Mechanical Technician's Handbook

List of Contributors

Robert S. Christie, Senior Engineer, Plasma Physics Laboratory, Princeton University *(Fabrication Techniques)*

Walter P. Heady, Chief Draftsman, Hoke Inc. *(Interpreting and Using Engineering Drawings)*

Robert C. Knauer, Late Instrumental Specialist, Plasma Physics Laboratory, Princeton University (*Selecting Measuring Instruments* and *Instrument Calibration and Standards*)

Maurice J. Webb, Vice President of Engineering, Victaulic Company of America (*Selecting Materials, Selecting Components and Equipment, Safety Techniques and Devices, Useful Formulas, Miscellaneous Techniques, Tables and Conversions,* and *Selected Properties of Materials and Physical Constants*)

Mechanical Technician's Handbook

Maurice J. Webb
Editor in Chief

Vice President of Engineering
Victaulic Company of America

A James Peter Book
James Peter Associates, Inc.

McGraw-Hill Book Company

New York St. Louis San Francisco Auckland Bogotá
Hamburg Johannesburg London Madrid Mexico
Montreal New Delhi Panama Paris São Paulo
Singapore Tokyo Toronto

Library of Congress Cataloging in Publication Data

Main entry under title:
Mechanical technician's handbook.

"A James Peter book."
Includes index.
1.Technicians in industry—Handbooks, manuals, etc. I.Webb, Maurice J.
TA158.M4 620 81-23652
ISBN 0-07-068802-8 AACR2

1234567890 898765432

ISBN 0-07-068802-8

The editors for this book were Patricia Allen-Browne, Alice Goehring, and James T. Halston, the designer was Mark E. Safran, and the production supervisor was Paul Malchow. It was set in Century Schoolbook by University Graphics, Inc.

Printed and bound by R. R. Donnelley and Sons

Contents

About the Editor in Chief

Maurice J. Webb is Vice President of Engineering in charge of product design, research and development, and quality control at Victaulic Company of America. He holds graduate degrees from both London and Princeton Universities, and he has been a member of Princeton University's research faculty for more than seven years as well as serving as a consultant to industry.

Preface

Mechanical technicians provide the engineering laboratories and research centers of industrial, educational, and research organizations with a specialized pool of talent for future development and growth. These men and women make, test, evaluate, and prove new items and materials in support of new products and technology and fundamental research. They analyze failures and the limitations of existing products and devices and verify the adequacy of production items. These technicians multiply enormously the effectiveness of the engineering and scientific work force by relieving them of mechanical tasks for which they often do not have the skills or knowledge to perform. The technicians' services give the engineers and scientists additional time to pursue activities more appropriate to their advanced education and training.

A career as a mechanical technician provides very unique opportunities. The work is far from routine and includes building models, prototypes, test fixtures, and apparatus using all available manufacturing techniques. The area of testing involves the most advanced and sophisticated instrumentation and the reward of seeing one's creation put to work as its performance is measured and its results analyzed. Often, this career is an opportunity to work at the forefront of technology and share in the excitement of discovery.

There is no one typical task for a mechanical technician. Depending upon the setting, the project, and the precision of the results desired, literally thousands of different operations are performed in the engineering or research laboratory. Most mechanical technicians are capable of developing a procedure to solve a problem they have never faced, but the process can be time-consuming and involve many errors. This book draws upon the knowledge, skills, and talents of top technical people in areas of mechanical engineering technology and presents the essentials that make work proceed faster and produce more dependable results.

We selected each contributor based upon unique personal experience and knowledge and asked each one to include in his chapter everything he did routinely in the laboratory as well as other, less-common tasks he had performed to solve problems for which there were no ready answers. In other words, we have included not only the basics of mechanical engineering technology but also the details worked out by a person who has faced many problems and then effectively solved them.

The guiding principle for selecting material in this handbook was a technician's three primary activities:

1. Building models and apparatus from specific instructions, primarily drawings
2. Selecting components and instruments for the assembly of apparatus
3. Safely conducting tests and recording results

Because there is such wide diversity between different industries and laboratories regarding the technicians' specific fields, this book sufficiently describes the various devices and components selected, to provide an understanding of their operating principles. These details will help technicians make the wisest selection of materials, components, and instruments for their particular requirements and aid them as they troubleshoot in the event of difficulties or malfunctions. In addition, understanding the functional principles will help technicians select alternate items that may be more suitable to the particular circumstances.

The International System (SI) of units is included with the still familiar English units. An exception occurs in Chapter 3, where, because of their continued wide usage, English units only are used. Because of the prevalence for material such as bar and rod to be supplied in "inches," for screw threads still to be manufactured predominantly in "threads per inch," and for rotational speeds to be measured in "revolutions per minute," it seems unwise to convert them to their SI equivalents if, in fact, the equivalents are not yet in common use.

Maurice J. Webb

c h a p t e r 

Interpreting and Using Engineering Drawings

Walter P. Heady

1.1 GRAPHIC INTERPRETATION

Of primary importance to the engineering technician is the development of the necessary skills for interpreting or reading drawings. Reading a drawing is essentially a two-step process; involving (1) obtaining an overall impression of the general shape of the part and (2) appreciating the details through reading the dimensions, symbols, and notes. The entire process is similar to that used by a golfer to read the green. At first glance, an accurate portrayal of the green is registered immediately in the golfer's mind, and with further study all the intricacies come into focus. Just as the golfer must learn to read the green, so the technician must learn to read a drawing. Proficiency is acquired only through concentrated and repeated practice once the principles have been learned. This proficiency must be developed if serious errors are to be avoided.

For consistency and uniformity, certain conventions and standards have been adopted by drafters, to eliminate ambiguity. Some standards, especially those in symbol form, generally are not used extensively enough in any one technical field, so that even though they were once learned, they are eventually forgotten or, even worse, misinterpreted. But the technician must be versatile, often working in different fields; this section of the handbook is a ready reference for and provides information on the fundamentals. The standards most commonly used and the ones adopted in this chapter are from the American National Standards Institute (ANSI) and designated ANSI Y14.

The mechanical technician commonly uses two principal types of drawings. The *engineering drawing* contains the necessary information for the manufacture of parts—the dimensions, tolerances, finishes, and, most likely, materials to be used for the part—and assemblies. Engineering drawings are prepared by the orthographic projection process, which provides a precise visual graphic representation of the actual part. The *schematic drawing* is used especially for plumbing and wiring hookups. Schematic drawings are not to scale, and components are indicated with standard symbols that represent each component's function.

1.2 ORTHOGRAPHIC PROJECTION

Orthographic projection shows exactly the shape of a three-dimensional object by projecting perpendiculars from the profile and other distinguishing features of an object onto various planes that are usually perpendicular to one another. The resulting "views" on the planes give a complete description of the part (Fig. 1.1). As far as possible, the object being drawn is oriented to the projection planes so that its principal surfaces and axes are parallel to the planes. There is thus a maximum number of dimensions appearing in true length on the projection planes, which enables greater direct utilization of the dimensions during manufacture. This technique lacks the more common pictorial visualization, but it eliminates distortion in the views and simplifies dimensioning, although the views are not always easy to interpret.

Principal Views

The technique of projecting perpendiculars to a plane to provide the shape of the three-dimensional object involves six planes, to give six principal views. Each plane is perpendicular to the adjacent planes. Therefore, the planes can be visualized as enclosing the object in a "box" (Fig. 1.1). To illustrate the part on a flat

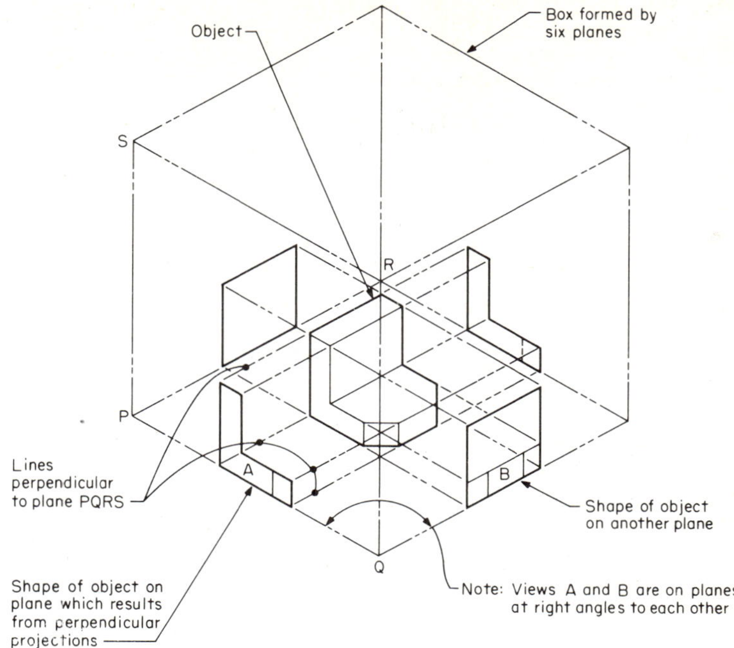

FIG. 1.1 Orthographic projection.

sheet of paper, certain edges of the box are cut, whereas others are folded or hinged so that the sides of the box can be displayed on a single plane, as in Fig. 1.2. The principle is illustrated for a simple object in Fig. 1.2*a*, which is considered suspended within the box in Fig. 1.1, and the views are projected to the sides and the box opened up in Fig. 1.2*b*. Note how the top, right-side, bottom, and left-side views are hinged to the front view and the back view is hinged to the left-side view. The views are then rotated until they are in the same plane as the front view (Fig. 1.3). The top view is directly above the front view, and the bottom view is directly below. All other views are in line with the front view. This convention is always adopted. Views are not normally labeled as in Fig. 1.3, but by convention are (called) front view, right-side view, and so on. The front view is chosen arbitrarily by the drafter.

This method of illustrating a part is known as *third-angle projection* and invariably is used for working drawings (that is, drawings from which parts are to be made). A working drawing, when necessary, shows dotted lines that represent surfaces, edges, and intersections not visible to the viewer from the direction of the plane under consideration. In Fig. 1.3, note that the bottom and left-side views show dashed lines because important surfaces are hidden from view. A working drawing shows only the views necessary to manufacture the part. In Fig. 1.3, all but the front and top views are eliminated because they are the only views needed for fabrication. Normally, the views shown on drawings are the front, right-side, and top views, with other views used as needed to clarify the part.

Sometimes only one view is necessary to describe a part (Fig. 1.4). Note that

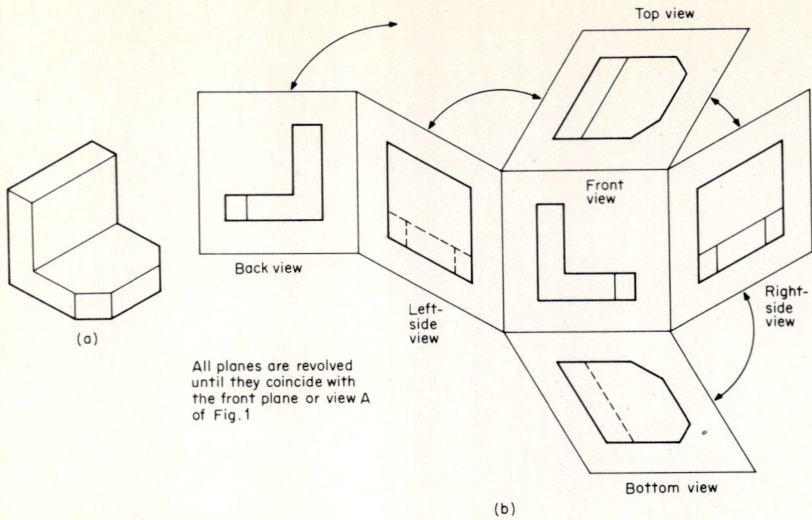

FIG. 1.2 The six principal views.

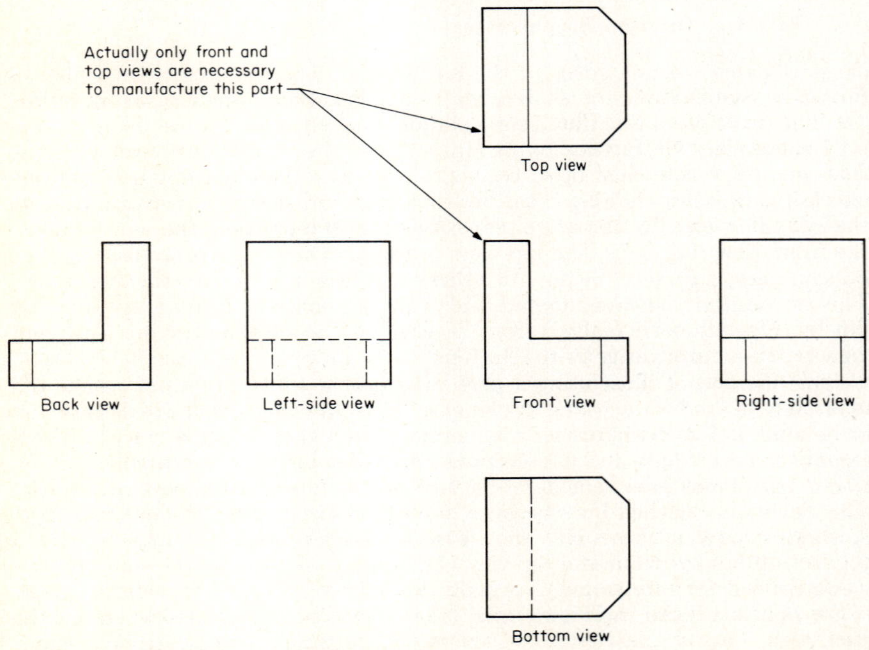

FIG. 1.3 Example of a six-view drawing.

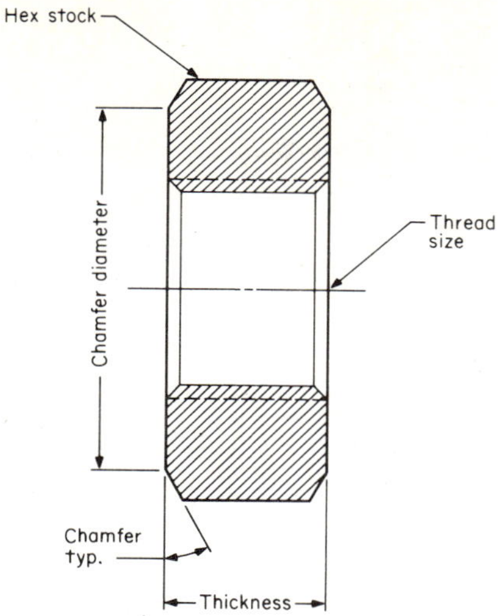

FIG. 1.4 Example of a single-view drawing.

all pertinent dimensions are shown, along with a brief word description. It is important not to make a determination from just a single view when multiple views are given; all views must be considered to evaluate the shape correctly.

Auxiliary Views

Auxiliary views clarify the shape. They are sometimes necessary to show true size and shape of a part. The view is parallel to the plane of projection.

The auxiliary view in Fig. 1.5 is hinged to the front view but could be taken from the top or side view. It is possible to have more than one auxiliary view on a drawing, as in Fig. 1.6. No other views are required because the front and two auxiliary views are all that are necessary to manufacture the part.

Sectional Views

When the inner surface of a part is difficult to illustrate and understand on a drawing that uses the conventional dashed or hidden line, a section is taken through the part. This is just an imaginary cut through the object to expose the inner details. The plane that cuts the object is represented by the *cutting-plane line,* which indicates where the cut is made. Arrows show the direction of viewing the object, and identification letters indicate where the view is located, as in Fig. 1.7. In many cases it is quite obvious where the section is taken and so no cutting lines are necessary (Fig. 1.8). Portions of the part cut by the plane are shaded or lined, to clearly indicate that a section has been made and to differentiate between cut and uncut surfaces.

Complicated drawings may not have sufficient room to show sectional views in their correct location. It is permissible to show these views elsewhere on the drawing as long as the view is shown parallel to the cutting plane and clearly labeled, as in Fig. 1.7.

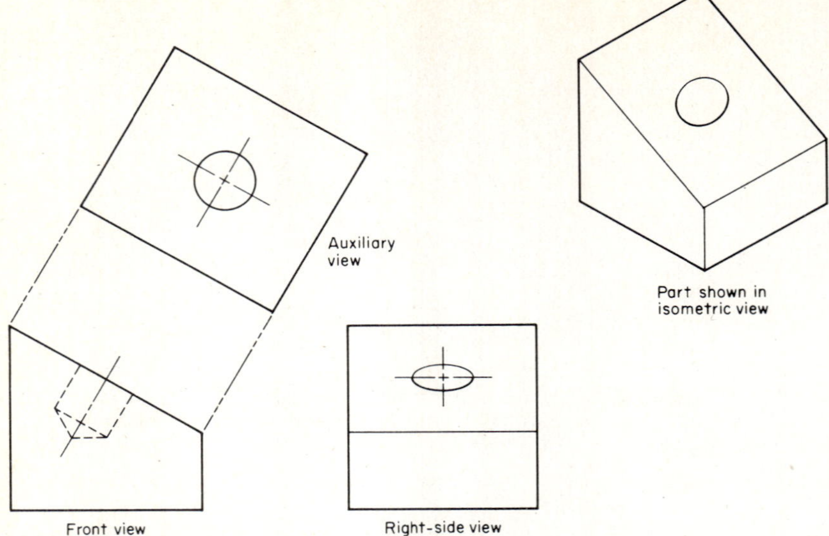

FIG. 1.5 Drawing with one auxiliary view.

Sometimes it is not necessary to section along a complete cutting plane; the section can be taken part way and then broken off. This technique is referred to as a *broken section*. A wavy line indicates where the section is broken. One side of the wavy line shows section lining, and the other side shows the view for the external shape. Figure 1.9 is typical of a broken section.

Rather than drawing a separate view, sometimes a section is made and rotated 90° and superimposed upon the single view (Fig. 1.10). This technique is used especially for long parts that have a uniform cross section. It is also used for webs. At times, instead of rotating the section and retaining relative positioning, the rotated section view may be removed and located close to the main view. A complex part may have several rotated sections.

Assembly drawings (drawings showing a complete unit comprised of more than one part) are often sectioned to show better how the assembly is built from its components. Section lining helps distinguish the individual parts and, in some

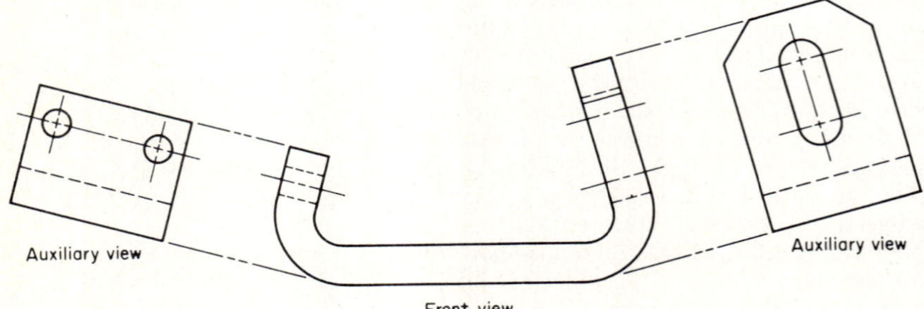

FIG. 1.6 Drawing with two auxiliary views.

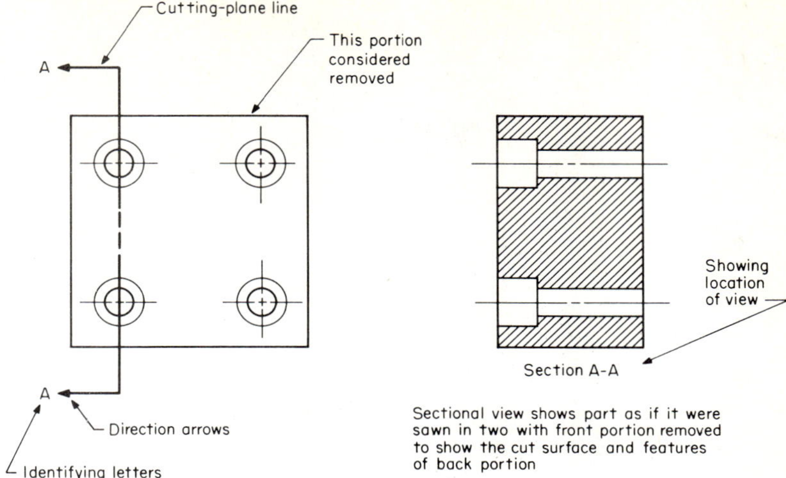

FIG. 1.7 Sectional view.

cases, identifies materials. Typical material linings or shading are shown in Fig. 1.11; Fig. 1.12 shows a valve assembly made from various metals. Actually, the section lining for material identifications is used primarily for installation drawings to show such items as foundations, insulation, and masonry work. Assembly drawings for mechanical or plastic components have insufficient section linings to depict the many different materials, so one must use a bill of materials, which

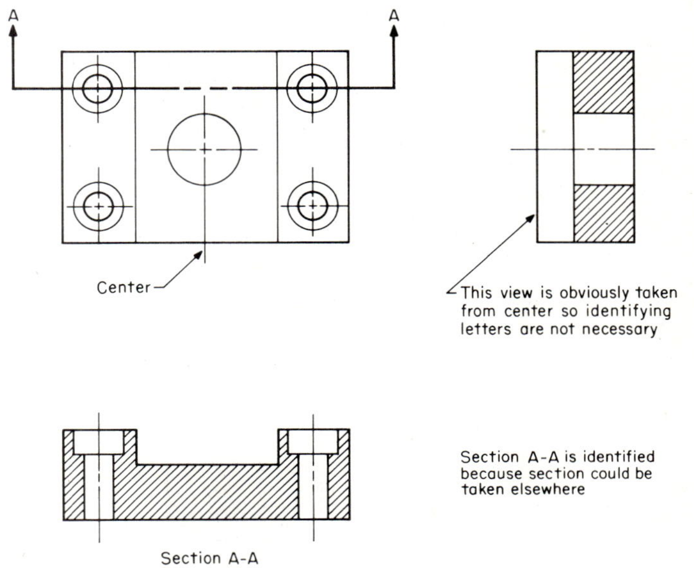

FIG. 1.8 Sectional views.

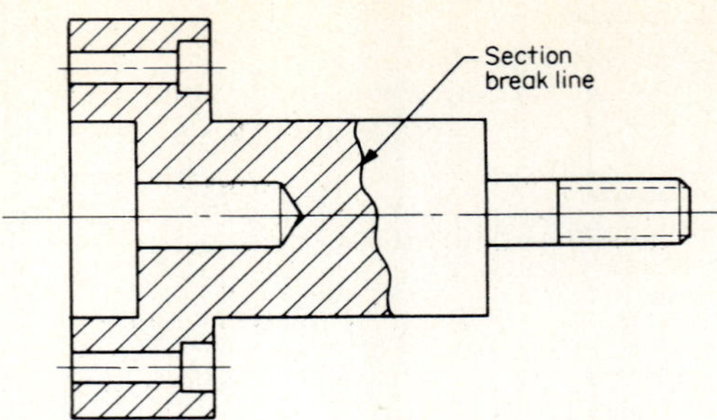

FIG. 1.9 Broken section.

is directly on the assembly drawing or on a separate sheet. Different components made from the same material are lined in different directions in any assembly drawing so that the reader can more easily distinguish the separate components.

By convention, certain items and parts of items are not shown sectioned, even though the section plane passes through them. This violation of the section principle is limited to small items of circular section (e.g., bolts, shafts, stems, pins) and thin sections (e.g., webs, ribs, and spokes). Figures 1.12 and 1.13 include examples of section violations.

Another convention that violates the principles of projection is illustrated in Fig. 1.14. To minimize the number of views required, the drafter sometimes shows portions of a part deliberately moved from their true positions. Most often this movement is confined to details rotated about a symmetric axis to a position where true dimensions are projected with respect to the axis. In Fig. 1.14*a* a hole has been rotated, whereas in Fig. 1.14*b* a spoke of a wheel has been moved. To

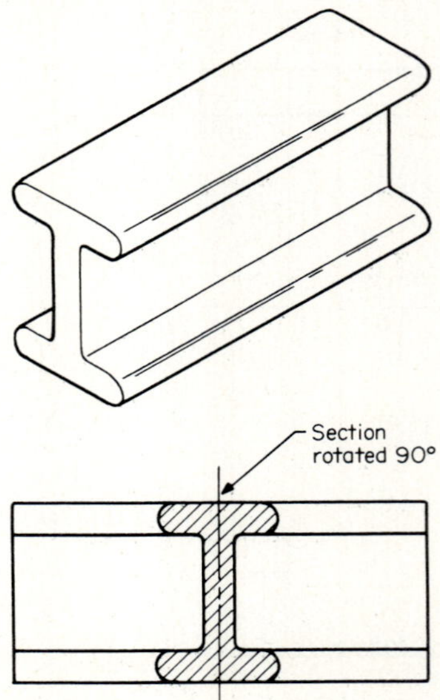

FIG. 1.10 Revolved section.

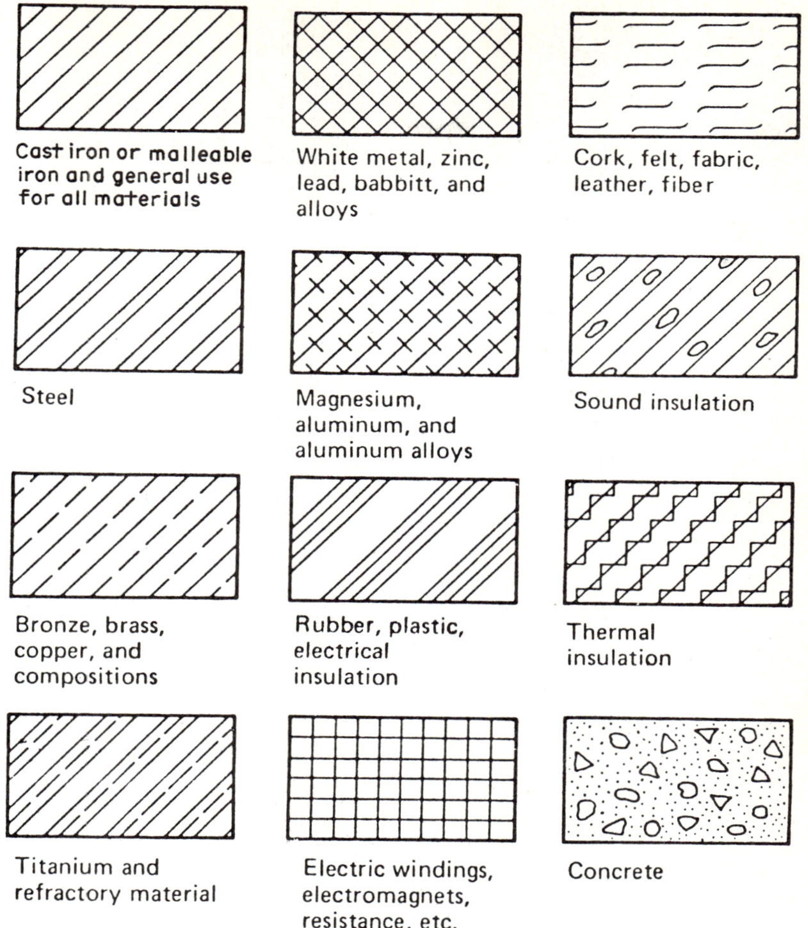

FIG. 1.11 Section linings for various materials.

help the reader of a drawing, a note, e.g., "rotated out of true position," is usually placed on the drawing.

Other special sections are shown for tube or pipe and bar. Figure 1.15 shows typical drawings.

1.3 DRAWING LINES

Different kinds of lines convey different meanings. Figure 1.16 illustrates the lines commonly used and describes their meanings. Because drawings are composed of lines, their meanings must be committed to memory for effective interpretation.

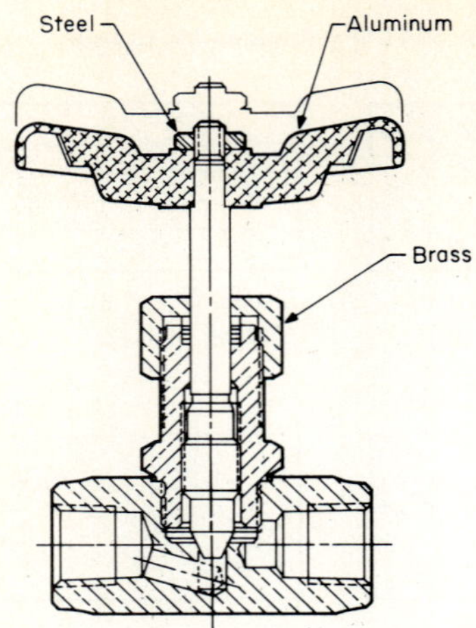

FIG. 1.12 Assembly drawing with various section linings.

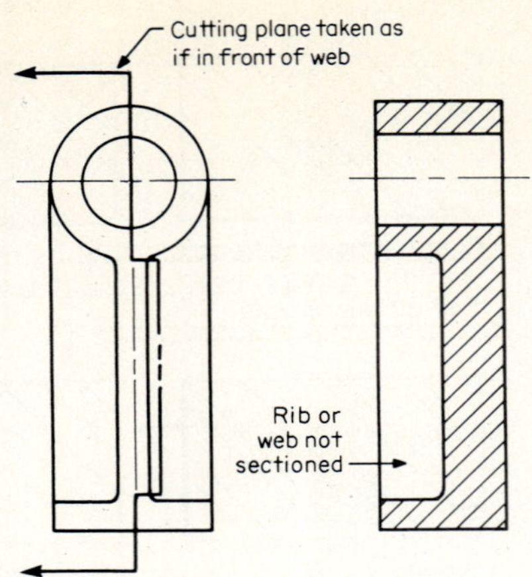

FIG. 1.13 Web in section (section violation).

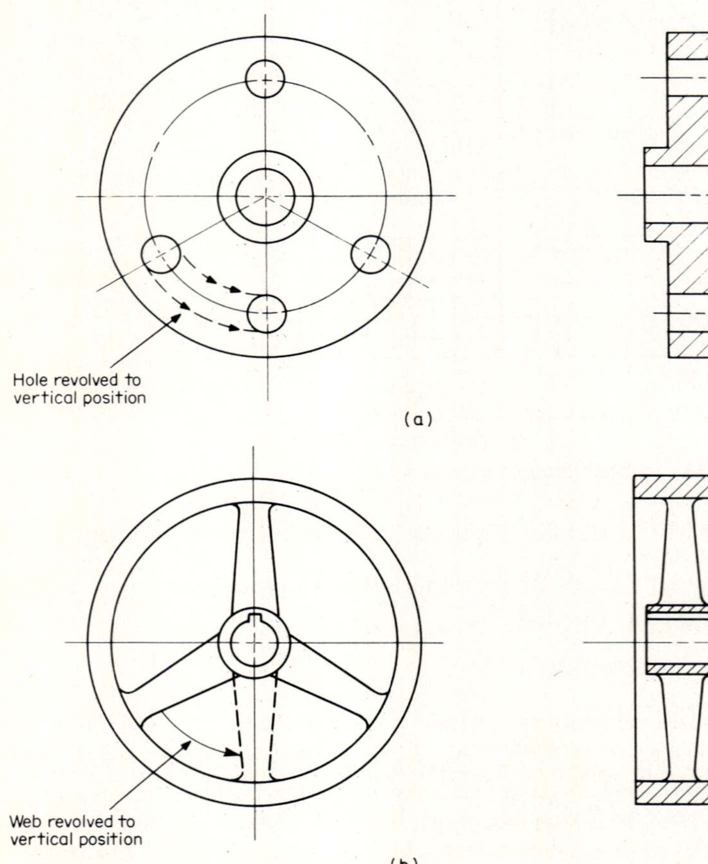

FIG. 1.14 Violation of projections.

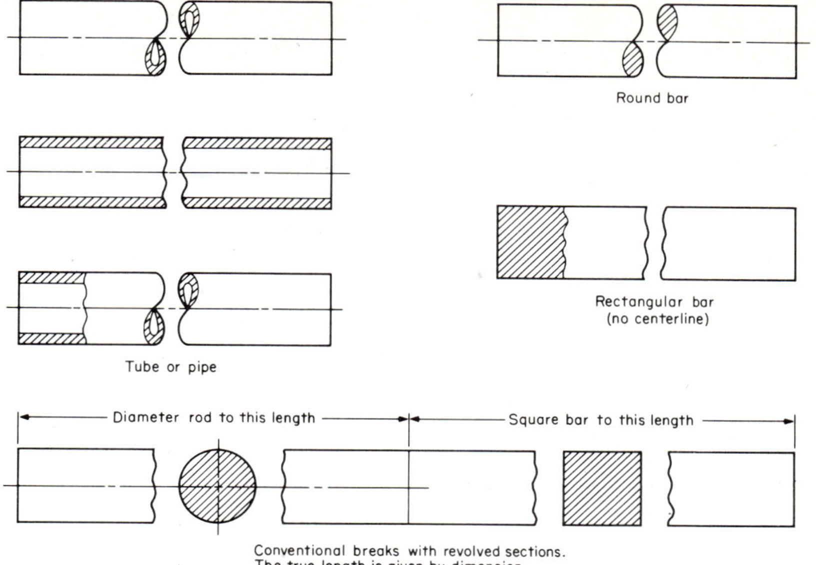

FIG. 1.15 Conventional breaks in parts.

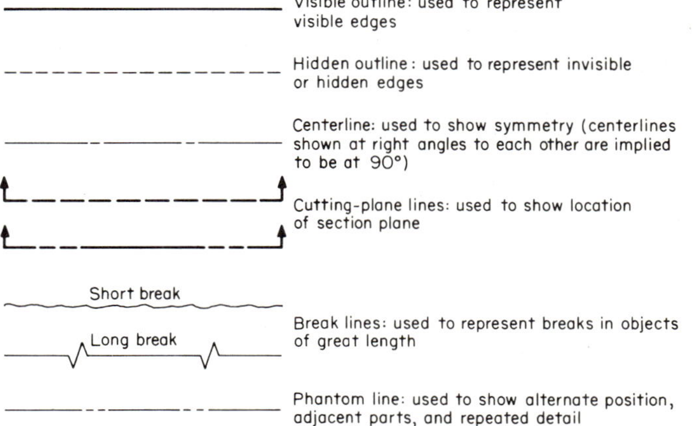

FIG. 1.16 Drawing lines.

1.4 THREAD REPRESENTATION

Generally, three conventions are used for representing screw threads on drawings: the *detailed,* the *schematic,* and the *simplified.* Figure 1.17 shows all three. The representation used by the drafter depends upon such factors as the purpose and use of the drawings, general work quality, and drafting time. No one presentation is more significant than another, so all three conventions may be on one drawing if the drafter believes they add clarity. For engineering drawings, the simplified presentation is most commonly used.

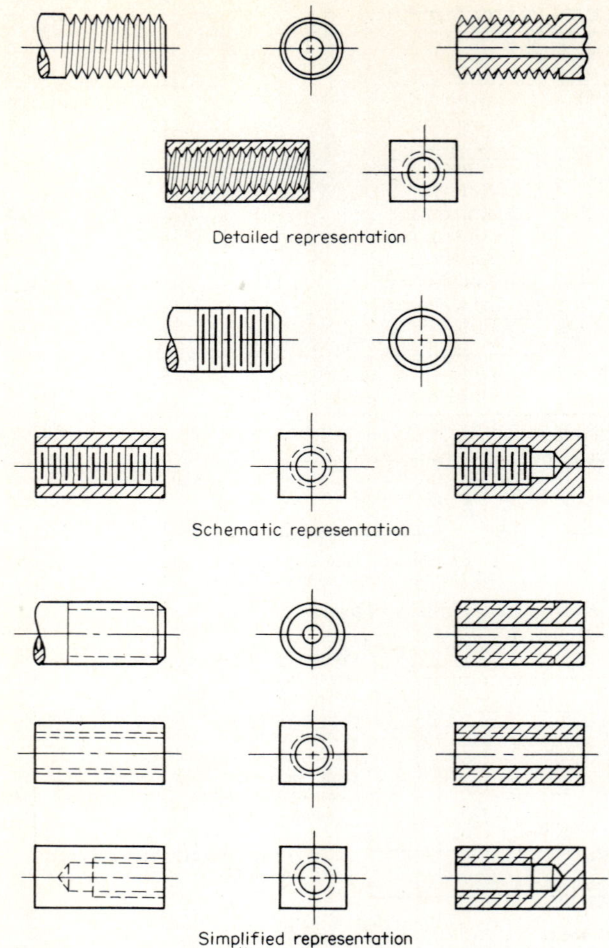

FIG. 1.17 Thread representations.

Thread Call-Out

Standard abbreviations are used to designate or call out thread size and type. Figures 1.18 and 1.19 illustrate how this is done for a straight thread and a taper thread, respectively. Table 1.1 lists the common thread abbreviations. Dimensions and thread forms for the various sizes and classes of thread are discussed in Sec. 3.9.

1.5 KNURLING REPRESENTATION

Knurling is a process in which a surface, usually cylindrical, is precisely roughened to create a distinct geometric pattern. Knurling is used for decorative purposes; for providing a gripping surface, as for a knob or handle; or for providing

TABLE 1.1 Thread Designations

Designation	Series
Acme-G	Acme threads, general-purpose (see also Stub-Acme)
L.H.*	Left-hand thread
8N	American National 8-thread
12N	American National 12-thread
16N	American National 16-thread
NC†	American National coarse-thread
NEF	American National extrafine-thread
NF†	American National fine-thread
NPT	American Standard taper pipe threads for general use
NPTF	Dryseal American Standard taper pipe threads
Stub-Acme	Stub-Acme threads
UNC	Unified National coarse-thread
UNEF	Unified National extrafine-thread
UNF	Unified National fine-thread

*All threads right hand unless designated otherwise.
†Superseded by Unified Thread Series.

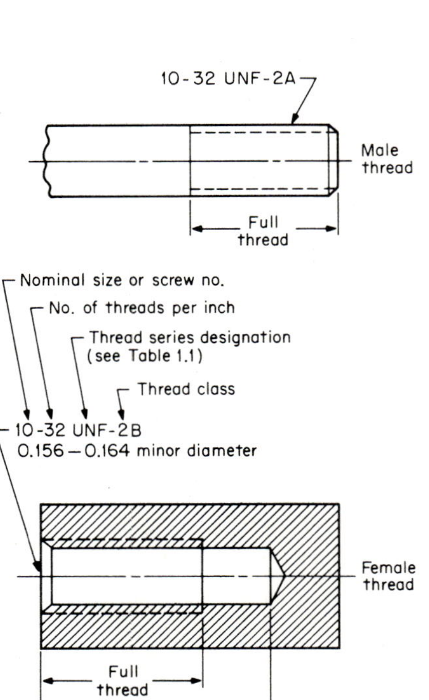

FIG. 1.18 Typical male and female straight-thread call-outs.

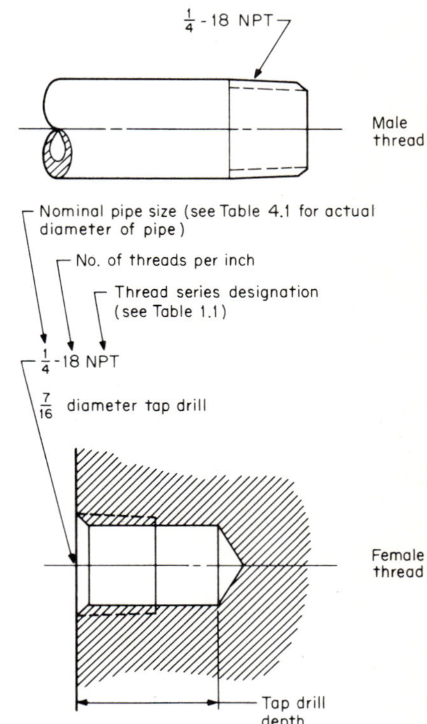

FIG. 1.19 Typical male and female American Standard taper pipe thread call-outs.

an interference fit to lock components together. Circular pitch and diametral pitch knurling are the two methods in use; the diametral pitch system is covered by ANSI 94.6, 1966, reaffirmed 1972.

Knurling is depicted on a drawing almost pictorially (Fig. 1.20). For clarity, it is generally spelled out along with other descriptive information that requires explanation. First, the type of knurl is described as *straight, diagonal,* or *diamond* knurling (Fig. 1.21). Diagonal knurling is either left hand (L.H.) or right hand (R.H.); the outside diameter (OD) is given before and after knurling; the pitch (teeth per inch—TPI—or outside diametral pitch—ODP) and length of knurl (Figs. 1.21 and 1.22) are also given. The approximate TPI is all that is necessary to call out when general-appearance knurls such as handwheels and knobs are used; see Fig. 1.23.

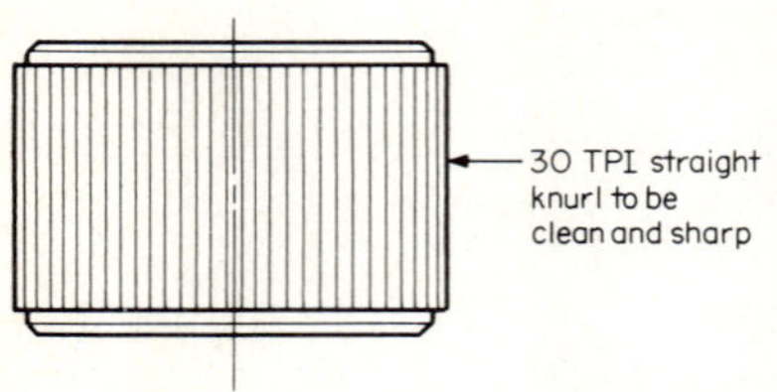

FIG. 1.20 Decorative handwheel knurl.

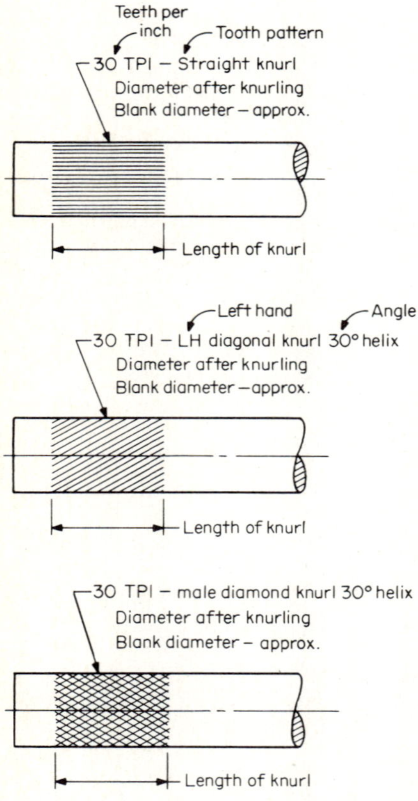

FIG. 1.21 Dimensioning of circular pitch knurling.

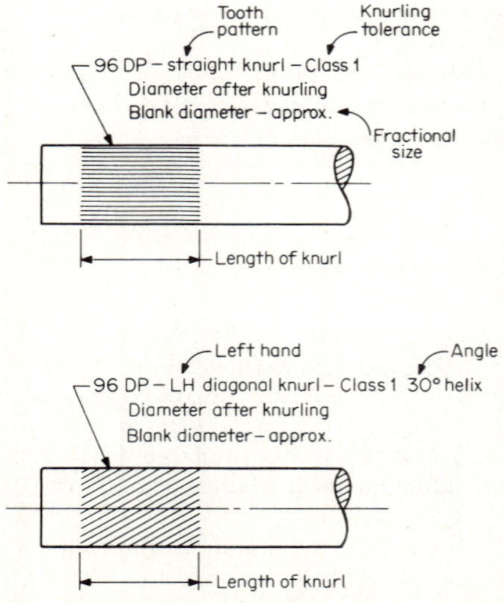

FIG. 1.22 Dimensioning of diametral pitch knurling.

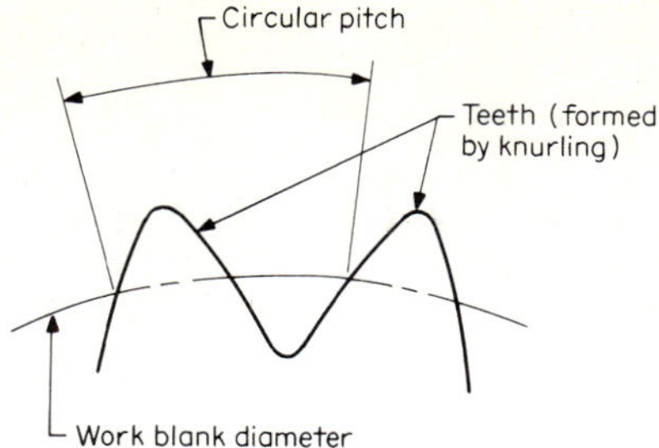

FIG. 1.23 Circular pitch for knurl.

Drawings may still be found that call out coarse, medium, or fine knurls. Coarse is normally considered 16 to 20 TPI, medium 25 to 35 TPI, and fine 40 to 80 TPI.

Circular Pitch Knurling

The distance between teeth on the circumference of the work is referred to as *circular pitch knurling* (Fig. 1.23) and is called out as TPI (Fig. 1.21). Blank diameters vary with the circular pitch of the knurling selected and should be specified only after the proper diameter of the blank has been determined by trial.

Diametral Pitch Knurling

Diametral pitch knurling uses fractional sizes as work blanks, in multiples of $\frac{1}{64}$ and $\frac{1}{32}$, depending upon the choice of pitch (Fig. 1.22). The four standard pitches are 64, 96, 128, and 160, with 96 the one generally preferred. The standards 96 and 160 are used for work blanks with multiples of $\frac{1}{32}$, and 64 and 128 with multiples of $\frac{1}{64}$.

Diametral pitch knurling is the result of the total number of teeth (Nw) divided by the work-blank diameter (Dw). It is called out as DP (Fig. 1.22).

$$\text{DP} = \text{diametral pitch} = \frac{Nw}{Dw}$$

1.6 SURFACE FINISH

The surface finish to be provided on a part is denoted on the drawing by a special symbol (like a check mark) surrounded by various numbers. The position of a number indicates a control with respect to a specific characteristic of finish. On a single part, different surfaces often require different types of finish. The system of finish marks has been very cleverly devised and conveys an enormous amount of information to the reader, particularly dictating the method of manufacture (see Sec. 3.12).

The finish symbol is easily recognized and cannot be confused with any other. Figure 1.24 shows the mark and how it is typically used to denote which surfaces are to be finished to the specified standard. If no finish marks appear around the part itself, there is usually a general remark regarding finish in the title block of the drawing. Specific finish marks always take precedence over general finish instructions. For example, finish marks may apply to only one surface of a part, the remaining surfaces being required to comply with only the general requirements.

Three important characteristics of finish are roughness, waviness, and lay. Roughness refers to the finer irregularities of the surface that are typically inherent in any machining process. Because the irregularities are related to the machining process, invariably they are similarly shaped and run parallel to one another. Figure 1.25 shows the fine structure of surface roughness. Waviness is an irregularity of much greater wavelength than roughness. Lay is the predominant direction of the pattern producing roughness.

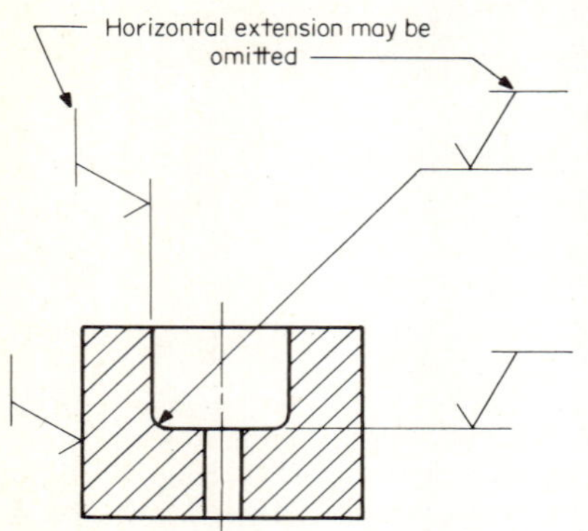

FIG. 1.24 The surface-finish symbol on a drawing.

In Fig. 1.24, the internal cavity of the part has three different finish call-outs: one for the cylindrical surface, one for the radius, and one for the flat bottom. These three call-outs might all call for the same finish or, in recognition of the manufacturing process and the function of the part, for different finishes. Only one external surface has a specific finish requirement that, if the part were circular, would apply to the entire cylindrical surface.

Roughness height is measured in microinches (a microinch is 0.000001 in) or micrometers. To measure the roughness, the average (arithmetic) value is taken over a certain distance called the *roughness-width cutoff.* If the roughness is measured over too great an area, it can include a contribution from waviness. The location of roughness values next to the finish symbol is shown in Fig. 1.26,

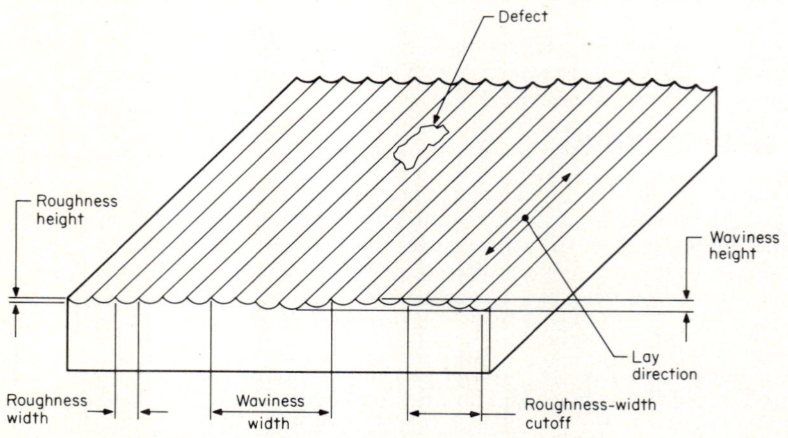

FIG. 1.25 Terms used to describe surface finish.

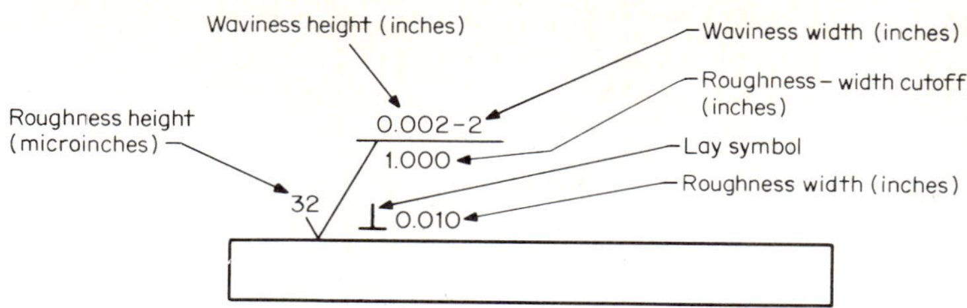

FIG. 1.26 Positioning of characters around surface-finish symbol.

which also shows the location for the roughness-width cut-off dimension. Typical values of roughness specified are 8, 16, 32, 63, 125, and 250 microinches for machined surfaces (0.2, 0,4, 0.8, 1.6, 3.2, and 6.4 μm).

Waviness, significantly larger in wavelength than roughness, is measured in inches. In the same way that it is necessary to specify a length for roughness measurement, so it is for waviness. The length is referred to as *waviness width.* Figure 1.26 shows the location by the finish-symbol dimensions of waviness.

Lay patterns are conveniently grouped into six categories. The six symbols for describing lay are in Fig. 1.27 and the location for lay symbol in Fig. 1.26. A surface roughness width is often used in conjunction with the lay symbol.

A surface flaw, such as a hole, scratch, or pit, is not included in surface-finish measurement unless specifically stated on the drawing by the drafter. Dimensions for surface finish are understood to be maximum values; smaller values are acceptable.

Often, only the roughness value is specified, in which case the roughness-width cutoff is then assumed as 0.030 in (0.762 mm). When neither waviness nor lay are specified, it is usually because the profile of the part has been adequately limited by the drafter through geometric characterization of such factors as flatness and straightness.

1.7 TOLERANCES

No matter how precise one wishes to be in the manufacture of a part, there is obviously a limit to the exactness to which the part and its characteristics can be measured. Common measuring instruments measure to one thousandth (0.001) of an inch (0.025 mm) and sometimes to tenths of thousandths (0.0001) of an inch (0.0025 mm). Beyond these values, very special instruments are required, and measurements must be made in areas under carefully controlled temperatures because the expansion and contraction of the part from temperature change can be a significant factor in the true size or shape. Recognizing these practical limits, engineering drawings ordinarily limit accuracy through design of the parts to require only the degree of exactness necessary to provide proper function and safety. These limits of accuracy are called *tolerances.*

Various types of tolerances are needed to describe the size and configuration of a part. The simplest one is a dimensional tolerance, but many others are used to describe shape of form and the positioning or location of one feature with respect to another. Just as other drafting descriptors use a shorthand notation, tolerances are characterized and standardized in a shorthand form for drawing convenience and precise description.

Lay Symbol	Meaning	Example Showing Direction of Tool Marks
=	Lay approximately parallel to the line representing the surface to which the symbol is applied.	 =
⊥	Lay approximately perpendicular to the line representing the surface to which the symbol is applied.	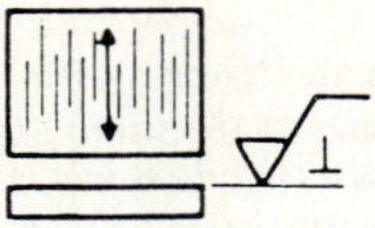 ⊥
X	Lay angular in both directions to line representing the surface to which the symbol is applied.	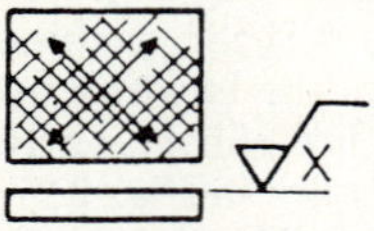 X
M	Lay multidirectional.	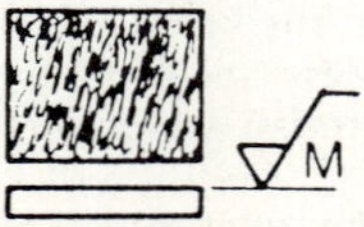 M
C	Lay approximately circular relative to the center of the surface to which the symbol is applied.	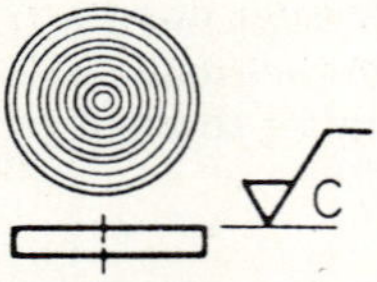 C
R	Lay approximately radial relative to the center of the surface to which the symbol is applied.	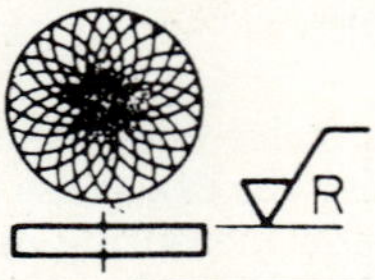 R
P	Lay particulate, nondirectional, or protuberant.	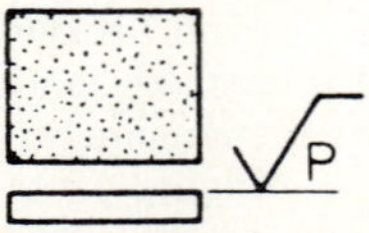 P

FIG. 1.27 Lay symbols used to describe surface finish. *(ANSI B46.1–1978.)*

Dimensional Tolerances

Dimensional tolerances are generally shown in three ways. The first involves the number of digits (usually the decimal digits) used to dimension the feature together with a note on the drawing that gives the key to the system being followed. The key is normally adjacent to or part of the drawing title block. Figure

1.28 illustrates a typical tolerance key: A dimension described to only two decimal places (for example, 13.10) is required to be maintained to ±0.010 of that dimension (that is, from 13.09 to 13.11) in inches or millimeters. Similarly, dimensions described to additional precision are required to be maintained to the smaller tolerances shown.

Dimensional tolerances are also shown by specifying the dimensional limits as part of the dimension, as illustrated in Fig. 1.30. The third way shows the nominal dimension with the tolerance (above and below nominal) directly following it. When more than one method is used on the same drawing, the tolerances expressed as part of the actual dimension take precedence over general tolerances described by a note or key.

Unless otherwise specified:

1. Dimensions are in inches

*2. Tolerances:

Decimals	0.0000	± 0.0005
	0.000	± 0.005
	0.00	± 0.01
Angles	________	± 3°

*3. Machined surface fin. $^{63}\surd$

*4. Geometric tolerances and positioning within 0.010 TIR

5. Symbols as per ANSI Y14.5

*6. Break all sharp edges 0.005 – 0.010

*Not applicable to final assembly drawings

FIG. 1.28 Typical key for general dimensional tolerances.

Reference Dimension

Occasionally on a drawing a dimension is given in parentheses or is followed by "REF," as shown in Fig. 1.29. Such dimensions are *reference dimensions*. They normally appear without tolerances because they are not meant for manufacturing or inspection. Just as the term implies, the dimensions are for reference only, and most often the drafter has placed them there for the technician's convenience. A reference dimension provides better overall perspective of the part because the reader does not have to compute such a dimension from other dimensions on the drawing.

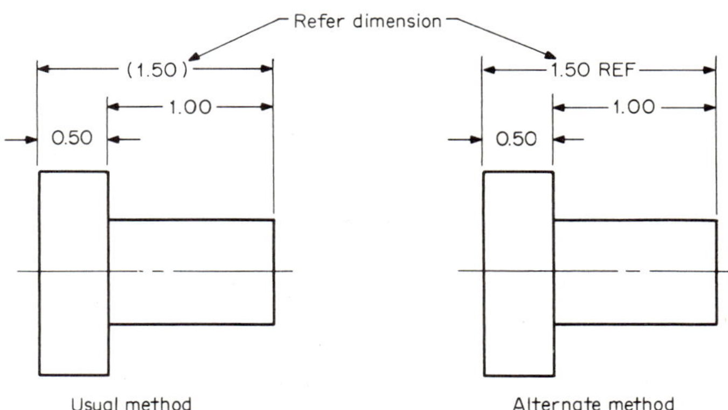

FIG. 1.29 Reference dimension.

Basic Dimension

Sometimes a dimension is enclosed in a rectangle or followed by "BSC" (Fig. 1.30). This is a *basic dimension*. Basic dimensions are theoretically exact (without any tolerance) and normally found in conjunction with other dimensions, which are toleranced, to describe contours and shapes.

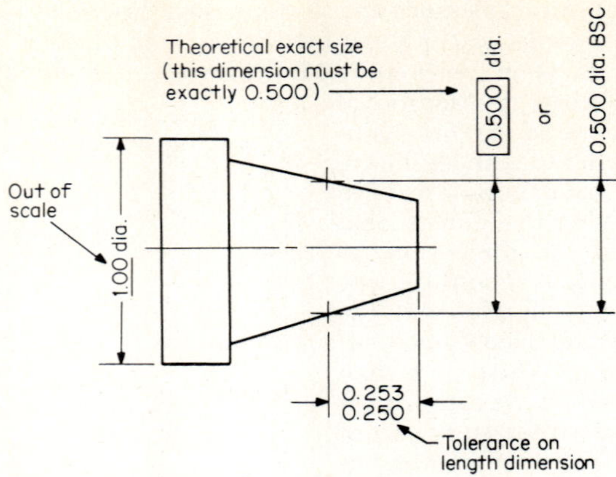

FIG. 1.30 Basic dimension.

Not-to-Scale Dimension

An underline (Fig. 1.30) denotes that a particular dimension is not to scale. Such dimensions must be watched for carefully because dimensions out of scale also mean the part is drawn out of scale; a wrong overall perspective of the part is thus easily conveyed. Although a dimension may be shown out of scale, the part must be made precisely to the dimension shown, including its tolerances.

Geometric Tolerances

Besides dimensional tolerances, *geometric tolerances,* which describe the geometry of the part, are used. Fourteen different characteristics, such as straightness and roundness, more precisely describe the features of a part. Each characteristic is described by a standard symbol or shorthand notation; Table 1.2 lists the characteristics with their corresponding symbols.

Figure 1.31 illustrates the manner in which these symbols are shown on a

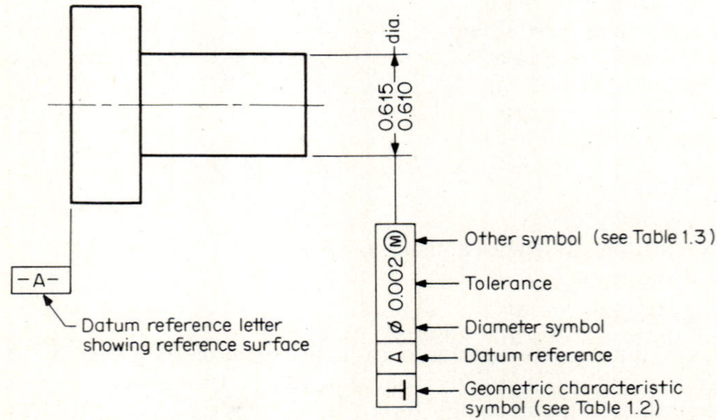

FIG. 1.31 Feature-control symbols incorporating datum references.

TABLE 1.2 Geometric Characteristic Symbols and Their Meanings

Symbol	Characteristic
—	Straightness
▱	Flatness
○	Roundness
⌭	Cylindricity
⌒	Profile of a line
⌓	Profile of a surface
∠	Angularity
⊥	Perpendicularity
//	Parallelism
⌖	Position
◎	Concentricity
⌯	Symmetry
↗	Circularity
↗	Total runout

drawing. In a rectangular box there are usually five positions, four for symbols and one for a number (the tolerance). All five positions are not required in every case, but at least two are (i.e., a geometric symbol plus the associated tolerance). Occasionally there may be more than five positions.

In Fig. 1.31, in addition to the symbols and tolerance itself, there is a *datum reference,* which relates one point, line, or surface to another. This is often necessary with geometric characteristics. Therefore, one line or surface is used as a reference, with the positioning and tolerance of the other surface with respect to it. The datum reference alphabetical letter always follows the geometric characteristic symbol associated with it in the rectangular box. The reference line or surface itself is indicated on the drawing by the alphabetical letter in a separate box, as shown. On a single drawing, usually datum references are assigned in alphabetical order. In more complex drawings, several datum references are not unusual. The datum reference is assumed to be exact for the purpose of establishing the relationship of features with respect to it. For example, in Fig. 1.31, the end surface is the datum reference, and the symbols tell the reader that the cylindrical portion (0.610 to 0.615 diameter) is to be perpendicular ($\perp$) to that surface within a specified tolerance.

To interpret the geometric tolerances, it is first necessary to recognize four additional abbreviations and symbols used in conjunction with geometric symbols and that may also appear in the rectangular box; see Table 1.3. Their interpretation will be apparent from the following examples used to explain the meanings of geometric tolerance notations. The examples are the ones used in ANSI Y14.5-1973 to describe the correct usage of these tolerances.

TABLE 1.3 Several Abbreviations Used with Geometric Tolerances

Abbreviation	Meaning	Symbol
DIA	Diameter	⌀
MMC	Maximum material condition (within stated tolerance)	Ⓜ
RFS	Regardless of feature size (within stated tolerance)	Ⓢ
FIM	Full indicator movement (as measured by a dial indicator, for example)	None

Straightness

Any feature that, in projection, appears as a straight line on a drawing may have a straightness tolerance applied to it. Figure 1.32*a* to *c* illustrates both the interpretation of the straightness characterization of the sides of a cylinder (a very common application) in three different forms and the use of additional symbols (see Table 1.3).

For simplicity, the examples show the allowable tolerance zone for an assumed single curvature. The conditions are equally applicable to a wavy centerline or profile, but the total allowable straightness tolerance zone still applies. In the case of a long shaft, examine the drawing carefully because the straightness conditions may be different from one section to another. If a straight line on the drawing represents a surface, then a flatness symbol is used.

Flatness

The flatness symbol and tolerance are shown in Fig. 1.33, which applies to a plane. The tolerance defines a zone within two parallel planes separated by the amount of the tolerance. All the surface undulations must fall within the zone.

Often a finish symbol is applied to the same surface. Ordinarily, "flatness" is a greater tolerance than "waviness" applied by the surface-finish requirements. Furthermore, the waviness width (Fig. 1.26) is typically too small to control flatness of a surface, except for small parts whose dimensions are of the same order as the waviness width.

Roundness or Circularity

Roundness or circularity may be applied to any circular section, such as cylinders, cones, and spheres. When applied, this tolerance specifies a radial zone within which a surface must fall (Fig. 1.34). It applies to the total length of the surface (cylinder or cone) unless noted otherwise. By itself it imposes no conditions on the straightness or concentricity of the surface, which are achieved by other controls.

The roundness tolerance cannot exceed one-half of the normal diametral tolerance. In the absence of roundness tolerance, the circularity or roundness of a part is controlled by the diametral tolerance alone. However, this is sometimes inadequate, and although a relatively large diametral tolerance is acceptable, the circularity of the surface may be maintained to higher precision.

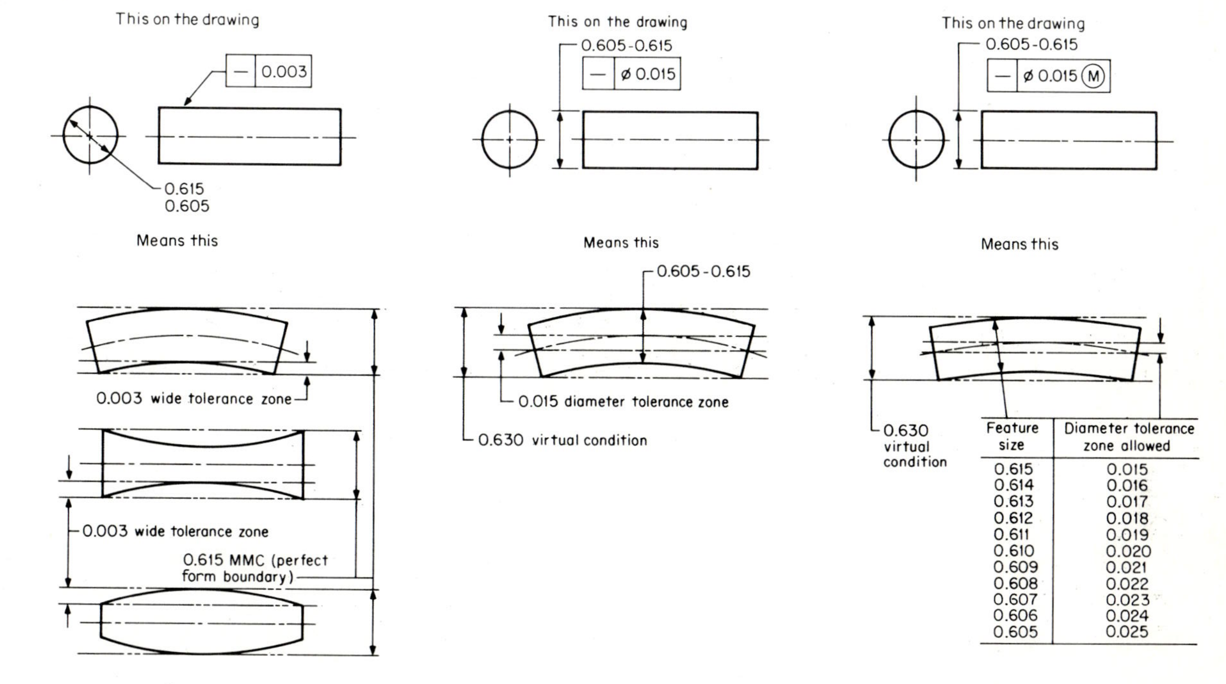

Feature size	Diameter tolerance zone allowed
0.615	0.015
0.614	0.016
0.613	0.017
0.612	0.018
0.611	0.019
0.610	0.020
0.609	0.021
0.608	0.022
0.607	0.023
0.606	0.024
0.605	0.025

The feature must be within the specified tolerance of size and the boundary of perfect form at MMC (0.615); each longitudinal element of the surface must lie between two parallel lines (0.003 apart) where the two lines and the nominal axis of the part share a common plane.

(a)

Each circular element of the feature must be within the specified tolerance of size. The derived axis or centerline of the actual feature must lie within a cylindrical tolerance zone of 0.015 diameter regardless of feature size.

(b)

Each circular element of the feature must be within the specified tolerance of size. The derived axis or centerline of the actual feature must lie within a cylindrical tolerance zone of 0.015 at MMC. As the feature departs from MMC, an increase in the straightness tolerance is allowed which is equal to the amount of such departure.

(c)

FIG. 1.32 Interpretation of the straightness symbol when used alone (*a*) and with other symbols (*b*) and (*c*). *(American Society of Mechanical Engineers.)*

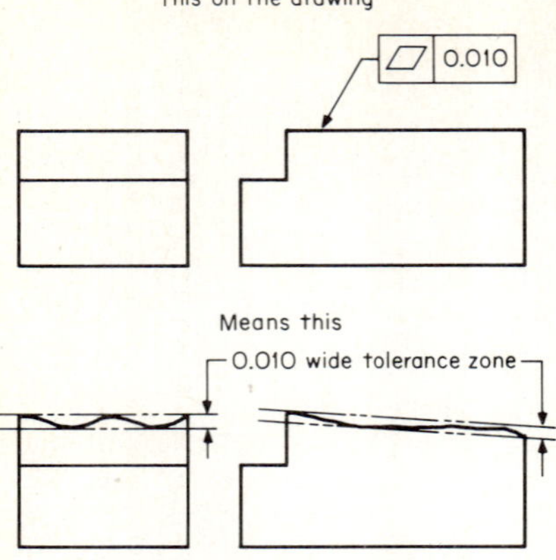

FIG. 1.33 Interpretation of the flatness symbol on a drawing. *(American Society of Mechanical Engineers.)*

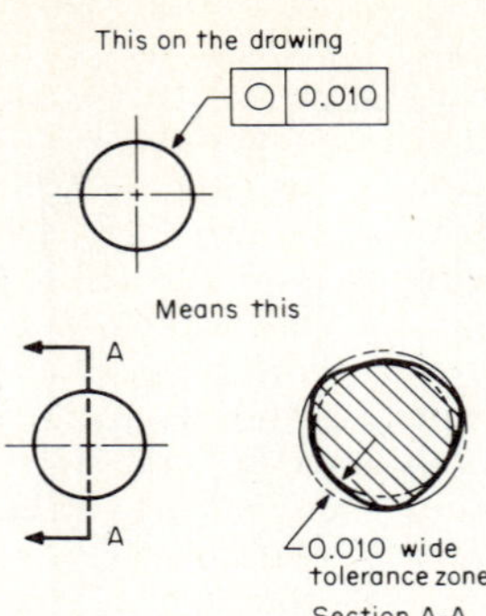

FIG. 1.34 Interpretation of the roundness symbol on a drawing. *(American Society of Mechanical Engineers.)*

Cylindricity

Cylindricity, a convenient combination of straightness and roundness that may be applied to cylindrical components, defines a tolerance zone bounded by two cylindrical surfaces considered generated from a perfectly straight axis. Figure 1.35 illustrates the application. Note that it is a radial tolerance and cannot exceed one-half the diametral tolerance.

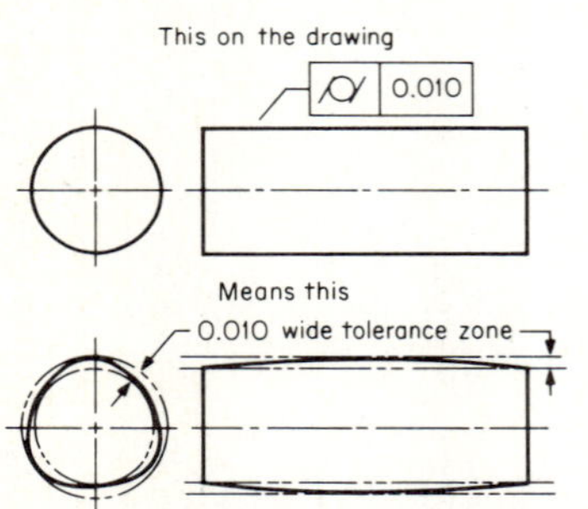

FIG. 1.35 Interpretation of the cylindricity symbol on a drawing. *(American Society of Mechanical Engineers.)*

This tolerance is rather restrictive and exacting and therefore normally applied only to a short cylinder. If it were applied to a long cylinder, it would impose high precision from various aspects, so the use of separate roundness and straightness tolerances are still required for many items.

Profile of a Line

The profile of a line is exactly what orthographic projection provides for each view. The profile tolerance is applied mostly to curved components of the view rather than to the surfaces and edges that appear as simple straight lines or circles. The profile itself is typically referenced to one or more projected straight lines because this is the most practical way of constructing (for the drawing) and reconstructing (for the machined part) the required form.

Because views are mostly of three-dimensional parts, there is much greater use of the profile-of-a-surface symbol (see the next section) than for a line. The profile of a line is used in sectional views that incorporate complex shapes and is especially useful when many sections are required to describe the object (e.g., sectional views of the hull of a ship).

The specified tolerance for a line profile (and also for a surface profile) is taken to be equally distributed on either side of the theoretically exact profile. However, the engineer may want the tolerance to be wholly within or wholly outside the exact profile. These three cases are illustrated in Fig. 1.36*b*, which shows the appropriate dimensioning.

Basic and reference dimensions are frequently used to describe profiles. Where the profile cannot be described by simple arcs and straight lines, it is established through a series of ordinates measured from a datum line or surface. Ordinate points are joined by a smooth curve, and the machinist must use some discretion when reproducing the profile. When checking the actual profile, only the ordinate positions can be used.

The profile tolerance may apply to only a short section of the object, in which case it is shown directly under the tolerance box by between X and Y, and points X and Y are found on the appropriate view. If the tolerance applies to the entire profile, the words All around are written below the tolerance box, as in Fig. 1.36*a*.

Profile of a Surface

The interpretation of the requirements for the profile of a surface are the same as for the profile of a line, with one important addition: the possibility of having the surface in the plane of the profile view, a datum surface. Obviously this reference surface must then be called out in another view, as in Fig. 1.36*a*. When this is done on the drawing, the additional requirement is essentially to impose a perpendicularity tolerance between the profile and the datum surfaces. Therefore, the full tolerance of the line profile is not available if any nonperpendicularity exists. If perpendicularity is perfect, the surface profile and line profile tolerances are equal.

Angularity Tolerance

Tolerance on angles is often expressed as a small deviation of the angle itself, for example, 30° ±0.1°. However, a different and additional angularity tolerance has been established that permits the angular tolerance to exist between two parallel planes or axes separated by a defined (tolerance) amount. Figure 1.37 illustrates clearly the use of this tolerance for both a plane surface *a* and the axis of a hole *b*.

Perpendicularity Tolerance

Like angularity, perpendicularity provides a linear tolerance dimension applicable to surface or axes over the length of the surface or axis. Figure 1.38*a* to *d* illustrates different situations where this tolerance may be found and its interpretation.

Parallelism Tolerance

The parallelism tolerance is found to establish parallelism between two surfaces, between a surface and an axis of a hole or cylinder, or between two axes. The specified linear tolerance is the separation between the two planes or axes between which the actual plane or axis must fall. This tolerance zone applies across the full length of the surface or axis under consideration (Fig. 1.39*a* and *b*).

This on the drawing

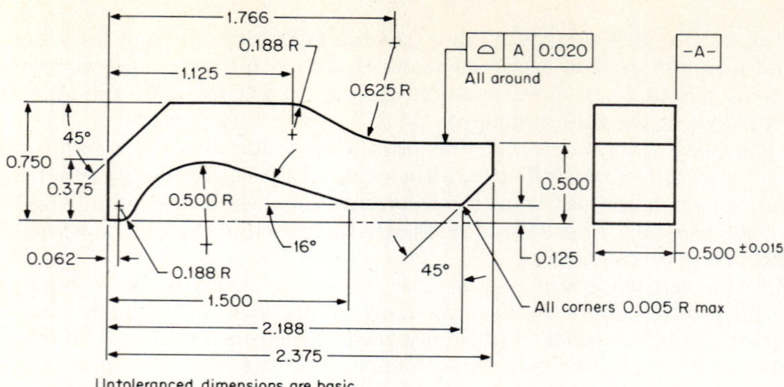

Means this

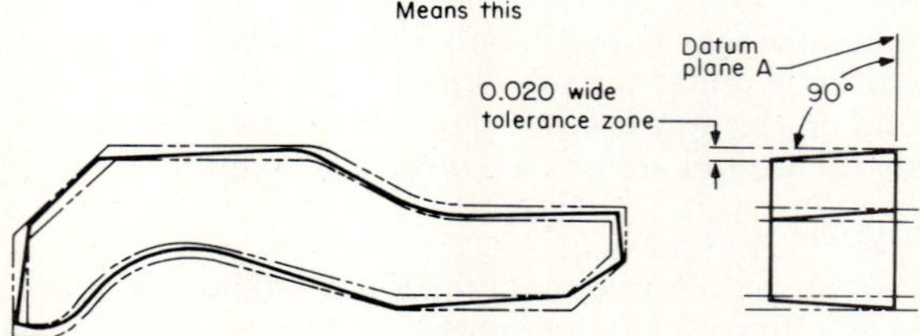

Surfaces all around must lie within two parallel boundaries 0.020 apart equally disposed about the true profile which are perpendicular to datum plane A. Radii of part corners must not exceed 0.005 R.

(a)

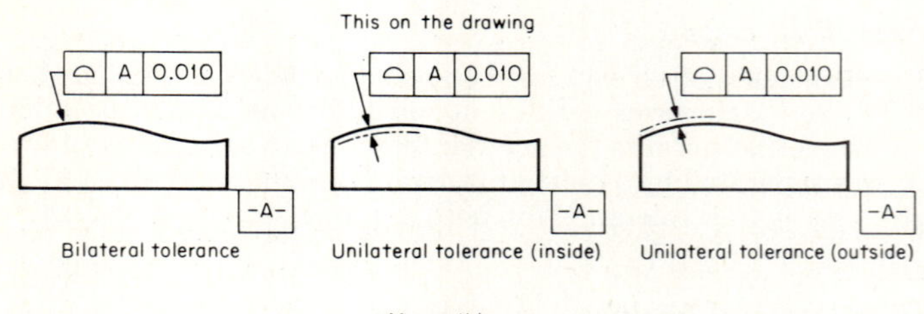

Means this

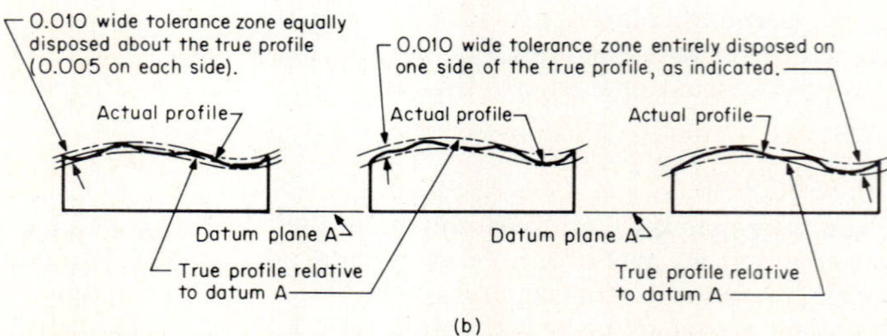

(b)

FIG. 1.36 Interpretation of the profile of a surface symbol on a drawing. *(American Society of Mechanical Engineers.)*

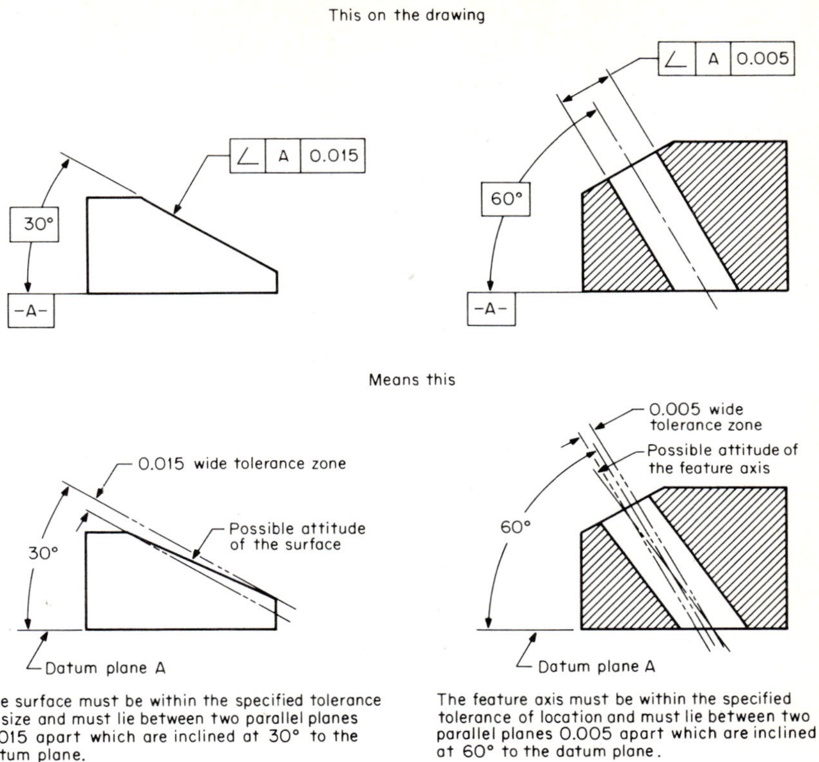

FIG. 1.37 Interpretation of the angularity symbol on a drawing. (*a*) Angularity for a plane surface; (*b*) angularity for an axis. *(American Society of Mechanical Engineers.)*

Positional Tolerance

A positional tolerance is usually found on drawings of flanges, cover plates, manifolds, and other devices where two components are to be assembled together and have a number of bolt holes that must be aligned between the two so all the bolts pass through all the holes. Basic dimensions are used extensively to describe the theoretically exact position of the hole centers, and the tolerance describes the variation permitted from that true position. The tolerance represents a circular or cylindrical zone in which the actual center of the hole must pass. The tolerance is the diameter of the circle or cylinder.

The maximum material condition (MMC) often accompanies the tolerance description (Fig. 1.40). For a hole, the MMC is the minimum diameter permitted by the dimensional tolerance of the hole, which will have been dimensioned separately. The position tolerance is permitted to increase as the hole size increases beyond the MMC up to the maximum permitted by the hole tolerance. Therefore, this maximum tolerance is allowed when the hole is of maximum diameter.

Another situation for positional tolerances is shown in Fig. 1.41, where the datum for positional tolerances is the center hole of the piece (0.735 to 0.740

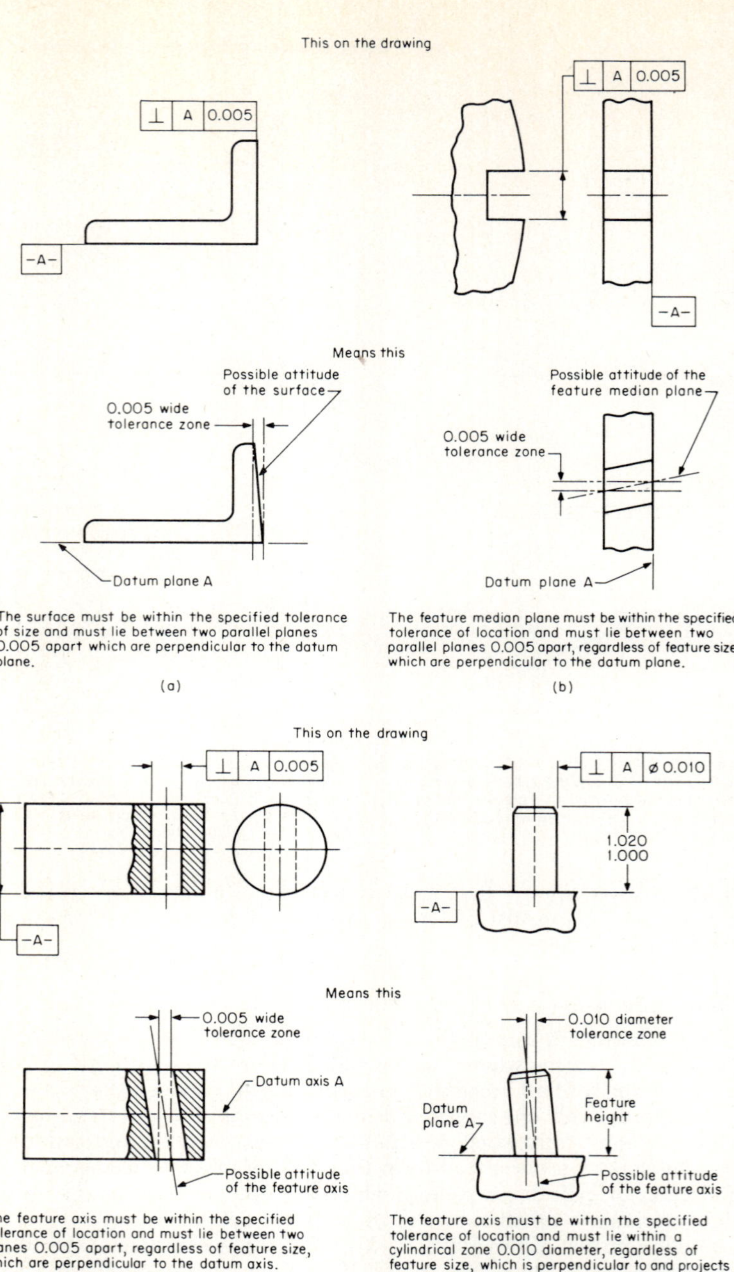

FIG. 1.38 Interpretation of the perpendicularity symbol for a(n) (*a*) plane surface, (*b*) median (middle) plane, (*c*) axis (hole), (*d*) axis (pin or boss). *(American Society of Mechanical Engineers.)*

This on the drawing

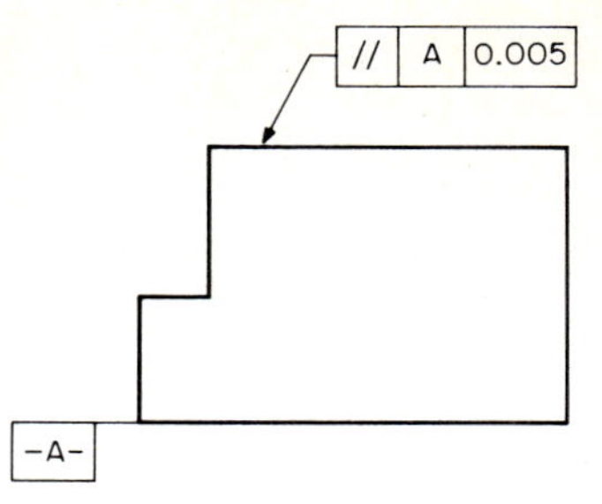

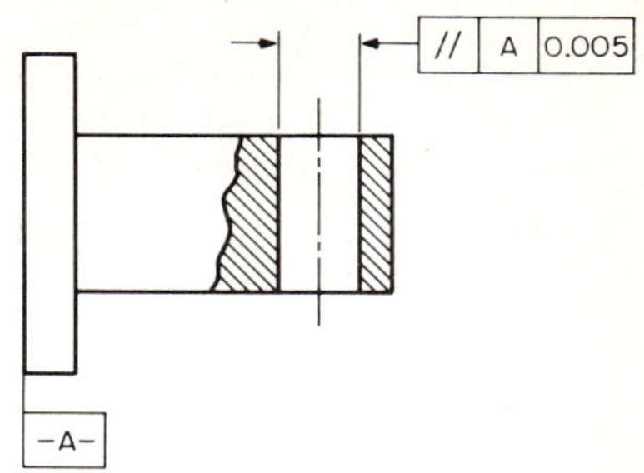

Means this

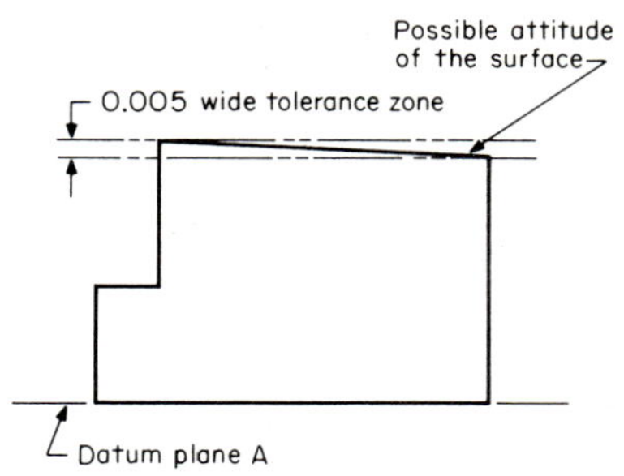

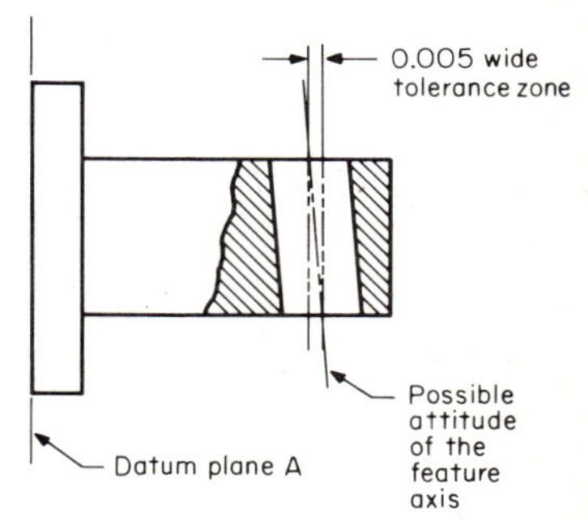

The surface must be within the specified tolerance of size and must lie between two planes 0.005 apart which are parallel to the datum plane.

(a)

The feature axis must be within the specified tolerance of location and must lie between two planes 0.005 apart regardless of feature size, which are parallel to the datum plane.

(b)

FIG. 1.39 Interpretation of the parallelism symbol. *(American Society of Mechanical Engineers.)*

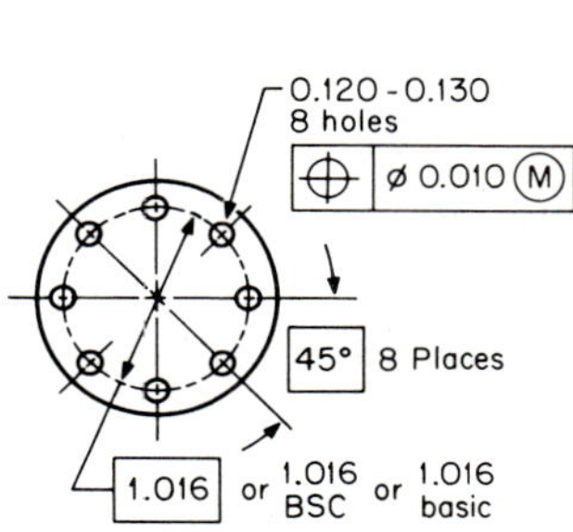

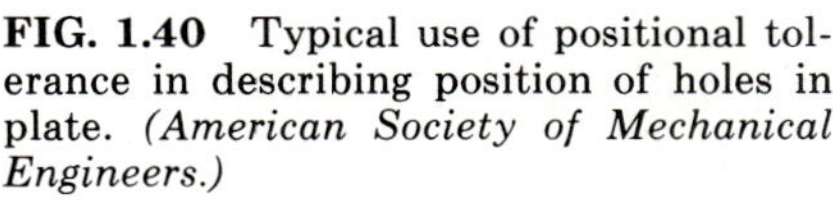

FIG. 1.40 Typical use of positional tolerance in describing position of holes in plate. *(American Society of Mechanical Engineers.)*

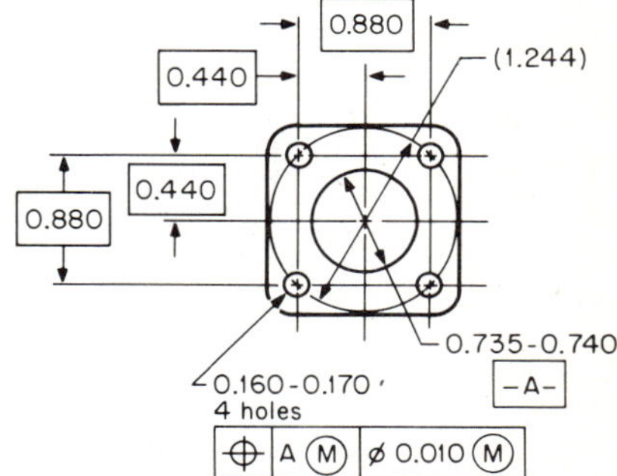

FIG. 1.41 Interpretation of positional tolerances. *(American Society of Mechanical Engineers.)*

diameter). The datum is for the MMC when the hole is of minimum diameter (0.735). The positional tolerance for each hole is given as 0.010 at MMC (i.e., for holes of 0.160 diameter). As the hole diameter increases (to 0.170), the positional tolerance may increase up to 0.020. An additional positional tolerance can also be used if the center hole increases above its MMC (i.e., above 0.735). In fact, another 0.005 (from 0.735 to 0.740) is available if the diameter is actually 0.740.

Positional tolerances may have more than one datum reference. It is common to include the surface in the plane of the paper (which requires another view on the drawing to show it), which essentially places a perpendicularity requirement on the holes.

Concentricity Tolerance

The concentricity tolerance controls the misalignment of two or more cylindrical surfaces. The axes of the datum cylinder and the feature cylinders are established by considering the entire length of the cylindrical surfaces. This is important because if one axis is at an angle to the other, one circular section may be

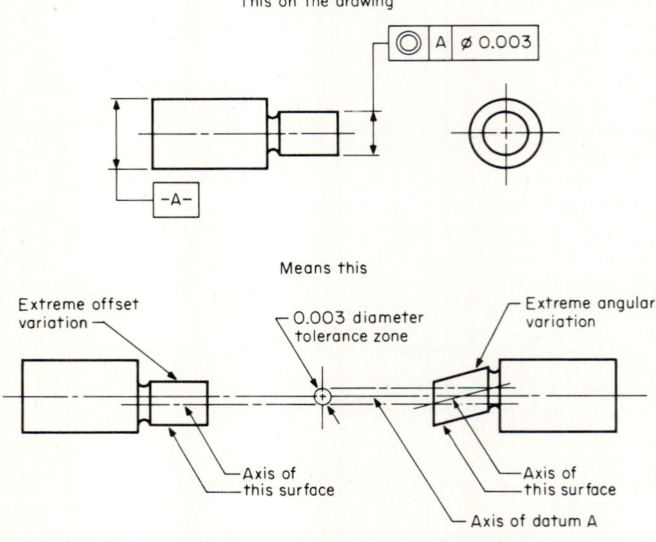

FIG. 1.42 Interpretation of the concentricity symbol on a drawing. *(American Society of Mechanical Engineers.)*

concentric, whereas the whole cylinder is not, as shown in Fig. 1.42. Therefore, when checking pieces for concentricity, more than one section must be measured for each cylindrical surface.

The concentricity tolerance is the total eccentricity along the entire length of the feature axis. It does not include eccentricity caused by other features, such as out-of-roundness. By simple measurement, using a dial indicator, it is impossible to segregate eccentricity from other factors (e.g., roundness). Therefore, although the runout tolerance is more often used, the concentricity tolerance, when called for, requires special consideration in measurement.

Symmetry Tolerance

The symmetry tolerance describes the exactness of a feature about a datum axis plus other possible surfaces. Figure 1.43 explains the use of this tolerance; a second datum, that of a surface, is used in addition to the axis of symmetry. As in some previous tolerances, this additional datum (plane *A* in Fig. 1.43) imposes a perpendicularity requirement on the feature.

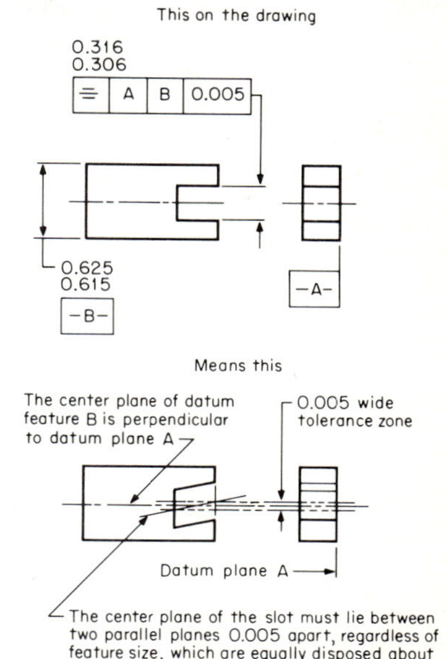

FIG. 1.43 Interpretation of the symmetry symbol on a drawing. *(American Society of Mechanical Engineers.)*

Runout Tolerance

A runout tolerance controls axisymmetric sections or surfaces, such as circular sections, cylinders, and cones, and also surfaces normal to an axis. The datum reference is either an axis, a surface normal to the axis, or both. Figure 1.44 illustrates typical features that may be controlled by runout tolerances.

Runout is checked by using a dial indicator. The tolerance is the full pointer movement measured normal to the feature surface unless otherwise noted. Evidently, when measuring runout in this manner, various attributes can contribute to the indicator movement, such as eccentricity and circularity.

Two different runout tolerances are used. One applies to circular sections (single position for the indicator) and is called *circular runout,* which is used mostly for short features. The other tolerance is *total runout* and requires that several positions of the indicator be used to measure runout along the entire length of the feature. Therefore, total runout includes not only eccentricity and out-of-roundness but also straightness and angularity. When used to control the runout on a surface normal to the datum axis, it will also include perpendicularity and flatness. Circular runout and total runout may be identified in Fig. 1.44.

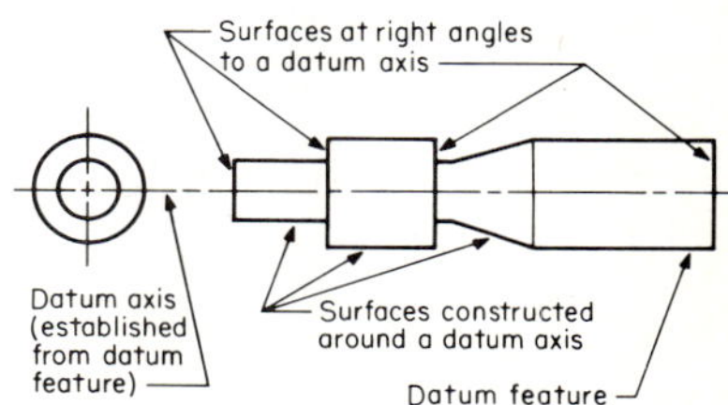

FIG. 1.44 Interpreting runout tolerances. *(American Society of Mechanical Engineers.)*

1.8 SCHEMATIC DRAWINGS

Schematic drawings are symbolic representations used especially for piping and wiring diagrams. An engineering drawing may be provided to show the exact locations of components, but the interconnecting piping and wiring are often left

TABLE 1.4 Graphical Symbols for Pipe Fittings, Valves, and Piping

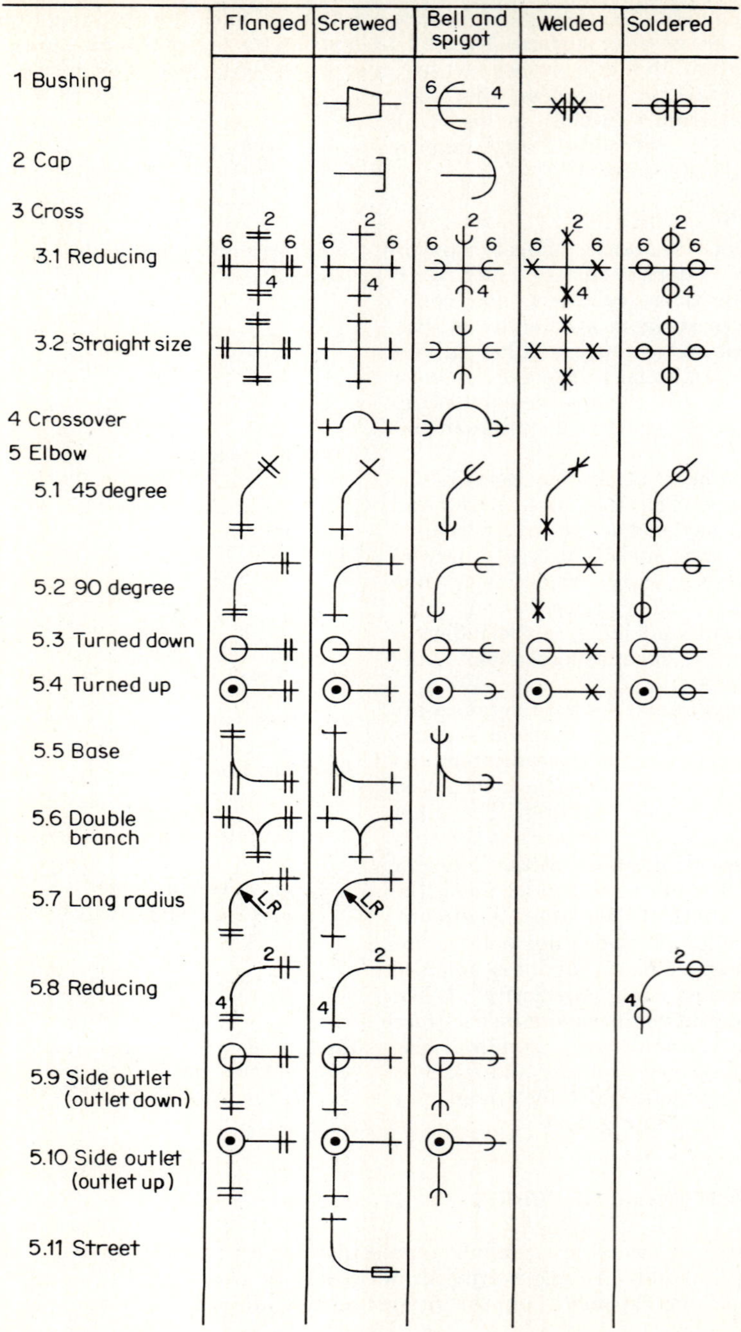

	Flanged	Screwed	Bell and spigot	Welded	Soldered
1 Bushing					
2 Cap					
3 Cross					
3.1 Reducing					
3.2 Straight size					
4 Crossover					
5 Elbow					
5.1 45 degree					
5.2 90 degree					
5.3 Turned down					
5.4 Turned up					
5.5 Base					
5.6 Double branch					
5.7 Long radius					
5.8 Reducing					
5.9 Side outlet (outlet down)					
5.10 Side outlet (outlet up)					
5.11 Street					

TABLE 1.4 (*Continued*)

	Flanged	Screwed	Bell and spigot	Welded	Soldered
6 Joint					
6.1 Connecting pipe					
6.2 Expansion					
7 Lateral					
8 Orifice flange					
9 Reducing flange					
10 Plugs					
10.1 Bull plug					
10.2 Pipe plug					
11 Reducer					
11.1 Concentric					
11.2 Eccentric					
12 Sleeve					
13 Tee					
13.1 Straight size					
13.2 Outlet up					
13.3 Outlet down					
13.4 Double sweep					
13.5 Reducing	2 6 4	2 6 4	2 6 4	2 6 4	2 6 4
13.6 Single sweep					
13.7 Side outlet (outlet down)					
13.8 Side outlet (outlet up)					

TABLE 1.4 (*Continued*)

	Flanged	Screwed	Bell and spigot	Welded	Soldered
14 Union					
15 Angle valve					
15.1 Check					
15.2 Gate (elevation)					
15.3 Gate (plan)					
15.4 Globe (elevation)					
15.5 Globe (plan)					
15.6 Hose angle	Same as	symbol	23.1		
16 Automatic valve					
16.1 Bypass	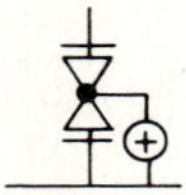				
16.2 Governor-operated					
16.3 Reducing	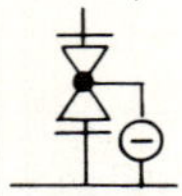				
17 Check valve					
17.1 Angle check	Same as	symbol	15.1		
17.2 Straight way					
18 Cock					
19 Diaphragm valve	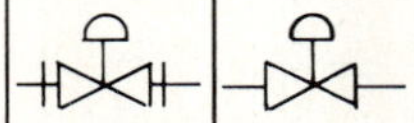				
20 Float valve					

TABLE 1.4 ***(Continued)***

	Flanged	Screwed	Bell and spigot	Welded	Soldered
21 Gate valve					
*21.1					
21.2 Angle gate	Same as	symbols	15.2 and 3		
21.3 Hose gate	Same as	symbol	23.2		
21.4 Motor-operated					
22 Globe valve					
22.1					
22.2 Angle globe	Same as	symbols	15.4 and 5		
22.3 Hose globe	Same as	symbol	23.3		
22.4 Motor-operated					
23 Hose valve					
23.1 Angle					
23.2 Gate					
23.3 Globe					
24 Lockshield valve					
25 Quick-opening valve					
26 Safety valve					
27 Stop valve	Same as	symbol	21.1		

*Also used for general Stop Value symbol when amplified by specification.
SOURCE: American Society of Mechanical Engineers, Publication ASA Z32.2.3–1949.

for the installer to assemble with discretion. It is important that components are connected properly, which is why schematic drawings are prepared.

In fluid flow and electric systems, certain components (e.g., valves and switches) are used so frequently that standardized symbols have been developed to denote them. These symbols not only reduce the drafting time but convey to the reader much more information than would otherwise be possible. Tables 1.4 and 1.5 show the common symbols in piping and wiring systems.

The pipes and wires themselves are usually represented by a single line,

TABLE 1.5 Symbols Used with Electrical Wiring Schematic Drawings

Symbols	Description
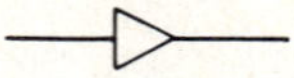	Amplifier
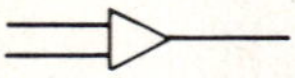	Amplifier with two inputs
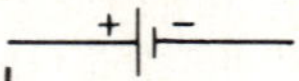	Battery
	Circuit breaker
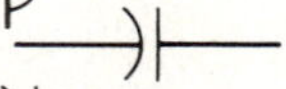	Capacitor
	Variable capacitor
	Photosensitive cell
or	Female contact
 or	Male contact
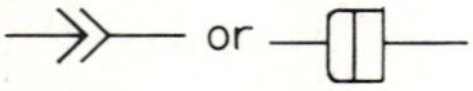 or	Separable connector
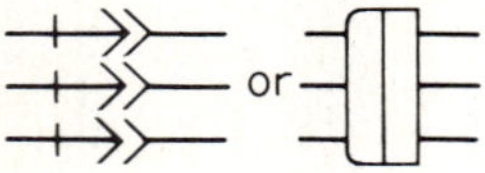 or	3-conductor connector
 or or	Fixed contact
	Fuse
	Fusible element
	Ground, chassis, or frame
	Lamp
	Motor
	Junction of connected wires
	Unconnected wires
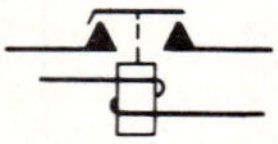	Relay, one pole, double contact, normally open
	Resistor
	Adjustable resistor
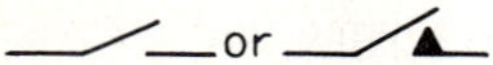 or	Switch
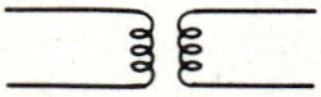	Transformer

regardless of actual size. However, the line may carry a sizing number or written description to indicate exactly what the designer requires. It is common also to identify each component with a letter and number. For example, all valves may be designated "V" and numbered sequentially on the schematic drawing (V-1, V-2, and so on). The bill of materials for the particular drawing then references these component designations with a full description of the component (e.g., size, pressure rating, manufacturer and catalog number).

In piping diagrams usually the style of end connection for the component is shown symbolically. However, many proprietary joining techniques have not been standardized. This information is found on the diagram, bill of materials, or part of the purchase specification.

chapter 2

Selecting Materials

Maurice J. Webb

2.1 GENERAL SELECTION CONSIDERATIONS

When determining which materials are well-suited for a particular purpose, technicians are primarily interested in two areas: (1) selecting materials for the items that they themselves will be constructing and (2) evaluating potential items for purchase. To aid in the latter, manufacturers nearly always list the construction materials in their technical bulletins, catalogs, and brochures. If this information is not readily available in these publications, it can always be obtained by contacting the manufacturer. Only the potential user can really select the appropriate materials suitable for the application; the manufacturer can only provide valuable guidance. The guidance is reliable but not a guarantee.

Unfortunately, material selection is not always simple and straightforward. Sometimes a material that gave perfectly satisfactory service in one application is unsatisfactory in another which is apparently the same. The problem is that the two applications are not entirely identical, and the presence of, say, different trace chemicals or slightly different temperatures create, for that particular situation, a very different set of circumstances which in general terms were considered to be "the same." Then, too, technicians are often working at the forefront of technology, where there is little or no prior experience upon which material selection can be firmly based. In this case, reference should be made to materials that most closely represent in chemical and physical behavior the ones of interest.

In constructing apparatus, the first consideration is adequate strength at operating conditions; the second is chemical compatibility of the materials with the internal and external environment. It usually rests with an engineer to analyze the components for strength and to specify materials and designs that are structurally sound and environmentally suitable. The technician, however, may well be expected to substitute a new part for one that has failed or to add a minor feature to apparatus. By referring to the mechanical properties of the original material, an improved material or adequate alternate can often be selected if failure has occurred. Where a suitable substitution cannot be found, then the physical properties may be reviewed. Often a change in a physical property can render a material mechanically suitable that, at first consideration, was unsuitable. For example, if a part fails structurally in high-temperature service, an alternate material, with a higher thermal conductivity but otherwise of no greater strength at elevated temperature, may well perform better. This is so because it is capable of conducting heat away more efficiently and therefore operates at a lower temperature where its strength is greater. Alternatively, it might be determined that a protective coating of some kind is the best approach. The coating may be used for protection against temperature, erosion, or chemical reaction.

Both the internal and external environment must be considered in the selection of materials. A strongly corrosible atmosphere, for example, salt water, might well pose a more severe external chemical compatibility problem than a chemical fluid passing internally through a pipe, valve, or filter. In addition to external atmospheric corrosion, other external conditions are heat, especially radiant heat, and nuclear radiation.

The chemical compatibility guide of Table 2.1 lists some of the more common chemicals and materials and broadly categorizes them into four groups ranging from excellent service (A) to unacceptable service (D), with two intermediate ratings (B and C). Also, many of the combinations are rated conditionally, as indicated by the numbers referring to the notes at the end of the table. The dif-

TABLE 2.1 A Guide to the Chemical Compatibility of Various Materials

Ratings: A. Excellent service, no appreciable reaction expected in many years of service

B. Good service, some reaction expected but adequate life in many applications

C. Limited service, reaction expected to limit life severely

D. Unacceptable service, not recommended for use

All solutions 100% room temperature unless stated otherwise.

Chemicals	Metals and alloys																	Elastomers					Plastics				
	Ductile iron	Bronze	Gold, 14K	Zirconium	Titanium	Tantalum	Stainless steel, Carpenter's 20-Cb3*	Stainless steel 430 F	Nickel	Stainless steel 316	Stainless steel 303 and 304	Carbon steel	Brass	Aluminum	Monel†	Inconel†	Hastelloy C-276‡	Buna	Silicone	Fluorocarbon	EPDM	SBR	Teflon§	Kel-F**	Nylon	Nylatron GS¶	Vespel§
Acetaldehyde	C		A 22	A 22	A 22	A 22	A 22	A	A 36	A 22	A 22	A 34	A 22	A 22	A 22	A 34	A 34	C	B	D	B	D	A		B 14	B 14	A
Acetamide																					A	C	A				
Acetic acid			A 22	A 22	A 22	A 22	A 34	C 3	A 18	A 11	A	D	D	B	B	B	A 30	B	B	D	B		A	A	C 1	C 1	
Acetic acid, 100%, hot	D		A	A	A	A	B 34	D	B	B	D	D	D	B	B	B	A	D 28	C 28		D 28		A	B			
Acetic anhydride	C	C	A 22	A 22	A 22	A 22	A	D	B	B	B	D	D	A	B	B	A	D	B	D	B	B	A	B 32			
Acetone	A 22	A 22	A 22	A 22	A 22	A 22	A 22	A 22	A 22	A 22	A 22	A 20	A 22	A 36	A 22	A 22	A 4	D	A	D	D	B	A	B	A	A	
Acetylene	A	B	A			A	A	A	A	A	A	A	D		A	A	A				B	A					
Alcohols	C	A	A		A		A	A	A	A	A	B	A	B	A	A	A						A	B	B		
Aluminum chloride	D	B	A	A 26	B	A	A	D	B	D 1	D 1	D 16	D	D 26	D	D	A	A	B	A	A	A	A	A			
Aluminum chloride, 10%, boiling			A	A	D	A	C	D	B 9								A						A	A 9			

TABLE 2.1 (*Continued*)

Chemicals	Metals and alloys																	Elastomers					Plastics				
	Ductile iron	Bronze	Gold, 14K	Zirconium	Titanium	Tantalum	Stainless steel, Carpenter's 20-Cb3*	Stainless steel 430 F	Nickel	Stainless steel 316	Stainless steel 303 and 304	Carbon steel	Brass	Aluminum	Monel†	Inconel†	Hastelloy C-276‡	Buna	Silicone	Fluorocarbon	EPDM	SBR	Teflon§	Kel-F**	Nylon	Nylatron GS¶	Vespel§
Aluminum fluoride	D																				A	A					
Aluminum sulphate	D	B	A 35	A 35	A 11	A 35	A 11	D 26	A 26	A 26	B 26	D					A	A	A		A	B					
Aluminum sulphate, <10%, boiling			A	A	A	A	A		D								B										
Aluminum sulphate, >10%, boiling			A	A	A	A	A	D 22	D	A 22	A 22						B										
Amines, mixed			A							A								D	B		B						
Ammonia, anhydrous	A	D	A	A	A	A	A	A	A	A	A	A	A	A	A	A	B	B	B	D	A		A	A	B 1		
Ammonia, gas, cold			A	A	A	A	A	A	A	A	A	A		A	A	A	A	A	A		A	A			C		
Ammonia, gas, hot			A	A	A	A	A	B	A	B	B	A		A	A	A	B	D	B		B						
Ammonium chloride, 10%	D	D	A		A	A	B	C	B	B	B	C	D	B	B	B	A	A		A	A	A	A		A		
Ammonium chloride, <10%, boiling			A		A	A	B		B	B	D				B	B	A						A				
Ammonium chloride, >10%, boiling			A		B	A	B	D 22	B	D 22	D 22	D			B	B	A 14						A	A			
Ammonium hydroxide, 10%	C		A															A	A		A	D	A	A			
Ammonium hydroxide, hot			A																				A	A			

Ammonium hydroxide, concentrated			A															D	A	A	A		A			
Ammonium nitrate		D	A	A	D	A	B	A 11	B	A 11	A 11	A	D	B	B		A 1	A 1	B 1		A 1	A	A			
Ammonium persulfate			A			A	B	A 2	B	A	A 2	A 2	D	D	D	B	A 1	D 1			A 1	D	A	A 9		
Ammonium phosphate, dibasic	D	C	A	A	A	A	B	A	B	B	B	B	D	D	B	B	A 2	A	A		A	A	A			
Ammonium sulfate			A	A	A	A	B	A 2	B	A 2	A 2	B	D	D	B	B	B 12	A		A	A	B	A			
Ammonium sulfate, >10%, boiling			A	A	A	A	B	D 22	B	A 22	D 22	D			B	B							A	A 9		
Ammonium sulfite	C	C	A			A	B	D 1	D	B 1	B 1	D	D	C	D	D	A 4									
Amyl acetate	C	B	A																		A	C				
Amyl alcohol	C	B	A																		A	B				
Aniline	B		A	A	A	A	A	B	B	B	B	A	D	B	B	B	B	D	D	A	B	D	A	A	C	C
Aniline dyes	C	C																			B	B				
Aniline hydrochloride			A	A	A	A	D	D	C	D	D	D	D	D	D	D	C	B	C		B	C				
Antimony trichloride	D	D	A			A		D	D	D	D	D	D	D	D		A								C 1	C 1
Aqua regia			D	A	A	D	D	D	D	D	D	D	D	D	D	D	C	D	D		C	D	A 30	A 30		
Arsenic acid	D	D	A	A		A	B		D	B	B	D	D	D	D	B	B	A	A		A	A		A 4		
Asphalt	B	A	A															B	D		D	D				
Barium chloride	C	B	A	A	A	A	B	B 1	B	B 1	B 1	A	D	B	B	B	B	A	A		A	A			A 1	A 1
Barium chloride, <5%, hot			A	A	A	A	B	D 22	B	B	D 22	D	D	B	B	B	A 18									

TABLE 2.1 ***(Continued)***

Chemicals	Metals and alloys																	Elastomers					Plastics				
	Ductile iron	Bronze	Gold, 14K	Zirconium	Titanium	Tantalum	Stainless steel, Carpenter's 20-Cb3*	Stainless steel 430 F	Nickel	Stainless steel 316	Stainless steel 303 and 304	Carbon steel	Brass	Aluminum	Monel†	Inconel†	Hastelloy C-276‡	Buna	Silicone	Fluorocarbon	EPDM	SBR	Teflon§	Kel-F**	Nylon	Nylatron GS¶	Vespel§
Barium hydroxide	B	B	A	A		A	A	B	A	A	B	B	D	D	B	B	B 30	A	A	A	A	A					
Barium nitrate			A	B		A	B	B 22	B	B	B 22	B	B	B		B	B	A	A		A						
Barium sulfate	C	C	A																		A	A					
Barium sulfide	C	C	A																		A	B					
Beer	D	B	A							A	A							A	A	A	A	A			B	B	
Beet sugar liquors	B	A	A															A	A		A	A					
Benzene	A	A	A	A	A	A	A 11	B 22	A 37	B 22	B 22	B	B	A	A 37	A 37	B 30	A	D	C	D	D	B	D 30	A	A	A
Benzonic acid			A	A	A	A	B	B 8	B	B 8	B 8	B	B	B	B	B	A 1	D	D		D		A 4,9	C 9	C 9		A
Benzylamine			A			A	B	B	B	B	B	A	D	D	B	B	B										C
Borax		A	A															B	A	A	B	B	A				C
Boric acid	D	A	A	A	A	A	A 1	B	B	A 1	A	C	B	A	B	B	A 30	A	A	A	A	A			A 1	A 1	
Bromic acid			A			A	D	D	D	D	D	D	D	D	D	D											
Bromine, liquid			D					D		D	D						A	D	D		D		A	A	D	D	
Bromine, dry gas	D 22	B 22	A		D 22	A 22		D 22	A 22	D 22	D 22	D 22	D 22	B 22	A 22	A 22	A 22			A 22			A 22				
Bromine, moist gas	D	B	C	A	D	A		D	A	D	D	D	D	D	D	D	A						A				

Bromine trifluoride			D	D		D	B	B	B	B	B	B	B		B	B		D	D		D	D	D				C
Butadiene	F																				C	D					
Butane	B	A	A																		D	D					
Butter—animal fat			A															A	B		A	D					
Buttermilk	D	D	A					A		A	A																
Butyric acid	D		A	A	A	A	A	A 2	C	A 2	A 2	D	D	A	B	C	A	D			B				B 18	B 18	A
Butyric acid, hot, concentrated			A	A	A	A	A	D	C	B 22	D 22			D	C	C	A 30										
Calcium bisulfite	D	B	A	A	A	A	B	D	B	B	B	D	D	B 1	B	B	B	A	A	A	D	D					
Calcium chloride	C	B	A	A	A	A	A 26	C	A 14	C	C	A	B	A	A	A	A	A	A	A	A	A	A	A	B 1	B 1	
Calcium hydroxide	C	A	A			A	A	B	B	B	B	B	B	D	B	B	A	A		A	A	A					
Calcium hydroxide, 10%, boiling			A				A	B	B	B	B	B	B		B	B	A			A							
Calcium hydroxide, 20%, boiling			A							A 22	A 22						A			A							
Calcium hydroxide, 30%, boiling			A							B 22	C 22						A 11			A							
Calcium hypochloride			A															D			A	D					
Carbolic acid (phenol)	D	B	A	A	A	A	A	B 8	A	B 8	B 8	A	A	A	A	A	A	D	D	A	B	D	A	A 2	D	D	
Carbon dioxide, dry	B	A	A	A		A	A	A	A	A	A	A	A	A	A	A	A	A	B		B	B					
Carbon dioxide, wet	B	A	A	A				A	B	A	A	D	D	A	A	A	A	A	B		B	B					
Carbon disulphide	A	C	A															D		A	D		A	A	A	A	
Carbon tetrachloride, dry, hot	C	C	A	A	A		A	A	A	B 8	B 8	A	B	A	A	A	A	B	D	A	D	D	B	D	A	A	A
Carbonic acid			A	A	A	A	A	B	B	B 8	B 8	B	D	A	C	A	A 8	B	A		A	B			A 1	A 1	

TABLE 2.1 ***(Continued)***

Chemicals	Metals and alloys																	Elastomers					Plastics				
	Ductile iron	Bronze	Gold, 14K	Zirconium	Titanium	Tantalum	Stainless steel, Carpenter's 20-Cb3*	Stainless steel 430 F	Nickel	Stainless steel 316	Stainless steel 303 and 304	Carbon steel	Brass	Aluminum	Monel†	Inconel†	Hastelloy C-276‡	Buna	Silicone	Fluorocarbon	EPDM	SBR	Teflon§	Kel-F**	Nylon	Nylatron GS¶	Vespel§
Castor oil	B	A	A				A	A	A	A	A	A	A	A	A	A	A				B	A					
Chloroacetic acid	D	B	A	A	A	A		D	B	D	D	D	D	D	C	C	A	D		D	B			C 4	D 1	D 1	
Chloric acid			A			A	B	D	D	D	D	D	D	D	D		A										
Chlorinated water, saturated			A	A	A	A	B	D		B	C	B		B		B	A	C			B		A		D 1	D 1	
Chlorine, dry			B 13	A	A		A	C	B	B	B	A	D	B		A	A	D	D	A	D	D	A	A 10			
Chlorine, moist gas (wet)	D	D	D	D	A		D	D	D	D	D	D	D	D		D	A 3			A	C	D	A				
Chlorine trifluoride			A		D		A	A	A	A	A	A	A	A	A	B		D	D	A	D	D	D				
Chromic acid	D	D	A	A 11	A 11	A	B	B 2	D	A 2	A 2	A	D	D	D	B	B	D 11	C 11	A 11	C 11	D	A	A	C 7	C 7	B
Chromic acid, <10%, boiling			A	A	A	A	D	C 1		B 1	B 1				D		D 1						A				
Chromic acid, >10%, boiling			A			A	D	D 22		C 22	D 22					D 22	D						A	A			
Chromic acid + sodium bisulfate			A			B				B	B																
Citric acid	D	B	A	A	A	A	A	A 2	B	A 2	A 2	D	D	B	A 26	A 14	A	A	A	A	A	A			C 1	C 1	
Citric acid, hot, concentrated			A	A	D	A	B	D	C	B	D			C	B	B	A 30										

Copper chloride	D	C																			A	A					
Copper nitrate	D		A	A		A	A	B 22	D	A 22	A 22	D	D	D	D	D	B 12	A	A		A						
Copper sulfate	D		A 22	A	A	A	A 26	B 22	B	A	A	D	D	D	B	B	A 30	A 12	A 12	A	A 12	B	A 4,9		B 1	B 1	
Creosote	B	A	A	A		A	A	B 22	A	B 22	B 22	A	B	B	A	A	A 4	A	D	A	D	D			A	A	
Cresylic acid	D	B																			D	D					
Cupric chloride			B	D	B	B	D		D			D	D		B		A	A	A	A	A			A 4,9	B 1	B 1	
Cupric chloride, 5%			B	D	B	B	D	B 33	D	B 33	C 33	D	D	D	B		A	A		A							
Diesel oil	B	B	A				A	A	A	A	A	A	A	A	A	A	A				D	D					
Diethylene glycol	A		A																		A	A					
Dimethylamine			A			A	A	A		A	A	A	D														
Ethyl acetate	C	C	A																		B	D					
Ethyl chloride	C		A	A	A	A	A	D	A	A	D	D	A 5	A 5	B	A 5	B 12	A	D	A	A	B					
Ethylene glycol	B	B	A	A	A	A	A	B		B	B	A	B	A	B	B	A	A	A	A	A	A	A	A 4,9	B 8	B 8	
Ethylene oxide	B	A	A							A	A		C			A	A				C						
Fatty acids		C	A	A	A	A	A	B	A	A	B	D	C	A	B	B	A 4	B			D	C					
Ferric chloride	D	D	A	B	A	A	D	D 14	D	D 14	D 14	D	D	D	D	C	B	A	B	C	A	A	A				B
Ferric chloride, <1%			A					B		A	B						A						A				
Ferric chloride, <1%, boiling			A					D		D	D						D						A				
Ferric chloride, >1%, boiling			A	D	A	A								D			D						A	A 9			

TABLE 2.1 ***(Continued)***

Chemicals	Metals and alloys																	Elastomers					Plastics				
	Ductile iron	Bronze	Gold, 14K	Zirconium	Titanium	Tantalum	Stainless steel, Carpenter's 20-Cb3*	Stainless steel 430 F	Nickel	Stainless steel 316	Stainless steel 303 and 304	Carbon steel	Brass	Aluminum	Monel†	Inconel†	Hastelloy C-276‡	Buna	Silicone	Fluorocarbon	EPDM	SBR	Teflon§	Kel-F**	Nylon	Nylatron GS¶	Vespel§
Ferric nitrate	D	D	A			A	A 35	A 2	D	A 2	A 2	D	D	D	D	D	A 1	A	B		A	A					
Ferric sulfate	D	D	A																		A	A	A				
Ferric sulfate, 10%			A 1	A	A	A	A 1	A	A 1	A	A	D	D	D	D	A	A 26										
Ferrous sulfate, 10%	D	B	A	A 14	A 11	A	B	B	D	B 22	B 22	D	D	A	C	B	B 30							A 4,9			
Fluorine, dry gas			A 32	B	A	A	A	A	A	A	A	A	B	A	A	A	B						D	A 29	D	D	
Fluorine, dry, 300°F			A 22	B	A	A	A	A	A	A	A	A	B	A	A	A							D				
Fluorine, moist gas			B						B				D	D			A										
Formaldehyde	C	A	A	A	A	A	A	A 22	A 14	A 22	A 22	A	A 11	A	A 18	A 18	B	C	B	A 23	B		A	B 4	B 14	B 14	
Formic acid, <50%		C	A	A	D	A	A	C	B	A	B	D	D	B	B	B	A						A	A	D 1	D 1	B
Formic acid, >50%		C	A	A	B	A	A	A	B	A	A	D	D	B	B	B	A				A	A	A				
Formic acid, <50%, boiling		D	A	A	D	A	A	D	B	D	A				D	D	B 18						A				
Formic acid, >50%, hot		D	A	A	D	A	A	D	B	A	A			D	C	D	B			D			A 30	C			
Freon-11,15																		B	D	B	D	D		C			

Freon-12			A	A	A	A	A	B	B	A	A	D 17	D 17	B	B	B		A	D	B	B	A		B	A	A	A
Freon-13			A	A	A	A	A	A	A	A	A	B	A	A	A	A	A	A	D	B	A	A					A
Freon-13B1																		A	D	B	A	A					
Freon-14																		A	D	A	A						
Freon-17			A	A	A	A	A	A	A	A	A	A	A	A	A	A	A										A
Freon-21																		D	D	D	D	D					
Freon-22			A	A	A	A	A	A	A	A	A	A	A	A	A	A	A	D	D	C	A	A		B			A
Freon-31																		D		D	A	B					
Freon-32																		A		D	A	A					
Freon-112																		B	D		D						
Freon-113			A	B	A	A	A	A	A	A	A	A	A	A	A	A	A	A	D	B	D	B		B			A
Freon-114																		A	D	A	A	A					
Freon-114B2																		B	D	A	D	C					
Freon-115																		A		A	A	A					
Freon-142b																					A	A					
Freon-152a																		A			A	A					
Freon-152b																		A			A						
Freon-218																					A	A					
Freon-C316																					A	A					
Freon-C318																				A	A	A					
Freon-502																		B			A	A					
Freon-BF																		B	D		D	D					
Freon-MF																		B	D		D	B					
Freon-PCA																		A	D		D						

TABLE 2.1 (*Continued*)

	Metals and alloys																	Elastomers					Plastics				
Chemicals	Ductile iron	Bronze	Gold, 14K	Zirconium	Titanium	Tantalum	Stainless steel, Carpenter's 20-Cb3*	Stainless steel 430 F	Nickel	Stainless steel 316	Stainless steel 303 and 304	Carbon steel	Brass	Aluminum	Monel†	Inconel†	Hastelloy C-276‡	Buna	Silicone	Fluorocarbon	EPDM	SBR	Teflon§	Kel-F**	Nylon	Nylatron GS¶	Vespel§
Freon-TF																		A	D		D	B		A	A		
Freon-T-WD602																					B	B					
Freon-TMC																					B	C					
Freon-T-P35																					A	A					
Freon-TA																					A	A					
Freon-TC																					B	A					
Fuel oil	B	A	A							A 4	A 4							A	D	A	D	D					
Furfural	B	A	A	A	A	A	B	B 22	A	B 22	B 22	A	B	B	B	B	B 30	D	D	D	B	C			A	A	
Gasoline, refined	B	B	A					A		A	A						A 4	A	D	A	D	D	A	B 30			A
Gelatin	D	A	A																		A	A					
Glucose	B	A	A																		A	A					
Glycerine	A	A	A			A	A	B	A	B	B	A	A	A	A	A	A 8				A	A					
Hydraulic oil (petroleum)	B	B	A																		D	D					
Hydrazine hydrate	C	C	A				A	B	A	B	A	C	C	A		A	A	B	B		A						
Hydrochloric acid, <1%			A														A						A		C	C	
Hydrochloric acid			A	A	B	A	C	D 12	B	D 12	D	D	D	D	B	B	A	C 1	D 1	A 18	A 1		A		D 1	D 1	B

Hydrochloric acid, >20%			A	A	D	A	D		D			D	D	D	D	D	A	D 8	D 8	A 23	C 8		A	A			
Hydrochloric acid, <0.5%, 175°F			A						A							A	A						A				
Hydrochloric acid, 0.5–2%, 175°F (hot)			A														A			A	C	D	A				
Hydrochloric acid, >2%, 175°F			A 22	A 23		A	D										C 1			A	C	D	A	A 23			
Hydrochloric acid, <0.25%, boiling			B																				A				
Hydrochloric acid, 0.25–1%, boiling			B																				A				
Hydrochloric acid, >1%, boiling			A 2	C		A																	A	A			
Hydrochloric acid, aerated			A	D	D	A	D	D	D	D	D	D	D	D	D	D	B										
Hydrofluoric acid, <40%			A	D	D			D	B	D	D	D	D	D	A	A	B	C	D	A 24	A		A		D	D	C
Hydrofluoric acid, <40%, hot			A												A	D	B						A				
Hydrofluoric acid, >40%			A	D	D		B	D	D	D	D	A	D	D	A	D	B	D	D	A	C		A				
Hydrofluoric acid, >40%, hot			A								D	B			B		B						A				
Hydrofluoric acid, boiling			A						D			D			B	D	B	D	D	C	D		A				
Hydrofluosilicic acid	D	A															B	B	D		A	B					
Hydrogen chloride, dry			A	A		A	A	D	A	A	A	A	B	D	A	B	A 4										
Hydrogen chloride, moist			A			A			D			D			D	D											
Hydrogen fluoride, dry			A	A	A		A	B	A	A	A	A	B	B	B	B	A 6						D				C
Hydrogen peroxide	D	C	A	A	A	A	B	B	B	B	B	B 1	D	A	B	B		B	A		A		A	A 26	D 7	D 7	

TABLE 2.1 (*Continued*)

Chemicals	Metals and alloys																	Elastomers					Plastics				
	Ductile iron	Bronze	Gold, 14K	Zirconium	Titanium	Tantalum	Stainless steel, Carpenter's 20-Cb3*	Stainless steel 430 F	Nickel	Stainless steel 316	Stainless steel 303 and 304	Carbon steel	Brass	Aluminum	Monel†	Inconel†	Hastelloy C-276‡	Buna	Silicone	Fluorocarbon	EPDM	SBR	Teflon§	Kel-F**	Nylon	Nylatron GS¶	Vespel§
Hydrogen peroxide, 90%			A	A		A	A			A	A			A				D	B	A	C	D	A				
Hydrogen peroxide, boiling			A	A	A	A	B	C 22	B	A 22	C 22	B		A	B	B				B			A				
Hydrogen sulfide, dry, cold	B	C	A	A	A		B	C	B	B	C	B	B	B	B	B	A	A	C		A						
Hydrogen sulfide, dry, hot			A	A	A		B	B	B	B	B	B	B	B	B	B	A	D	C	B	A			A			
Hydrogen sulfide, wet, cold	D	D	A																		A	D	A				
Hydrogen sulfide, wet, hot			A																		A	D	A				
Iodine, dry			A		A	A	B	D	A	D	D	B	D	A	A	D	B	B			B		A	D	D		B
Iodoform	C	C	A																		A		A				
Isopropanol			A	A	A	A	B	B	B	B	B	B	B	B	B	B	A 1	B	A	A	A		A	B	B		
Kerosene	B	A	A	A	A	A	A	A	A	A	A	A	A	A	A	A	A	A	D	A	D	D	A	A			
Lacquers and lacquer solvents	C	A																			D	D					
Lactic acid, 5%	D		A					B		A	A						A	A		A	A		A				
Lactic acid, 10%	D		A	A	A	A	B	D	B	B	B	D	C	B	D	B	A	A		A 25	A		A	C	C		B
Lactic acid, 5%, boiling	D		A															D			D		A				

Lactic acid, 10%, boiling	D		A	A	A	A	B	D 22	D	B 22	B 22		D	D	D	B	B	D			D		A				
Lead acetate			A	A	A	A	B	B	B	B	B	A	D	D	B	B	B	B	D		A		A	B	B		B
Linseed oil	A	B	A	A	A	A	A	A	A	A	A	A	A	A	A	A	A				B	D	A	A			
Liquid petroleum gas	B	A	A																		D	D					
Lithium, molten			D	B	C	A	A	A	B	A	A	A	D	D	D	B							D				B
Lubricating oils (petroleum)	A	B	A	A	A	A	A	A	A	A	A	A	A	A	A	A	A				D	D	A	A			
Magnesium chloride	D	B	A	A	A	A	A 14	B 11	A 14	B 11	B 11	A	B	A 18	A 1	A 26	A	A	A	A	A	A		A 1	A 1		
Magnesium hydroxide	D		A	A			A	A	A	A	A	A	A	B	B	A	A	B		A	A	B		A 1	A 1		
Magnesium sulfate	A	B	A	A 11	A	A	A 35	B	A 35	B	A	B	A 26	A 14	A 14	A 14	B 11	A	A		A	B		A 1	A 1		
Magnesium sulfate, boiling			A	A 11	A	A	A	B 22	A	B 22	B 22	B	A	A	A	A	B 11										
Maleic acid		B																			C	B					
Malic acid	D	B																			D	B					
Mercuric chloride, 10%	D	D	A	A	A	A		D	C	D	D	D	D	D	D	D	B	A 9		A 9	A 9	A		C 3	C 3		B
Mercury	A	D	D	A	A	A	A	B	B	B	B	A	D	D	B	B	A 4	A		A	A	A	A	A	A		
Mercuric cyanide			D				B	C 1		B	B	B	D	D			B										
Methyl alcohol	A	A																			A	A	A	A	A	A	A
Methyl cellosolve	B	A																			B	D					
Methyl chloride	A	A	A		A	A	A	B	B	A	A 5	A	A 5	C	B	B	B	D	D		C	D					
Methyl ethyl ketone	A	A																			A	D					
Methyl formate	C	A																			B	D					

TABLE 2.1 ***(Continued)***

Chemicals	Metals and alloys																	Elastomers					Plastics				
	Ductile iron	Bronze	Gold, 14K	Zirconium	Titanium	Tantalum	Stainless steel, Carpenter's 20-Cb3*	Stainless steel 430 F	Nickel	Stainless steel 316	Stainless steel 303 and 304	Carbon steel	Brass	Aluminum	Monel†	Inconel†	Hastelloy C-276‡	Buna	Silicone	Fluorocarbon	EPDM	SBR	Teflon§	Kel-F**	Nylon	Nylatron GS¶	Vespel§
Methanol			A	A	A	A	A	B	A	B	B	A	B	B	A	A	A 17	A	A	C	A		A	B	B		
Milk	D	A						A		A	A							A	A		A	A		A	A		
Molasses	C	A						A		A	A																
Naptha	B	B	A	A	A	A	A	B	A	B	B	A	A	A	A	A	A	B	D	A	D	D		A			
Naphthalene	B	B																			D	D					
Natural gas	A	A																			D	C					
Nickel chloride	D	D	A	A	A	A	B	D 26	D	B	B	D	D	D	B	B	A	A	A		A	A					
Nickel sulfate	D	D	A	A		A	B	A	B	A 1	A 1	D	C	D	B	B	B 11	A	A		A	B			A 1	A 1	
Nitric acid, 10%	D	D	A			A	A	B 29	D	A 29	A 29	D	D	D	D	B	A	D	D	A	B		A	A	D	D	D
Nitric acid, red fuming			C	A	A	A	A	A	D	A	A	C	D	A	D	B	B	D	D	C	D	D	A	A			
Nitric acid, white fuming			C	A	A	A	B	D	D	B	B	D	D	A	D	B		D	D		D						
Nitric acid, boiling, 20%			A			A	D	B		A	A						B	D	D		D		A	B			
Nitric acid, boiling, 65%			A			A	D	D		B	B						B	D	D		D		A	A			
Nitric acid, boiling, concentrated			B			A	C	D		D	D			D			B	D	D	D	D	D	A				
Nitrobenzene	B	D																			D	D					

Nitrous acid			A			A	B	B	D		B	D	D	C	D	D											
Nitrogen	A	A																			A	A					
Nitrogen tetroxide			A		A	A	A		A	A	A	B	D	A	A	A		D	D	C	D	D	A				
Oleic acid	C	B																			B	C					
Oxalic acid 100%	D	B	A	A		A	B	D	B	D	B	D	C	B	B	B	B	B	B		A	B					
Oxalic acid, 10%			A	A	B	A	B	B		B	B	D	B	B	B	A	B								C	C	
Oxalic acid, boiling, 10%			A	A	D	A	B	D 22	C	D 22				D	B	B	B							B			
Oxalic acid, boiling, 50%			A	A		A	B	D 22	C	D 22	D 22			D		B	B										
Phosphoric acid (ortho), <10%	D	D	A 22				B	A 2	B	A 2	A 2	D	B	D	B	B	A	D	B		A		A		D	D	
Phosphoric acid (ortho), 10–50%	D	D	A				B		B	B	B	D	B	D	B	B	A				B	D	A				B
Phosphoric acid (ortho), concentrated	D	D	A		D		B	D	B	B	D	D	B	D	A	B	A	D	C		B	D	A				
Phosphoric acid (ortho), <20%, 175°F	D	D	A				B		D	B	B		B		B	B	A					D	A				
Phosphoric acid (ortho), >20%, 175°F	D	D	A				B			B	B		B		B	B	B					D	A				
Phosphoric acid (ortho), <10%, boiling	D	D	A				B	D	D	B	B		D		B	B	B					D	A	A			
Phosphoric acid (ortho), 85%, boiling	D	D	A				B		D	B	D		D		D		B			A 22		D	A	A			
Phosphoric acid, +20–200 ppm chloride	D	D	A								A	D															
Picric acid, H_2O solution			A			A	B	B	B	B	B	D	D	A	D	B	B	A			B	B					
Picric acid, molten			A															B	D	A	B	B					
Potassium, molten			D		A	A			B		B												D				B
Potassium acetate																					A						

TABLE 2.1 ***(Continued)***

Chemicals	Metals and alloys																	Elastomers					Plastics				
	Ductile iron	Bronze	Gold, 14K	Zirconium	Titanium	Tantalum	Stainless steel, Carpenter's 20-Cb3*	Stainless steel 430 F	Nickel	Stainless steel 316	Stainless steel 303 and 304	Carbon steel	Brass	Aluminum	Monel†	Inconel†	Hastelloy C-276‡	Buna	Silicone	Fluorocarbon	EPDM	SBR	Teflon§	Kel-F**	Nylon	Nylatron GS¶	Vespel§
Potassium bisulfite	D	C																									
Potassium bromide	D	C	A	A	A	A	B	B	B	B 22	B	C	B	D	B	B	A 4	A	A		A				A 1	A 1	
Potassium carbonate	B	B	A	A	A	A	B	B	B	B	B	B	B	D	B	B	B	A	A		A				A 35	A 35	
Potassium chlorate	B	B	A	A	A	A	B 26	B 22	C	B 22	B 26	A	B	B	C 1	C 1	B	A	A		A						
Potassium chloride	B	C	A	A	A	A	A	B 1	A 26	B 1	B 1	A	B	D	A 26	A 26	A 26	A	A		A				A 17	A 17	
Potassium chloride, hot			A	A	A	A	B	D	B	B 22	B 22	D			B	B	A										
Potassium cupro cyanide																					A	A					
Potassium cyanide	A		D		D	A	B	B	B	B	B	A	D	D	B	B	B	A	A		A	A					
Potassium dichromate			A	A	A	A	A 26	B	B	A 26	A 26	B	B	A	B	B	B	A	A	A	A	B		A 4,9	C 2	C 2	
Potassium diphosphate	A	B																									
Potassium ferricyanide, 10%	C	D	D	A		A	B	A 2	B	A 2	A 2	B	B	B	B		B								A 26	A 26	
Potassium ferrocyanide, 10%			D	A		A	B	A	B	A	A	B	B	A	B	B	B								A 26	A 26	
Potassium hydroxide, 50%	B		A	A	A		B	B	A	B	B	B	D	D	A	B	B	B	C	A	A	B	A	A 1	A 1	C 11	

Potassium hydroxide, 30%, 175°F			A	A	A 1		B	B	A	B	B	B	D		A	B	B			A			A	A		
Potassium hydroxide, 50%, 175°F			A	A			B	B	A	B	B	B	D		A	B	B			A			A			
Potassium hydroxide, 30%, boiling			A	A	D		B	B 22	A	B 22	B	B	D		A	B	B			A			A			
Potassium hydroxide, 50%, boiling			A	A	D		D	B 22	B	B 22	B	B	D		A	B	B			A			A	A		
Potassium hypochlorite			A		A	A	B	D	D	B	D	D	D	D	A	B	B 12	A	A		A					
Potassium iodide	C	D																								
Potassium nitrate																					A	A				
Potassium permanganate, diluted	B	B	A 1			A 1	B	A 2	B	A 2	A 2	B	B	A	C	B	A			C 26			A	A 9	D	D
Potassium sulfate, 10%	C	B	A		A	A	A 1	B	A	B	B	B	C	A	A	A	A	A	A		A	B			A 8	A 8
Potassium sulfate, diluted, boiling			A			A	A	B 1	A	B 1	B 22	D	B	A	A	B	A									
Potassium sulfide, saturated			A	A 1	A 1	A	B	B 1	D	B	B	B	C	A	B	A 1	A 1								A 17	A 17
Potassium triphosphate	A																									
Propane, liquid and gas	B	A	A	A	A	A	A	A	A	A	A	A	A	A	A	A	A	A	D	A	D	D			A 10	A 10
Propyl acetate																					B	D				
Propyl alcohol			A				A	A	A	A						A					A	A	A	A		
Propyl nitrate																					B					
Pyrogallic acid	C	B	A			A	B	B	B	B	B	C	B	B	B	B	B							A 4,9		
Rosin, molten			A			A	B	B	A	B	B	B	B	B	B	A	A 4									
Salicylic acid	D	C	A			A	B	B	B	B	B	D	B	B	B	B	A 4	B			A	B		A	A	A

TABLE 2.1 (*Continued*)

Chemicals	Metals and alloys																	Elastomers					Plastics				
	Ductile iron	Bronze	Gold, 14K	Zirconium	Titanium	Tantalum	Stainless steel, Carpenter's 20-Cb3*	Stainless steel 430 F	Nickel	Stainless steel 316	Stainless steel 303 and 304	Carbon steel	Brass	Aluminum	Monel†	Inconel†	Hastelloy C-276‡	Buna	Silicone	Fluorocarbon	EPDM	SBR	Teflon§	Kel-F**	Nylon	Nylatron GS¶	Vespel§
Salt water/spray			A		A	A															A	A					
Sewage																					B	B					
Silver bromide, 10%			A		A	A	B	D	B	D	C	D	B	D	B		A										
Silver chloride, 10%			A		A	A	B	D		D	D	D	D	D			B										
Silver nitrate	D	D	A	A	A	A	A 35	B 22	D	B 22	B 22	D	D	D	D	B	A	B	A		A	A			A	A	
Soap solutions	B	A																			A	B					
Sodium, molten			D	A	A	A	A	B	A	A	A	A	A	A	A	A	B			A			D				B
Sodium acetate	C	B	A	A	A	A	A 1	B	B	B	B		B	A	B	B	B	B	D		A	C			B 35	B 35	C
Sodium bicarbonate	C	B																			A	A	A	A			
Sodium bisulfite	D	B																			A	B	A				
Sodium bisulfate	D	B	A	A	A	A	A	B	B	A 1	D	D	D	B	B	B	B 12	A	A		A						
Sodium bisulfate, 140°F			A	A	A	A	A 1	B 22	B	A 1	A 22	D			B	B	B										
Sodium borate	C	B																			A	A					
Sodium bromide, diluted	D	B	A			A	B	D	B	D	A	B	C	D	B	B	B 12								A 1	A 1	
Sodium carbonate, 10%	B	C	A	A	A	A	A	B	A	B	A	B	D	D	A	A	A	A	A		A			A 7	A 11	A 11	

Sodium chlorate	C	B																									
Sodium chloride	C	B	A	A	A	A	B	B 1	A	B 1	B 1	B	B	C	A	A	A 4	A	A	A	A	A		A 4,9	A 17	A 17	
Sodium chloride, saturated, boiling					A			D	A		A				A	A				A							
Sodium cyanide	A	D	D		A	A	A 1	A	B	A 1	A 1	A	D	D	B	A		A	A		A	A			A 1	A 1	
Sodium fluoride	D	C	A			A	B	C 1	B	B	C 1	A 1		A	B												
Sodium hydroxide	A		A	A	A		A	A	A	A	A	A	D	D	A	A	A	B 1	A 1	A 18	A	A	A	A 1	A 1	C 11	
Sodium hydroxide, 40%, 175°F			A	A	A		A	A 34	A	A 34	A 34	B	D		A	A	B						A				
Sodium hydroxide, 40–80%, 175°F			A	B	B		B	A	A	A	A	C	D		A	A	B						A				
Sodium hydroxide, < 30%, boiling			A	A	A		B	B 22	A	A 22	A 22	C	D		B	B	B						A	B 26			
Sodium hydroxide, > 30%, boiling			A	B	B		C	D 22	A	B 22	B 22	D	D		B	B	B			C			A	A 11			
Sodium hydroxide, molten			A	D	D		D	D	D	D	C	D	D	D	D	D											
Sodium hypochlorite (still)	D	D	A	A	A	A	B	D	D	B	D	D	D	D	A	D	A 1	B 18	B	A 2	B	C	A		D	D	
Sodium hyposulfite			A			A	B	C	B	B	B	D	C		B		B 12										
Sodium metaphosphate																					A	A					
Sodium monophosphate	D	B																									
Sodium nitrate	A	B	A		A	A	A 35	A	B	A	B	B	C	A 17	B	A	B 26	B	D		A	B			A 11	A 11	
Sodium perborate, 10%	C	B	A			A	B	B	B	B	B	B		D	B	B	B	B 9	A 9		A 9	B			B	B	C

TABLE 2.1 (*Continued*)

Chemicals	Metals and alloys																	Elastomers					Plastics				
	Ductile iron	Bronze	Gold, 14K	Zirconium	Titanium	Tantalum	Stainless steel, Carpenter's 20-Cb3*	Stainless steel 430 F	Nickel	Stainless steel 316	Stainless steel 303 and 304	Carbon steel	Brass	Aluminum	Monel†	Inconel†	Hastelloy C-276‡	Buna	Silicone	Fluorocarbon	EPDM	SBR	Teflon§	Kel-F**	Nylon	Nylatron GS¶	Vespel§
Sodium peroxide, 10%	C	D	A			A	B	B	B	B	B	B	D	B	B	B	B	B 9	D 9	A 9	A 9	B	A 9				
Sodium phosphate, tribasic	C	C	B	A	A	B	B	B	B	B	A	B	B	D 1	B	B	B	A	A		A	A		A 4,9	A 17	A 17	
Sodium potassium (NaK)			D			A				A	A						B 31										
Sodium silicate	A	A	A	A	A	A	B	B	B	B	B	B	B	A 38	B	B	B	A			A	A	A	A	A	A	
Sodium sulfate (all concentrations)	B	B	A		A	A	A	D	B	A	A	B	B	A	A 1	B	B 26	A	A		A	B	A		A 17	A 17	
Sodium sulfate, hot			A		A	A	A	D	B	B	B	B	B	A	A 1	B	B										
Sodium sulfide	B	D	A		A	A	A 1	D	A 1	B 1	D 11	C 1	D	D	B 1	B 1		A	A		A		A				
Sodium sulfite, 10%	A	B	A 9		B	A 9	A	B	B	A	A	B	D	B	B	B	A 4	A 9	A 9		A 9						
Sodium thiosulfate	C	C	D			D	B 18	B 18	B	A	A	D		A	B		A	B	A		A	B	A	A	A 1	A 1	
Soybean oil	C	B																			C	D	A				
Stannic chloride	D	C	A	A	A	A	D	D	D	D	D	D	D	D	D	D	B	A	B	A	B	A	A				B
Stannic chloride, 5%	D	D	A	A	A	A	D	D 14	D	D 14	D 14	D	D	D	D	D	B 11	A 11	B		A		A	A 4,9	C 1	C 1	
Stannic chloride, boiling	D	D	A							D	D												A				

Stannous chloride	D	D	A	A 14		A	A 1	C	B	A 1	C	B	D	D	B	B	B	A 19	B	A	B	A	A				
Steam, 212°F	A	A	A			A	A	A	B	A	A	A		A	B	A		D 20	C 20	C 27	A 20	D 13					
Steam, 600°F	A	A					A	A	B	A	A	A		D	D	A		D 21	D 21	D 21	C 21	D 21	D	D	D	D	
Stearic acid	C	C																			B	B	A				
Sucrose solution	B	A																			A	A	A				
Sulfite liquors																	A	B	D		B	B	A				
Sulfur	B	C	A		A	A	A	A	A	A	A	A	C		A	A	A				A	D	A				
Sulfur, molten, 266°F	B	D	A		A	A	A	A	A	A	A	A	D	A	A	A	A	D	C	A	C		A		A	A	
Sulfur chloride	D	A	A			A	B	B	B 5	B	B	C	D	D	D	B	B	D	C		D	D					
Sulfur dioxide, 250°F, dry	B	B	A				B	B	B	B	B	A	C	B	B	B	B	D	B		A	C		A	B	B	
Sulfur dioxide, moist		D					B	D	D	A	D	D			D	B		D	B	A	A	C					A
Sulfur hexofluoride																				A	A						
Sulfur trioxide, dry	C	A																		B	D						
Sulfuric acid, 2%			A							A					A		A						A	A	C	C'	
Sulfuric acid, 2–40%	D	C	A	B	D	A	A	C 1	B	B 1	D 1	D	C	D	A	B 1	A	D 1	D		B		A	A	D 2	D 2	
Sulfuric acid, 40%	D	C	A	B	D	A	A		B			D	C	D	A	D	A			A			A	A			
Sulfuric acid, 93–100% (concentrated)	B	B	A	D	D	A	A	C	D	B	C	C	B	D	A	C	A	D	D	A	D	D	A	A			
Sulfuric acid, 10%, boiling			A			A	D	D	D	D	D		D		B	D	B 1			A			A	A			
Sulfuric acid, 10–80%, boiling			A			A	D		D						D					A			A	A			
Sulfuric acid, concentrated, boiling			B			A	B		D								B 3,25			A			A	A			

TABLE 2.1 (*Continued*)

Chemicals	Metals and alloys																	Elastomers					Plastics				
	Ductile iron	Bronze	Gold, 14K	Zirconium	Titanium	Tantalum	Stainless steel, Carpenter's 20-Cb3*	Stainless steel 430 F	Nickel	Stainless steel 316	Stainless steel 303 and 304	Carbon steel	Brass	Aluminum	Monel†	Inconel†	Hastelloy C-276‡	Buna	Silicone	Fluorocarbon	EPDM	SBR	Teflon§	Kel-F**	Nylon	Nylatron GS¶	Vespel§
Sulfurous acid	D	D	A	B 1	A	A	B	D	D	B	D	D	D	B	D	B	B	B	D	A	B	B			D 1	D 1	B
Sulfuric acid and nitric acid			A			A	B		D				D	D	D	D	A 23										
Tannic acid, 10%	C	B	A	A	A	A	B	B	B 9	B	B	D	C 9	B	B 9	B 9		A	B	A	A	B					
Tar, hot	A	A															A	B	B		D	D			B	B	A
Tartaric acid	D	A	A	A	A	A	A 11	A 1	B	B 1	A 1	C	A	B	B	B	B 12	A	A	A	B	B			B 1	B 1	A
Tetraethyl lead	C	B								B	B										D	D					
Toluene	A	A	A	A	A	A	A	A	A	A	A	A	A	A	A	A	A	D	D	A	D	D	B	B	A	A	A
Trichlorethylene	C	A	A	A	A	A	B	B 22	A	B 22	B 22	A	B	A	A	A	A	C	D		D	D	A		B	B	A
Triethylamine		A	A			A	A	A		A	A	A	D										A	B			
Tung oil	C	C								A	A										D	D					
Turpentine	A	C	A			A	B	B	B	B	B	B	B	B	B	B	B	A	D	A	D	D	A		A	A	
Uranium hexafluoride					A		A		A	A	A	B		A	A												
Varnish	C	A						A		A	A							B	D		D	D					
Vegetable oils								B		A	B						A	A	A		C	D			A	A	
Vinegar	D	B						A		A	A							B	A		A	B			C	C	

Water, acid mine	D	C	A			A				A	A	D		A							A						
Water, boiler feed			A			A	A	A	A	A	A	B	B	A	A	A					A						
Water, cold		A								A	A	A								A	A	A					
Water, distilled			A	A		A	A	A	A	A	A	A	A	B	A	A	A 4	A	A	A	A	A	A		A	A	
Water, hot (150°F+)	A	A	A	A	A	A	A	A	A	A	A	A	A	A	A	A	A		A	A	A		A	A		A	A
Water, salt sea			A		A	A	A	C	A	A	A	C	C	B	A	A	A						A	A			
Whiskey and Wine	D	B								A	A	D						A	A		A	A	A	A	B	B	
Xylene	B	A	A	A	A	A	B	B	A	B	B	B	A	A	A	A	A 4	D	D	A	D	D	A	D	A	A	
Zinc acetate																					A	C					
Zinc chloride	C	D	A	A	A	A	A 26	A 11	B	A 11	A 11	A 5	D	D	B	B	B	A			A	A	A		C 1	C 1	B
Zinc chloride, 5%, boiling								B 11		B 11	B 11									A			A				
Zinc hydrosulfite	B	C								A	A	A											A				
Zinc sulfate	D	B	A			A	A	A 2	A 26	A 2	A 2	D	D	D	B	A 18	A	A	A		A	B		B 4,9	A 1	A 1	B

Notes:

1. 10%
2. 5%
3. 160°F
4. Hot
5. Dry
6. Wet–moist
7. Dilute
8. Concentrated
9. Saturated
10. Gas
11. 50%
12. All concentrations
13. 145°F (up to)
14. 40%
15. Type of Freon must be known
16. Up to 140°F
17. 90%
18. 20%
19. 15%
20. Below 350°F
21. Above 350°F
22. Up to 212°F
23. 37%
24. 48%
25. 85%
26. 30%
27. 260°F
28. High pressure
29. 185°F
30. Boiling
31. Considered good for long-time use
32. 198°F (up to)
33. Agitated
34. Up to 125°F
35. 60%
36. Up to 600°F
37. Up to 70%
38. 0 at 10%

*Registered trademark of the Carpenter Technology Corporation.
†Registered trademark of Huntington Alloys, Inc.
‡Registered trademark of the Cabot Corporation.
§Registered trademark of the Du Pont Company.
**Registered name of the 3M Company (for Kel-F).
¶Registered name of Polymer Corporation (for Nylatron GS).

ficulty in being precise is evident. What is "good service" in one application may be unsatisfactory in another because one user, for example, expects many more years of troublefree service than another and the difference in service life is tied to other factors anyway.

There are few absolutes in material selections, and compromise figures prominently in a decision. It is fortunate, however, that there are so many materials from which to choose, for they open up many possibilities for the solution of a problem. Added to the materials themselves are the various combinations and techniques for using them. The limits to a solution are, perhaps, one's own ingenuity rather than the materials themselves.

2.2 METALS AND ALLOYS

The majority of the naturally occurring elements are metals, and most of them have properties sufficiently unique that they are practically all refined for specific uses. However, only a few metals find engineering application in the pure state, and these for the most part are alloyed with one or more other metals and other elements to provide a fantastic array of materials with various attributes of strength, corrosion resistance, high-temperature service, and so on. Because of its low cost and versatility, the metal iron is the most important from which engineering alloys are produced. Two general categories of iron alloys will be included here: cast iron and steels. The other base metals that provide important alloys are aluminum, copper, beryllium, magnesium, lead, nickel, tin, and zinc. For the ultimate in high-temperature service there are the refractory alloys from the metals tungsten, molybdenum, tantalum, and columbium (niobium).

Cast Irons and Steels

Cast Irons

Iron alloys are frequently cast to complex shapes, which minimizes machining, reduces weight, saves metal, and provides specific mechanical properties not otherwise economically available. The common cast irons are called *white cast iron, gray cast iron, malleable iron,* and *ductile iron.*

The white cast irons, named for the white appearance of a fresh fracture, have limited use because of their brittleness. They are very hard, with Brinell hardness numbers (BHN) to 600, so that they provide good abrasion and wear resistance and are used, for example, for balls in ball mills. The white cast irons are not used structurally.

The gray cast irons, also named for the appearance of their fractures, are not as brittle, are relatively soft (BHN 120 to 200), and have low ultimate tensile strength (Table 2.2). They are used for their resistance to corrosion and wear for components that are neither highly stressed nor subject to impact loads.

Malleable cast irons are available in a range of strengths and ductility, are easily machined, and have a good cast appearance. Table 2.2 shows how ductility (elongation) varies with strength for several alloys. While the ductility decreases markedly with increasing strength, there are many alloys with properties that permit them to be used in highly stressed components where the loading may be applied rapidly.

Ductile cast irons provide somewhat better strength and ductility than malleable, but are similar in the way strength and ductility characteristics vary with

TABLE 2.2 Typical Mechanical Properties of Some Cast Irons

Name	Specification	Class, type, or grade	Tensile strength, psi × 10^3 (MPa) min.	Yield strength, psi × 10^3 (MPa) min.	Elongation, %	Compressive strength, psi × 10^3 (MPa) min.	Hardness, BHN
Gray cast iron	ANSI A48-74	Class 30	30 (207)	*	*	115 (793)	195
	ANSI A436-72	Type 2b	30 (207)	*	*	115 (793)	195
	ANSI A278-75	Class 50	50 (345)	*	*	150 (1034)	238
Malleable iron	ANSI A47-74	Grade 32510	50 (345)	32 (221)	10	208 (1434)	133
	ANSI A220-76	Grade 45006	65 (448)	45 (310)	6	235 (1620)	180
	ANSI A220-76	Grade 60004	80 (552)	60 (414)	4	226 (1558)	242
	ANSI A220-76	Grade 90001	105 (724)	90 (621)	1	265 (1827)	242
Ductile iron	ANSI A536-72	Grade 65-45-12	65 (448)	45 (310)	12	140 (965)	180
	ANSI A536-72	Grade 120-90-02	120 (827)	90 (621)	2	270 (1862)	280

*No specific value required.

one another (Table 2.2). Malleable and ductile cast irons may replace components where forgings were once the only practical approach. They may be selectively hardened, for example, on wear surfaces, which further extends their use.

Steels

Steels have less carbon than the cast irons and are generally supplied in the form of rods, bars, sheets, and plate extrusions that are rolled and drawn to the desired cross section. While the rods, bars, sheets, and plate extrusions may be used in that form for fabricating parts, often they are the raw material from which parts are machined, especially small parts. Steels are also used for casting and have better ductility (elongation) than the cast irons.

Steels are generally categorized as either carbon steels, alloy steels, tool steels, or stainless steels. The carbon steels typically contain the elements sulfur, silicon, phosphorus, and manganese in small amounts, whereas the alloy steels contain additional elements, especially larger amounts of manganese and also small amounts of molybdenum, chromium, and nickel. Tool steels are a special group of alloy steels that are invariably heat-treated to give them specific properties of hardness and strength. Stainless steels, too, are a group of alloy steels with larger amounts of alloying elements and of such importance as to be considered separately.

Carbon Steels • Carbon steels typically contain from 0.15 to 1.0 percent carbon, the strength increasing with the carbon content. Carbon-steel bars are supplied in one of three forms: hot-rolled, cold-rolled, and cold-drawn. The hot-rolled bars are characterized by a black oxide coating or mill scale, whereas the cold-drawn bars are clean. Another distinguishing feature is the precision of the dimensions, the cold-drawn being more precise. Cold drawing is a process of drawing a bar, previously hot-rolled and descaled, through a die. A degree of cold

TABLE 2.3 Mechanical Properties of Carbon Steels

AISI designation and condition	Yield strength, psi × 10³ (MPa)	Tensile strength, psi × 10³ (MPa)	Elongation, %
1015:			
Hot-rolled	40 (276)	65 (448)	25
Cold-drawn	51 (352)	61 (421)	15
1035:			
Hot-rolled	42 (290)	76 (524)	18
Cold-drawn	71 (490)	85 (586)	12
1095:			
Hot-rolled	66 (455)	130 (896)	9
Cold-drawn and annealed*	76 (524)	99 (683)	13
12L15:			
Hot-rolled	34 (234)	57 (393)	22
Cold-drawn	70 (483)	80 (552)	14
1118:			
Hot-rolled	40 (276)	69 (476)	28
Cold-drawn	55 (379)	74 (510)	17
1144:			
Hot-rolled	51 (352)	94 (648)	15
Cold-drawn	90 (621)	100 (689)	10

*Annealed at 1450°F (1061 K).

working is performed, which improves the material strength but with some sacrifice in ductility. Table 2.3 gives some typical properties.

The common designation for carbon steels is their AISI (American Iron and Steel Institute) number. The last two digits of the four-digit number show the minimal carbon content (hundredths of a percent). The first two digits indicate the general kind of steel. A letter "L" between the second and third digit indicates the addition of lead, which improves the machinability of the steel (for example, 12L13 is a leaded steel with 0.13 percent carbon).

Because the machinability of carbon steels is very important for reasons of manufacturing economy, two grades, the 1100 and 1200 series, have been specially developed for ease of machining. Further machining improvement is possible for some of these with the addition of lead. Selection between one steel and another is primarily based upon required strength, machinability, and cost.

Alloy Steels • Alloy steels provide improved strength over the carbon steels especially as a result of heat treating. There are two general types: those with low carbon content and intended for surface hardening through carburizing[1] and those with high carbon content for through hardening (Table 2.4). The carburizing grades are used primarily where a hard surface is required for improved

TABLE 2.4 Mechanical Properties of Alloy Steels

AISI designation and condition	Yield strength, psi × 10^3 (MPa)	Tensile strength, psi × 10^3 (MPa)	Elongation, %
	Carburizing grades		
4320:			
Annealed	61 (421)	84 (579)	29
Carburized	105 (724)	148 (1020)	17
8620:			
Annealed	55 (379)	77 (531)	31
Carburized	80 (552)	124 (855)	19
E9310:			
Annealed	63 (434)	119 (820)	17
Carburized	120 (758)	155 (1069)	15
	Through-hardening grades		
1340:			
Annealed	63 (434)	102 (703)	25
Hardened and tempered	100 (689)	115 (793)	21
4130:			
Annealed	52 (359)	81 (558)	28
Hardened and tempered	115 (793)	130 (896)	115
6150:			
Annealed	59 (407)	96 (662)	23
Hardened and tempered	130 (896)	145 (1000)	15

[1]Carburizing is a process that adds additional carbon to the material to a small depth only (usually 0.010–0.050 in, or 0.25–1.3 mm) to provide a steel composition at the surface different from that throughout the core of the piece.

wear, although some improvement in strength is usually obtained because the outer layers of an item are often stressed more highly than the core material. The through-hardening grades are brought to increased strength and hardness by heat treating (hardening and tempering). Increased strength is accompanied by increased hardness in a consistent manner, and hardness measurements may be used to verify the strength indirectly.

Alloy steels are supplied in the annealed (soft) condition so that they may be heat-treated as desired. Because heat treating usually causes some distortion of a part, machining to approximate dimensions is often done first while the material is easier to work. After heat treating, final dimensions are obtained by other machining techniques such as grinding.

Tool Steels • Tool steels are specially controlled, higher-grade alloy steels. They contain alloying elements that provide extreme hardness, toughness, and strength to permit them to give long life as cutting tools and consistent performance. Some grades are specially developed to retain their properties at elevated temperatures because they often get very hot in their intended cutting-tool role. However, these same attributes are important in the design of other items, so tool steels do see other service as well.

Tool steels may be obtained in common shapes (round, square, rectangle) to very close dimensional tolerances. For this reason, they are used, for example, as keys and pins to secure gears and pulleys to shafts.

Stainless Steels • Stainless steel is not a single alloy of iron but a whole family of alloys providing wide ranges of chemical resistance and strengths. The primary alloying element which makes these steels "stainless" is chromium. When chromium is present to the extent of 12 percent or greater, the steel is generally called a *stainless steel.* This amount of chromium is sufficient to prevent rusting from ordinary atmospheric exposure. Larger amounts of chromium and other elements, particularly nickel, are added to provide increased corrosion resistance to a large number of chemicals. Like all steels, the stainless steels contain a small amount of carbon (less than 1 percent). In general, the stronger the steel the more carbon it contains and the less resistant it is to corrosion. Figure 2.1 illustrates how the more common stainless-steel alloys relate to each other on a basis of strength and corrosion resistance.

The most common stainless steel is type 304, which contains 18 percent chromium and 8 percent nickel and is typical of 18/8 stainless steels. Its corrosion resistance is good and it does not rust, although a powdery rust-colored oxide may form on the surface under severe atmospheric exposure, but this may be wiped off. Another 18/8 stainless, type 302, is very similar and is often used for such items as springs, belleville washers, and fasteners. When the term "stainless steel" is used and the type is not specified, it is often type 302. Where greater corrosion resistance is required, then use type 316, and if that is not adequate then use an alloy such as Carpenter's 20Cb-3 Stainless. For less corrosion resistance, use type 430 or type 405. None of these alloys is heat-treatable to improve their mechanical properties, but they are improved through cold working (e.g., machining). They may be annealed, and this may be required if extensive cold forming is planned (e.g., stamping, drawing).

The type 440 stainless steels are specifically heat-treatable to a high hardness [about Rockwell C (abbreviated as Rc) 50 to 60]. Carpenter's Custom 450 and Custom 455 alloys (Carpenter Technology Corp.) provide a combination of good corrosion resistance and high strength and may be heat-treated to a range of hardnesses and strengths to suit a wide variety of applications.

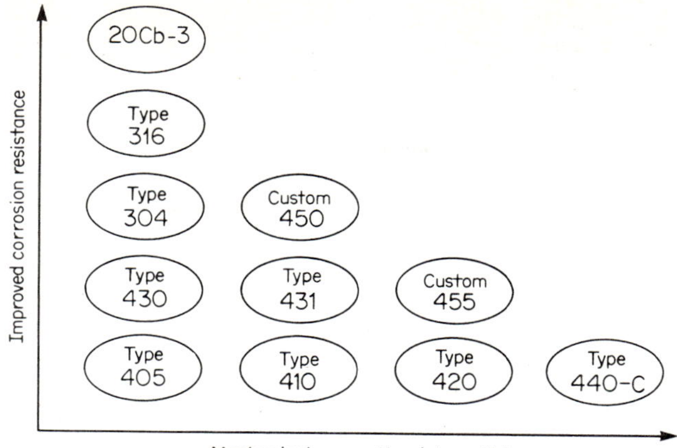

FIG. 2.1 Variation of corrosion resistance and strength of various stainless-steel alloys. *(Carpenter Technology Corp.)*

Table 2.5 summarizes the key characteristics of some common stainless alloys. The balance of material not specified in the tabulated compositions is iron.

Two other types of stainless steel in common use are types 304L and 316L. The "L" is for low carbon. The carbon content is 0.03 percent maximum. These grades are specified where it could be difficult to anneal properly. Annealing types 304 and 316 alloys is done rapidly by quenching in water to prevent the precipitation of carbides. The presence of these carbides substantially decreases the corrosion resistance. In certain large forgings and in welding it is often impossible to cool rapidly. For these applications it is usual to specify the L grades so that the likelihood of carbides forming is significantly reduced because so little carbon is present. If welded parts permit annealing and rapid cooling following the welding process, then there is no need to use the L grades. Small assemblies can usually be annealed, but on-site fabrication usually precludes it.

The corrosion resistance of most stainless steels is enhanced by passivation. To passivate, immerse the parts in a solution of nitric acid and sodium dichromate.

Various case-hardening processes can be used to surface-harden the stainless steels that are not heat-treatable (e.g., types 302, 304, 316). However, these processes change the alloy composition at the surface, and the normal corrosion-resistant properties are lost.

Aluminum

Aluminum is one of the lightest metals, whitish gray in color, relatively inexpensive, and widely used. It may be polished to a fine luster with excellent reflectivity. Like most pure metals it is relatively weak, but a large number of alloys are readily available to fill almost any need where temperatures do not exceed about 500°F (530 K). These alloys are used for castings, rods, bars, sheet, and plate and may be formed, shaped, machined, and joined by all conventional techniques. Aluminum and its alloys readily oxidize and react with most acids and alkalis. However, the oxide coating that forms rapidly in the atmosphere at room temperature frequently inhibits further corrosion. Anodized coatings are especially adherent and tough.

TABLE 2.5 Composition and Typical Properties of Some Stainless Steels

	Grade						
	302	304	316	410	440-C	Carpenter's 20Cb-3 (bar)	17-4PH
Composition, %:							
Chromium	17–19	18–20	16–18	11.5–13.0	16–18	19–21	15.5–17.5
Nickel	8–10	8–10.5	10–14	...	...	32.5–35.0	3– 5.0
Carbon	0.15 max.	0.08 max.	0.08 max.	0.15 max.	0.95–1.20	0.06 max.	0.07 max.
Molybdenum	...	...	2.0–3.0	...	0.75 max.	2.0–3.0	
Copper	...	...	...	...	...	3.0–4.0	3.0–5.0
Manganese	2.00 max.	2.00 max.	2.00 max.	1.00 max.	1.00 max.	2.0 max	1.0 max.
Silicon	1.00 max.	1.00 max.	1.00 max.	0.5 max.	1.0 max.	1.0 max.	1.0 max.
Phosphorus	0.045 max.	0.045 max.	0.045 max.	0.04 max.	0.040 max.	0.035 max.	0.04 max.
Sulfur	0.03 max.	0.030 max.	0.03 max.	0.03 max.	0.03 max.	0.035 max.	0.03 max.
Columbium and tantalum	...	...	...	...	...	8 × C min.–1.0 max.	0.15–0.45

Mechanical properties:								
Condition	Annealed	Annealed	Annealed	Annealed	Hardened at 1850°F; tempered at 500°F	Hardened at 1900°F; tempered at 600°F	Annealed	Condition H-900
Yield strength, 0.2% offset, psi × 10^3 (MPa)	35 (241)	38 (262)	34 (234)	40 (276)	158 (1089)	285 (1896)	40 (276)	185 (1275)
Ultimate tensile strength, psi × 10^3 (MPa)	85 (586)	84 (579)	82 (565)	75 (517)	193 (1331)	275 (1965)	90 (621)	200 (1379)
Elongation, %	60	60	60	35	17	2	40	14
Hardness	. . .	. . .	. . .	. . .	Rc 43	Rc 56	Rb 90	Rc 45
Remarks	Cannot be heat-treated. Cold working improves strength, especially yield.			Tempering at higher temperatures provides lower strength and hardness and increased ductility.		Intended for use in hardened condition.	Hardened only by cold working, not by heat treating.	Higher-temperature heat treating provides lower hardness and strength with better ductility.

TABLE 2.6 Composition and Typical Properties of Aluminum Alloys

	Alloy designation							
	1100-0	2014-T6	2024-T4	5052-0	6061-0	6061-T6	6063-T6	7075-T62
Composition, %:								
Silicon, max.		0.50–1.20	0.5		0.4–0.8	0.4–0.8	0.2–0.6	0.4
Iron, max.		0.7	0.5		0.7	0.7	0.35	0.5
Silicon + Iron (bracketed)	1.0			0.45				
Copper, max.	0.05–0.20	3.9–5.0	3.8–4.9	0.10	0.15–0.40	0.15–0.40	0.10	1.2–2.0
Manganese, max.	0.05	0.4–1.2	0.3–0.9	0.10	0.15	0.15	0.10	0.30
Magnesium, max.	...	0.2–0.8	1.2–1.8	2.2–2.8	0.8–1.2	0.8–1.2	0.45–0.9	2.1–2.9
Chromium, max.	...	0.10	0.10	0.15–0.35	0.04–0.35	0.04–0.35	0.10	0.18–0.35
Zinc, max.	0.10	0.25	0.25	0.10	0.25	0.25	0.10	5.1–6.1
Titanium, max.	...	0.15	0.15	...	0.15	0.15	0.10	0.20
Aluminum	99.0 min.	Remainder	Remainder	Remainder	Remainder	Remainder	Remainder	Remainder
Mechanical properties:								
Density, lb/in^3 (kg/m^3)	0.098 (2624)	0.10 (2768)	0.10 (2768)	0.097 (2685)	0.098 (2624)	0.098 (2624)	0.099 (2740)	0.101 (2796)
Tensile strength, min., psi $\times 10^3$ (MPa)	11 (76)	60 (414)	60 (414)	25 (172)	22 (152)	38 (262)	30 (207)	81 (558)
Yield strength, min., psi $\times 10^3$ (MPa)	3 (21)	53 (365)	44 (303)	10 (69)	16 (110)	35 (241)	25 (172)	72 (496)
Elongation, min., %	25	7	12	...	16	8	8	7
Modulus of elasticity, psi $\times 10^6$ (MPa)	10.0 (68,950)	10.6 (73,084)	10.6 (78,084)	10.2 (70,327)	10.0 (68,950)	10.0 (68,950)	10.6 (73,084)	10.4 (71,706)

Wrought-aluminum alloys are most frequently identified by a four-digit number (for example, 5052) that specifies the alloy composition. That number is followed by a dash and a letter or number which specifies the material condition, such as strain-hardened or heat-treated (for example, 2024–T6). Consult manufacturers' data for the various aluminum alloys and their mechanical properties. Table 2.6 gives these data for a few common alloys. Alloy 1100 is commercially pure aluminum containing a minimum of 99 percent aluminum. The fully annealed condition (–0) is very soft and ductile but rapidly work-hardens.

White-aluminum alloys provide some of the highest strength-to-weight ratios of all the metals and are therefore used extensively for lightweight structures, but they have many other uses. Their high electric and thermal conductivities lead to their application as electric wires, air-cooled gasoline engine cylinders, and cooking utensils.

Beryllium

Beryllium has a number of unique properties. Its density is low, 0.0665 lb/in^3 (1840 kg/m^3), and comparable to that of magnesium. Its strength (90,000 psi, or 620 MPa, tensile at room temperature) is higher than other light metals, and it retains 50 percent of its room-temperature yield strength to 1000°F (810 K). It is resistant to oxidation to 1500°F (1090 K). It has a modulus of elasticity higher than most metals (40×10^6 psi, or 2.76×10^5 MPa) and, therefore, can provide rigid structures. Its very low neutron cross section combined with its high strength and temperature capabilities make it useful for nuclear reactor applications. There are two factors, however, which limit its use. The first is cost because in finished form it approaches that of the precious metals. The second is toxicity because the airborne particles are dangerous to certain individuals. Consequently the pure metal is machined in special facilities. In the pure form its use is limited to a few applications in the aerospace and nuclear industries where its special properties are costwise still advantageous.

The commercial uses for beryllium are as an alloying element, especially with copper and nickel, where it imparts high strength and hardness to the alloys. Although beryllium is a minor constituent, the metals so alloyed are usually referred to as *beryllium-bronze* and *beryllium-nickel alloys.* Table 2.7 gives the

TABLE 2.7 Composition and Mechanical Properties of Beryllium and Copper Alloys

	Alloy 170	Alloy 172
Composition, %:		
Beryllium	1.60–1.79	1.80–2.00
Nickel Cobalt	0.2 min.	0.2 min.
Iron	0.4 max.	0.4 max.
Copper + others above	99.5 min.	99.5 min.
Mechanical properties:		
Yield strength, quarter-hard, drawn, and heat-treated, psi $\times 10^3$ (MPa)	135 (931)	150 (1030)
Tensile strength, psi $\times 10^3$ (MPa):		
Quarter-hard, drawn	75–88 (517–607)	75–80 (517–552)
Quarter-hard, drawn, and heat-treated	160–190 (1103–1310)	175–205 (1206–1413)
Elongation, quarter-hard, drawn, %	10	10

composition and properties of two of these alloys. Beryllium alloys generally have excellent fatigue resistance and are used for items such as springs, dies for drawing alloy steels, electric contacts, and molds for plastics. They combine good thermal and electrical conductivity with exceptional strength.

Beryllium-bronze alloys are normally supplied in the annealed and cold-worked condition. Various amounts of cold working (rolling) are available. Depending upon the amounts of cold working, they are specified as *annealed, quarter-hard, half-hard,* and *hard.* The quarter-hard condition is readily machined, having a hardness about Rockwell B (abbreviated Rb) 80. The full strength of the material is obtained by a precipitation-hardening heat treatment, which is usually done following machining. After precipitation hardening, the material is much improved (tensile strength), and hardnesses are typically in the range Rc 35–40.

Copper

Copper is used in the pure form because it has very high electric and thermal conductivities, the highest of all the commercial metals. Pure copper is not a strong material, but it is used in a wide range of alloys that provide good strength characteristics and corrosion resistance. Although copper alloys generally provide good thermal and electric conductivity, they do not compare with the pure metal in this regard even when the alloying elements are present only in a small amount.

Many of the alloys of copper are called either *brass* or *bronze.* Brass is an alloy of copper and zinc with, perhaps, other minor elements to improve machinability (e.g., lead) or mechanical properties. Bronzes have other alloying elements, with copper the primary constituent. For example, aluminum bronze is an alloy of copper (90 percent approx.) and aluminum (10 percent approx.). The word "bronze" by itself denotes an alloy of copper and tin. Some of the more common alloys and their properties are shown in Table 2.8.

Copper and its alloys are readily cold-worked, although they do work-harden easily and may have to be annealed frequently. Cold working improves the mechanical properties, and drawn bars are available in varying hardnesses depending upon the amount of annealing and drawing. Typical hardnesses range from soft to half-hard and hard. For this reason a wide range of strengths is given in Table 2.8 for most alloys, the lower values applying to the soft material and the higher to the hard.

Whereas the color of copper is a reddish orange, many of the alloys have a color similar to gold. While these alloys are relatively corrosion-resistant, they are attacked by mineral acids and salts. Mercury forms an amalgam with many of the alloys, causing especially severe intergranular corrosion. Such alloys may disintegrate when exposed to either the liquid metal or its vapor, even after short exposure times.

Lead

Lead is not often used in the pure state because it has a low melting point (621°F, or 600 K), low strength, and poor creep characteristics. It is resistant to atmospheric corrosion and many weak acids. Capable of absorbing vibration, it is used for mechanical damping and noise control. Another use is for attenuating high-energy radiation (e.g., x-rays, gamma rays). As an alloying element it is often a minor constituent, added to improve machinability.

TABLE 2.8 Composition and Properties of Copper and Copper Alloys

	Common name and ASTM designation					
	Pure copper, B111-75	Free machining brass, B16-74	Aluminum bronze, B150-74	Naval brass, B21-74	Phosphor bronze, B139-71	Leaded red brass, B140-74
Composition, %:						
Copper	99.95 min.	60.0–63.0	88	59.0–62.0	Remainder	87.5–90.5
Lead	...	2.5–3.7	...	0.2 max.	0.05 max.	1.3–2.5
Iron	...	0.35 max.	3	0.1 max.	0.1 max.	0.1 max.
Aluminum	...	...	9	...	...	
Nickel	...	...	...	...	...	0.7 max.
Manganese	...	0.50 max.	...	...	...	
Phosphorus	0.005 max.	...	...	...	0.03–0.35	
Tin	...	...	...	0.50–1.0	9.0–11.0	
Zinc	...	Remainder	...	Remainder	0.20 max.	Remainder
Mechanical properties:						
Yield strength, psi × 10^3 (MPa)	20–40 (140–275)	15–45 (100–310)	30–50 (210–340)	20–45 (140–310)	...	10–30 (70–210)
Tensile strength, psi × 10^3 (MPa)	30–45 (210–310)	40–80 (275–550)	75–90 (520–620)	50–67 (350–460)	60–160 (420–1100)	35–50 (240–350)
Modulus of elasticity, psi × 10^6 (MPa)	...	14 (96.5 × 10^3)	16 (110 × 10^3)	14 (96.5 × 10^3)	15 (103 × 10^3)	14 (96.5 × 10^3)
Density, lb/in^3 (kg/m^3 × 10^3)	0.32 (8.85)	0.31 (8.58)	0.28 (7.75)	0.31 (8.58)	0.32 (8.85)	0.32 (8.85)

Other elements generally used in lead alloys are tin, bismuth, antimony, and cadmium. They all have low melting temperatures. A solder used extensively for electric connections is a lead alloy containing 52 percent lead and 48 percent tin, whereas other solders used for filling and "wiping" contain more lead and provide a wide temperature range over which the alloy remains soft or mushy so that it may be shaped. Antimony added to lead-tin alloys produces alloys that expand upon solidification. Castings of these alloys provide very accurate reproductions of fine mold details and are used for printer's type and toys. Other lead alloys may be formulated to provide a wide range of melting points from 63 to 350°F (257 to 450 K). The low-melting-point alloys are used for temperature-responsive devices (e.g., fire alarms).

Because lead is a weak structural material, it may be applied as a coating or cladding on supporting materials such as steel and plastic in sheet form. Its applications in noise-, vibration-, and radiation-absorbing structures are often facilitated through these backing materials.

Magnesium

Magnesium is even lighter than aluminum and has a density of only 0.063 lb/in^3 (1750 kg/m^3). It is usually alloyed with aluminum, manganese, zinc, and zirconium to provide a number of high-strength-to-weight structural materials. These alloys are available as bars, plates, forgings, and castings. They are extremely notch-sensitive, and stress concentrations, sharp corners, and rapid changes of section must be avoided in structural parts. Magnesium alloys work-harden rapidly, but apart from this characteristic, they are readily fabricated by all normal methods. Magnesium alloys are generally applied below about 300°F (420 K).

Magnesium and its alloys are resistant to alkalies and seawater. Improved corrosion resistance is obtained by anodizing and with various proprietary coatings. These alloys burn fairly easily in air and emit a brilliant white light. Shavings and chips, such as those produced during machining, are most easily ignited and require special care in handling. The fire risk is substantially reduced when very small amounts of beryllium and lithium are added to the alloys. Molten magnesium reacts vigorously with water, obviating the use of water to extinguish magnesium fires.

Its light weight makes it useful for portable equipment such as housings for hand tools and ladders and for aerospace and marine applications. Properties of some of the magnesium alloys are shown in Table 2.9.

Nickel

Nickel is one of the most important commercial metals because it finds considerable use in the pure form and extensive use as an alloying element. In the latter it imparts exceptional corrosion resistance, high strength from cryogenic to elevated temperatures, and permits fabrication by most common techniques. The widely used alloys of nickel are Duranickel, Monel, Inconel (all trademarks of Huntington Alloys, Inc.), and Hastelloy (trademark of the Cabot Corp.). Some important properties of these materials are summarized in Table 2.10.

Commercially pure nickel (Nickel 200 with 99 percent minimum nickel) is a relatively strong pure metal that may be used from cryogenic temperatures (−400°F, or 33 K) to elevated temperature (800°F, or 700 K), where it still retains much of its room-temperature strength. Nickel has a thermal conductivity over twice that of stainless steel, a factor which can sometimes be used to

TABLE 2.9 Composition and Properties of Magnesium Alloys

	Use and alloy designation					
	Sand casting		Forging,	Die casting,	Extruded shapes, rod, bar, and wire	
	AM100A-T6	ZK61A-T6	AZ80A-T5	AZ91A	AZ61A-F	ZK60A-T5
Composition, %:						
Magnesium	Remainder	Remainder	Remainder	Remainder	Remainder	Remainder
Aluminum	9.3–10.7	. . .	7.8–9.2	8.3–9.7	5.8–7.2	
Manganese, min.	0.1	. . .	0.12	0.13	0.15	
Zinc	0.3 max.	5.5–6.5	0.2–0.8	0.35–1.0	0.4–1.5	4.8–6.2
Zirconium	. . .	0.6–1.0	. . .	. . .	. . .	0.4 min.
Silicon, max.	0.3	. . .	0.1	0.5	0.1	
Copper, max.	0.1	0.1	0.05	0.10	0.05	
Nickel, max.	0.01	0.01	. . .	0.03	. . .	
Mechanical properties:						
Density, lb/in³ (kg/m³ × 10³)	0.065 (1.80)	0.066 (1.83)	0.066 (1.83)	0.065 (1.80)	0.065 (1.80)	0.064 (1.77)
Tensile strength, min., psi × 10³ (MPa)	35 (241)	40 (276)	42 (290)	34 (234)	43 (297)	46 (317)
Yield strength, min., psi × 10³ (MPa)	17 (117)	26 (179)	28 (198)	23 (159)	28 (193)	38 (262)
Elongation, %	. . .	5	2	3	8	4
Modulus of elasticity, psi × 10⁶ (GPa)	6.5 (44.8)	6.5 (44.8)	6.5 (44.8)	6.5 (44.8)	6.5 (44.8)	6.5 (44.8)

TABLE 2.10 Composition and Properties of Nickel and Nickel Alloys

	Nickel 200		Duranickel 301		Monel 400	
Composition, %:						
Nickel	99.0 min.		93.0 min.		63.0 min.	
Copper	0.25 max.		0.25 max.		28.0–34.0	
Chromium	. . .		. . .		. . .	
Aluminum	. . .		4.00–4.75		. . .	
Iron	0.40 max.		0.60 max.		2.5 max.	
Molybdenum	. . .		. . .		. . .	
Manganese, max.	0.35		0.50		2.0	
Silicon, max.	0.35		1.00		0.5	
Sulfur, max.	0.01		0.01		0.02	
Phosphorus	. . .		. . .		. . .	
Columbium plus tantalum	. . .		. . .		. . .	
Carbon, max.	0.15		0.30		0.30	
Titanium	. . .		0.25–1.00		. . .	
Cobalt	. . .		. . .		. . .	
Mechanical properties:						
Condition	Cold drawn and annealed		Cold drawn and annealed	Aged	Cold drawn and annealed	
Temperature, °F (K)	70 (294)	1000 (811)	70 (294)	1000 (811)	70 (294)	1000 (811)
Yield strength, psi × 10^3 (MPa)	15–30 (103–207)	13.5 (93)	30–60 (207–414)	99 (683)	55–100 (379–689)	23 (159)
Tensile strength, psi × 10^3 (MPa)	55–75 (379–517)	31 (214)	90–120 (621–827)	120 (827)	84–120 (579–827)	55 (379)
Elongation, %	55–40	69	55–35	7	40–22	38
Hardness	45–70 Rb	. . .	75–90 Rb	. . .	85 Rb–20 Rc	. . .
Density, lb/in^3 (kg × 10^4/m^3)	0.32 (0.88)		0.30 (0.83)		0.32 (0.88)	
Thermal conductivity at 70°F (294 K), Btu/(ft^2·h·in·°F) [W/(m·K)]	485 [69.8]		165 [23.7]		151 [21.7]	

advantage where high-temperature containment and corrosive environments are present because nickel is more easily cooled. Nickel provides good corrosion resistance to mineral acids (except nitric acid) and to most salts, organic acids, and alkalies. It may also be used for dry fluorine, chlorine, and hydrogen chloride but should first be treated with hydrofluoric acid to provide a protective layer of nickel fluoride on the exposed surfaces. Nickel may be hot- and cold-formed, conventionally machined, and joined by welding, brazing, and soldering.

Duranickel (alloy 301) is a high-nickel alloy (93 percent) whose main additive is aluminum. It possesses the excellent corrosion resistance of nickel but can be heat-treated to greater strength and hardness (Table 2.10). Tensile strengths in excess of 200,000 psi (1400 MPa) are achievable, with hardnesses in the range Rc 40–45. Duranickel may be cold-worked and machined, but special care must be used for welding, brazing, and soldering because of the temperature-hardening characteristics of this alloy.

The monel metals are alloys of nickel and copper plus other specific elements to provide special characteristics. Monel 400, one of the most common, is used extensively for marine (seawater) environments and other corrosive applications. The room-temperature strength of Monel 400 is comparable to pure nickel. Its strength improves markedly to very low temperatures while retaining much of its ductility and impact resistance so that it is an excellent choice for cryogenic structural applications, even for liquid hydrogen. It has good resistance to sulfuric, hydrochloric, and hydrofluoric acids but is not recommended for nitric acid.

Monel K-500		Inconel 600		Inconel 625		Hastelloy-C-4	
63.0 min.		72.0 min.		58.0 min.		Balance (~62)	
27.0–33.0		0.50 max.		...			
...		14.0–17.0		20.0–23.0		14.0–18.0	
2.30–3.15		...		0.40 max.			
2.0 max.		6.00–10.00		5.0 max.		3.0 max.	
...		...		8.0–10.0		14.0–17.0	
1.5		1.00		0.50		1.00	
0.5		0.5		0.50		0.08	
0.1		0.01		0.01		0.03	
...		...		0.01 max.		0.04 max.	
...		...		3.15–4.15			
0.25		0.15		0.10		0.015	
0.35–0.85		...		0.40 max.		0.70 max.	
...		...		...		2.00 max.	
Cold drawn and annealed	Aged	Cold drawn and annealed	As drawn	Cold drawn and annealed		Aged	Aged
70 (294)	70 (294)	70 (294)	1000 (811)	70 (294)	1200 (867)	70 (294)	1200 (867)
40–60 (276–414)	160 (1103)	25–50 (172–345)	100 (690)	60–110 (414–758)	60 (414)	55 (379)	37 (255)
90–110 (621–758)	200 (1380)	80–100 (552–689)	110 (758)	120–160 (827–1103)	130 (90)	115 (793)	87 (600)
50–25	...	55–38	13	60–30	30	56	56
85–90 Rb	48 Rc	70–85 Rb	...		...	92 Rb	
0.31 (0.86)		0.30 (0.83)		0.30 (0.83)		0.31 (0.86)	
121 [17.4]		103 [14.8]		68 [9.8]		70 [10.1]	

Another monel alloy, which contains a small amount of aluminum, is Monel K-500. This alloy provides almost the same corrosion resistance as Monel 400 but may be heat-treated (aged) to greater strength and hardness. It can be hardened to Rc 40, with tensile strength to 200,000 psi (1400 MPa).

Two important high-temperature alloys are Inconel 600 and 625. These are nickel alloys containing principally chromium, about 20 percent, and other minor elements (Table 2.10). The addition of chromium extends the corrosion resistance beyond that of the monels, especially in oxidizing atmospheres at elevated temperatures. These alloys have been used to temperatures up to 2000°F (1370 K). They may be hot- or cold-formed, machined by conventional techniques, and joined by welding, brazing, and soldering. The fatigue characteristics, retained strength at elevated temperature, and corrosion resistance make the 625 alloy a prime choice for many structural components in hostile environments. These two alloys gain greater strength and hardness through cold work (e.g., drawing), not heat treatment.

The Hastelloy alloys C-4 and C-276 are nickel-chromium-molybdenum alloys having outstanding resistance to chemical corrosion. These alloys have good strength even at elevated temperatures (Table 2.10). They are available in sheet, plate, bar, wire, pipe, and forging stock and ordinarily supplied in the solution-heat-treated condition.

The chemical resistance of these alloys allows them to be used in air to 1900°F (1300 K) and with hot solutions of hydrochloric, nitric, and sulfuric acids and with wet chlorine, seawater, and brine. Consult manufacturers' data for specific corrosion rates when the alloys are exposed to a wide range of oxidizing and reducing chemicals.

The Hastelloy alloys' utility is enhanced by its fabricability. All the normal fabrication techniques may be used. It does work-harden but can still be deep-drawn and spun. All the common electric welding techniques can be used, and no special treatment is required to retain the corrosion-resistant properties of the heat-affected zones; thus welded assemblies may be used in the as-welded condition. This is of special benefit when fabricating specialty equipment for chemical processing.

Tin

Tin is a silvery-white metal with a bluish hue. It melts at 450°F (506 K), is soft and ductile, and may be extensively cold-worked without hardening. At temperatures of 60°F (290 K) and below, the metal breaks up into a gray powder; avoid low-temperature service. It is resistant to atmospheric corrosion but otherwise is quite chemically reactive. In the pure form it is used most extensively as a coating for steel in tin plate. Its alloys are used for die casting, fusible metals, bearing metals, solder, and pewter. None of the alloys has any great strength.

Typical solders are alloys of tin and lead (Table 2.11). An alloy of 63 percent tin and 37 percent lead is eutectic and has the lowest melting point. A typical

TABLE 2.11 Composition and Properties of Tin Alloys

	Alloy and designation				
	Solder		Bearing alloy, ASTM B23No3	Die-casting alloy, ASTM B102	Pewter
	63A	50A			
Composition, %:					
Tin	63	50	84	65	89.3
Lead	37	50	...	18	
Bismuth	...	...	...	...	1.8
Antimony	...	...	8	15	7.1
Copper	...	...	8	2	1.8
Melting temperature:					
Solidus, °F (K)	361 (456)	361 (456)	464 (513)		
Liquidus, °F (K)	361 (456)	421 (489)	792 (696)		
Ultimate compressive strength at 70°F (294 K), psi (MPa)	...	...	17,600 (121)		

commercial solder is a 50/50 alloy of tin and lead, and while it melts at 361°F (456 K) it is not completely fluid until a temperature of 421°F (489 K) is reached.

A white metal-bearing alloy, often known as *babbitt metal,* is a high-tin alloy. A typical composition is shown in Table 2.11. These alloys are used for plain bearings. They are easily cast into the supporting journals, which assures intimate contact and therefore good support for the soft and relatively weak alloy.

Pewter is used for eating utensils and decorative artwork. The composition for fine pewter is given in Table 2.11. Many variations of pewter alloy exist, and often they contain lead. Lead-containing pewter alloys must not have less than 65 percent tin to obviate the possibility of lead contamination of foodstuffs.

Zinc

Zinc is another low-melting-temperature metal (788°F, or 693 K). It is bluish white in color. There are three primary uses for this metal. One is for coating steel and iron, i.e., galvanizing (see Sec. 2.6). The second is as a constituent in brass (see "Copper," Sec. 2.2), and the third is as the major constituent in zinc-base die-casting alloys. The pure metal is very ductile to a temperature as low as 250°F (400 K) and is resistant to atmospheric corrosion because of the formation of a protective carbonate film, hence its use as a protective coating. The wide use of zinc alloys for die casting results from their low cost, low melting temperature, and good casting characteristics.

Many of the zinc-base die-casting alloys are essentially zinc-aluminum alloys with only about 4 percent aluminum. Other metals, especially copper and magnesium, may also be present in small, but controlled, amounts. A typical alloy is shown in Table 2.12. These alloys are not strong (40,000–50,000 psi, or 275–345 MPa, tensile strength) but are adequate for many functional as well as decorative purposes. They are easily machined and may be plated (e.g., chromium-plated) to provide added attractiveness. These alloys cannot be used above 200°F (370 K) because they become brittle and lose strength.

TABLE 2.12 Typical Zinc Base Die-Casting Alloy

Designation	ASTM AC41A
Composition, %:	
Copper	0.75–1.25
Aluminum	3.5–4.3
Magnesium	0.03–0.08
Iron, max.	0.100
Lead, max.	0.005
Cadmium, max.	0.004
Tin, max.	0.003
Zinc	Remainder
Mechanical properties:	
Tensile strength, psi (MPa)	47,000 (324)
Elongation, %	7

Refractory Metals and Alloys

While refractory metals are considered to be those with melting points above 3600°F (2250 K), only four are of engineering interest: columbium (niobium), molybdenum, tantalum, and tungsten. Unfortunately, while these metals do have high melting points and provide good strength to high temperatures, they all suffer from poor resistance to oxidation in air. A protective coating is required to take advantage of their other benefits unless vacuum (space) or inert-atmosphere service is contemplated. The refractory metals and alloys have been used for liquid-metal containment, lamp filaments and components, surgical implants, electronic components, and electrodes.

Tungsten has the highest melting temperature (Table 2.13) and greatest strength but oxidizes in air at a temperature of only 375°F (464 K). It can be worked by most of the common methods, but welding is best done by the electron-beam process to exclude air. Because of its high thermal conductivity and low specific heat, temperature spreads rapidly from the point of weld fusion.

TABLE 2.13 Some Properties of Refractory Metals and Alloys

	Tungsten	Tantalum	90% Ta + 10% W*	Columbium	Molybdenum
Melting point, °F (K)	6152 (3673)	5425 (3269)	...	4470 (2789)	4730 (2883)
Density, lb/in³ (kg/m³ × 10³)	0.70 (19.4)	0.60 (16.6)	...	0.31 (8.6)	0.37 (10.2)
Specific heat	0.113	0.034	...	0.065	0.065
Yield strength, psi × 10³ (MPa):					
At 70°F (294 K)	200 (1379)	33 (228)	88 (607)	15 (103)	97 (669)
At 1800°F (1256 K)	70 (488)	23 (159)	75 (517)	7 (48)	27 (186)
Tensile strength, psi × 10³ (MPa):					
At 70°F (294 K)	220 (1517)	35 (241)	98 (676)	25 (172)	110 (758)
At 1800°F (1256 K)	80 (552)	20 (138)	63 (434)	11 (76)	30 (207)
Elongation at 70°F (294 K), %	3	42	26	25	42
Modulus of elasticity, psi × 10⁶ (MPa × 10³)	58 (400)	27 (186)	...	14.3 (99)	47 (324)

*Alloy 90% tantalum and 10% tungsten.

Tantalum is the most versatile of the refractory metals. It provides oxidation resistance to 500°F (530 K) in air and is strongly resistant to many chemicals at lower temperatures. It is completely inert to body fluids. It is available in foil, sheet, plate, rod, and tube forms. It may be worked by most conventional methods except grinding. Its tendency to tear or gall with cutting tools can be appreciably reduced with the generous use of a cutting lubricant such as trichlorethylene. Welding is done by either the TIG (tungsten-inert-gas) or electron-beam techniques. With TIG, adequate inert-gas protection of the postweld heat-affected zones must be provided.

The yield strength of tantulum decays rapidly above room temperature. However, alloys of tantalum with about 10 percent tungsten have excellent high-temperature strength characteristics (Table 2.13), even to temperatures as high as 2400°F (1590 K).

Columbium, whose internationally accepted name is niobium, has good chemical corrosion characteristics but not as good as tantalum. It is used as an alloying agent for steels, especially stainless steels, and with other refractory metals (including hafnium) to produce higher-strength materials. It may be fabricated in a manner similar to tungsten and tantalum and is available in the same forms. Like tantalum, columbium loses its strength rapidly with increasing temperature, but alloys containing zirconium, hafnium, and tungsten provide good strength at high temperature and are somewhat easier to machine.

Molybdenum is in many respects similar to tungsten, although it is easier to machine and has greater ductility (Table 2.13); witness its ability to be drawn into fine wires. Its principal use is as an alloying element in steels, especially those for high-temperature service. Molybdenum has good corrosion resistance (even to hydrofluoric acid) and is coated on other metals by plasma spraying or vapor deposition.

2.3 CERAMICS

The ceramic materials or nonmetallics provide a variety of mechanical and physical properties that include extreme hardness, chemical inertness, and very high temperature resistance, and electrical and thermal properties that range from good conductors to excellent insulators. Table 2.14 summarizes some of the important properties.

The oxides of aluminum (alumina), beryllium (beryllia), silicon (silica), and magnesium (magnesia) are especially good electrical insulators, may be used to high temperatures even in an oxidizing (air) atmosphere, and except for beryllia are good thermal insulators. Because beryllia has a fairly high thermal conductivity and excellent electrical insulating properties, it is used in electric systems requiring dissipation of heat.

The carbides of boron and silicon are very hard and provide excellent resistance to abrasion. The nitrides of these same elements provide excellent resistance to thermal shock and are inert at high temperatures.

The properties of the ceramics given in Table 2.14 are only approximate and vary considerably with the method used to manufacture an item. The oxides may be spray-coated or slip-cast and kiln-fired. The carbides and nitrides are usually cast or hot-pressed, and final dimensions, if accuracy is required, are obtained by grinding. Consult the manufacturer for property values of its product.

Graphite is an excellent structural material for high-temperature applications provided that oxygen or other oxidizing agents are kept from direct contact. A

TABLE 2.14 Some Properties of Ceramics

	Alumina	Beryllia	Boron carbide	Boron nitride	Graphite	Pyroltic graphite	Magnesia	Silica	Silicon carbide	Silicon nitride
Bulk density, g/cm^3 (or kg/m^3 × 10^3)	3.89	2.85	2.41	2.30	1.85	2.02	3.58	2.30	2.9	2.8
Melting point, °C (K)	2016 (2289)	2530 (2803)	2450 (2723)	3000 (3273) sublimes	3650 (3923) sublimes	3650 (3923) sublimes	2800 (3073)	1400 (1673)	2250 (2523) dissociates	1890 (2163) dissociates
Maximum useful temperature, °C (K)	1700 (1970)	600 (870)	600 (870)	1650 (1920)	2300 (2570) nonoxidizing	2800 (3070) nonoxidizing	2300 (2570) reducing, 1700 (1970) oxidizing	1750 (2020)	1850 (2120)	1650 (1920)
Specific heat	0.21	0.31	. . .	. . .	0.39	. . .	0.28	0.28	0.29	0.29
Thermal conductivity at 20°C (293 K), g·cal/(s·cm^2·°C/cm) [W/(m·K)]	0.08 (33.4)	0.48 (201)	0.063 (26.3)	0.065 (27.2)	0.075 (31.4)	0.082 (34.3)	0.006 (2.5)	0.0049 (2.0)	0.037 (15.5)	0.038 (15.9)
Coefficient of linear expansion at 20°C (293 K), 10^{-6}/°C (or 10^{-6}/K)	7.2	7.6	1.8	2.4	7.8	0.6	13.5	8.6	4.4	4.4
Thermal shock resistance	Fair	Excellent	Fair	Excellent	Good	Excellent	Fair	Excellent	Excellent	Excellent
Flexural strength, psi × 10^3 (MPa)	55 (379)	40 (276)	44 (304)	. . .	5 (35)	. . .	. . .	. . .	110 (759)	130 (900)
Modulus of elasticity, psi × 10^6 (GPa)	54 (372)	51 (352)	65 (447)	7 (48)	1.3 (9)	. . .	. . .	. . .	62 (427)	25 (172)
Abrasion resistance	Good	Good	Excellent	Poor	Poor	Poor	Poor	Fair	Excellent	Excellent
Electrical resistivity at 20°F (267 K), Ω·cm (Ω·m)	10^{14} (10^{12})	10^{17} (10^{15})	0.6 (0.006)	10^{13} (10^{11})	10 (0.1)	4 (0.04)	4 × 10^6 at 800°C (4 × 10^4 at 1070 K)	10^8 (10^6)	700 (7)	10^{15} (10^{13})

special form of graphite, pyrolytic graphite, is made so that the crystals are aligned, and the resulting properties differ markedly from one direction to another (i.e., anistropic). For example, it may be used as a thermal or electric conductor in one direction but as a resistor in a direction 90° away. This unique capability affords ingenious uses for pyrolytic graphite.

2.4 PLASTICS

There are many plastic materials that one may choose to fill a specific need. However, unlike the metals, the choice is not generally as simple for several reasons. First, since they are plastic materials they tend to *cold-flow;* that is, even at low temperatures (room temperature) permanent deformation results from mechanical stress. Stress values must be kept small to prevent distortion and failure. This phenomenon accelerates as the temperature increases. The stress values given by manufacturers (and those in Table 2.15) are for short-term standard testing, and if the test time were increased, then in most cases the failure stress would be much reduced. In some cases creep data are available for a range of temperatures to aid the design engineer, but the analysis is often complex when dealing with materials in the plastic-flow regime.

Second, many plastics are prone to stress cracking. Material subject to stress becomes increasingly notch-sensitive with time, and surface scratches or discontinuities in thickness become sites for the propagation of cracks.

Third, weathering (temperature, humidity, water, and sunlight) and aging affect the properties of plastics so that what was initially satisfactory might fail several months or years later because of changes in the basic material properties. Usually plastics become hard and brittle. They may also discolor or fade, which is a disadvantage if selected for their original decorative value.

Manufacturers have become adept at modifying the basic properties of generic types of plastics by adding plasticizers, stabilizers, fillers, and lubricants that not only modify the properties but also allow for easier fabrication. Therefore the data from one manufacturer can be, and often are, different from another for the same general type of plastic material. The properties listed in Table 2.15 are only a guide, and the specific manufacturer should be consulted for actual values. Substitutions of one manufacturer's product for another of the same generic type should be done with caution.

The maximum useful temperature capability shown in Table 2.15 is for a zero-stress condition, which seldom exists. When plastic is used as a coating for metal, thermal stresses are often introduced because of the large difference between the expansion of plastic and the metal; accordingly the reinforcing value of a base material (metal) is a detriment rather than an asset. Reinforcement by introducing fiberglass fibers into the plastic is more successful in extending the strength capability, thereby increasing the possibility that the listed value for maximum temperature may be more closely approached.

Teflon (Du Pont Company) is one plastic that deserves special mention because of its outstanding chemical inertness. Actually, Teflon is made from either of two resins, TFE (polytetrafluroethelyne) or FEP (fluorinated ethelyne propylene). There are almost no differences in the properties of the two except that TFE has a slightly higher temperature capability. There is a difference, however, in the processibility of the two materials.

Teflon is so inert that it is better considered from a list of substances with which it does react, for this list is very short. There are two groups of substances

TABLE 2.15 Typical Properties of Selected Plastics

	ABS	Acrylic	Epoxy	Nylon	Phenolic	Polyester	Poly-ethylene	Poly-propylene	Poly-styrene	Polyimide (Vespel)	PVC	Teflon	
												FEP	TFE
Specific gravity	1.01	1.19	1.12	1.13	1.4	1.4	0.9	0.9	1.0	1.36	1.4	2.16	2.2
Tensile strength, psi × 10^3 (MPa)	6 (41.4)	10.5 (72.4)	6 (41.4)	12 (82.7)	6.5 (44.8)	8 (55.2)	2 (13.8)	5 (34.5)	10 (68.9)	10.5 (72.4)	7 (48.3)	1.8 (12.4)	1.3 (8.96)
Elongation, %	10	6	7	60	3	10	100+	12	2	8	50	350	275
Modulus of elasticity, psi × 10^5 (GPa)	3.3 (2.3)	4.3 (3.0	4 (2.8)	4.2 (2.9)	12 (8.3)	2.8 (1.9)	0.4 (0.3)	1.6 (1.1)	6 (4.1)	...	4 (2.8)	0.6 (0.4)	0.5 (0.3)
Flexural strength, psi × 10^3 (MPa)	10.5 (72.4)	16 (110)	13 (89.6)	20 (138)	10 (68.9)	13 (89.6)	...	...	15 (103)	5 (34.5)	10 (68.9)	...	
Impact strength, ft·lb/in (N·m/m)	6.5 (347)	0.4 (21.4)	0.8 (42.7)	1.0 (53.4)	0.3 (16.0)	0.5 (26.7)	3+ (160+)	2.0 (107)	0.3 (16.0)	...	3 (160)	No break	3 (160)
Hardness	103R	92M	100M	120R	80E	65M	15R	90R	70M	68M	70D	55D	55D
Maximum useful temperature, °F (K)	175 (353)	200 (367)	250 (394)	225 (381)	300 (422)	250 (394)	220 (378)	225 (381)	150 (339)	500 (533)	175 (353)	400 (478)	500 (533)
Coefficient of thermal expansion, 10^5/°F (10^5/K)	3.0 (5.4)	2.2 (4.0)	3.2 (5.8)	4.5 (8.1)	2.2 (4.6)	3.8 (6.8)	9 (16)	5.0 (9.0)	4.5 (8.1)	2.8 (5.0)	4 (7.2)	5.2 (9.4)	5.5 (9.9)
Water absorption, %/24h	0.3	0.3	0.1	1.5	0.6	0.1	<0.10	0.01	0.06	0.24	0.1	0.01	<0.01

for which it is not suitable: molten alkali metals (e.g., molten sodium) and strong fluorinating compounds (e.g., fluorine or chlorine trifluoride).

In addition to its chemical inertness, it has excellent electrical resistance properties and a very low coefficient of friction.

The plastics may generally be classified in one of two categories: thermoplastic or thermoset. Thermoplastic materials are formed by heating the material to a soft and pliable state (often molten) that can be shaped, extruded, etc. The thermosets are polymerized or hardened by heat. Actually many plastics are available in either form, and some thermosets may be set chemically. A few thermosets (epoxies, polyesters, and polyurethanes) are of special interest to the technician because they permit simple laboratory use for casting or for otherwise forming complex shapes. With simple chemical additives, these materials can be poured and brushed and left to set or cure, mostly without heat. They are especially useful for holding small mechanical assemblies in position for, say, cutaway inspection, as well as for forming an actual item by using a mold or lay-up technique.

2.5 ELASTOMERS

Many types of elastomers or rubber compounds are available. Table 2.16 lists some of the more common generic types along with their properties. Natural rubber has given way to many synthetic elastomers because it lacks chemical resistance and has poor weathering characteristics. Many nitrile elastomers are called *Buna;* the fluorocarbons are better known under manufacturers' trade names, such as Viton (Du Pont) and Fluorel (3M). The properties listed are typical; consult the manufacturer for more specific data. Within these generic types there is considerable latitude for individual compounding to produce specific physical properties. Substitution of one manufacturer's product for another, even of the same generic type, can give different results. However, most manufacturers will compound to a variety of needs.

The useful life of elastomers decreases with increasing temperature, sometimes dramatically so. However, applications differ depending upon the elastomer property of prime importance. For example, the performance of an O ring is strongly dependent upon the compression-set characteristics. Compression set may be minimal at 200°F (367 K) and significant, especially after several months of service at, say, 250°F (394 K), while other properties suffer less change. Therefore the listed temperature capability of the elastomer needs more detailed examination for the specific use. By special compounding one property can be traded for another within limits to provide the most desirable characteristics.

The chemical compatibility chart (Table 2.1) is a good starting point when selecting an elastomer. Following that, review the important physical properties at the operating temperature. Where several choices exist, one will naturally opt for the least expensive.

It is not unusual for elastomers to swell in contact with various fluids. If the swelling is minimal (a few percent) and no other degradation occurs, it is usual to rate it for satisfactory service (e.g., Table 2.1). In most applications a small amount of swelling is no detriment, but there may be designs where it would be troublesome. Therefore, while it might be rated as satisfactory for chemical compatibility, it could give mechanical problems because of small dimensional changes.

Not included in Table 2.1 is the elastomer Kalrez (Du Pont), a relatively new

TABLE 2.16 Some Properties of Natural Rubber and Selected Elastomers

Property	Natural rubber	Chlorinated butyl	EPDM	Buna	Polyurethane	Silicone	Flourocarbon
Specific gravity	0.092	0.092	0.086	1.00	1.20	1.00	1.85
Tensile strength, psi (MPa)	4000 (28)	3000 (21)	3000 (21)	4000 (28)	10,000 (69)	1200 (8)	2500 (17)
Elongation, %	700	700	600	600	700	700	300
Hardness range, Shore A	30–90	40–80	30–90	40–90	30–80	30–85	60–95
Compression set	Excellent	Good	Good	Good	Good	Excellent to poor	Good
Maximum temperature long-term use, °F (K)	150 (340)	200 (370)	230 (380)	180 (360)	160 (345)	350 (450)	300 (420)
Resistance to abrasion	Excellent	Poor	Good	Good	Excellent	Fair	Good
Sunlight aging	Poor	Good	Excellent	Poor	Good	Excellent	Excellent
Weathering	Poor	Excellent	Excellent	Poor	Excellent	Excellent	Excellent
Gas permeability	Fair	Excellent	Fair	Fair	Poor	Poor	Excellent
Resistance to tearing	Excellent	Good	Fair	Good	Excellent	Fair	Fair

material that has chemical resistance similar to Teflon. However, formaldehyde does cause some swelling. While Kalrez is presently expensive, it is worthwhile considering for those otherwise difficult situations for O-ring seals and the like. It provides good temperature service to 550°F (560 K).

Of special interest to the technician are the castable polyurethane elastomers because they lend themselves to prototype work. Single-cavity molds machined from aluminum alloys permit the design of an elastomeric part to be confirmed before committing to more expensive molds for production items. There is also usually considerable time saving by following this route. These elastomers provide a wide range of properties so that the intended production elastomer may be closely duplicated in most respects and therefore provide reliable preproduction data. Additionally, where special parts are required, they provide excellent means for producing one-of-a-kind items.

2.6 COATINGS

Coating parts is a common way to provide protection against corrosion, chemical attack, temperature, erosion, and wear. The coatings may be metallic, ceramic, elastomeric, or plastic. They are not used as a last resort but are planned as part of the manufacturing process. The methods for applying coatings are usually sufficiently simple that they also lend themselves for corrective measures once a problem has arisen. Table 2.17 lists the common coating techniques and typical base materials or substrates on which they are used.

TABLE 2.17 Coating Processes and Applications

Coating process	Coating material	Typical substrate upon which coating applied
Painting	Oils, plastics	Metal, plastics, wood
Conversion	Oxides, phosphates, and chromates	Metal (iron, steel, aluminum, magnesium beryllium)
Electroplating	Nickel, silver, gold, chromium, tin, zinc, copper, cadmium, rhodium, platinum	Metal
Electroless plating	Nickel, copper, silver	Metal, glass, plastic, ceramics
Fluidized beds	Plastics	Metal
Spray coatings	Metals, ceramics	Metal
Galvanizing	Zinc	Iron and steel
Rubberizing	Rubber	Metals

Painting

The preserving and protective properties of paints are well-known. Of particular interest are the epoxy paints, which can provide complete encapsulation of parts and afford excellent protection against salt water and chemically corrosive atmospheres, acids, alkalies, and some solvents. While the coatings are somewhat brittle, it is of little consequence in most applications because they attain excellent adhesion to the substrate and usually gain adequate support from it so as not to be subject to undue flexing and cracking. Apart from their protective qualities, epoxy coatings have excellent electrical resistance characteristics, provide very good decorative finishes, and are available in a multiplicity of colors.

Conversion Coatings

Protective coatings may be formed by chemically converting the surface atoms of metals to compounds that provide special characteristics for various attributes. These attributes are improved corrosion resistance, better wearing surfaces, and enhanced lubrication. It is usually not just the newly formed surface compound alone which provides these features but that compound in combination with other treatments. The compound is typically strongly adherent to the base metal and provides a strongly retentive surface for treatment with paints, sealants, oil, waxes, etc.

Most of the coating compounds are phosphates, chromates, and oxides. Use phosphate coatings mainly on steel. Chemically remove dirt, grease, and scale before immersing the workpiece in the phosphate bath. After thorough rinsing and drying, give the item its next treatment. Apply chromate coatings to ferrous and nonferrous metals in much the same manner as for phosphates. While the chromate surfaces are usually more corrosion-resistant than the phosphate, they do not provide such good retention of oils and waxes. For added protection cadmium- and zinc-coated surfaces are often chromated. With the exception of aluminum oxide, oxide coatings are generally not as good as phosphate or chromate. Aluminum oxide is strongly adherent and hard. While aluminum readily oxidizes in air, a tougher, more dependable coating is formed by anodizing.

To anodize, immerse the workpiece in an electrolyte and make it the anode in an electric circuit. Control coating thickness by the exposure time and the magnitude of the electric current. Because the resulting oxide coatings are porous, it is usual to seal the surface with a secondary operation that often contains a dye. Many different colors for anodized coatings are produced with dyes that may be used for decorative or identification purposes.

Electroplating

The electrodeposition of metals is limited to a few (copper, silver, nickel, cadmium, chromium, gold, rhodium, platinum). The deposit is a very pure form of the metal and of very low porosity. Therefore as chemically protective coatings they are excellent. Copper, silver, nickel, and chromium are used extensively for decorative finishes; nickel, chromium, gold, and platinum are used for chemical resistance, and rhodium is used for its excellent electric conductivity and nonoxidizing characteristics (electric contacts). Both nickel and chromium may be deposited as a hard coating to provide resistance to wear and abrasion and for rebuilding worn or incorrectly machined parts. Silver is used as an antigalling coating, especially on stainless steel.

Electroplating is done in a bath containing a solution (electrolyte) that must have metallic ions required to be plated. Maintain the concentration of ions by using an anode of the metal to be deposited. Material is removed from the anode as it is deposited on the workpiece (cathode). Connect the anode and cathode to a direct-current electric supply. For best results the electrolyte comprises several salts in solution, maintained at a specific temperature and constantly filtered. The thickness of the deposited material may vary considerably over the surface of the workpiece because of varying current density. The electroplater can minimize this by reducing the overall current and therefore the rate of deposition,

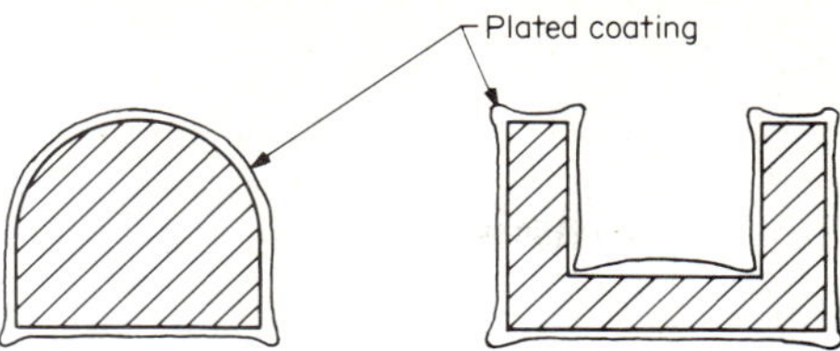

FIG. 2.2 The variation in electrodeposited coating thickness for two items with different sections.

and by the location and shape of the anode, but there remains limits to thickness uniformity. These limits depend greatly on the shape of the part to be plated. Sharp corners are especially difficult to deal with because of the high current density. Figure 2.2 shows for various part configurations how the thickness of the deposited material can be expected to vary. The thicker the deposit the more pronounced the variation. Portions of the workpiece can be rendered nonconductive by insulating them with coatings of wax or lacquer. Areas insulated in this manner, or masked, are protected against plating so that only selected areas are plated. This procedure is especially important for rebuilding specific portions of a workpiece.

The finish of the deposited materials depends upon two principal factors. One is the finish of the substrate; the higher the polish of the unplated surface the better the finish on the plated surface. The other is the thickness of the deposit; the heavier the deposit the poorer the finish. Sometimes in rebuilding parts to uniform thickness and good finish it is necessary to remachine the coating before full thickness is reached. Remachining corrects geometry and restores the finish.

The best corrosion resistance and adherence of plating is often achieved by using different layered coatings. For example, for chromium-plated steel, best results are obtained by first plating with copper, then nickel, and finally chromium. If repeated thermal shock is to be experienced by the coating and the final layer is of expansion coefficient different from the substrate, apply layers of plating of intermediate expansion characteristics.

Electroless Plating

Electroless plating is a simple process where the surface to be plated is first sensitized and then immersed in a plating solution. No electric current is used. The thickness of the plating depends upon the time the workpiece is left in the solution. The solution becomes depleted and requires replenishment from time to time.

Two particular advantages are offered by the electroless technique. First, the deposit is of uniform thickness regardless of the shape of the workpiece. Holes and internal cavities are plated as easily as simple exterior surfaces. Second, the workpiece does not have to be electrically conductive; thus plastics, glass, and ceramics may be plated with proper sensitization of the surface. In fact, an electroless-deposited layer on an otherwise nonconductive surface makes it possible to electrodeposit subsequently.

Coatings formed in this manner are often brittle and porous. Cleanliness, proper solution temperatures, and adequate replenishment of solutions minimize these problems.

Fluidized Bed

The fluidized-bed technique offers the technician a wide variety of plastic coatings that provide not only for versatile chemical compatibility but also for decorative finishes. Plastics that can be applied this way include epoxies, PVC, polyethylene, nylon, Teflon, polypropylenes, and polyurethanes. Heavy coatings to $\frac{1}{16}$ in (1.6 mm) can be applied this way. Small quantities of parts are easily handled. Internal surfaces, where complete flow-through is possible, may be coated uniformly.

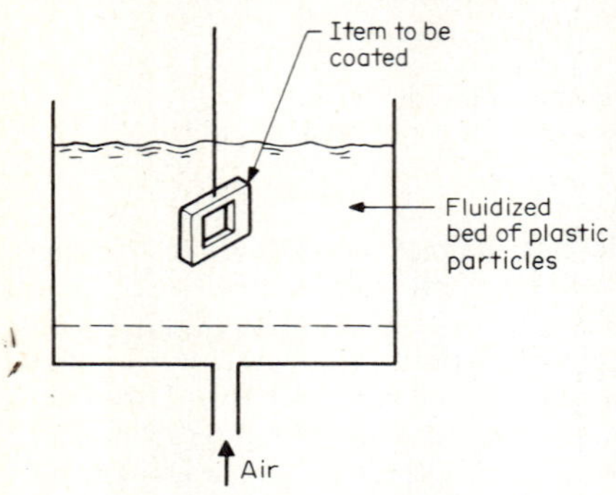

FIG. 2.3 Schematic arrangement for plastic coating in a fluidized bed.

Figure 2.3 illustrates schematically the fluidized-bed equipment. The powder-plastic material is maintained in suspension by the flow of air and is called the *fluidized bed* because it behaves very much like a fluid. To coat an item, heat it in an oven to a temperature from 300 to 750°F (420 to 670 K), depending upon the plastic. The suspended particles adhere, melt, and fuse together to form the coating. The longer the exposure to the particles the heavier the coating. The temperature requirements do not normally present a problem for metal parts but keep them in mind. The hanger used to suspend the part sometimes leaves an impression or thin spot, and if this is likely, then select its location to present no detriment.

Spray Coatings

Spray coatings of metals, alloys, ceramics, and mixtures of these may be applied to metallic surfaces. These coatings are tailored to meet a wide variety of applications and provide a unique way to obtain improved serviceability and performance for many items. Some of the commonly used coatings are for wear resistance, heat resistance, corrosion resistance, improved electric conductivity, electrical insulation, and for the repair and salvage of worn or incorrectly machined hardware. The coating materials themselves are either in a powder or wire form and are heated to a high temperature before being sprayed onto the base material. Special spray guns are used.

There are two general types of gun. One uses combustible gases, e.g., oxygen and acetylene, to heat and propel the particles of material being sprayed. The other uses electricity to heat and melt the spray material. Either an electric arc is used between electrodes of the material to be deposited or a plasma arc is used, which, in turn, heats a powder for spraying. A spray gun for combustible gases and powder is shown in Fig. 2.4.

The use of spray coatings may be considered for two reasons. The first is for the correction of a problem that materializes during testing. For example, higher than anticipated corrosion or heat-transfer problems can often be tackled quickly by coating existing hardware with a suitable protective layer. The cost of redesign and new hardware is eliminated, as are lengthy downtimes. Since the physical properties of the coatings are known, improved design data are frequently obtained from improved knowledge of the environment which caused the problem. This is an additional benefit for the engineer.

FIG. 2.4 Oxyacetylene powder spray gun. *(Metco Inc.)*

The second reason for spray coating is to consider it as part of the original design to be incorporated as a regular manufacturing step. For technicians, this, in all probability, will require an evaluation of the coating along with other aspects of the equipment during prototype or pilot testing.

Galvanizing

The application of zinc as a coating for iron and steel is called *galvanizing.* It is applied specifically to reduce atmospheric corrosion. The protection is provided by two separate mechanisms. One is the corrosion resistance of the zinc itself, and the other is by galvanic or electrolytic protection when both iron and zinc are exposed together. Such dual exposure might occur, for example, if a bolt hole has been drilled into a galvanized part. The galvanic action occurs because two dissimilar metals are in close proximity, and normal acids in the atmosphere, especially industrial and ocean atmospheres, create a local galvanic cell. This galvanic cell causes faster corrosion. In the case of zinc and iron, the cell is weak and corrosion-slow; furthermore the zinc is corroded faster than the iron, which, in effect, reduces the rate of iron corrosion. Therefore, the degree of protection afforded by galvanizing depends upon the zinc-coating thickness and results from considerations of both the zinc corrosion alone plus its sacrificial role in the galvanic process.

Galvanizing may be done by electroplating zinc or by hot dipping. Hot dipping permits heavier coatings to be applied more quickly. In the hot-dip process, thoroughly clean the iron or steel parts, usually by degreasing and acid pickling, before immersing them in a bath of molten zinc. Maintain the zinc at a temperature of about 850°F (730 K). The thickness of the coating depends upon the duration of immersion in the bath, the temperature of the bath, and the rate of withdrawal of the part from the bath.

Rubberizing

Rubberizing or rubber coating is mostly done to provide protection to iron and steel. In particular, it is often placed on internal surfaces of pipes and pumps to protect against acids and erosion caused by abrasion of solid particles, particularly in slurries. Rubber is also applied to exterior metal surfaces where a degree of resilience is required and a precise shape maintained (e.g., typewriter platens).

Thoroughly clean the surface to be coated, and apply special coatings before the rubber is vulcanized to the surface. Temperatures of 300–400°F (420–480 K) are required in the process.

chapter 3

Fabrication Techniques

Robert S. Christie

3.1 GENERAL CONSIDERATIONS

Fabrication is the act or process of making something. To fabricate is to construct or build; to form by art and labor; to manufacture. To manufacture is to make wares by hand or by machinery. Here we consider the techniques, methods, or details of procedures essential in the fabrication process. The "expertise" is provided by the technician in producing parts and assemblies of parts. We describe the various procedures and operations and some of the machines and tools usually available in the technician's shop for the performance of these activities.

The engineering technician must perform many operations to make models, prototype assemblies, and parts. The technician has to start with raw materials, such as plate, bar, or sheet stock, and use operations or techniques like cutting, milling, turning, drilling, reaming, grinding, lapping, threading, brazing, and soldering to produce the end product. When making a single piece, not all these techniques are necessarily used; but for an assembly of a prototype machine, assembly, or model, all are involved at one time during the various phases of the work.

FIG. 3.1 Assembly.

The material specified on the drawings of the parts to be made affects the techniques used to produce the part. Not all material can be worked by all these procedures. For instance, ceramics do not lend themselves to turning or milling, except in their "green" state, but they can be readily ground by the proper grinding wheel. On the other hand, copper is difficult to grind. The material also affects the time it takes to make a part. For example, it takes as much as three times longer to machine a piece from stainless steel as it does to machine the same piece out of aluminum or brass.

Figure 3.1 shows a part consisting of two flanges and a tube; one flange is round, the other rectangular. Most of the just-mentioned techniques were used to produce this part. The round flange was made by cutting, turning, drilling, and threading operations. The rectangular flange was produced by cutting, milling, drilling, and threading. Cutting and milling produced the tube. The various pieces were all brazed to produce the finished part.

Each operation we mentioned is discussed in detail in the following sections. The tables, drawings, and photographs will help the technician perform the various tasks involved.

3.2 CUTTING

Cutting is the penetration with a sharp-edged instrument, to separate from the main body. Cutting is divided into two general sawing categories: cutoff and contour. Both operations are performed with band saws; various hacksaws, circular saws, abrasive saws, and radial saws are primarily cutoff machines.

"Contour band sawing is a process for cutting metal and other materials by means of a power-driven saw-band, to produce workpieces of desired contour."[1] Cutoff sawing more closely fits the dictionary description of cutting. Any machine tool basically performs a cutting operation. Cutting, as used here, refers to the processes by which a workpiece is roughed out, prior to its being further worked upon by some other machining or finishing operation.

Hacksaws

Power hacksaws are used for cutting off bars, extruded and rolled shapes, pipe, and so on. The blade is stretched in a frame that is reciprocated (by an appropriate drive) above the work. The work is held in a vise located on the saw bed. The weight of the frame causes the blade to feed into the work; the frame weight varies according to the size of the machine. The pressure required between the blade and the work for various materials is affected by adding weights to the frame, by spring tension, by positive screw feed or a friction screw feed, and in some cases by hydraulic feed mechanisms.

Blades for power hacksaws are made of tungsten alloy steel, tungsten high-speed steel, molybdenum steel, and molybdenum high-speed steel. The best service is from tungsten high-speed steel blades, which are available in lengths up to 36 in and in pitches from 2½ to 18 teeth per inch (TPI). Thickness varies from 0.032 to 0.100 in. Table 3.1 suggests blade teeth per inch and strokes per minute for various materials.

Generally, with power hacksaws use a blade only as long as is needed to cover the workpiece size, wide and thick enough to take the feed pressure, and with at least three teeth in contact with the work at all times. Use coarser teeth on the softer materials, finer teeth on hard materials.

Cutting speeds range from 150 to 60 strokes per minute. The higher speeds are generally used for softer materials and lower speeds on harder and tougher ones. When cutting soft materials and light cross sections, use moderate feed pressure. The harder materials and heavier sections require greater feed pressure.

In all cases, rigidly clamp the work. When cutting more than one piece of stock at a time, securely hold all pieces, to help prevent damage to saw and blade and to provide straight cuts.

[1]From *Metals Handbook,* 8th ed., vol. 3, "Machining," American Society for Metals, Metals Park, Ohio, copyright 1967.

Use coolants of soluble oil in water or sulfur-base cutting oil diluted in kerosene when power hacksawing all ferrous metals except cast iron. This greatly aids chip removal and helps prolong blade life. Figure 3.2 shows a typical power hacksaw; Fig. 3.3 is a close-up view of work properly clamped in the saw vise.

Hand hacksaw frames are made to take the standard 8-, 10-, and 12-in-long hacksaw blades readily available on today's market. The blades are 7/16 to 9/16 in wide and 0.025 in thick, in four pitches: 14, 18, 24, and 32 TPI. Blades are made of tungsten alloy steel, carbon steel, and high-speed steel. Table 3.2 is a guide for the selection of the right blade for various purposes and materials.

In general, use a 14-tooth blade for cutting stock equivalent to 1 in round or more and relatively soft material requiring maximum chip clearance. For materials 1/4 to 1 in in diameter and for general shop use, an 18-tooth blade is recommended. Use the 24-tooth blades on materials 1/8 to 1/4 in thick. Cut materials less than 1/8 in thick with a 32-tooth blade. A blade should have a minimum of three consecutive teeth in contact with the workpiece at all times.

When using a hand hacksaw, develop the technique of a long, steady stroke at a rate of about 40 to 60 strokes per minute. Apply pressure only on the forward stroke, and raise the blade slightly on the return stroke to prevent dulling the

TABLE 3.1 Recommended Blade Teeth and Cutting Speeds for Power Hacksaws

Material	Teeth per inch	Strokes/min
Ferrous material:		
Drill rod	10	90
Forging stock:		
Alloy	4–6	90
Mild	3–4–6	120
Iron:		
Cast	6–10	90–120
Malleable	6–10	90
Rails	6–10	60–90
Steel:		
Alloy	4–6	60–90
Carbon tool	4–6	90–120
High-speed	6–10	60–90
Machinery	3–4–6	90–120
Stainless	3–4–6	60–90
Die blocks	4–6	60–90
Pipe	6–10	120
Tubing:		
Thick-wall	6–10	120
Thin-wall	14	120
Nonferrous material:		
Aluminum	3–4–6	120
Babbitt	4–6	120
Brass castings:		
Hard	4–6	90–120
Soft	4–6	120
Bronze castings	4–6–10	90
Bronze-manganese	6–10	60–90
Copper bars	3–4–6	90
Copper tubing	10	120
Monel metal	4–6	60–90

SOURCE: By permission of Nicholson, USA.

FIG. 3.2 Power hacksaw.

teeth. Moderate pressure is usually adequate; excessive pressure causes a crooked cut.

It is particularly important when hacksawing stainless steel to lift the blade on the return stroke. In addition to keeping the teeth sharp, this technique avoids the work-hardening effect of dragging the teeth backward through the work. For example, for chromium-nickel stainless steels, the work hardening caused by drag may be severe. A lubricant and coolant are recommended when hacksawing stainless steel. "Stick"-type lubricants provide sufficient lubrication in most cases. A hand hacksaw frame and blades are shown in Fig. 3.4.

FIG. 3.3 Close-up of power hacksaw vise with workpiece.

TABLE 3.2 Recommended Blade Teeth and Cutting Speeds for Hand Hacksaws

Material	Teeth per inch	Strokes/min
Ferrous material:		
BX	32	60
Conduit, rigid	24	60
Drill rod	18–24	40
Iron, cast	14	60
Pipe	24	60
Rails	14	40
Sheet metal	24–32	60
Steel:		
Machinery	14–18	60
Tool	18–24	50
Structural shapes		
Heavy	18	60
Light	24	60
Tubing, light	32	60
Nonferrous material:		
Aluminum	14	60
Brass and bronze	14–24	60
Brass tubing	24	60
Copper	14	60
Structural shapes	14–24	60
Nonmetals:		
Asbestos	14	60
Fiber	14	60
Slate	14	50

SOURCE: By permission of Nicholson, USA.

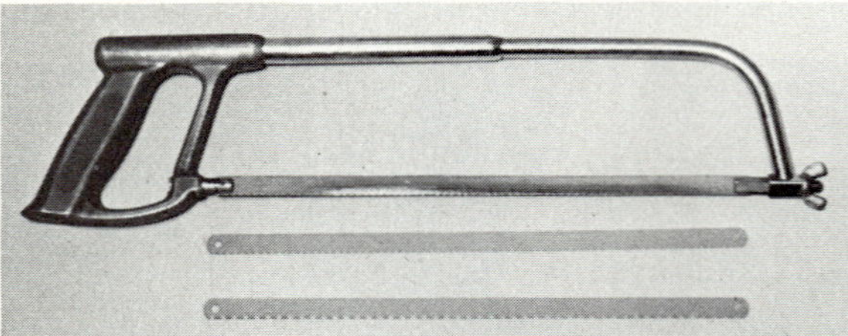

FIG. 3.4 Hand hacksaw and blades.

Band Saws

The two types of band saws are horizontal and vertical. Cutoff band sawing is usually accomplished on the horizontal machines; contour band sawing is done on the vertical machines. The vertical machine is also handy for cutoff operations, especially for smaller, hand-held size work.

Two common vertical band saws in the technician's shop are shown in Figs. 3.5 and 3.6. The larger machine is equipped with blade cutting and welding attachments. The machines have a C-shaped frame and a continuous band that moves vertically down at the point of cutting. The size of the workpiece that can be handled depends upon the depth of throat or height between the band-saw guides. The bigger and more powerful the machine, the larger the work that can be accommodated.

Figure 3.7 is a typical horizontal band saw found in job and the technician's shop areas. This machine is used for cutoff sawing and in most shops is replacing the power hacksaw for this type of work. Its advantages are uninterrupted cutting (the band is continuous), even band wear over the entire length, high cutting rate, less frequent band changing, and less band binding.

Bands for both vertical and horizontal machines are made of various materials. Carbon-steel bands are most commonly used, with fixed table machines usually found in the technician's area. These bands are used satisfactorily on carbon and low-alloy steels and the more easily machined nonferrous alloys. High-speed steel bands are used for stainless steels, tool steels, and some nonferrous alloys. Bands with tungsten-carbide inserts welded or brazed to a carbon-steel band are used for hard-to-machine alloys and today's more exotic materials.

Several saw bands are used for cutting metal, plastic, and wood; Fig. 3.8 shows five types:

1. Regular-tooth or precision-tooth
2. Skip-tooth or buttress-tooth
3. Claw-tooth or hook-tooth

FIG. 3.5 Vertical band saw (large).

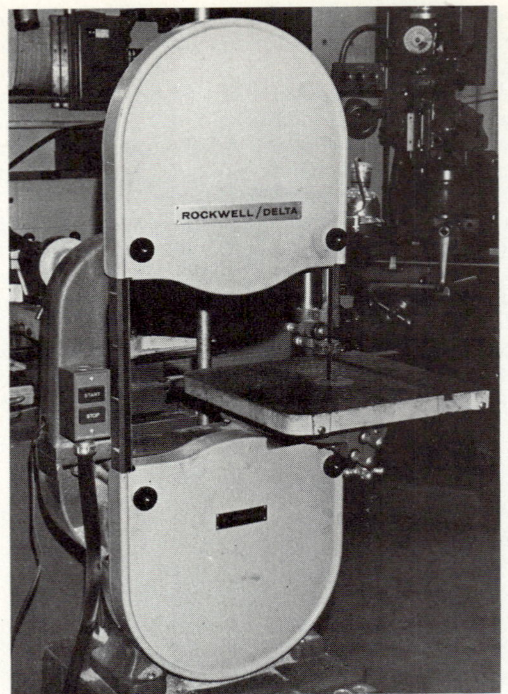

FIG. 3.6 Vertical band saw (small).

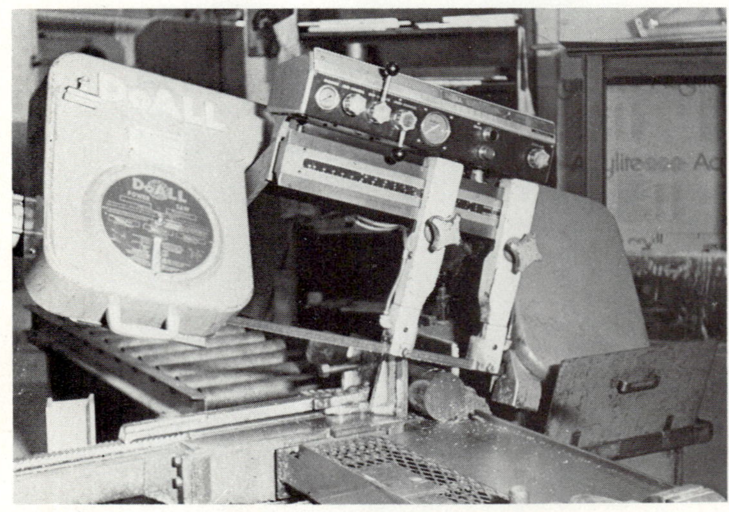

FIG. 3.7 Horizontal band saw.

4. Friction
5. Abrasive

Figure 3.8 also shows the standard band-tooth nomenclature. Precision- and buttress-tooth forms are the same, except the buttress- or skip-tooth form has a lengthened gullet for larger chip loads. The claw-tooth form has a positive rake angle of 10° and a lengthened gullet.

Types 1 to 3 are usually available in three tooth sets (see Fig. 3.8): raker set, wave or wavy set, and alternate or straight set. In the raker set, one tooth is set to left, one set to right, and one not set. The wave set has a group of teeth set alternately to left and right. The straight set has individual, alternately set teeth, one to right, one to left, and so on.

Friction bands are used in *friction band sawing,* which differs from other band-saw operations. It is more like torch cutting in that heat, generated by the blade moving at high speed (from 6000 to 15,000 fpm), is built up in the workpiece at the contact point between band and workpiece, causing softening or melting of the metal, which is then removed by the teeth. Friction bands are usually made of carbon steel with specially heat-treated teeth, are thicker than standard bands, and have a wider set.

Abrasive bands have a cutting edge of diamond, silicon-carbide, or aluminum oxide fused or bonded to the edge of a steel band. Diamond-edge band saws can

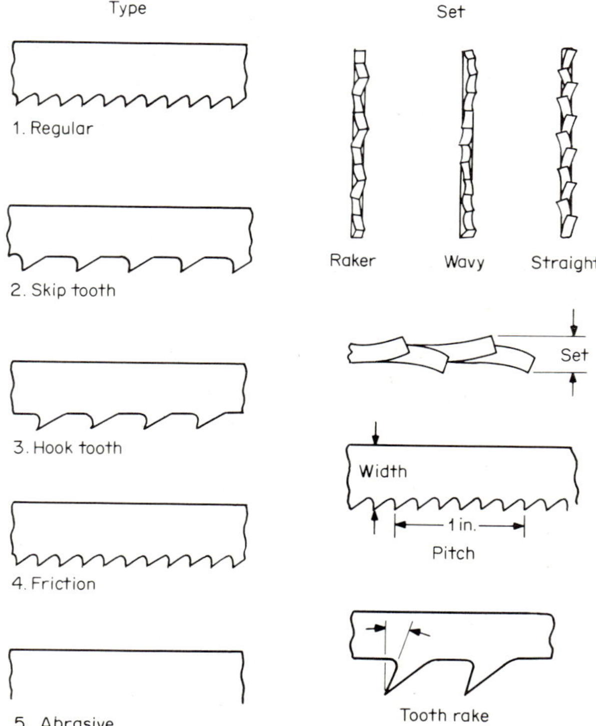

FIG. 3.8 Types of band saw blades and tooth set.

be used for contour or straight cutting such materials as glass, ceramics, silicon, and barium-titanate. Band saws edged with tungsten- or silicon-carbide are used to cut or "fine grind" ceramics, glass, quartz, and vitrified materials. Those edged with aluminum oxide are used on such materials as heat-treated alloy steels, heat-resistant steels, and wear-resistant nonferrous alloys. All these bands generate considerable heat, so coolants should be used. Dry cutting is possible if an air blast or strong vacuum-suction system is used to keep the abrasive edge from loading down and preventing powdered material from becoming airborne. To sharpen an abrasive band, dress it by moving a diamond dressing stick lightly over the cutting edge while running the band.

The pitch of the band is the number of teeth per inch. The pitch required for a given job is determined primarily by the thickness of the material to be cut. Remember that at least two teeth should be engaged in the material at all times. Light- or thin-wall sections require fine teeth; heavy or large sections require coarse teeth. When contour sawing, use a raker-set band. Also, use the widest blade capable of cutting the smallest radius to be cut (see Table 3.3).

Because of the many factors involved in band sawing, it is practical to express the feed pressures to use only in general terms. The fewer the teeth, the less feed pressure required. Even pressure without forcing the work generally gives the best results. For straight, accurate cuts, use moderate pressure. Examining the chips may help determine whether the right feed is being used. If the chips are fine and powdery, the feed is too light. Too heavy a feed results in discolored, blued, or straw-colored chips. The ideal feed usually results in free-cut curling chips.

Proper cutting speed is very important in band sawing. Table 3.4 gives speeds and feeds for various metals; Table 3.5 recommends speeds for various nonmetals, plastics, and woods. For those who have machines capable of the higher cutting speeds, Table 3.6 gives recommended friction cutting speeds for common metals. Table 3.7 provides information related to friction sawing stainless steels.

Whenever feasible, use coolant or lubricants in band sawing, except with cast iron, which is always cut dry. Paraffin is a common lubricant for cutting aluminum. Use light cutting oil and soluble oils for ferrous metals.

One of the more recent materials available on the market is Machinable Glass Ceramic, sometimes called MGC or Macor (registered trade names of Corning

TABLE 3.3 Recommended Band Width for Various Radii

Band width, in	Radius to be cut, in
1/16	Square or less than 1/16
3/32	1/16
1/8	1/8
3/16	5/16
1/4	5/8
3/8	1 7/16
1/2	2 1/2
5/8	3 3/4
3/4	5 7/16
1	7 1/4
1 1/4	12
1 1/2	21
2	28

TABLE 3.4 Recommended Band Saw Cutting Speeds and Feeds

AISI type	Description	High-speed-steel bands		Carbon-steel band saw	
		Saw speed, fpm	Cutting rate, in^2/min	Saw speed, fpm	Cutting rate, in^2/min
		Group I—Easily machinable			
1010–1035	Straight carbon steels	350	12	175	1.5
1040–1050	Medium carbon steel	250	12	125	1.5
1108–1130	Free-cutting low carbon steels	350	12	200	1.5
1137–1150	Free-cutting medium carbon steels	270	12	200	1.5
		Group II—Moderately difficult to machine			
1065–1095	High carbon steels	150	7	100	0.7
1320–1345	Manganese steels	200	7	100	0.7
2317–2517	Nickel steels	175	6	90	0.6
3115–3315	Nickel-chrome steels	200	7	90	0.7
4017–4068	Moly steels	250	7	125	0.7
4130–4150	Chrome-moly steels	225	7	75	0.7
4317–4340	Nickel-chrome moly steels	200	6	75	0.6
4608–4820	Nickel-moly steels	200	6	75	0.6
5045–5160	Chromium steels	200	6	75	0.6
50100–52100	Carbon-chrome steels	150	6	100	0.6
6117–6152	Chrome-vanadium steel	175	7	75	0.7
8615–8645	Nickel-chrome moly steels	175	6	65	0.6
8647–8750	Moly steels	175	6	65	0.6
9255–9262	Silicon steels	150	5	100	0.5
9310–9850	Nickel-chrome moly steels	150	5	50	0.5
A–2, O–1, O–2, O–6	Tool and die steels	175	5	100	0.5
H–2, H–24	Hot-work tool steels	150	4	50	0.4
S–1, S–5	Shock-resisting tool steels	150	5	50	0.5
		Group III—Machined with difficulty			
2515	5% nickel steel	150	3	75	0.3
T–1, T–2	High-speed steel	100	3	70	0.3
T–4, T–5	High-speed steel	100	2.5	60	0.25
T–6, T–8, T–15	High-speed steel	75	2	55	0.25
M–1, M–2	High-speed steel	130	5	90	0.5
M–3, M–10	High-speed steel	100	3	60	0.3
D–2, D–3	Die steel	100	4	60	0.4
D–7	Die steel	75	2	50	0.2
308, 309, 310, 314, 316, 317, 330, 420, 430, 446	Stainless steel	75	3		
302, 304, 321, 347, 440	Stainless steel	100	4		
303	Free-machining	150	6		
420F	Stainless steel	125	5		
430F, 440F		125	5		

TABLE 3.4 (*Continued*)

AISI type	Description	High-speed-steel bands: Saw speed, fpm	High-speed-steel bands: Cutting rate, in^2/min	Carbon-steel band saw: Saw speed, fpm	Carbon-steel band saw: Cutting rate, in^2/min
	Group IV—Nonferrous				
	Aluminum			2000	
	Aluminum-bronze	225	14		
	Beryllium-copper	300	8	100	0.8
	Manganese-bronze	400	15	125	1.5
	Phosphor-bronze	200	10	135	1
	Silicon	325	15	200	1.5
	Bronze-titanium	85	1.5		

SOURCE: *Metal Cutting Manual*, Henry G. Thompson Co., Branford, Conn., 1968.

Glass Works). It is the first available glass ceramic material that is machinable with ordinary tools. The recommended band-saw band for MGC is ¼-in-wide carbide, with a 100-fpm band speed. Experience has shown that a tungsten-carbide abrasive band, used with suction to remove the powder and help prevent loading of the band, gives excellent results. Figure 3.9 shows a piece of MGC being cut with a tungsten-carbide abrasive band. Note the use of a vacuum system to remove the powder.

TABLE 3.5 Recommended Cutting Speeds for Nonmetals, Plastics, and Woods

Material	Carbon band speed*, fpm	Material	Carbon band speed*, fpm
Asbestos	1000–1500	Formica	600–3300
Backer board	6000–8000	Insurock	2800–3800
Brake lining	1000–1500	Lamacoid	2200–3800
Builders board	1500–4500	Lucite	2000–3500
Carbon	2000–5000	Lustron	3800–4500
Celotex	3000–4000	Micarta	2500–4300
Chipboard	2000–3500	Neoprene	3500–4000
Cork	3000–5000	Phenolite	2500–4000
Fiber board	2500–4000	Plaskon	3500–4500
Fire brick (soft)	90–200	Plexiglas	2200–3500
Insulation (glass)	4000–6000	Teflon	2500–3000
Insulite	4000–6000	Tenite	1500–3500
Sponge	3500–6000	Texolite	2500–3500
Transite	90–200	Vinylite	4000–4500
Bakelite	1500–4000	Hardwood	2500–4000
Catalin	1800–4000	Softwood	3000–4000
Farlite	3000–4000	Plywood	2000–3500
Fibestos	2500–3800		

*When cutting larger work sections, select a slower speed within the recommended range.

SOURCE: *Metal Cutting Manual*, Henry G. Thompson Co., Branford, Conn., 1968.

TABLE 3.6 Average Recommended Friction Cutting Speeds for Common Metals

Metal	Cutting speed, fpm*
Armor plate	7000–13,000
Carbon steel	6000–12,000
Cast steel	7000–13,000
Chromium steel	8000–15,000
Chromium-vanadium steel	8000–15,000
Free-machining steel	6000–12,000
Gray cast iron	7000–13,000
Malleable cast iron	7000–13,000
Manganese steel	6000–12,000
Moly steel	8000–15,000
Molybdenum	8000–15,000
Nickel-chromium steel	8000–15,000
Nickel steel	8000–15,000
Silicon steel	8000–15,000
Stainless steel	7000–13,000
Tungsten steel	8000–15,000

*Low range for thin sections, high speed for maximum areas.
SOURCE: *Metal Cutting Manual,* Henry G. Thompson Co., Branford, Conn., 1968.

TABLE 3.7 Stainless-Steel Friction Sawing Data

Dimension	Saw pitch	Saw velocity, fpm	Cutting rate
Sheet and plate:			
3/32	14	6000	72 in/min
5/8	14	9000	32 in/min
1/8	10	5000	29 in/min
1/4	10	7000	4.5 in^2/min
1/8	14	5000	24 in/min
18 gage	10	5000	22 in/min
3/4	10	12,000	16.8 in/min
3/8	10	8000	5.16 in^2/min
1	10	14,000	12.8 in/min
7/32	10	8000	32 in/min
1/16	14	6000	125 in/min
1/32	18	5000	396 in/min
Tubing:			
6 × 3/8 wall	10	10,000	20 cuts/min
1 × 0.046 wall	10	5000	0.02 min/cut
1½ × 3/64 wall	14	8000	30 cuts/min
3 × 3/32 wall	10	9000	20 cuts/min
2 7/16 × 3/16 wall	14	8000	0.04 min/cut
Gates and risers:			
9/16 × 7/16	10	10,000	0.10 min/cut
7/8 × 7/8	10	10,000	0.15 min/cut
1 × 5/8	10	10,000	0.19 min/cut
1½ × 7/16	10	10,000	0.23 min/cut
1¾ × 1½	10	10,000	0.33 min/cut

SOURCE: "Stainless Steel Fabrications," Allegheny Ludlum Steel Corporation, Pittsburgh, 1957.

FIG. 3.9 Band sawing machinable glass ceramic.

Other Saws

In the technician's shop, radial, table, and hand circular saws are used primarily to cut wood. However, the larger and higher-powdered radial and table saws can be used to cut aluminum and brass, particularly plate stock up to perhaps 1 in thick. Because those machines are usually not multiple-speed, the only way to change the cutting speed is to change the blade diameter (the larger the diameter blade, the higher the surface feet per minute for a given revolution per minute), and it is difficult to specify speed and feed information. As an example, when cutting aluminum, a 10-in-diameter blade with 10 or 12 carbide-tipped teeth and running at 4200 to 4500 rpm very nicely cuts up to 1-in stock. Use a light, steady feed pressure. Apply stick paraffin or grease to the blade frequently to help clear chips from work and blade. A worthwhile trick when using a table saw is to adjust the blade so it protrudes no more than ⅛ in above the material. Among other things, this helps prevent chips from flying up and hitting the operator in the face. In this and similar operations, wear a safety shield or glasses and observe all safety precautions. Blades of this type are available in sizes from 8 to 14 teeth made to fit different size arbors. All such blades have a maximum-rated speed of revolutions per minute that should not be exceeded. Figure 3.10 is a photograph of such a blade.

There are numerous types and sizes of blades for cutting woods, hardboard, plywoods, veneers, plastics, and other materials. The most common are combination saw blade, ripsaw blade, crosscut blade, and miter saw blade. Figure 3.11 shows examples of the variety of blades available from a number of sources.

Abrasive cutoff saws use rubber- or bakelite-bonded abrasive wheels running at speeds as high as 16,000 fpm. They are very useful for cutting ceramic and quartz tubing, hardened high-speed steel, stellite, and like materials as well as the more common materials. They come in both "wet" and "dry" versions, in sizes having wheel diameters from 3 to 12 in and over. Wet machines incorporate a coolant pumping system and provide smooth cutting with little burr and long wheel life. Dry machines usually include an attached dust-collecting system; they cut faster than wet machines.

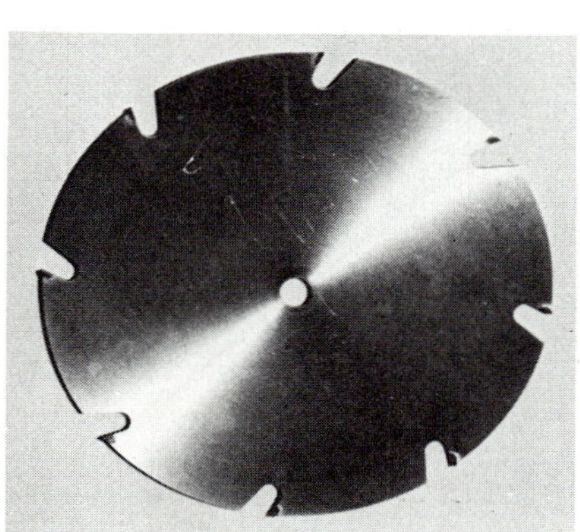

FIG. 3.10 Carbide-tipped circular saw blade.

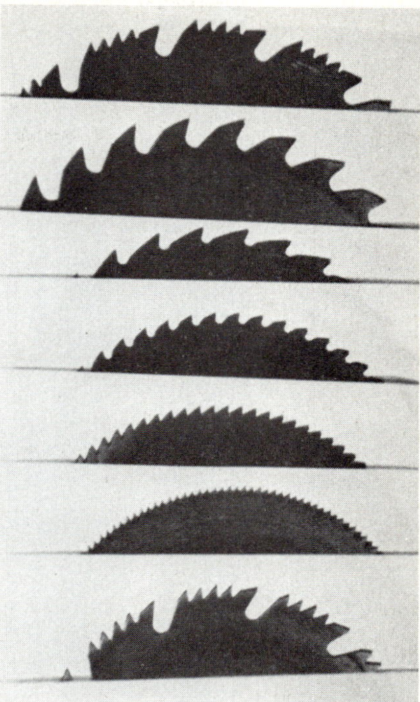

FIG. 3.11 Circular saw blade profiles. *(Sears, Roebuck & Co.)*

3.3 DRILLING

The drilling process is piercing or boring a hole in something. A drill or drill bit is a shaftlike tool with two or more cutting edges for making holes in firm material, especially by rotation; and a tool, especially a hand tool, for holding and operating a shaftlike tool.

Most machine tools capable of rotating a tool or workpiece have been used at one time for drilling, including lathes, milling machines, jig borers, and boring mills. However, the most common drilling machine for the technician's and job shop work is the hand-feed drill press, such as in Fig. 3.12. In some of the larger shops or technician's areas, a power-feed drill press may be used.

Hand-feed drill presses consist of a (1) table to hold the workpiece, (2) vertical-driven drill chuck, and (3) mechanism to lower the chuck by hand and feed the drill into the work. Hand feeding lets the operator control the rate of feed and use a sense of feel in the operation. These machines range in size from bench presses to floor models capable of drilling 1½-in diameter or larger holes.

Another machine frequently found in the technician's area is the portable magnetic drill. Such drills are available in many sizes, with capacities to accommodate 1¼-in drills. The magnetized base attaches to steel surfaces with sufficient force to allow horizontal and vertical drilling and even overhead drilling. It is an especially useful tool when drilling must be done accurately and the workpiece is much too large to be placed in a drill press. Figure 3.13 shows such a tool.

FIG. 3.12 Drill press.

FIG. 3.13 Portable drill press.

The portable electric hand drill is legion with all technicians who work with metals. The ¼- and ⅜-in drills are in most tool collections. There are single-speed as well as variable-speed models. Larger sizes are available but are bulkier, heavier, and not as easily handled. Where accuracy of hole location and hole size are not too important, portable electric hand drills are very useful tools.

Drills

Twist drills, the most common tools used in drilling operations, are made in many sizes and lengths. Straight-shank twist drills are available in numbered sizes, 1 through 80; lettered sizes, A through Z; and by fractions of an inch, in 1/64-in increments up to ½ in. Taper-shank drills come in sizes from ⅛- to 1¾-in diameter by 1/64 increments, then by 1/32-in increments to 2¼-in diameter, and up to 3½-in diameter by 1/16-in increments. Figure 3.14 shows the design features and terminology of typical straight- and taper-shank standard twist drills. Straight-shank twist-drill sizes are in Table 3.8; Table 3.9 lists taper-shank twist-drill sizes.

The recent move in the United States to adopt the metric system has made metric-size drills more readily available. Both straight- and taper-shank types can be obtained, in sizes ranging from 0.04- to 23-mm diameter in straight shank and 5- to 100-mm diameter in tapered (Morse) shank.

The usual point angle of a twist drill is 118°, and the helix angles of the flutes vary from 0 to 48°. Some materials are more easily drilled by using different point angles. Recommended point and helix angles for various materials are in Table 3.10.

Speeds for drilling materials largely depend upon the machine used, condition of the drill, and the operator's experience. It is best to start with moderate speed and feed and increase one or the other or both as the work progresses and the action and condition of the drill can be observed. Table 3.11 is a guide to suggested speeds for high-speed-steel drills. Run carbon-steel drills at speeds approximately 40 to 50 percent of those given in the table. Feed is best determined by drill size (Table 3.12).

Wherever possible, use a coolant when drilling because most hand-feed drill presses are not equipped with coolant systems. It is possible to brush on a sufficient amount of coolants and/or lubricants, and this procedure should be followed. Kerosene, soluble oils, and sulfurized oil are typical lubricants that can be used.

To facilitate changing from feet per minute cutting speeds to revolutions per minute, use Tables 3.13 to 3.15.

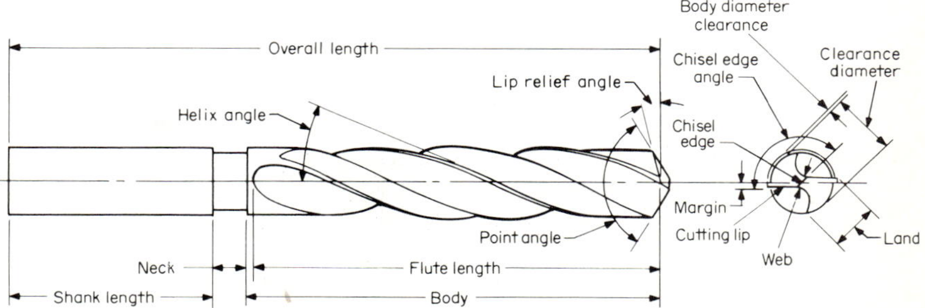

FIG. 3.14 Twist-drill terminology.

TABLE 3.8 Straight-Shank Twist-Drill Sizes to ½ In

Drill no. or letter	Fraction, in	Diameter, in	Drill no. or letter	Fraction, in	Diameter, in
80		0.0135	33		0.113
79		0.0145	32		0.116
	1/64	0.0156	31		0.120
78		0.016		1/8	0.1250
77		0.018	30		0.1285
76		0.020	29		0.136
75		0.021	28		0.1405
74		0.0225		9/64	0.1406
73		0.024	27		0.144
72		0.025	26		0.147
71		0.026	25		0.1495
70		0.028	24		0.152
69		0.0292	23		0.154
68		0.031		5/32	0.1562
	1/32	0.0312	22		0.157
67		0.02	21		0.159
66		0.033	20		0.161
65		0.035	19		0.166
64		0.036	18		0.1695
63		0.037		11/64	0.1719
62		0.038	17		0.173
61		0.039	16		0.177
60		0.040	15		0.180
59		0.041	14		0.182
58		0.042	13		0.185
57		0.043		3/16	0.1875
56		0.0465	12		0.189
	3/64	0.0468	11		0.191
55		0.052	10		0.1935
54		0.055	9		0.196
53		0.0595	8		0.199
	1/16	0.0625	7		0.201
52		0.0635		13/64	0.2031
51		0.067	6		0.204
50		0.070	5		0.2055
49		0.073	4		0.209
48		0.076	5		0.2055
	5/64	0.0781	4		0.209
47		0.0785	3		0.213
46		0.081		7/32	0.2187
45		0.082	2		0.221
44		0.086	1		0.228
43		0.089	A		0.234
42		0.0935		15/64	0.2344
	3/32	0.0937	B		0.238
41		0.096	C		0.242
40		0.098	D		0.246
39		0.0995	E	1/4	0.250
38		0.1015	F		0.257
37		0.104	G		0.261
36		0.01065		17/64	0.2656
	7/64	0.1094	H		0.266
35		0.110	I		0.272
34		0.111	J		0.277

TABLE 3.8 (*Continued*)

Drill no. or letter	Fraction, in	Diameter, in	Drill no. or letter	Fraction, in	Diameter, in
K		0.281	U		0.368
	9/32	0.2812		3/8	0.375
L		0.290	V		0.377
M		0.295	W		0.386
	19/64	0.2969		25/64	0.3906
N		0.302	X		0.397
	5/16	0.3125	Y		0.404
O		0.316		13/32	0.4062
P		0.323	Z		0.413
	21/64	0.3281		27/64	0.4219
Q		0.332		7/16	0.4375
R		0.339		29/64	0.4581
	11/32	0.3437		15/32	0.4687
S		0.348		31/64	0.4844
T		0.358		1/2	0.500
	23/64	0.3594			

TABLE 3.9 Taper-Shank Twist-Drill Sizes

Drill sizes, in	ASA or Morse taper no.	Drill sizes, in	ASA or Morse taper no.
1/8–15/32 × 1/64	1	1 33/64–1 3/4 × 1/64	5
31/64–25/32 × 1/64	2	1 25/32–2 1/4 × 1/32	5
51/64–1 1/16 × 1/64	3	2 5/16–3 × 1/16	5
1 5/64–1 1/2 × 1/64	4	3 1/16–3 1/2 × 1/16	6

TABLE 3.10 Twist-Drill Point and Helix Angles

Material	Point angle	Helix angle
Aluminum alloys	140	24–48
Bakelite	90–118	10–20
Brass and bronze	118	0–20
Carbon steel	118	20–32
Cast iron, soft	90–118	32
Copper	100–110	40
Inconel	118	30
Magnesium alloy	118	20–45
Magnesium alloy sheet	60	10
Molybdenum	118	30
Monel alloys	118	30
Nickel alloys	118	30
Plastic, molded	60–90	10–20
Stainless steels	130–140	24–32
Steel, forged	118–140	20–35
Titanium alloy	135–150	29
Tungsten	118°/90°*	30
Wood	70	30–40
Zinc die castings	118	30

*90° extending from margin up one-third of lip; notched point, 10 to 12° lip clearance solid carbide.

TABLE 3.11 Recommended Drill Speeds

Material	Cutting speed, fpm
Aluminum alloy	200–300
Bakelite	100–150
Brass and bronze	200–300
Carbon steel	80–100
Cast iron	70–100
Copper	75–175
Inconel	30–45
Magnesium alloys	150
MGC	20–30
Molybdenum	100–125
Monel alloys	40–60
Nickel alloys	40–60
Steel, forged	50–60
Stainless steel:	
AISI 300 series	20–80
AISI 400 series	20–100
Tantalum	40–50
Tungsten	200–250*
Wood	300–400
Zinc die, castings	200–300

*Use solid carbide drill accurately ground and tight fitting backup plate of steel. Do not drill closer than one drill diameter from any edge. Use large quantity of cutting fluid.

TABLE 3.12 Drill Feed

Drill diam., in.	Feed, in/rev
Under ⅛	0.001–0.002
⅛–¼	0.002–0.004
¼–½	0.004–0.007
½–1	0.007–0.015
1 and over	0.015–0.025

TABLE 3.13 Cutting Speeds and Equivalent rpm for Numbered Size Drills

	Cutting speed, fpm										
Size no.	30	40	50	60	70	80	90	100	110	130	150
1	503	670	838	1005	1173	1340	1508	1675	1843	2179	2513
2	518	691	864	1037	1210	1382	1555	1728	1901	2247	2593
4	548	731	914	1097	1280	1462	1645	1828	2010	2376	2741
6	562	749	936	1123	1310	1498	1685	1872	2060	2434	2809
8	576	768	960	1151	1343	1535	1727	1919	2111	2495	2879
10	592	790	987	1184	1382	1579	1777	1974	2171	2566	2961
12	606	808	1010	1213	1415	1617	1819	2021	2223	2627	3032
14	630	840	1050	1259	1469	1679	1889	2099	2309	2728	3148
16	647	863	1079	1295	1511	1726	1942	2158	2374	2806	3237
18	678	904	1130	1356	1582	1808	2034	2260	2479	2930	3380
20	712	949	1186	1423	1660	1898	2135	2372	2610	3084	3559
22	730	973	1217	1460	1703	1946	2190	2433	2676	3164	3649
24	754	1005	1257	1508	1759	2010	2262	2513	2764	3267	3769
26	779	1039	1299	1559	1819	2078	2338	2598	2858	3378	3898
28	816	1088	1360	1631	1903	2175	2447	2719	2990	3534	4078
30	892	1189	1487	1784	2081	2378	2676	2973	3270	3864	4459
32	988	1317	1647	1976	2305	2634	2964	3293	3622	4281	4939
34	1032	1376	1721	2065	2409	2753	3097	3442	3785	4474	5162
36	1076	1435	1794	2152	2511	2870	3228	3587	3945	4663	5380
38	1129	1505	1882	2258	2634	3010	3387	3763	4140	4892	5645
40	1169	1559	1949	2339	2729	3118	3508	3898	4287	5067	5846
42	1226	1634	2043	2451	2860	3268	3677	4085	4494	5311	6128
44	1333	1777	2221	2665	3109	3554	3999	4442	4886	5774	6662
46	1415	1886	2358	2830	3301	3773	4244	4716	5187	6130	7074
48	1508	2010	2513	3016	3518	4021	4523	5026	5528	6534	7539
50	1637	2183	2729	3274	3820	4366	4911	5457	6002	7094	8185
52	1805	2406	3008	3609	4211	4812	5414	6015	6619	7820	9023
54	2084	2778	3473	4167	4862	5556	6251	6945	7639	9028	10417

SOURCE: Erik Oberg, Franklin D. Jones, and Holbrook L. Horton (eds.), *Machinery's Handbook*, 20th ed., Industrial Press, New York, 1975.

TABLE 3.14 Cutting Speeds and Equivalent rpm for Letter Size Drills

	Cutting speed, fpm										
Size	30	40	50	60	70	80	90	100	110	130	150
A	491	654	818	982	1145	1309	1472	1636	1796	2122	2448
B	482	642	803	963	1124	1284	1445	1605	1765	2086	2407
C	473	631	789	947	1105	1262	1420	1578	1736	2052	2368
D	467	622	778	934	1089	1245	1400	1556	1708	2018	2329
E	458	611	764	917	1070	1222	1375	1528	1681	1968	2292
F	446	594	743	892	1040	1189	1337	1486	1635	1932	2229
G	440	585	732	878	1024	1170	1317	1463	1610	1903	2195
H	430	574	718	862	1005	1149	1292	1436	1580	1867	2154
I	421	562	702	842	983	1123	1264	1404	1545	1826	2106
J	414	552	690	827	965	1103	1241	1379	1517	1793	2068
K	408	544	680	815	951	1087	1223	1359	1495	1767	2039
L	395	527	659	790	922	1054	1185	1317	1449	1712	1976
M	389	518	648	777	907	1036	1166	1295	1424	1683	1942
N	380	506	633	759	886	1012	1139	1265	1391	1644	1897
O	363	484	605	725	846	967	1088	1209	1330	1571	1813
P	355	473	592	710	828	946	1065	1183	1301	1537	1774
Q	345	460	575	690	805	920	1035	1150	1266	1496	1726
R	338	451	564	676	789	902	1014	1127	1239	1465	1690
S	329	439	549	659	769	878	988	1098	1207	1427	1646
T	320	426	533	640	746	853	959	1066	1173	1387	1600
U	311	415	519	623	727	830	934	1038	1142	1349	1557
V	304	405	507	608	709	810	912	1013	1114	1317	1520
W	297	396	495	594	693	792	891	989	1088	1286	1484
X	289	385	481	576	672	769	865	962	1058	1251	1443
Y	284	378	473	567	662	756	851	945	1040	1229	1418
Z	277	370	462	555	647	740	832	925	1017	1202	1387

SOURCE: Erik Oberg, Franklin D. Jones, and Holbrook L. Horton (eds.), *Machinery's Handbook*, 20th ed., Industrial Press, New York, 1975.

Drill Grinding or Sharpening

Drill grinding or sharpening is done by hand or by machine. When sharpening drills, it is important that (1) the point have both cutting edges of the same length and be inclined at the same angle with the axis, (2) the angle of clearance back of the cutting edge or the lip clearance is correctly ground, and (3) there is the correct angle between the chisel edge and the lips. The standard drill-point angle is 118°; the lip-clearance angle should be 12° at the periphery of the drill and increase toward the center until the angle made by the straight line across the web and the cutting edges is 135°. For soft materials, the clearance angle can be increased up to 15°; for harder materials, decrease it slightly. However, the angles given are recommended for most work (see Fig. 3.15). These dimensions are the ones ground on the drill by most drill-sharpening machines. One additional point: As a drill is ground back, the web increases in thickness, which decreases chip clearance. Therefore, thin the web whenever the web exceeds approximately one-eighth of the drill diameter (see Fig. 3.16).

3.4 TURNING

Turning is a process in which a piece of metal, wood, or other material is shaped into round form with a cutting tool while being rotated. It is a machining process

TABLE 3.15 Cutting Speeds and Equivalent rpm for Fractional Size Drills

Diam., in	Cutting speed, fpm												
	30	40	50	60	70	80	90	100	110	120	130	140	150
1/16	1833	2445	3056	3667	4278	4889	5500	6111	6722	7334	7945	8556	9167
1/8	917	1222	1528	1833	2139	2445	2750	3056	3361	3667	3973	4278	4584
3/16	611	815	1019	1222	1426	1630	1833	2037	2241	2445	2648	2852	3056
1/4	458	611	764	917	1070	1222	1375	1528	1681	1833	1986	2139	2292
5/16	367	489	611	733	856	978	1100	1222	1345	1467	1589	1711	1833
3/8	306	407	509	611	713	815	917	1019	1120	1222	1324	1426	1528
7/16	262	349	437	524	611	698	786	873	960	1048	1135	1222	1310
1/2	229	306	382	458	535	611	688	764	840	917	993	1070	1146
5/8	183	244	306	367	428	489	550	611	672	733	794	856	917
3/4	153	203	255	306	357	407	458	509	560	611	662	713	764
7/8	131	175	218	262	306	349	393	436	480	524	568	611	655
1	115	153	191	229	267	306	344	382	420	458	497	535	573
1 1/8	102	136	170	204	238	272	306	340	373	407	441	475	509
1 1/4	92	122	153	183	214	244	275	306	336	367	397	428	458
1 3/8	83	111	139	167	194	222	250	278	306	333	361	389	417
1 1/2	76	102	127	153	178	204	229	255	280	306	331	357	382
1 5/8	70	94	117	141	165	188	212	235	259	282	306	329	353
1 3/4	65	87	109	131	153	175	196	218	240	262	284	306	327
1 7/8	61	81	102	122	143	163	183	204	224	244	265	285	306
2	57	76	95	115	134	153	172	191	210	229	248	267	287
2 1/4	51	68	85	102	119	136	153	170	187	204	221	238	255
2 1/2	46	61	76	92	107	122	137	153	168	183	199	214	229
2 3/4	42	56	69	83	97	111	125	139	153	167	181	194	208
3	38	51	64	70	89	102	115	127	140	153	166	178	191

SOURCE: Rupert LeGrand (ed.), *New American Machinists Handbook*, McGraw-Hill, New York, 1955.

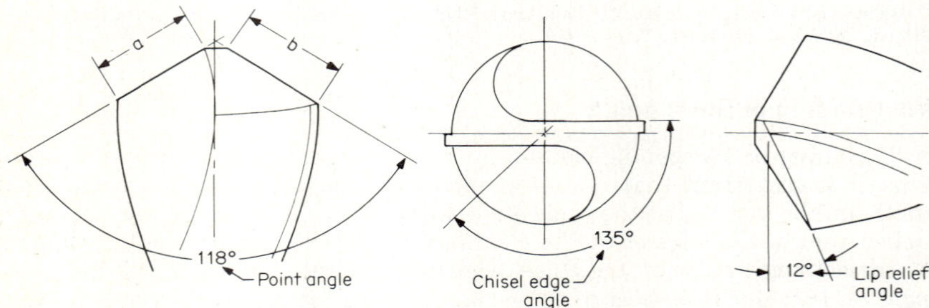

FIG. 3.15 Twist-drill sharpening (I)

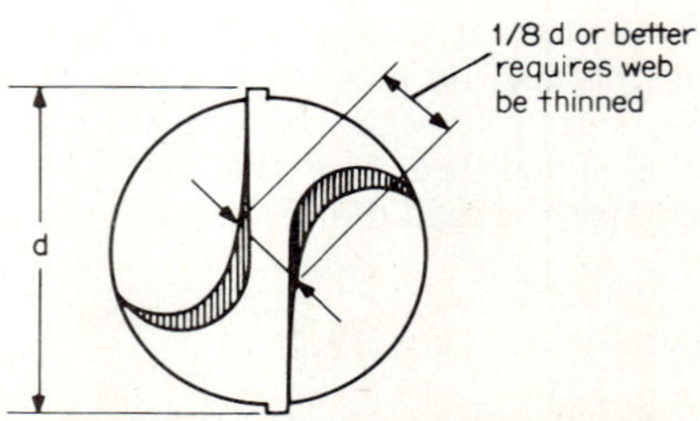

FIG. 3.16 Twist-drill sharpening (II)

FIG. 3.17 Lathe.

for generating external surfaces of revolution. When this process is used to produce internal surfaces of revolution, it is called *boring*.

Turning is most often done on a lathe. All the numerous types of lathes are modifications of the basic machine, which consists of a bed (with ways or guides), headstock, tailstock, carriage, cross slide, toolholder, source of power to rotate the work, and a lead screw arrangement to move the carriage along the ways. The modifications are too numerous to mention but include all sorts of tool room and production machines.

The technician is most likely to use the hollow-spindle lathe, which is constructed with a hollow headstock spindle to accommodate lengths of bar or tubular stock. Collets grip, hold, and center the workpiece for rotation. The machine also takes a jawed chuck for clamping and centering a workpiece. The size of a lathe is specified by giving the swing (diameter) over the bed and the distance between centers when the tailstock is flush with the end of the bed. A third figure often given indicates the overall length of the bed. Lathes are man-

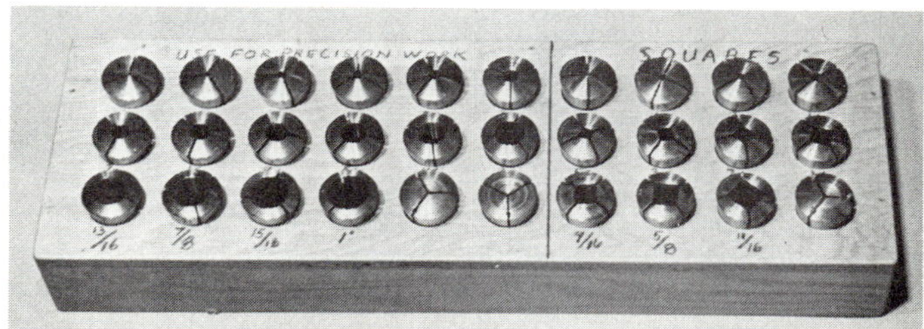

FIG. 3.18 Lathe collets.

FIG. 3.19 Three-jaw chuck.

ufactured in sizes ranging from 6-in swing up to as much as 72 in. The smaller sizes are most common in job and technicians' shops. A typical lathe is 14-in swing, 36 in between centers, with a bed 5 or 6 ft long. Modern machines have multiple-speed controls to drive the spindle at the various speeds required for the different jobs that can be performed. A typical lathe is shown in Fig. 3.17. A typical set of collets is shown in Fig. 3.18, and a three-jaw chuck in Fig. 3.19.

Besides turning, many operations can be performed on a lathe, including boring, facing, tapering (with appropriate attachment), drilling, reaming, tapping and threading, knurling, grinding (with attachment), and chamfering.

Tools

In turning and boring, *the single-point tool* removes metal. The common single-point tools are the *solid tool* and the *tipped tool.* The solid tool is produced by grinding the appropriate cutting edge on a bar of tool steel. Tipped tools are made by inserting a tip of cutting material, such as carbide, into a shank of appropriate material. The insert is brazed in place. Tools ground from square cutting materials are clamped in a holder, as in Fig. 3.20, which shows both a modern toolholder and the type used until recent years.

Figure 3.21 illustrates several of the more common shapes of tools ground from solid bars. Standard nomenclatures and abbreviations for tool angles are in Fig. 3.22.

To maintain an inventory of tools or to store them so they can be readily

FIG. 3.20 Toolholders, old and new styles.

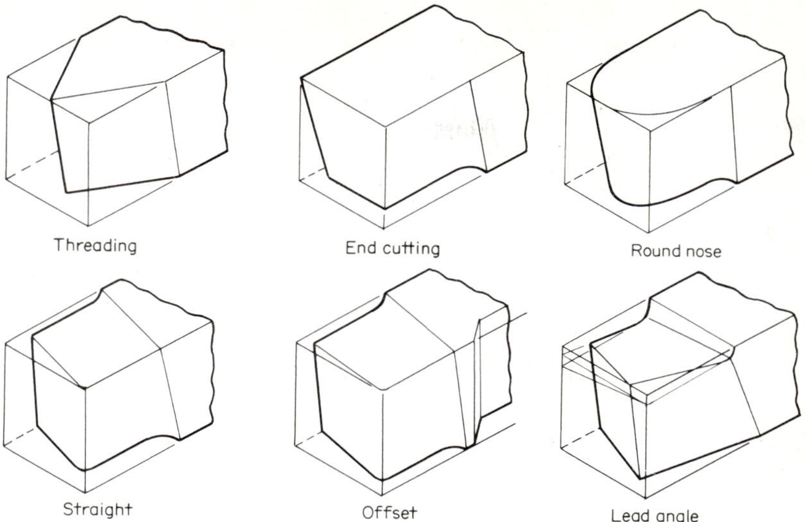

FIG. 3.21 Single-point tool shapes ground from solid bars.

identified, use the standard way of designating tool geometry, in which a series of numbers represents the various tool angles. The conventional order of listing tool angles (in degrees) of single-point tools is:

Tool designation (degrees)—8-14-6-6-20-15-3/16

- Back-rake angle (BR) — 8
- Side-rake angle (SR) — 14
- End-relief angle (ER) — 6
- Side-relief angle (SR) — 6
- End-cutting-edge angle (ECEA) — 20
- Side-cutting-edge angle (SCEA) — 15
- Nose radius (in) (NR) — 3/16

Square and rectangular tools are made in numerous sizes. Rectangular tools are made in various depths, from $1\frac{1}{2}$ to 2 times the widths of $\frac{1}{4}$, $\frac{5}{16}$, $\frac{3}{8}$, $\frac{1}{2}$, $\frac{5}{8}$, $\frac{3}{4}$, $\frac{7}{8}$, 1, and $1\frac{1}{2}$ in. Square tools come in the following sizes: $\frac{1}{4}$, $\frac{5}{16}$, $\frac{3}{8}$, $\frac{1}{2}$, $\frac{5}{8}$, $\frac{3}{4}$, $\frac{7}{8}$, 1, $1\frac{1}{4}$, and $1\frac{1}{2}$ in^2. Solid tool stock is available in carbon steel, high-speed steel, and stellite. Tipped tools are also available in all the shank sizes indicated. Tips are produced of stellite and carbides of tungsten, tantalum, and titanium. Carbide tips are made in the styles shown in Fig. 3.23.

Styles 0000 through 4000 in Fig. 3.23 come in sizes from $\frac{1}{16} \times \frac{1}{8} \times \frac{5}{8}$ in long to $\frac{1}{2} \times \frac{3}{4} \times 1\frac{1}{2}$ in long. Style 5000 is made in sizes $\frac{1}{16} \times \frac{1}{4} \times \frac{5}{16}$ in to $\frac{1}{4} \times \frac{3}{4} \times 1$ in; style 6000 is available in sizes from $\frac{3}{32} \times \frac{5}{16} \times \frac{3}{8}$ in long to $\frac{3}{16} \times \frac{3}{4} \times \frac{3}{4}$ in long. Figure 3.24 shows a variety of carbide-tipped tools commonly used in lathe operations.

There are a variety of cutoff blades for cutoff operations on a lathe. The standard configurations are indicated in Figure 3.25 and the dimensions in Table 3.16.

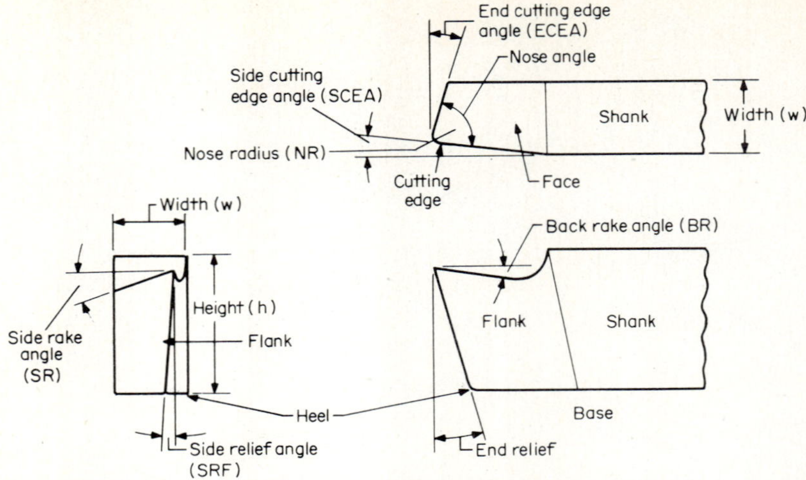

FIG. 3.22 Standard nomenclatures and abbreviations for tool angles.

Tool sharpness is very important for good work because a sharp tool produces better surface finish and accuracy. To produce the proper cutting edge on high-speed-steel tools, first grind with an aluminum oxide wheel, 50 to 60 grit. Finish grind with a 320-grit shellac-bonded wheel. The finish grinding should not remove more than 0.001 in of material, less if possible. Do grinding wet. As a guide to grinding high-speed-steel tools for performing turning operations on a lathe (especially since the technician will most likely use high-speed-steel tools), Table 3-17 lists cutting angles for various materials to be machined. To help select feeds and speeds to be used in turning, Table 3.18 gives recommended

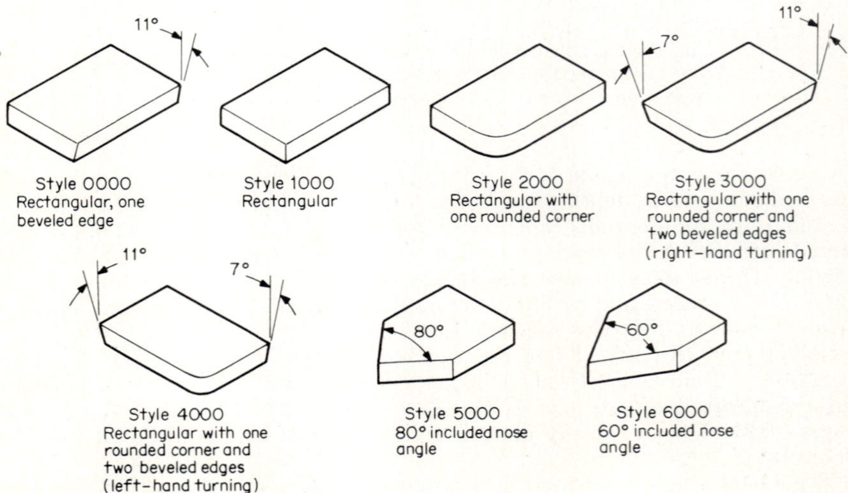

FIG. 3.23 Carbide-tipped shapes.

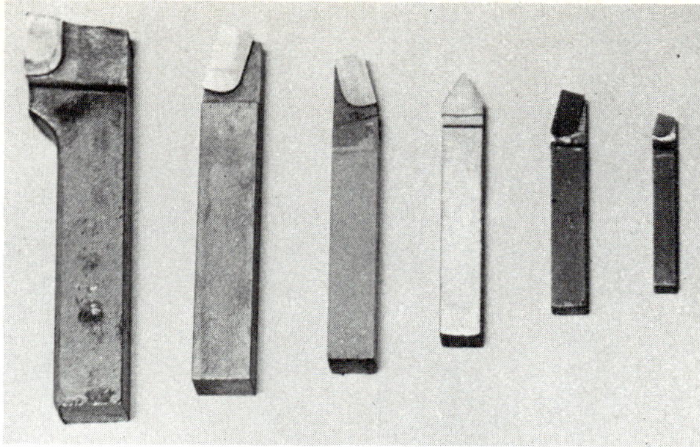

FIG. 3.24 Carbide-tipped tools.

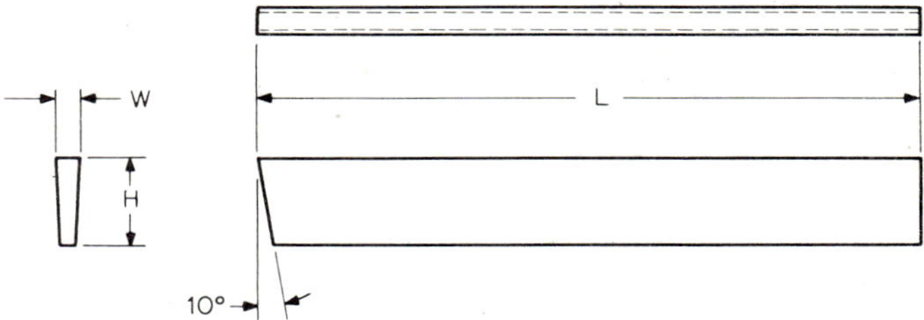

FIG. 3.25 Cutoff blade.

TABLE 3.16 Dimensions of Straight Cutoff Blades

Width (w)	Height (h) × length (l)				
1/16	½ × 4½	11/16 × 5	13/16 × 6		
5/64	½ × 4½				
3/32	½ × 4½	11/16 × 5	13/16 × 6		1 × 6½	
1/8	½ × 4½	11/16 × 5	13/16 × 6		1 × 6½	
5/32		11/16 × 5	13/16 × 6			
3/16		11/16 × 5	13/16 × 6	7/8 × 6	1 × 6½	1⅛ × 6½
1/4			11/16 × 6	7/8 × 6	1 × 6½	1⅛ × 6½

speeds and feeds for various materials when using both high-speed-steel and carbide-tipped tools.

Tables 3.19 and 3.20, useful for converting surface feet per minute (sfpm) of the workpiece into revolutions per minute (rpm), have been developed from the following formula:

$$S = \frac{3.1416 \times d \times \text{rpm}}{12}$$

TABLE 3.17 High-Speed-Steel Tool Cutting Angles

Material	Back-rake angle	Side 1 rake angle	End-relief angle	Side-relief angle
Aluminum	35	15	8	12
Bakelite and molded plastics	0	0	8	12
Brass	0	1–5	8	10
Bronze	0	0–5	8	10
Bronze-phosphor	10	0	12	10
Cast iron	5	12	8	10
Copper	16	20	12	14
Copper alloys	0	0	8	10
Inconel	8	25	8	8
Monel	8	14	13	15
Nickel	8	14	13	15
Steel:				
Alloy	8–16	12–22	8	10–12
Carbon	16	14	8	12
Stainless	10	15–20	8	10
Tool	8	12	8	10

or

$$\text{rpm} = \frac{S \times 12}{3.1416 \times d}$$

where S = cutting speed, sfpm
d = diameter of workpiece, in

The tables give revolutions per minute of workpieces with diameters up to 12 in and speeds up to 800 sfpm. Obviously the revolutions per minute above the spindle speed of the lathe being used should be ignored. Most lathes do not have spindle speeds of more than 4000 rpm. The values in the tables are to the nearest integer, regardless of fractions, but this is sufficient accuracy for use when turning. Set up the lathe to give a spindle speed as near as possible to that in the tables.

Lathe Set-Up

When setting up a lathe for turning operations, securely clamp the workpiece in a collet, a chuck, or between centers or a chuck and one center. The toolbit and holder must be solidly and securely fastened down in the toolpost. During turning operations, the tool is set in the holder or toolpost in a horizontal and level position, and the point is even with the center of the work (see Fig. 3.26). A convenient way to set the point on center is to line it up with the tailstock's center point. The angles in Fig. 3.26 are called working angles and vary for different tools and how they are ground. In most work being done by the technician, the tool or toolholder shank is always parallel to the lathe bed and the carriage's slide movement. In this case the back-rake and true rake angles are the same, as are the end-relief and working-relief angles. If after setting up and taking a cut it is determined, because of chip form, that the rake angle should be increased or decreased, raise or lower the tool in the toolpost so it is positioned slightly above or below the center of the workpiece. This effectively changes the back-

TABLE 3.18 Cutting Speeds and Feeds for Various Materials

Material	High-speed steel		Carbide	
	Speed, fpm	Feed, ipr	Speed, fpm	Feed, ipr
Metals:				
Aluminum alloys	700	0.012	4000	0.015
Cast iron:				
Hard	75	0.010	375	0.015
Soft	100	0.010	325	0.015
Copper alloys:				
Soft	400	0.010	1000	0.010
Medium	275	0.010	750	0.010
Hard	125	0.010	600	0.010
Steel:				
Alloy, AISI 4340	80	0.010	400	0.015
Maraging	50	0.007	225	0.003
Plain carbon	160	0.010	550	0.015
Stainless:				
Type 410	100	0.008	350	0.012
Type 416	110	0.005	350	0.010
Type 420	75	0.008	300	0.010
Type 430	100	0.008	350	0.012
Type 431	75	0.008	300	0.010
Type 440	50	0.008	275	0.010
Type 302	70	0.008	275	0.012
Type 303	100	0.005	350	0.010
Type 316	75	0.008	300	0.012
Type 321	70	0.008	275	0.012
Nickel alloys:				
Soft	65	0.030	*	*
Hard	185	0.008	*	*
Monel alloys	95	0.010	325	0.008
Monel alloy 405	125	0.010	325	0.008
Incoloy alloys	55	0.010	350	0.008
Inconel alloys	20	0.008	45	0.008
Ni-Span-C:				
Aged	20	0.008	45	0.008
Unaged	65	0.010	250	0.008
Nonmetals:				
MGC	30	0.002	50	0.005
Hard rubber	200	0.015	300	0.025
Acrylic*	65	0.010	†	†
Polystyrene*	600	0.005	†	†
Plastics, glass-reinforced	†	†	400	0.010
Phenolic:				
Laminates	750	0.030		
Cast	800	0.010		

*Set tool 60° to work axis.
†Not recommended.

TABLE 3.19 Workpiece rpm and Cutting Speeds

Diam., in	Cutting speed, sfpm												
	20	30	40	50	60	70	80	90	100	110	120	130	140
0.063	1222	1833	2444	3055	3666	4278	4889	5500	6111	6722	7333	7945	8556
0.125	611	916	1222	1527	1833	2139	2444	2750	3055	3361	3666	3972	4278
0.188	407	611	814	1018	1222	1426	1629	1833	2037	2240	2444	2648	2852
0.250	305	458	611	763	916	1069	1222	1375	1527	1680	1833	1986	2139
0.313	244	366	488	611	733	855	977	1100	1222	1344	1466	1589	1711
0.375	203	305	407	509	611	713	814	916	1018	1120	1222	1324	1426
0.438	174	261	349	436	523	611	698	785	873	960	1047	1135	1222
0.500	152	229	305	381	458	534	611	687	763	840	916	993	1069
0.563	135	203	271	339	407	475	543	611	679	746	814	882	950
0.625	122	183	244	305	366	427	488	550	611	672	733	794	855
0.688	111	166	222	277	333	388	444	500	555	611	666	722	777
0.750	101	152	203	254	305	356	407	458	509	560	611	662	713
0.813	94	141	188	235	282	329	376	423	470	517	564	611	658
0.875	87	130	174	218	261	305	349	392	436	480	523	567	611
0.938	81	122	162	203	244	285	325	366	407	448	488	529	570
1.000	76	114	152	190	229	267	305	343	381	420	458	496	534
1.250	61	91	122	152	183	213	244	275	305	336	366	397	427
1.500	50	76	101	127	152	178	203	229	254	280	305	331	356
1.750	43	65	87	109	130	152	174	196	218	240	261	283	305

2.000	38	57	76	95	114	133	152	171	190	210	229	248	267
2.250	33	50	67	84	101	118	135	152	169	186	203	220	237
2.500	30	45	61	76	91	106	122	137	152	168	183	198	213
2.750	27	41	55	69	83	97	111	125	138	152	166	180	194
3.000	25	38	50	63	76	89	101	114	127	140	152	165	178
3.250	23	35	47	58	70	82	94	105	117	129	141	152	164
3.500	21	32	43	54	65	76	87	98	109	120	130	141	152
3.750	20	30	40	50	61	71	81	91	101	112	122	132	142
4.000	19	28	38	47	57	66	76	85	95	105	114	124	133
4.500	16	25	33	42	50	59	67	76	84	93	101	110	118
5.000	15	22	30	38	45	53	61	68	76	84	91	99	106
5.500	13	20	27	34	41	48	55	62	69	76	83	90	97
6.000	12	19	25	31	38	44	50	57	63	70	76	82	89
6.500	11	17	23	29	35	41	47	52	58	64	70	76	82
7.000	10	16	21	27	32	38	43	49	54	60	65	70	76
7.500	10	15	20	25	30	35	40	45	50	56	61	66	71
8.000	9	14	19	23	28	33	38	42	47	52	57	62	66
8.500	8	13	17	22	26	31	35	40	44	49	53	58	62
9.000	8	12	16	21	25	29	33	38	42	46	50	55	59
9.500	8	12	16	20	24	28	32	36	40	44	48	52	56
10.000	7	11	15	19	22	26	30	34	38	42	45	49	53
10.500	7	10	14	18	21	25	29	32	36	40	43	47	50
11.000	6	10	13	17	20	24	27	31	34	38	41	45	48
11.500	6	9	13	16	19	23	26	29	33	36	39	43	46
12.000	6	9	12	15	19	22	25	28	31	35	38	41	44

TABLE 3.20 Workpiece rpm and Cutting Speeds

	Cutting speed, sfpm							
Diam., in	100	200	300	400	500	600	700	800
0.063	6111	12223	18334	24446	30557	36669	42780	48892
0.125	3055	6111	9167	12223	15278	18334	21390	24446
0.188	2037	4074	6111	8148	10185	12223	14260	16297
0.250	1527	3055	4583	6111	7639	9167	10695	12223
0.313	1222	2444	3666	4889	6111	7333	8556	9778
0.375	1018	2037	3055	4074	5092	6111	7130	8148
0.438	873	1746	2619	3492	4365	5238	6111	6984
0.500	763	1527	2291	3055	3819	4583	5347	6111
0.563	679	1358	2037	2716	3395	4074	4753	5432
0.625	611	1222	1833	2444	3055	3666	4278	4889
0.688	555	1111	1666	2222	2777	3333	3889	4444
0.750	509	1018	1527	2037	2546	3055	3565	4074
0.813	470	940	1410	1880	2350	2820	3290	3760
0.875	436	873	1309	1746	2182	2619	3055	3492
0.938	407	814	1222	1629	2037	2444	2852	3259
1.000	381	763	1145	1527	1909	2291	2673	3055
1.250	305	611	916	1222	1527	1833	2139	2444
1.500	254	509	763	1018	1273	1527	1782	2037
1.750	218	436	654	873	1091	1309	1527	1746
2.000	190	381	572	763	954	1145	1336	1527
2.250	169	339	509	679	848	1018	1188	1358
2.500	152	305	458	611	763	916	1069	1222
2.750	138	277	416	555	694	833	972	1111
3.000	127	254	381	509	636	763	891	1018
3.250	117	235	352	470	587	705	822	940
3.500	109	218	327	436	545	654	763	873
3.750	101	203	305	407	509	611	713	814
4.000	95	190	286	381	477	572	668	763
4.500	84	169	254	339	424	509	594	679
5.000	76	152	229	305	381	458	534	611
5.500	69	138	208	277	347	416	486	555
6.000	63	127	190	254	318	381	445	509
6.500	58	117	176	235	293	352	411	470
7.000	54	109	163	218	272	327	381	436
7.500	50	101	152	203	254	305	356	407
8.000	47	95	143	190	238	286	334	381
8.500	44	89	134	179	224	269	314	359
9.000	42	84	127	169	212	254	297	339
9.500	40	80	120	160	201	241	281	321
10.000	38	76	114	152	190	229	267	305
10.500	36	72	109	145	181	218	254	291
11.000	34	69	104	138	173	208	243	277
11.500	33	66	99	132	166	199	232	265
12.000	31	63	95	127	159	190	222	254

and true rake angles. Raising the tool increases the rake angle, and vice versa. Be careful here because the working-relief angle also changes, but in the reverse direction of the rake angle. Raising the tool decreases the relief angle. See Fig. 3.27. This is possible on small-diameter pieces, but on larger diameters it may reduce the clearance sufficiently to cause trouble. Setting the tool off center causes excessive wear on the tool's cutting edge. If the tool is not on the center-

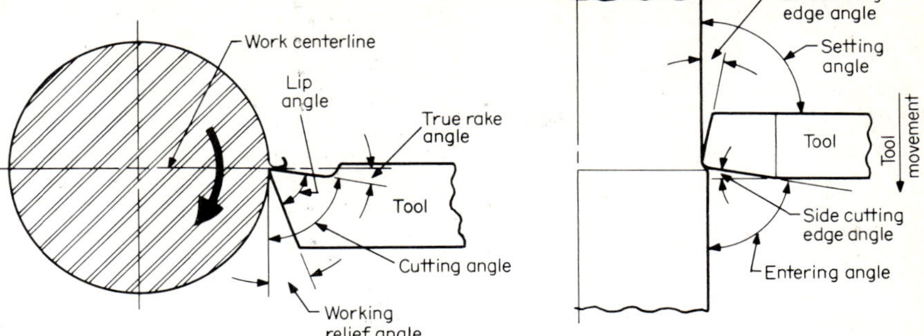

FIG. 3.26 Working-tool angles.

line of the workpiece, feed in will take off more or less material than indicated on the dial of the slide. Never let the toolbit protrude beyond the slide more than needed for all parts to clear the workpiece. Excess tool overhang causes tool chatter, and if it is extreme, it may break the tool. Cutting tool shapes and the types of cuts they are used for are shown in Fig. 3.28.

The lathe must be set up carefully and neatly. If a piece is to be held in a chuck, be sure the chuck is clean and the jaws of the chuck—both three- and four-jaw chucks—have been inserted following the proper procedure. Make sure the scroll is in the correct position when inserting each jaw in numerical order in the corresponding numbered slot. Be sure the chuck is not too small for the work. If the jaws protrude too far beyond the diameter, there may be interference with other parts of the machine or the jaws may not be engaged enough to prevent their flying out when rotated at higher speeds. Always remember to remove the chuck key before operating the machine.

It is always good practice after setting up to turn the workpiece over by hand to be sure it clears all parts of the machine. The workpiece should be balanced, especially with odd-shaped pieces fastened to a faceplate for turning. An unbalanced workpiece is dangerous for the operator and can damage the machine.

Support long pieces with the tailstock center and/or a steady rest. When using centers, be sure to clean the centers as well as the center holes they will be inserted into.

If a finish-machined piece must be chucked, protect it from being marred by the chuck jaws. Use collets only on smooth material because collets are sized to

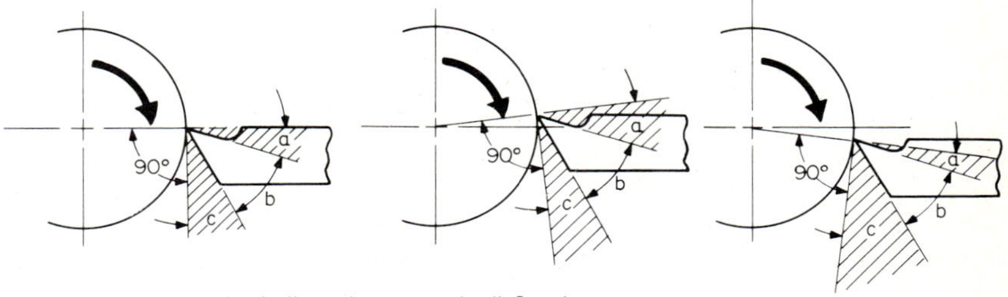

FIG. 3.27 Effect of tool position on tool angles.

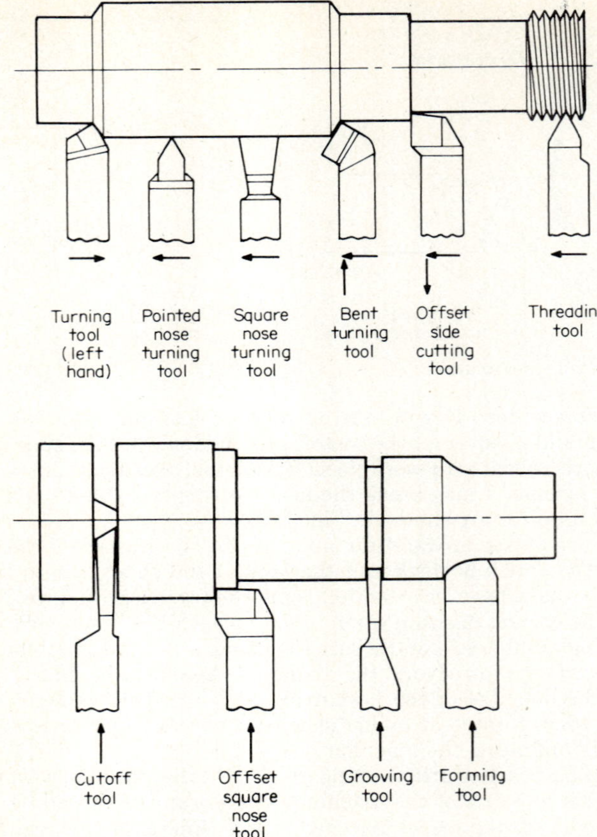

FIG. 3.28 Tool shapes and cuts.

take a specific size of smooth round stock. Scale and the like damage the collets. Do not force material into a collet; use the right-size collet.

Thread Cutting

Thread cutting with a single-point tool is an operation frequently performed on a lathe, particularly when accuracy is required. All types and sizes of screw threads can be turned on a lathe. Do not turn external threads less than ¼ in in diameter and internal threads less than ½ in in diameter; they are made better with threading dies and taps—see Sec. 3.9 for thread types and sizes.

To thread on the lathe, set up the toolpost with the tool set at a 60° angle, as indicated in Fig. 3.29, and then feed in in successive steps until the required thread depth is reached. Alternately, if a 60° point tool is used, set up as indicated in Fig. 3.30. In either case, the fact that the slide is at angle means that when the tool is advanced x inches, the actual feed into the workpiece is x cos 30°. Specifically, for every 0.010-in advance on the dial, the actual depth of cut

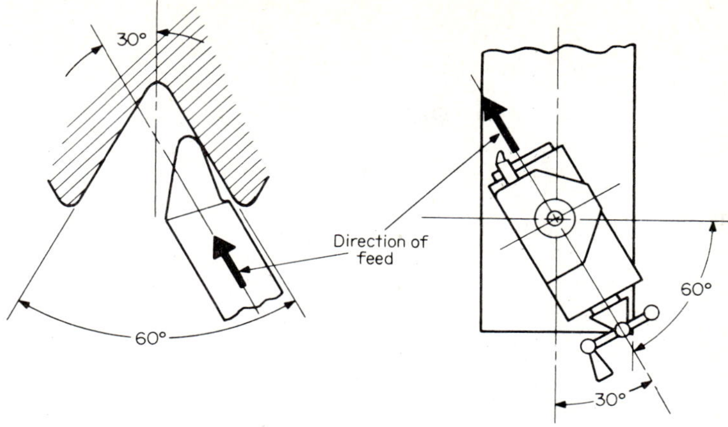

FIG. 3.29 Threading setup on lathe, first method.

is 0.0087 in. A further word of caution: Although the depth of cut is 0.0087 in for every 0.010-in advance of feed dial, after the cut is made the diameter has been reduced twice that depth, or 0.0174 in.

A third method of cutting threads on a lathe is to first set up with a 60° point tool perpendicular to the axis of the workpiece and then proceed as follows. Advance the feed for the first cut. After making the cut, return the carriage and slide to the starting point. Advance in the feed again, and also move the carriage to the left 0.58 times the amount of the feed. Repeat this procedure until the full depth of thread is obtained. See Fig. 3.31.

As a guide to cutting threads, Fig. 3.32 and Tables 3.21 to 3.26 contain the dimensions of the six series of United and American screw thread sizes. Table 3.27 and Fig. 3.33 contain metric standard screw thread proportions. Acme

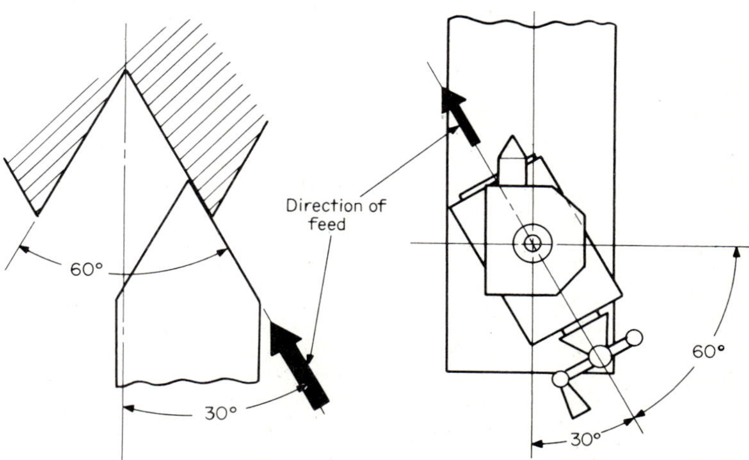

FIG. 3.30 Threading setup on lathe, second method.

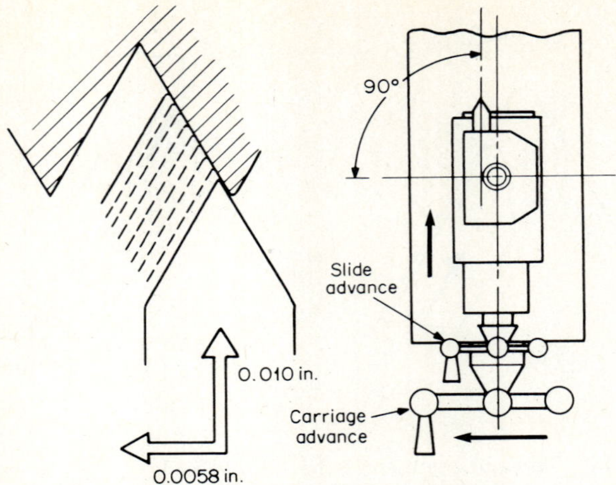

FIG. 3.31 Threading setup on lathe, third method.

Thread Series dimensions and proportions are in Figs. 3.34 to 3.37 and Table 3.28.

3.5 MILLING

In milling, a machining process that produces plain or shaped surfaces, the material is removed by a rotating cutter with multiple teeth. Each tooth removes

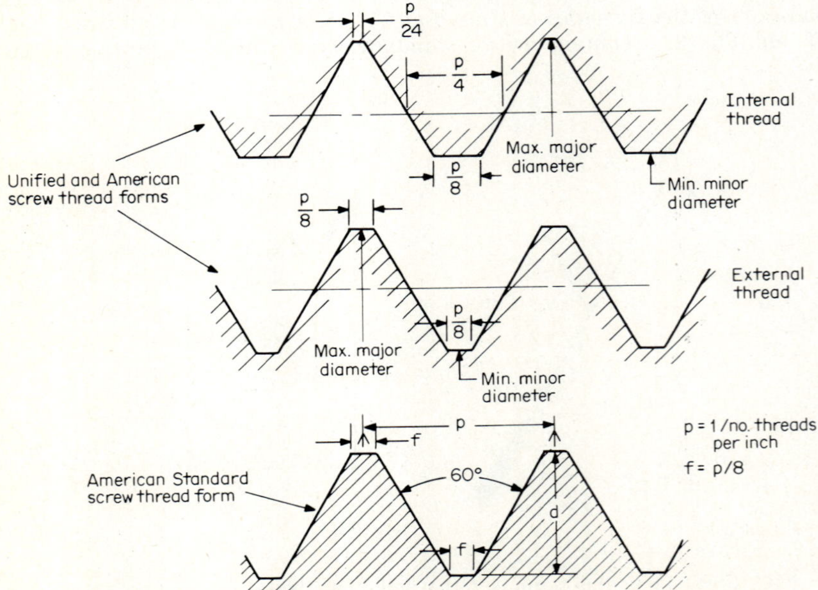

FIG. 3.32 Screw thread forms.

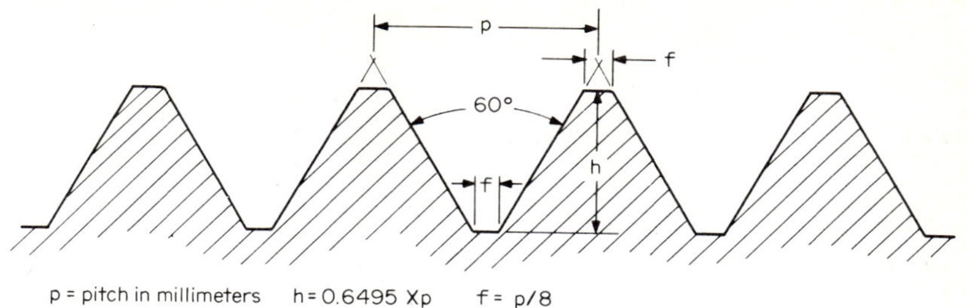

FIG. 3.33 Metric screw thread profile.

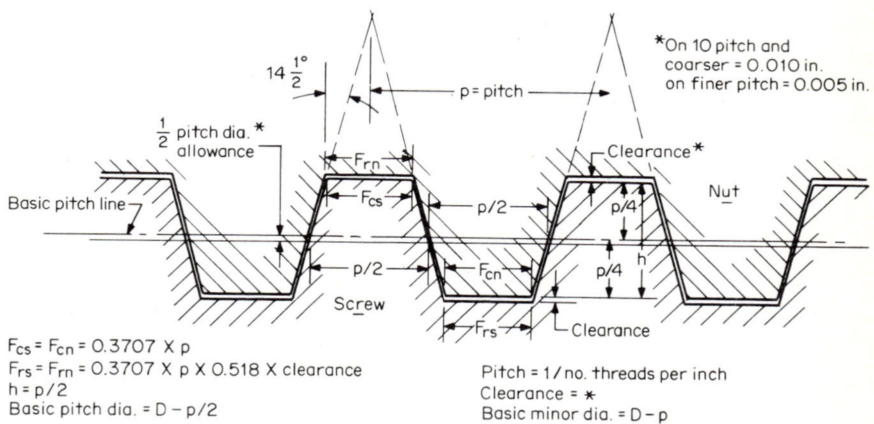

FIG. 3.34 General-purpose Acme Thread.

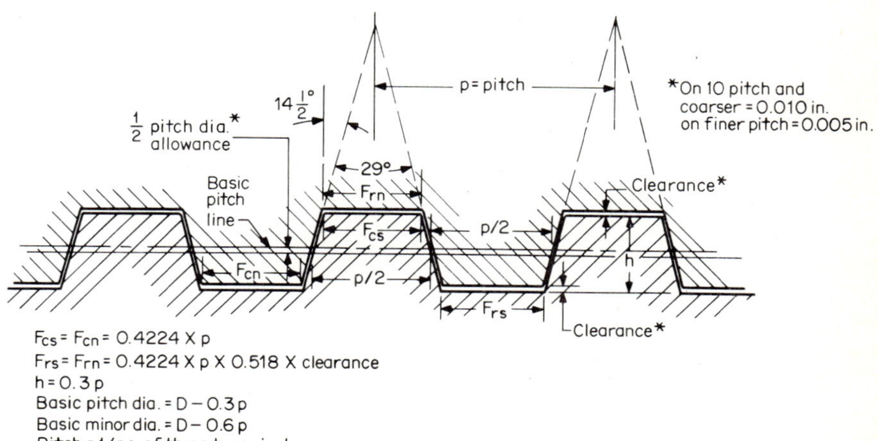

FIG. 3.35 29° Stub-Acme Thread.

TABLE 3.21 Coarse-Thread Series, UNC and NC

All dimensions in inches

Identification		Basic screw diameters			External threads			Internal threads		Areas of sections	
Size	Threads/in, n	Major diam. D^* max.	Pitch diam. E max.	Basic minor diam. K_s	Allowances,† classes 1A and 2A	Class 1A, major diam. tolerances‡	Classes 2A and 3A, major diam. tolerances‡	Minor diam. K_n min.	Minor diam. tolerances, classes 1B, 2B, and 3B§	Basic min. minor diam., in²	Stress area, in²
1(0.073)	64	0.0730	0.0629	0.0538	0.0006	...	0.0038	0.0561	0.0062	0.0022	0.0026
2(0.086)	56	0.0860	0.0744	0.0461	0.0006	...	0.0041	0.0667	0.0070	0.0031	0.0036
3(0.009)	48	0.0090	0.0085	0.0734	0.0007	...	0.0045	0.0764	0.0081	0.0041	0.0048
4(0.112)	40	0.1120	0.0958	0.0813	0.0008	...	0.0051	0.0849	0.0090	0.0050	0.0060
5(0.125)	40	0.1250	0.1088	0.0943	0.0008	...	0.0051	0.0979	0.0083	0.0067	0.0079
6(0.138)	32	0.1380	0.1177	0.0997	0.0008	...	0.0060	0.1042	0.0098	0.0075	0.0090
8(0.164)	32	0.1640	0.1437	0.1257	0.0009	...	0.0060	0.1302	0.0087	0.0120	0.0139
10(0.190)	24	0.1900	0.1629	0.1389	0.0010	...	0.0072	0.1449	0.0106	0.0145	0.0174
12(0.216)	24	0.2160	0.1189	0.1649	0.0010	...	0.0072	0.1709	0.0098	0.0206	0.0240
¼	20	0.2500	0.2175	0.1887	0.0011	0.0122	0.0081	0.1959	0.0108	0.0269	0.0317
5/16	18	0.3125	0.2764	0.2443	0.0012	0.0131	0.0087	0.2524	0.0106	0.0454	0.0522
⅜	16	0.3750	0.3344	0.2983	0.0013	0.0142	0.0094	0.3073	0.0109	0.0678	0.0773

7/16	14	0.4375	0.3911	0.3499	0.0014	0.0155	0.0103	0.3602	0.0115	0.0933	0.1060
1/2	13	0.5000	0.4500	0.4056	0.0015	0.0163	0.0109	0.4167	0.0117	0.1257	0.1416
1/2	12	0.5000	0.4459	0.3978	0.0015	0.0172	0.0114	0.4098	0.0125	0.1205	0.1374
9/16	12	0.5625	0.5084	0.4603	0.0016	0.0172	0.0114	0.4273	0.0120	0.1620	0.1816
5/8	11	0.6250	0.5660	0.5135	0.0016	0.0182	0.0121	0.5266	0.0125	0.2018	0.2256
3/4	10	0.7500	0.6850	0.6273	0.0018	0.0194	0.0129	0.6417	0.0128	0.3020	0.3340
7/8	9	0.8750	0.8028	0.7387	0.0019	0.0208	0.0139	0.7547	0.0134	0.4193	0.4612
1	8	1.0000	0.9188	0.8466	0.0020	0.0225	0.0150	0.8647	0.0150	0.5510	0.6051
1 1/8	7	1.1250	1.0322	0.9497	0.0022	0.0246	0.0164	0.9704	0.0171	0.6931	0.7627
1 1/4	7	1.2500	1.1572	1.0747	0.0022	0.0246	0.0164	1.0954	0.0171	0.8898	0.9684
1 3/8	6	1.3750	1.2667	1.1705	0.0024	0.0273	0.0182	1.1946	0.0200	1.0541	1.1538
1 1/2	6	1.5000	1.3917	1.2955	0.0024	0.0273	0.0182	1.3196	0.0200	1.2938	1.4041
1 3/4	5	1.7500	1.6201	1.5046	0.0027	0.0308	0.0205	1.5335	0.0240	1.7441	1.8983
2	4 1/2	2.0000	1.8557	1.7274	0.0029	0.0330	0.0220	1.7594	0.0267	2.3001	2.4971
2 1/4	4	2.2500	2.1057	1.9774	0.0029	0.0330	0.0220	2.0094	0.0267	3.0212	3.2464
2 1/2	4	2.5000	2.3376	2.1933	0.0031	0.0357	0.0238	2.2294	0.0300	3.7161	3.9976
2 3/4	4	2.7500	2.5876	2.4433	0.0032	0.0357	0.0238	2.4794	0.0300	4.6194	4.9326
3	4	3.0000	2.8376	2.6933	0.0032	0.0357	0.0238	2.7294	0.0300	5.6209	5.9659
3 1/4	4	3.2500	3.0876	2.9433	0.0033	0.0357	0.0238	2.9794	0.0300	6.7205	7.0992
3 1/2	4	3.5000	3.3376	3.1933	0.0033	0.0357	0.0238	3.2294	0.0300	7.9183	8.3268
3 3/4	4	3.7500	3.5876	3.4433	0.0034	0.0357	0.0238	3.4794	0.0300	9.2143	9.6546
4	4	4.0000	3.8376	3.6933	0.0034	0.0357	0.0238	3.7294	0.0300	10.6084	11.0805

Values based upon a length of engagement equal to the nominal diameter.
*Major diameter at intersection of rounded root with flanks of threads.
†Allowances apply to external threads, classes 1A and 2A only. The allowance for class 3A threads is zero.
‡Major diameter of internal threads may extend to a P/24 flat.
§Minor diameter of external threads may extend to a P/8 flat.
SOURCE: Theodore Baumeister (ed.), *Standard Handbook for Mechanical Engineers*, McGraw-Hill, New York, 1967.

TABLE 3.22 Fine-Thread Series, UNF and NF

All dimensions in inches

Identification		Basic screw diameters			External threads			Internal threads		Areas of sections	
Size	Threads/in, n	Major diam. D* max.	Pitch diam. E max.	Basic minor diam. K_s	Allowances,† classes 1A and 2A	Class 1A, major diam. tolerances‡	Classes 2A and 3A, major diam. tolerances‡	Minor diam. K_n min.	Minor diam. tolerances, classes 1B, 2B, and 3B§	Basic min. minor diam., in^2	Stress area, in^2
0(0.060)	80	0.0600	0.0519	0.0447	0.0005	...	0.0032	0.0465	0.0049	0.0015	0.0018
1(0.073)	72	0.0730	0.0640	0.0560	0.0006	...	0.0035	0.0580	0.0055	0.0024	0.0027
2(0.086)	64	0.0860	0.0759	0.0668	0.0006	...	0.0038	0.0691	0.0062	0.0034	0.0039
3(0.099)	56	0.0990	0.0874	0.0771	0.0007	...	0.0041	0.0797	0.0068	0.0045	0.0052
4(0.112)	48	0.1120	0.0985	0.0864	0.0007	...	0.0045	0.0894	0.0074	0.0057	0.0065
5(0.125)	44	0.1250	0.1102	0.0971	0.0007	...	0.0048	0.1004	0.0075	0.0072	0.0082
6(0.138)	40	0.1380	0.1218	0.1073	0.0008	...	0.0051	0.1109	0.0077	0.0087	0.0101

8(0.164)	36	0.1640	0.1460	0.1299	0.0008	. . .	0.0055	0.1339	0.0077	0.0128	0.0146
10(0.190)	32	0.1900	0.1697	0.1517	0.0009	. . .	0.0060	0.1562	0.0079	0.0175	0.0199
12(0.216)	28	0.2160	0.1928	0.1722	0.0010	. . .	0.0065	0.1773	0.0084	0.0026	0.0257
¼	28	0.2500	0.2268	0.2062	0.0010	0.0098	0.0065	0.2113	0.0077	0.0326	0.0362
5⁄16	24	0.3125	0.2854	0.2614	0.0011	0.0108	0.0072	0.2674	0.0080	0.0524	0.0579
⅜	24	0.3750	0.3479	0.3239	0.0011	0.0108	0.0072	0.3299	0.0073	0.0809	0.0876
7⁄16	20	0.4375	0.4050	0.3762	0.0013	0.0122	0.0081	0.3834	0.0082	0.1090	0.1185
½	20	0.5000	0.4675	0.4387	0.0013	0.0122	0.0081	0.4459	0.0078	0.1486	0.1597
9⁄16	18	0.5625	0.5264	0.4943	0.0014	0.0131	0.0087	0.5024	0.0082	0.1888	0.2026
⅝	18	0.6250	0.5889	0.5568	0.0014	0.0131	0.0087	0.5649	0.0081	0.2400	0.2555
¾	16	0.7500	0.7094	0.6733	0.0015	0.0142	0.0094	0.6823	0.0085	0.3513	0.3724
⅞	14	0.8750	0.8286	0.7874	0.0016	0.0155	0.0103	0.7977	0.0091	0.4805	0.5088
1	12	1.0000	0.9459	0.8978	0.0018	0.0172	0.0114	0.9098	0.0100	0.6245	0.6624
1⅛	12	1.1250	1.0709	1.0228	0.0018	0.0172	0.0114	1.0348	0.0100	0.8118	0.8549
1¼	12	1.2500	1.1959	1.1478	0.0018	0.0172	0.0114	1.1598	0.0100	1.0237	1.0721
1⅜	12	1.3750	1.3209	1.2728	0.0019	0.0172	0.0114	1.2848	0.0100	1.2602	1.3137
1½	12	1.5000	1.4459	1.3978	0.0019	0.0172	0.0114	1.4098	0.0100	1.5212	1.5799

Values based upon a length of engagement equal to the nominal diameter.

*Major diameter at intersection of rounded root with flanks of threads.

†Allowances apply to external threads, classes 1A and 2A only. The allowance for class 3A threads is zero.

‡Major diameter of internal threads may extend to a P/24 flat.

§Minor diameter of external threads may extend to a P/8 flat.

SOURCE: Theodore Baumeister (ed.), *Standard Handbook for Mechanical Engineers*, McGraw-Hill, New York, 1967.

TABLE 3.23 Extrafine-Thread Series, UNEF and NEF

All dimensions in inches

Identification		Basic screw diameters			External threads			Internal threads		Areas of sections	
Size	Threads/in, n	Major diam. D* max.	Pitch diam. E. max.	Basic minor diam. K_s	Allowances,† classes 1A and 2A	Class 1A, major diam. tolerances‡	Classes 2A and 3A, major diam. tolerances‡	Minor diam. K_n min.	Minor diam. tolerances, classes 1B, 2B, and 3B§	Basic min. minor diam., in^2	Stress area, in^2
12(0.216)	32	0.2160	0.1957	0.1777	0.0009	...	0.0060	0.1822	0.0073	0.0242	0.2269
¼	32	0.2500	0.2297	0.2117	0.0010	...	0.0060	0.2162	0.0067	0.0344	0.0377
5⁄16	32	0.3125	0.2922	0.2742	0.0110	...	0.0060	0.2787	0.0060	0.0581	0.0622
⅜	32	0.3750	0.3547	0.3367	0.0010	...	0.0060	0.3412	0.0057	0.0878	0.0929
7⁄16	28	0.4375	0.4143	0.3937	0.0011	...	0.0065	0.3988	0.0063	0.1201	0.1270
½	28	0.5000	0.4768	0.4562	0.0011	...	0.0065	0.4613	0.0063	0.1616	0.1695
9⁄16	24	0.5625	0.5354	0.5114	0.0012	...	0.0072	0.5174	0.0070	0.2030	0.2134
⅝	24	0.6250	0.5979	0.5739	0.0012	...	0.0072	0.5799	0.0070	0.2560	0.2676

11/16	24	0.6875	0.6604	0.6364	0.0012	...	0.0012	0.0072	0.6424	0.3151	0.3280
3/4	20	0.7500	0.7175	0.6887	0.0013	...	0.0081	0.6959	0.0078	0.3685	0.3855
13/16	20	0.8125	0.7800	0.7512	0.0013	...	0.0081	0.7584	0.0078	0.4388	0.4573
7/8	20	0.8750	0.8425	0.8137	0.0013	...	0.0081	0.8209	0.0078	0.5153	0.5352
15/16	20	0.9375	0.9050	0.8762	0.0014	...	0.0081	0.8834	0.0078	0.5979	0.6194
1	20	1.0000	0.9675	0.9387	0.0014	...	0.0081	0.9459	0.0078	0.6866	0.7095
1 1/16	18	1.0625	1.0264	0.9943	0.0014	...	0.0087	1.0024	0.0081	0.7702	0.7973
1 1/8	18	1.1250	1.0889	1.0568	0.0014	...	0.0087	1.0649	0.0081	0.8705	0.8993
1 3/16	18	1.1875	1.1514	1.1193	0.0015	...	0.0087	1.1274	0.0081	0.9770	1.0074
1 1/4	18	1.2500	1.2139	1.1818	0.0015	...	0.0087	1.1899	0.0081	1.0895	1.1216
1 5/16	18	1.3125	1.2764	1.2443	0.0015	...	0.0087	1.2524	0.0081	1.2082	1.2420
1 3/8	18	1.3750	1.3389	1.3068	0.0015	...	0.0087	1.3149	0.0081	1.3330	1.3684
1 7/16	18	1.4375	1.4014	1.3693	0.0015	...	0.0087	1.3774	0.0081	1.4640	1.5010
1 1/2	18	1.5000	1.4639	1.4318	0.0015	...	0.0087	1.4399	0.0081	1.6011	1.6397
1 9/16	18	1.5625	1.5264	1.4943	0.0015	...	0.0087	1.5024	0.0081	1.7444	1.7846
1 5/8	18	1.6250	1.5889	1.5568	0.0015	...	0.0087	1.5649	0.0081	1.8937	1.9357
1 11/16	18	1.6875	1.6514	1.6193	0.0015	...	0.0087	1.6274	0.0081	2.0493	2.0929
1 3/4	16	1.7500	1.7094	1.6733	0.0016	...	0.0094	1.6823	0.0085	2.2873	2.2382
2	16	2.0000	1.9594	1.9233	0.0016	...	0.0094	1.9323	0.0085	2.8917	2.9501

Values based upon a length of engagement equal to the nominal diameter.
*Major diameter at intersection of rounded root with flanks of threads.
†Allowances apply to external threads, classes 1A and 2A only. The allowance for class 3A threads is zero.
‡Major diameter of internal threads may extend to a P/24 flat.
§Minor diameter of external threads may extend to a P/8 flat.
SOURCE: Theodore Baumeister (ed.), *Standard Handbook for Mechanical Engineers,* McGraw-Hill, New York, 1967.

TABLE 3.24 8-Thread Series, 8N

All dimensions in inches

Identification		Basic screw diameters		External threads				Internal threads		Areas of sections	
Size	Threads/in, n	Major diam. D* max.	Pitch diam. E max.	Basic minor diam. K_s	Allowances,† classes 1A and 2A	Class 1A, major diam. tolerances‡	Classes 2A and 3A, major diam. tolerances‡	Minor diam. K_n min.	Minor diam. tolerances, classes 1B, 2B, and 3B§	Basic min. minor diam., in²	Stress area, in²
1⅛	8	1.1250	1.0438	0.9716	0.0021	...	0.0150	0.9897	0.0150	0.7277	0.7896
1¼	8	1.2500	1.1688	1.0966	0.0021	...	0.0150	1.1147	0.0150	0.9290	0.9985
1⅜	8	1.3750	1.2938	1.2216	0.0022	...	0.0150	1.2397	0.0150	1.1548	1.2319
1½	8	1.5000	1.4188	1.3466	0.0022	...	0.0150	1.3647	0.0150	1.4052	1.4899
1⅝	8	1.6250	1.5438	1.4716	0.0022	...	0.0150	1.4897	0.0150	1.6801	1.7723
1¾	8	1.7500	1.6688	1.5966	0.0023	...	0.0150	1.6147	0.0150	1.9796	2.0792

1⅞	8	1.8750	1.7938	1.7216	0.0023	...	0.0150	1.7397	0.0150	2.3036	2.4107
2	8	2.0000	1.9188	1.8466	0.0023	...	0.0150	1.8647	0.0150	2.6521	2.7665
2⅛	8	2.1250	2.0438	1.9716	0.0024	...	0.0150	1.9897	0.0150	3.0252	3.1469
2¼	8	2.2500	2.1688	2.0966	0.0024	...	0.0150	2.1147	0.0150	3.4228	3.5519
2½	8	2.5000	2.4188	2.3466	0.0024	...	0.0150	2.3647	0.0150	4.2917	4.4352
2¾	8	2.7500	2.6688	2.5966	0.0025	...	0.0150	2.6147	0.0150	5.2588	5.4164
3	8	3.0000	2.9188	2.8466	0.0026	...	0.0150	2.8647	0.0150	6.3240	6.4957
3¼	8	3.2500	3.1688	3.0966	0.0026	...	0.0150	3.1147	0.0150	7.4874	7.6738
3½	8	3.5000	3.4188	3.3466	0.0026	...	0.0150	3.3647	0.0150	8.7490	8.9504
3¾	8	3.7500	3.6688	3.5966	0.0027	...	0.0150	3.6147	0.0150	10.1088	10.3249
4	8	4.0000	3.9188	3.8466	0.0027	...	0.0150	3.8647	0.0150	11.5667	11.7995
4¼	8	4.2500	4.1688	4.0966	0.0028	...	0.0150	4.1147	0.0150	13.1228	13.3683
4½	8	4.5000	4.4188	3.3466	0.0028	...	0.0150	4.3647	0.0150	14.7771	15.0372
4¾	8	4.7500	4.6688	4.5966	0.0029	...	0.0150	4.6147	0.0150	16.5295	16.8042
5	8	5.0000	4.9188	4.8466	0.0029	...	0.0150	4.8647	0.0150	18.3802	18.6694
5¼	8	5.2500	5.1688	5.0966	0.0029	...	0.0150	5.1147	0.0150	20.3290	20.6330
5½	8	5.5000	5.4188	5.3466	0.0030	...	0.0150	5.3647	0.0150	22.3760	22.6945
5¾	8	5.7500	5.6688	5.5966	0.0030	...	0.0150	5.6147	0.0150	24.5211	24.8541
6	8	6.0000	5.9188	5.8466	0.0030	...	0.0150	5.8647	0.0150	26.7645	27.1118

Values based upon a length of engagement equal to the nominal diameter.

*Major diameter at intersection of rounded root with flanks of threads.

†Allowances apply to external threads, classes 1A and 2A only. The allowance for class 3A is zero.

‡Major diameter of internal threads may extend to a P/24 flat.

§Minor diameter of external threads may extend to a P/8 flat.

SOURCE: Theodore Baumeister (ed.), *Standard Handbook for Mechanical Engineers*, McGraw-Hill, New York, 1967.

TABLE 3.25 12-Thread Series, 12UN and 12N

All dimensions in inches

Identification		Basic screw diameters			External threads			Internal threads		Areas of sections	
Size	Threads/in, n	Major diam. D* max.	Pitch diam. E max.	Basic minor diam. K_s	Allowances,† classes 1A and 2A	Class 1A, major diam. tolerances‡	Classes 2A and 3A, major diam. tolerances‡	Minor diam. K_n min.	Minor diam. tolerances, classes 1B, 2B, and 3B§	Basic min. minor diam., in²	Stress area, in²
½	12	0.5000	0.4459	0.3978	0.0016	...	0.0114	0.4098	0.0127	0.1205	0.1374
⅝	12	0.6250	0.5709	0.5228	0.0016	...	0.0114	0.5438	0.0115	0.2097	0.2319
11/16	12	0.6875	0.6334	0.5853	0.0016	...	0.0114	0.5973	0.0112	0.2635	0.2883
¾	12	0.7500	0.6959	0.6478	0.0017	...	0.0114	0.6598	0.0109	0.3234	0.3508
13/16	12	0.8125	0.7584	0.7103	0.0017	...	0.0114	0.7223	0.0106	0.3895	0.4195
⅞	12	0.8750	0.8209	0.7728	0.0017	...	0.0114	0.7848	0.0108	0.4617	0.4943
15/16	12	0.9375	0.8834	0.8353	0.0017	...	0.0114	0.8743	0.0090	0.5000	0.5753
1 1/16	12	0.0625	1.0084	0.9603	0.0017	...	0.0114	0.9723	0.0110	0.7151	0.7556
1 3/16	12	1.1875	1.1334	1.0853	0.0017	...	0.0114	1.0973	0.0100	0.9147	0.9604
1 5/16	12	1.3125	1.2584	1.2103	0.0017	...	0.0114	1.2223	0.0100	1.1389	1.1898
1 7/16	12	1.4375	1.3834	1.3353	0.0018	...	0.0114	1.3473	0.0100	1.3876	1.4438
1⅝	12	1.6250	1.5709	1.5228	0.0018	...	0.0114	1.5348	0.0100	1.8067	1.8701
1¾	12	1.7500	1.6959	1.6418	0.0018	...	0.0114	1.6598	0.0100	2.1168	2.1853

1⅞	12	1.8750	1.8209	1.7728	0.0018	. . .	0.0114	1.7848	0.0100	2.4514	2.5250
2	12	2.0000	1.9549	1.8978	0.0018	. . .	0.0114	1.9098	0.0100	2.8106	2.8892
2⅛	12	2.1250	2.0709	2.0228	0.0018	. . .	0.0114	2.0348	0.0100	3.1943	3.2779
2¼	12	2.2500	2.1959	2.1478	0.0018	. . .	0.0114	2.1598	0.0100	3.6025	3.6914
2⅜	12	2.3750	2.3209	2.2728	0.0019	. . .	0.0114	2.2848	0.0100	4.0353	4.1291
2½	12	2.5000	2.4459	2.3978	0.0019	. . .	0.0114	2.4098	0.0100	4.4927	4.5916
2⅝	12	2.6250	2.5709	2.5228	0.0019	. . .	0.0114	2.5348	0.0100	4.9745	5.0784
2¾	12	2.7500	2.6959	2.6478	0.0019	. . .	0.0114	2.6598	0.0100	5.4810	5.5900
2⅞	12	2.8750	2.8209	2.7728	0.0019	. . .	0.0114	2.7848	0.0100	6.0019	6.1259
3	12	3.0000	2.9459	2.8978	0.0019	. . .	0.0114	2.9098	0.0100	6.5674	6.6865
3⅛	12	3.1250	3.0709	3.0228	0.0019	. . .	0.0114	3.0348	0.0100	7.1475	7.2714
3¼	12	3.2500	3.1959	3.1478	0.0019	. . .	0.0114	3.1598	0.0100	7.7521	7.8812
3⅜	12	3.3750	3.3209	3.2728	0.0019	. . .	0.0114	3.2848	0.0100	8.3812	8.5152
3½	12	3.5000	3.4459	3.3978	0.0019	. . .	0.0114	3.4098	0.0100	9.0349	9.1740
3⅝	12	3.6250	3.5709	5.5228	0.0019	. . .	0.0114	3.5348	0.0100	9.7132	9.8570
3¾	12	3.7500	3.6959	3.6478	0.0019	. . .	0.0114	3.6598	0.0100	10.4159	10.4649
3⅞	12	3.8750	3.8209	3.7728	0.0020	. . .	0.0114	3.7848	0.0100	11.1433	11.2970
4	12	4.0000	3.9459	3.8978	0.0020	. . .	0.0114	3.9098	0.0100	11.8951	12.0540
4¼	12	4.2500	4.1959	4.1478	0.0020	. . .	0.0114	4.1598	0.0100	13.4725	13.6411
4½	12	4.5000	4.4459	4.3918	0.0020	. . .	0.0114	4.4098	0.0100	15.1480	15.3265
4¾	12	4.7500	4.6959	4.6478	0.0020	. . .	0.0114	4.6598	0.0100	16.9217	17.1099
5	12	5.0000	4.9459	4.8978	0.0020	. . .	0.0114	4.9098	0.0100	18.7936	18.9916
5¼	12	5.2500	5.1959	5.1478	0.0020	. . .	0.0114	5.1598	0.0100	20.7636	20.9717
5½	12	5.5000	5.4459	5.3978	0.0020	. . .	0.0114	5.4098	0.0100	22.8319	23.0496
5¾	12	5.7500	5.6959	5.6478	0.0021	. . .	0.0114	5.6598	0.0100	24.9983	25.2257
6	12	6.0000	5.9459	5.8978	0.0021	. . .	0.0114	5.9098	0.0100	27.2628	27.4988

Values based upon a length of engagement equal to the nominal diameter.

*Major diameter at intersection of rounded root with flanks of threads.

†Allowances apply to external threads, classes 1A and 2A only. The allowance for class 3A is zero.

‡Major diameter of internal threads may extend to a P/24 flat.

§Minor diameter of external threads may extend to a P/8 flat.

SOURCE: Theodore Baumeister (ed.), *Standard Handbook for Mechanical Engineers,* McGraw-Hill, New York, 1967.

TABLE 3.26 16-Thread Series, 16UN and 16N

All dimensions in inches

Identification		Basic screw diameters			External threads			Internal threads		Areas of sections	
Size	Threads/in, n	Major diam. D* max.	Pitch diam. E max.	Basic minor diam. K_s	Allowances,† classes 1A and 2A	Class 1A, major diam. tolerances‡	Classes 2A and 3A, major diam. tolerances‡	Minor diam. K_n min.	Minor diam. tolerances, classes 1B, 2B, and 3B§	Basic min. minor diam., in^2	Stress area, in^2
13/16	16	0.8125	0.7719	0.7358	0.0015	. . .	0.0094	0.7448	0.0085	0.4200	0.4429
7/8	16	0.8750	0.8344	0.7983	0.0015	. . .	0.0094	0.8073	0.0085	0.4949	0.5197
15/16	16	0.9375	0.8969	0.8608	0.0015	. . .	0.0094	0.8698	0.0085	0.5759	0.6025
1	16	1.0000	0.9594	0.9233	0.0015	. . .	0.0094	0.9323	0.0085	0.6630	0.6916
1 1/16	16	1.0625	1.0219	0.0958	0.0015	. . .	0.0094	0.9948	0.0085	0.7563	0.7867
1 1/8	16	1.1250	1.0844	1.0483	0.0015	. . .	0.0094	1.0573	0.0085	0.8557	0.8880
1 3/16	16	1.1875	1.1469	1.1108	0.0015	. . .	0.0094	1.1198	0.0085	0.9612	0.9955
1 1/4	16	1.2500	1.2094	1.1733	0.0015	. . .	0.0094	1.1823	0.0085	1.0729	1.1090
1 5/16	16	1.3125	1.2719	1.2358	0.0015	. . .	0.0094	1.2448	0.0085	1.1907	1.2287
1 3/8	16	1.3750	1.3344	1.2983	0.0015	. . .	0.0094	1.3073	0.0085	1.3147	1.3545
1 7/16	16	1.4375	1.3969	1.3608	0.0016	. . .	0.0094	1.3698	0.0085	1.4448	1.4865
1 1/2	16	1.5000	1.4594	1.4233	0.0016	. . .	0.0094	1.4323	0.0085	1.5810	1.6246
1 9/16	16	1.5625	1.5219	1.4858	0.0016	. . .	0.0094	1.4948	0.0085	1.7234	1.7687
1 5/8	16	1.6250	1.5844	1.5483	0.0016	. . .	0.0094	1.5573	0.0085	1.8719	1.9191
1 11/16	16	1.6875	1.6469	1.6108	0.0016	. . .	0.0094	1.6198	0.0085	2.0265	2.0757
1 13/16	16	1.8125	1.7719	1.7358	0.0016	. . .	0.0094	1.7448	0.0085	2.3542	2.4070
1 7/8	16	1.8750	1.8344	1.7983	0.0016	. . .	0.0094	1.8073	0.0085	2.5272	2.5819

1 15/16	16	1.9375	1.8969	1.8608	0.0016	. . .	0.0094	1.8698	0.0085	2.7062	2.7269
2 1/16	16	2.0625	2.0219	1.9858	0.0016	. . .	0.0094	1.9948	0.0085	3.0831	3.1434
2 1/8	16	2.1250	2.0844	2.0483	0.0016	. . .	0.0094	2.0573	0.0085	3.2807	3.3427
2 3/16	16	2.1875	2.1469	2.1108	0.0016	. . .	0.0094	2.1198	0.0085	3.4844	3.5483
2 1/4	16	2.2500	2.2094	2.1733	0.0016	. . .	0.0094	2.8123	0.0085	3.6943	3.7601
2 5/16	16	2.3125	2.2719	2.2358	0.0017	. . .	0.0094	2.2448	0.0085	3.9103	3.9708
2 3/8	16	2.3750	2.3344	2.2983	0.0017	. . .	0.0094	2.3073	0.0085	4.1324	4.2018
2 7/16	16	2.4375	2.3969	2.3608	0.0017	. . .	0.0094	2.3696	0.0085	4.3606	4.4319
2 1/2	16	2.5000	2.4594	2.4233	0.0017	. . .	0.0094	2.4323	0.0085	4.4950	4.6682
2 5/8	16	2.6250	2.5844	2.5483	0.0017	. . .	0.0094	2.5573	0.0085	5.0082	5.1790
2 3/4	16	2.7500	2.7094	2.6733	0.0017	. . .	0.0094	2.6823	0.0085	5.5940	5.6745
2 7/8	16	2.8750	2.8344	2.7983	0.0017	. . .	0.0094	2.8073	0.0085	6.1303	6.2143
3	16	3.0000	2.9594	2.9233	0.0017	. . .	0.0094	2.9323	0.0085	6.6911	6.7789
3 1/8	16	3.1250	3.0844	3.0483	0.0017	. . .	0.0094	3.0573	0.0085	7.2765	7.3678
3 1/4	16	3.2500	3.2094	3.1733	0.0017	. . .	0.0094	3.1823	0.0085	7.8864	7.9814
3 3/8	16	3.3750	3.3344	3.2983	0.0017	. . .	0.0094	3.3073	0.0085	8.5209	8.6194
3 1/2	16	3.5000	3.4594	3.4233	0.0017	. . .	0.0094	3.4323	0.0085	9.1799	9.2821
3 5/8	16	3.6250	3.5844	3.5483	0.0017	. . .	0.0094	3.5573	0.0085	9.8634	9.9691
3 3/4	16	3.7500	3.7094	3.6733	0.0017	. . .	0.0094	3.6823	0.0085	10.5715	10.6809
3 7/8	16	3.8750	3.8344	3.7983	0.0018	. . .	0.0094	3.8073	0.0085	11.3042	11.4170
4	16	4.0000	3.9594	3.9233	0.0018	. . .	0.0094	3.9323	0.0085	12.0614	12.1779
4 1/4	16	4.2500	4.2094	4.1733	0.0018	. . .	0.0094	4.1823	0.0085	13.6494	13.7730
4 1/2	16	4.5000	4.4594	4.4233	0.0018	. . .	0.0094	2.4323	0.0085	15.3355	15.4662
4 3/4	16	4.7500	4.7094	4.6733	0.0018	. . .	0.0094	4.6823	0.0085	17.1199	12.2575
5	16	5.0000	4.9594	4.9233	0.0018	. . .	0.0094	4.9323	0.0085	19.0024	19.1470
5 1/4	16	5.2500	5.2094	5.1733	0.0018	. . .	0.0094	5.1823	0.0085	20.9831	31.1350
5 1/2	16	5.5000	5.4538	5.4233	0.0018	. . .	0.0094	5.4323	0.0085	23.0620	23.2208
5 3/4	16	5.7500	5.7094	5.6733	0.0019	. . .	0.0094	5.6823	0.0085	25.2390	25.4047
6	16	6.0000	5.9594	5.9233	0.0019	. . .	0.0094	5.9323	0.0085	27.5142	27.6868

Values based upon a length of engagement equal to the nominal diameter.
*Major diameter at intersection of rounded roots with flanks of threads.
†Allowances apply to external threads, classes 1A and 2A only. The allowance for class 3A is zero.
‡Major diameter of internal threads may extend to a P/24 flat.
§Minor diameter of external threads may extend to a P/8 flat.
SOURCE: Theodore Baumeister (ed.), *Standard Handbook for Mechanical Engineers*, McGraw-Hill, New York, 1967.

TABLE 3.27 Metric Standard Screw Threads

	Pitch, mm					Pitch, mm			
Nominal diam., mm	French std.	German and Swiss std.	Pitch diam., mm	Root diam., mm	Nominal diam., mm	French std.	German and Swiss std.	Pitch diam., mm	Root diam., mm
2	0.40	0.40	1.740	1.48	24	3.00	3.00	22.051	20.10
3	...	0.50	2.675	2.35	26	3.00	...	24.051	22.10
3	0.60	...	2.610	2.22	28	3.00	...	26.051	24.10
4	...	0.70	3.545	3.09	30	3.50	3.50	27.727	25.45
4	0.75	...	3.513	3.03	32	3.50	...	29.727	27.45
5	...	0.80	4.480	3.96	34	3.50	...	31.727	29.45
5	0.90	...	4.415	3.83	36	4.00	4.00	33.402	30.80
10	1.50	1.50	9.026	8.05	38	4.00	...	35.402	32.80
12	1.50	...	11.026	10.05	40	4.00	...	37.402	34.80
12	...	1.75	10.863	9.73	42	4.50	4.50	39.077	36.15
14	2.00	2.00	12.701	11.40	44	4.50	...	41.077	38.15
16	2.00	2.00	14.701	13.40	46	4.50	...	43.077	40.15
18	2.50	2.50	16.376	14.75	48	5.00	...	44.752	41.50
20	2.50	2.50	18.376	16.75	50	5.00	...	46.752	43.50
22	2.50	2.50	20.376	18.75					

SOURCE: Theodore Baumeister (ed.), *Standard Handbook for Mechanical Engineers,* McGraw-Hill, New York, 1967.

TABLE 3.28 Acme Thread Series

All dimensions in inches

Dimensions	Acme-G	29° Stub-Acme	60° Stub-Acme	10° Modified Square-Acme
Thickness of thread	$0.5p$*	$0.5p$	$0.5p$	$0.5p$
Basic depth of thread, h	$0.5p$	$0.3p$	$0.433p$	$0.5p$¶
Basic width of flat, F_1†	$0.3707p$	$0.4224p$	$0.250p$	$0.4563p$**
Width of flat, F_2‡	$F_1 - (0.52 \times \text{clearance})$	$F_1 - (0.52 \times \text{clearance})$	$0.227p$	$F_1 - (0.17 \times \text{clearance})$
Basic pitch diameter	D§ $- 0.5p$	$D - 0.3p$	$D - 0.433p$	$D - 0.5p$
Basic minor diameter	$D - p$	$D - 0.6p$	$D - 0.866p$	$D - p$

*p = pitch.

†$F_1 = F_{cs} = F_{cn}$

‡$F_2 = F_{rs} = F_{rn}$

§D = outside diameter.

¶A clearance of at least 0.010 in is added to h on threads of 10-pitch and coarser, and 0.005 in on finer pitches to produce extra depth, thus avoiding interference with threads of mating parts at minor or major diameters.

**Measured at crest of screw thread.

SOURCE: Theodore Baumeister (ed.), *Standard Handbook for Mechanical Engineers*, McGraw-Hill, New York, 1967.

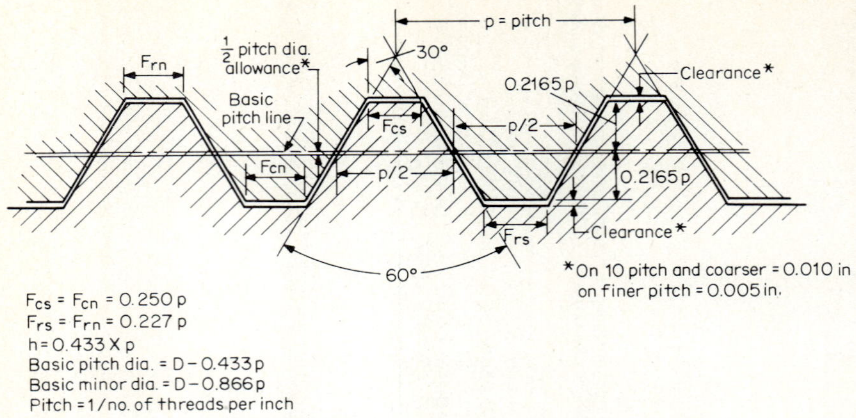

FIG. 3.36 60° Stub-Acme Thread.

an individual, small amount of material. Generally, the cutter is revolved about a vertical or horizontal axis as the workpiece is moved past it perpendicular to the cutter axis.

Milling machines are classfied by either type of construction or spindle orientation. A spindle-oriented milling machine is either horizontal or vertical. Type-of-construction classifications are (1) knee-and-column, (2) bed, and (3) special. Machines range in size from bench millers of less than 1 hp and 6-in travel to floor models with 50-hp motors and 60-in table travel. The type most likely found in the technician's area or shop is the *vertical spindle* knee-and-column machine, similar to the one in Fig. 3.38. A workpiece on the table of this very flexible and versatile machine can be moved up and down, in and out, left and right through movement of the knee, saddle, and table, respectively. The

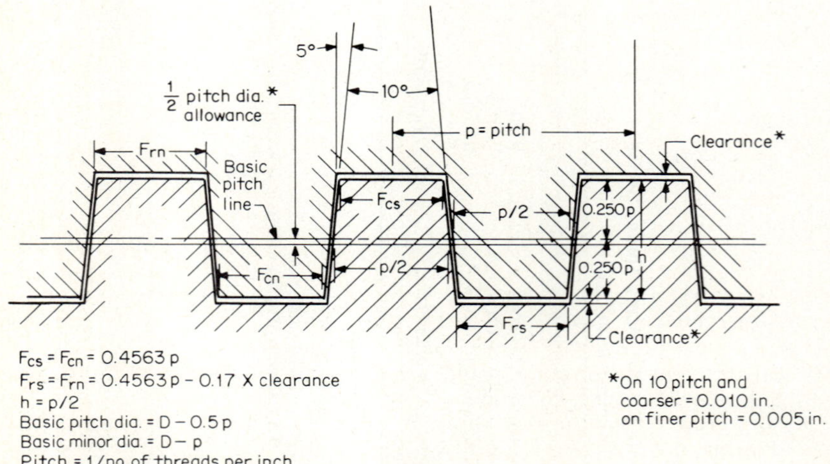

FIG. 3.37 10° Modified Square Acme Thread.

FIG. 3.38 Milling machine, floor model.

drive train, spindle, and motor are mounted on a horizontal slide at the top of the column so that the spindle can be moved in and out. The slide can be swiveled in a horizontal plane about a vertical axis through the column of the machine.

Figure 3.39 shows a typical small bench-type milling machine. Although this machine is not as flexible and powerful as the other kind and far from as large, it is very useful for small-parts work because it can be set up quickly and is easy to operate. On this machine the motor and spindle move vertically on a slide mounted vertically on the column. The table moves left and right, and the saddle moves in and out.

Other types of common milling machines, more likely found in volume machine shops and production shops, are the horizontal spindle knee-and-column, horizontal bed, vertical spindle bed, and manufacturing. The bed machines have a heavy bed, saddle, and table; the table is at a fixed vertical location. The spindle carrying the cutter can be moved vertically. Manufacturing-type machines are used for production work.

Useful attachments for milling machines are dividing heads and rotary tables. Dividing heads are usually secured on the machine table and used to accurately index the workpiece to a desired angle or angles. Rotary tables rotate the work for graduating, milling helixes, and indexing.

Two types of milling are usually employed: (1) *peripheral* and (2) *face*. In peripheral milling, the shaped surface is generated by teeth on the periphery of

FIG. 3.39 Milling machine, bench model.

the cutter. In face milling, the surface is produced by the action of cutting edges on both the periphery and the face of the cutter. Peripheral milling can be done by two methods: (1) *conventional* or (2) *climb*. In conventional milling, recommended for most work the technician will be performing, the cutter rotates *against* the direction of feed. In climb milling, the cutter rotates in the direction of feed (see Fig. 3.40).

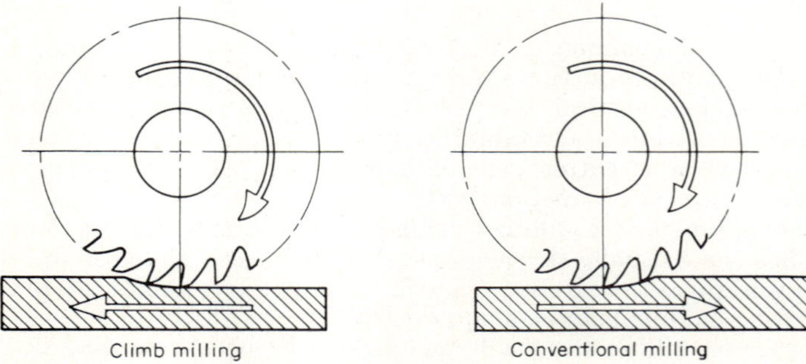

FIG. 3.40 Conventional and climb milling.

Milling Cutters

Milling cutters are available in countless sizes and shapes and can be classified rather broadly as (1) peripheral, (2) face, (3) end, and (4) special. *Peripheral mills* (see Fig. 3.41) cut with teeth that are on the periphery of the tool. They are generally used with their axis of rotation parallel to the surface being milled. *Face mills* (see Fig. 3.41) produce a milled surface on a workpiece by both the action of cutters on the face of the cutter and cutting edges on the outside diameter. This type cutter is usually driven by a spindle on an axis perpendicular to the work surface. *End mills* have cutting edges on the face and periphery of the tool and thus may be used as either of the types mentioned or in combination (Fig. 3.41). They are available with a straight and tapered shank. *Special mills* are special in design for a particular operation or cut.

Most useful for the technician's work are the peripheral mill (in the form of plain milling cutters), side milling cutters and slitting saws, and end mills. Plain milling cutters (teeth on periphery only), where the face is less than ¾ in wide, usually have straight teeth; wider ones have helical teeth. Side milling cutters have straight or staggered teeth on both the periphery and sides or the periphery and one side. End mills have helical or straight teeth on both the cylindrical surface and the end. Frequently both ends have cutting teeth with a straight shank between.

Materials used in milling cutters are the same as those used in single-point turning tools, namely, carbon steel, high-speed steels, and carbides. Carbides are inserts fastened into a body material; in some large carbide-tipped milling cutters the tips are removable for sharpening. The parts and tooth angles of a plain and side milling cutter are shown in Fig. 3.42.

There are too many types and sizes to list here specific milling cutters to be used on all the possible materials that may require milling. As a general guide to selecting cutters, keep in mind a few things. The cutter chip space should be sufficiently ample so that the chips will not cause binding and overheating. Use

FIG. 3.41 Peripheral, face, and end mills.

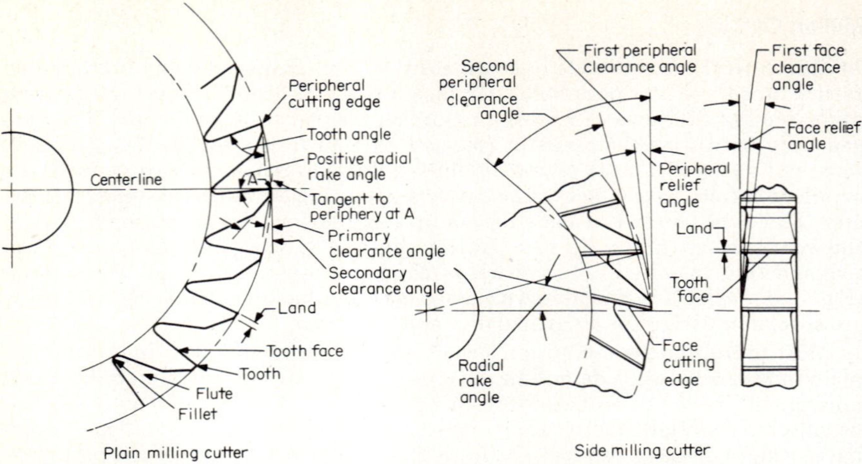

FIG. 3.42 Tooth parts, nomenclature, and angles of plain and side milling cutters.

cutters with the least number of teeth that will do the job required; this is usually the number of teeth that will ensure uninterrupted cutting action on the workpiece. The material being worked also influences the number of teeth selected for a given job. For brittle materials, which produce small or fine chips, more teeth are satisfactory because the space for chip clearance need not be as great. However, the opposite is true for stringy or gummy materials. Tooth spacing is not the only factor affecting chip space. A helical-tooth plain cutter provides the uninterrupted contact requirement, with fewer teeth for greater chip clearance. Fewer teeth can be tolerated on face milling cutters because each tooth is in contact with the work for a longer time than the teeth on a peripheral cutter. To further help select a cutter, Table 3.29 gives several tooth angles for high-speed-steel cutters for 12 common metals.

When setting up to perform milling, establish and/or work out several factors before operating the machine. Know what material is to be worked so a surface

TABLE 3.29 Tooth Angles for High-Speed-Steel Cutters

Workpiece material	Rake angle	Relief angle	Secondary clearance angle
Aluminum	20–35	3–7	7–12
Brass	10–20	3–4	6–10
Bronze	10–15	2–4	6–10
Copper	0–15	5–15	6–12
Inconel	5–15	7–8	12–14
Iron, cast	10–15	4–6	6–10
Nickel	5–15	7–8	12–14
Monel	5–15	7–8	12–14
Steel:			
Carbon	15	5	10–20
Stainless	5–10	5–12	10–20
Titanium	0	3–12	6–12
Magnesium	15–20	8–10	15–20

speed consistent with accepted practice can be selected. Cutting speeds for milling relate to the hardness of the material being worked; the harder the material, the slower the speed. Cutting speeds also correlate to tensile strength, again, inversely: The higher the strength, the lower the speed (see Fig. 3.43). A cutter must be selected, and some idea of the metal-removal rate that will or can be accomplished has to be determined. Determine metal removal rate in cubic inches per minute by multiplying the feed per tooth by the number of teeth on the cutter by depth and width of anticipated cut by rpm of cutter (rpm is found from the cutter diameter and the cutting speed in surface feet per minute):

$$\text{Metal removal rate (in}^3\text{/min)} = f\ n\ d\ w\ \text{rpm}$$

where f = feed per tooth,
n = number of teeth
d = depth of cut, in
w = width of cut, in
rpm = revolution per minute of cutter

To find the horsepower required at the cutter, multiply the removal rate by what has been called the K factor, or, more recently, "unit horsepower," for the material. This factor varies with hardness and type of material. K equal to one

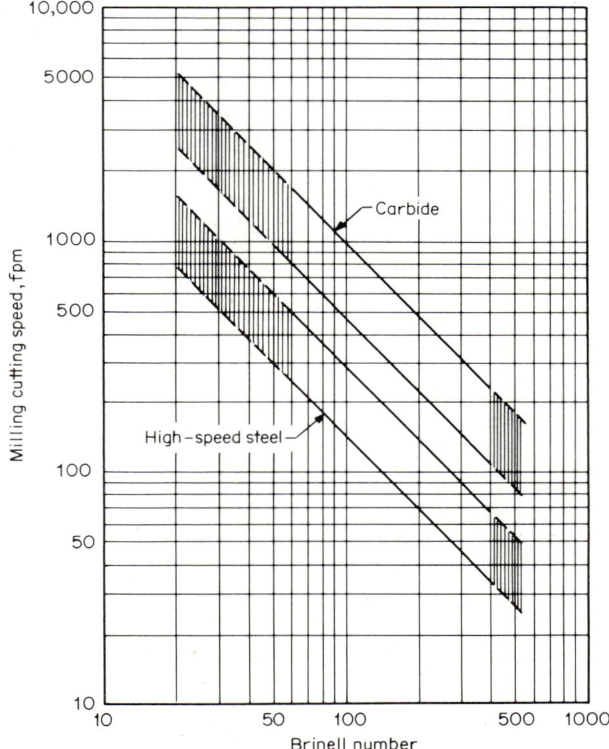

FIG. 3.43 Milling cutting speed vs. Brinell number.

unit of horsepower is the power required to cut at a rate that will remove 1 in^3 of material per minute. Table 3.30 lists some of the latest accepted values. Finally, to determine the required machine horsepower needed to deliver the required cutter horsepower, divide the cutter horsepower by the machine's overall efficiency. Most 5-hp or under machines (a 5-hp machine is probably the largest one that will be used by technicians) is considered 45 to 50 percent efficient for these purposes. By varying the many parameters involved in the previous determinations, the power required for a given job can be made to fit the available machine. The table speed to be used is the product of the feed per tooth, number of teeth in the cutter, and spindle revolutions per minute. As with most machining operations, use lubricants and cutting fluid in milling operations wherever possible. Suggested feeds per tooth and speeds for milling different materials are in Table 3.31.

Sharpen worn cutters when 1/64- to 1/32-in wear is evident. The tooth angles of main concern on a side cutting miller are the radial rake, relief (and clearance), and side-relief; axial rake and radial rake on a face mill. The width of the land is of concern on all cutters, so regrind as close as possible to its original width.

On peripheral cutters, grind the land with either a cup grinding wheel or a disk wheel. The cup produces a straight surface, the disk a slightly concave one. Using a cup wheel is the simplest procedure. Set the centers of the wheel and the cutter in line, then raise the cutter center (keeping the cutter tooth edge at the wheel center) an amount equal to the cutter radius times the sine of the clearance angle:

$$\frac{\text{Cutter diameter}}{2} \times \sin \alpha = h$$

where α = clearance angle
h = height to raise center of cutter
See Fig. 3.44.

Sharpening face milling cutters requires grinding relief on the periphery, face, and chamfer of the individual cutter teeth. The process is similar to sharpening a single-point tool. Teeth comprised of removable inserts make the sharpening easier.

End mills with helical flutes require varying depths ground along the axial width of the tooth. Produce relief along the entire length of the tooth by moving the cutter along its axis while rotating it with the tooth helix.

TABLE 3.30 ***K* Factor or Unit Horsepower**

Material	Hardness, BHN				
	150–175	176–200	201–250	251–300	301–350
Brass and bronze	0.60	0.72	0.84		
Iron:					
Gray cast	0.36	0.40	0.50	0.60	
Malleable	0.30	0.42	0.60	0.80	
Steel:					
Alloy	0.60	0.72	0.84	0.96	1.1
Carbon	0.72	0.85	0.95	1.05	1.2
Cast	0.60	0.73	0.80		
Stainless	0.60	0.72	0.88		
Tool	. . .	0.75	0.84	0.96	1.1

TABLE 3.31 Suggested Feeds and Speeds for Milling Various Materials

Material	Hardness, BHN	High-speed steel		Carbide	
		Speeds, fpm	Feed, in/tooth	Speeds, fpm	Feed, in/tooth
Aluminum alloys	. . .	900	0.005–0.020	Max.	0.005–0.020
Brass, soft	80–100	200–300	0.020	500–1000	0.012
Bronze:					
Hard	150–250	100–200	0.016	500–1000	0.010
Soft	100–150	200–300	0.018	500–1000	0.012
Cast iron:					
Soft	160	70–80	0.018	200–350	0.012
Hard	200	60–70	0.015	100–250	0.010
Copper	120–200	150–525	0.008–0.020	550–1200	0.010–0.018
Gray iron	140–200	75–140	0.010–.012	350–500	0.012–0.015
MGC	. . .	20–35	0.002	20–35	0.002
Malleable iron	120–220	75–150	0.010–0.012	200–500	0.012–0.015
Magnesium	. . .	600–1500	0.020	1000–1500	0.020
Nickel alloys	75–100	50–65	0.005–0.010	80–100	0.002–0.004
Monel	140–200	50–65	0.005–0.010	60–80	0.002–0.008
Steel:					
Alloy	200–300	40–100	0.002–0.008	200–300	0.003–0.014
Carbon	140	60–120	0.003–0.012	200–450	0.004–0.016
Stainless:					
Type 302	135–190	35–65	0.003–0.005		
Type 304	135–185	35–85	0.003–0.005		
Type 316	135–185	40–85	0.003–0.005		
Type 321	135–190	35–65	0.003–0.005		
Type 347	135–185	35–65	0.003–0.005		
Type 410	135–185	70–105	0.003–0.005		
Type 416	135–185	100–125	0.003–0.005		
Type 420	135–185	35–70	0.003–0.005		
Type 431	225–275	70–105	0.003–0.005		
Titanium (99.5)	110–170	125–150	0.001–0.007	325–375	0.001–0.006
Inconel	100–150	. . .	. . .	10–20	0.001–0.004
Incoloy	140–200	. . .	. . .	30–40	0.002–0.006
Hastelloy X	200–275	. . .	. . .	15–20	0.005
Zirconium	150–165	. . .	. . .	150–250	0.002–0.010
TZM molybdenum	220–290	90–110	0.001–0.010	225–275	0.004–0.005

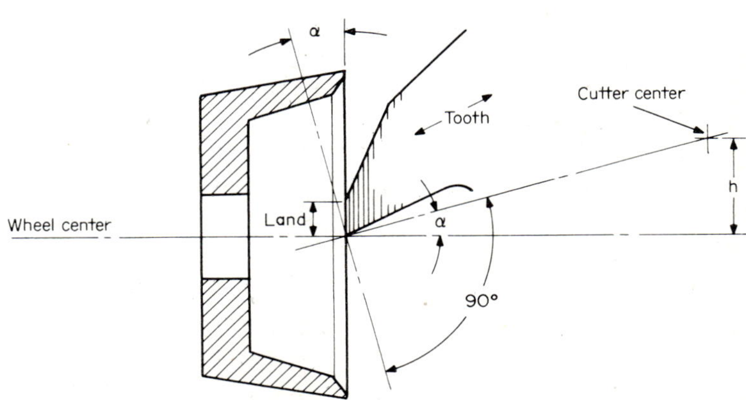

FIG. 3.44 Milling cutting speed.

Milling cutter sharpening is a job for an expert and best left to the professional if at all practical.

3.6 REAMING

Reaming is enlarging to desired size by using a reamer, which is a rotary tool with helical or straight flutes for finishing or enlarging holes. Most reamers have flutes that provide teeth for cutting and grooves for chips. Because the spiral cutting edges bridge irregularities better, they minimize chatter and surface finish is greatly improved. Both helical- (or spiral-) flute reamers are available with straight or taper shanks. Chucking reamers come in diameters of $\frac{3}{64}$ to $1\frac{1}{2}$ in in steps of $\frac{1}{64}$ in, in wire gage, letter, and decimal sizes. Wire-gage sizes range from No. 60 (0.0400-in diameter) to No. 1 (0.2280 in). Letter sizes are A to Z (0.2340 to 0.4130 in), and decimal sizes go from 0.124 to 0.501 in. Hand reamers are available in essentially the same sizes. A hand reamer differs from a chucking (sometimes called machine) reamer in that it has a square on the end of its shank to facilitate use of a tap wrench or vise and a much longer cutting edge (see Fig. 3.45).

Reaming is not a process that removes large amounts of material. However, if too little stock is removed, the reamer burnishes the work rather than cutting it, which damages both tool and work surfaces. The removal of 0.008 in per pass on soft materials and holes of $1\frac{1}{2}$-in diameter or less is considered minimum. This figure can be reduced to 0.005 in for harder materials. The need to remove more than 0.025 in indicates that other cutting processes should be considered.

Reamers are made in both solid and expansion forms. The expansion form is split and has a tapered plug that can be screwed into the tool to expand it slightly

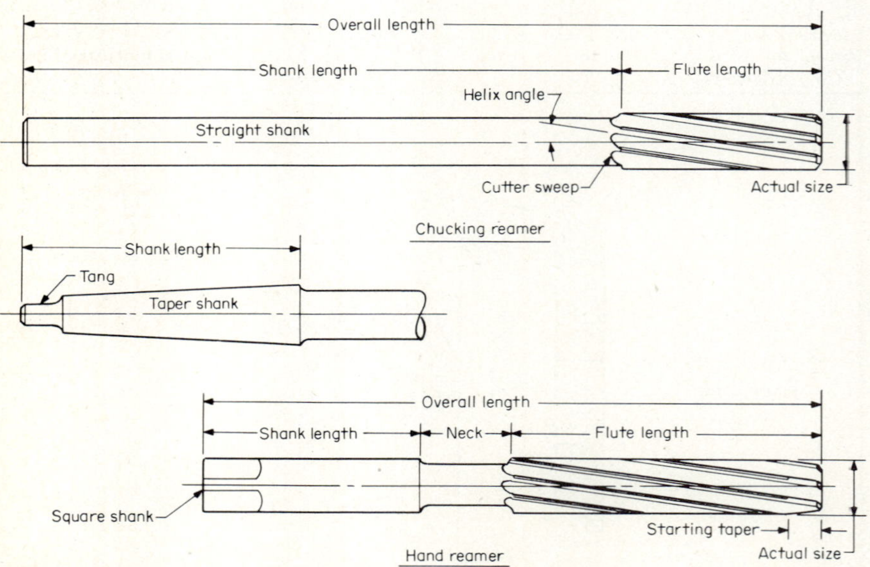

FIG. 3.45 Chucking and hand reamer.

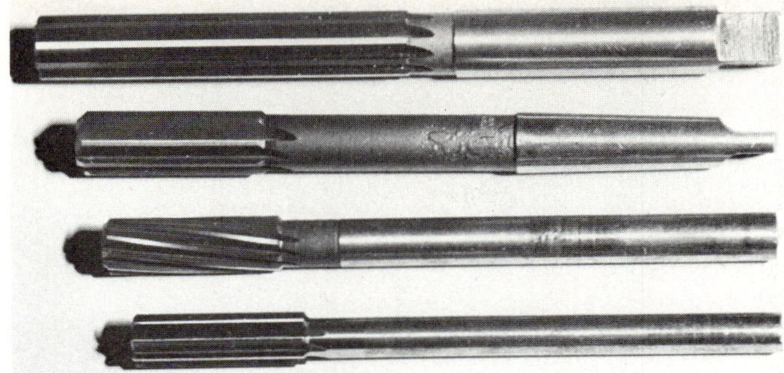

FIG. 3.46 Straight- and spiral-flute reamers.

(the amount of expansion depends upon the diameter). One-inch expansion reamers can be increased effectively in size by from 0.005 to 0.008 in. Expansion reamers are especially useful when an already reamed hole must be further enlarged. Typical straight- and spiral-flute reamers are shown in Fig. 3.46; Fig. 3.47 illustrates terms applying to reamers. Other types of reamers available but less likely found in the technician's shop are shell, stub screw machine, taper, taper pipe, die maker's, and taper pin reamers.

Hand reamers are usually made from carbon or low-alloy steel; chucking reamers (machine reamers) are made of high-speed steel or a tool steel shank with inserted carbide-tipped cutting edges. Carbide inserts are used on larger reamers. The reamers most likely used by the technician are made from high-speed or carbon steel.

In the technician's shop, reaming is mostly done on either a drill press or a milling machine. In either case, make sure the work is located correctly and securely fastened to the table or held in a heavy vise, to help produce a better reamed hole. A properly sized reamer, a coolant, and the correct selection of cutting speed helps immeasurably. Recommended cutting speeds for reaming materials are in Table 3.32.

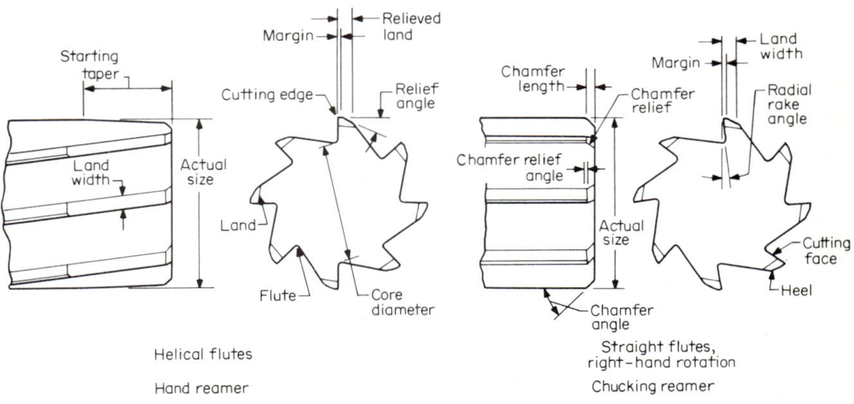

FIG. 3.47 Reamer terms.

TABLE 3.32 Recommended Cutting Speeds for Reaming Various Materials

Material	Hardness, BHN	Cutting speed, fpm	
		High-speed steel	Carbide
Aluminum alloy	. . .	250–300	700–850
Brass	. . .	160–175	320–360
Bronze-phosphor	. . .	50–60	180–200
Iron:			
Cast	100–320	20–70	90–275
Malleable	110–280	25–90	110–240
Magnesium alloys	. . .	300	850
Steel:			
Alloy	150–425	10–50	60–200
Maraging	250–325	30	100
Plain carbon	100–275	30–60	140–250
Stainless:			
AISI 300 series	135–185	25–30	100–140
AISI 400 series	135–425	10–45	50–200

When a hole is to be finished by reaming, first drill it undersized so there is material left for the reamer to remove. Drill holes up to 1½ in approximately 0.005 to 0.025 in undersized. The smaller the hole, the smaller the allowance for reaming.

Sharpen a reamer as soon as there is even slight evidence of its cutting undersized or poorly finished holes. When sharpening a chucking reamer, grind only the cutting edge (or point) to the proper angle and clearance. A hand reamer requires grinding a longer cutting edge because of the tapered lead in. See Fig. 3.48.

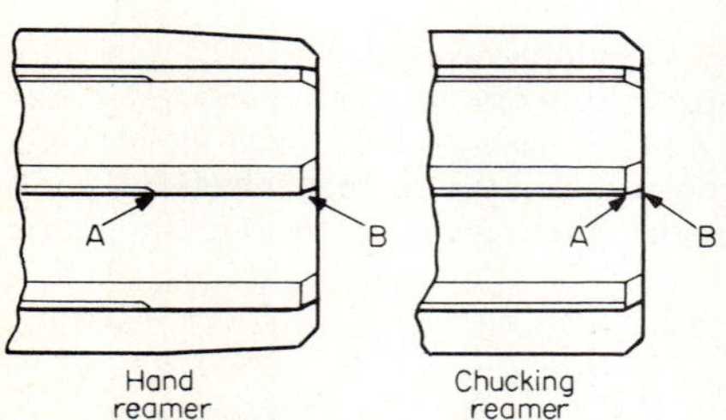

FIG. 3.48 Sharpening reamers.

As in most machining operations, use cutting fluid in reaming wherever possible and practical. Soluble oil is suitable for most materials; ream cast iron dry. When using no fluid, use an air-stream jet to cool the tool and remove chips.

3.7 GRINDING

In the grinding process, abrasive grains—either natural or manufactured—are used to remove metal from a workpiece. The more common grains used today are aluminum oxide, silicon-carbide, and diamond. Many methods are used to grind workpieces, including surface, cylindrical, centerless, internal, form, belt, and offhand grinding. Each method uses a different type of machine; the grinding methods most often used in the technician's shop are surface and offhand. Surface grinding is done on a machine similar to the one in Fig. 3.49, which is a

FIG. 3.49 Surface grinder.

FIG. 3.50 Bench grinder.

small bench-mounted precision surface grinder. It has a horizontal spindle with a reciprocating table and magnetic chuck. Other types of surface grinders have horizontal spindles with rotary tables or vertical spindles with reciprocating or rotary tables. The bench grinder is used to offhand grind parts and tools. A typical bench grinder is shown in Fig. 3.50.

Grinding Wheels

All machines use a grinding wheel composed of the abrasive grains held or bonded together by a binder material. The four main bond types are *vitrified, resinoid, rubber,* and *metal.* Vitrified bonds are used more than the other types because they can stand greater heat, are unaffected by water or fluids used in grinding, and are available in a wide range of porosity and strength. Resinoid bond is a more flexible bond; use it if the wheel will be subjected to impact or sudden loads. Resinoid bonded wheels are often used on portable grinders. Rubber is even more flexible than resinoid and is used for wheels that produce a very smooth finish. Metal bond is used for diamond wheels that grind hard materials.

The wheel's structure can be either dense or open, with many gradations. In a dense structure the grains of abrasive are closely packed. In an open structure the abrasive grains and bond material do not fill the entire wheel; there are openings or voids that provide for heat removal and cooling.

Open-structure wheels are generally used for tough materials and fine finishes. However, open-structure wheels are used for grinding cemented carbides and some other powdered, metallurgic materials. Most grinding operation done by the technician on a surface grinder falls in the middle- or medium-structure range.

Grinding wheels are identified by a standard marking system of numbers and letters in a specific sequence. Figure 3.51 shows the marking system used on grinding wheels other than diamond wheels.

When selecting a grinding wheel for a particular task, consider the factors of hardness, amounts of metal to be removed, and the finish desired. Generally, the harder the workpiece, the softer the wheel, and vice versa. Here soft refers to the "grade" of the wheel; grade is the ability of the bond material to hold the abrasive. Soft grade indicates less bond strength; hard grade indicates greater strength (see Fig. 3.51). Coarse grit removes more metal more rapidly, but fine grit produces a better finish.

Most surface grinding in the technician's shop can be done at a cutting or wheel speed of approximately 6000 fpm and table speeds between 40 and 60 fpm.

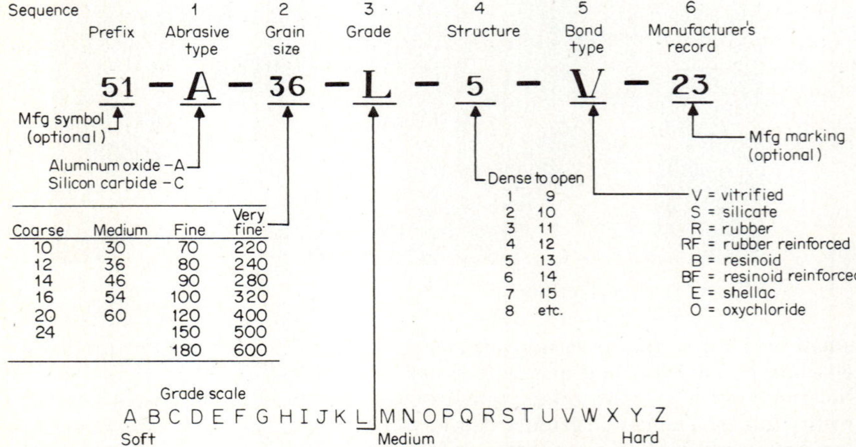

FIG. 3.51 Grinding wheel marking system (except diamond). (ANSI B74.13–1977.)

TABLE 3.33 Recommended Grinding Wheels for Various Materials

Material	Hardness	Wheel designations
Aluminum alloys	...	A-46-J-8-V
Copper alloys	120–180 BHN	A-60-K-8-V or
	180–300	A-46-J-8-V
Iron:		
Gray cast	54 Rc	C-36-H-8-V
Malleable cast	300 BHN	C-36-H-8-V
Magnesium alloys	...	A-46-J-8-V
Nickel alloys	150–230 BHN	A-46-I-8-V
Steel:		
Alloy	500 BHN	A-46-J-8-V
Plain carbon	500 BHN	A-46-J-8-V
Stainless:		
400 series	150–220 BHN	A-46-I-8-V
300 series	150–220 BHN	A-46-J-8-V
Precipitation-hardening	160–440 BHN	A-46-H-8-V
Tool	56–65 Rc	A-46-I-8-V or
		A-60-I-8-V
Titanium alloys	150–320 BHN	A-46-J-8-V
Zirconium	76–89 Rb	C-46-J-8-V
Molybdenum	250 BHN	A-60-H-8-V
Molybdenum, alloy (TZM)	220–300	A-46-F-9-V

SOURCE: "Precision Surface Grinding," Wilkie Brothers Foundation, Delmar Publishers, Albany, N.Y., 1964.

Generally, use down feeds of 0.001 in per pass for rough grinding and 0.0005 in per pass for finishing and cross feeds between 0.015 and 0.030 in per pass. Table 3.33 recommends grinding wheels for different materials. Diamond wheels have a layer of diamond grains held in a bond and applied to the operation surface of a nonabrasive core. Both their shape and composition are specified. Shapes are numerous, and the one used depends upon the operation to be performed.

Diamond wheels are designated differently than are other types of wheels. Figure 3.52 shows the method used for composition of the wheel. Diamond wheels are used by the technician for grinding hard materials like ceramics and tungsten-carbide tools. Table 3.34 lists recommended diamond wheel compositions for surface and tool grinding.

In Table 3.34, the type of wheel shape used for surface grinding is the normal

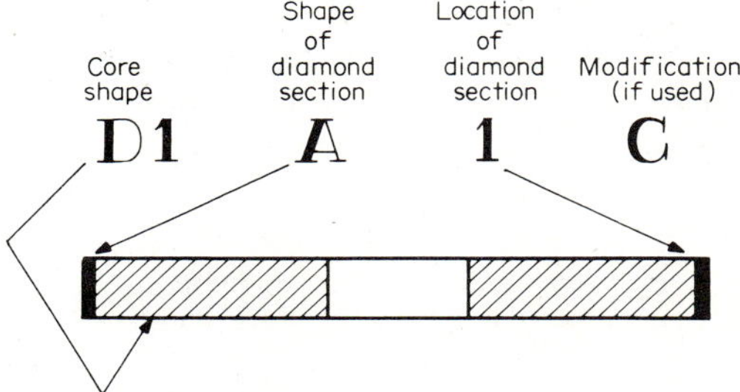

FIG. 3.52 Diamond-wheel marking system.

TABLE 3.34 Recommended Diamond Wheels

Operation	Wheel composition marking
Surface grinding	MD 120-N100-B ⅛ (rough) MD 240-P100-B ⅛ (finish)
Tool grinding (offhand)	MD 100-N100-B ⅛ (rough) MD 220-P75-B ⅛ (finish)

wheel, with diamond on the rim's surface. For tool grinding, a wheel with a layer of grit on its face is used. See Fig. 3.53.

Grinding Procedure

All grinding operations require safety precautions. Because grinding wheels can break during operation, they must be carefully mounted and used. The following precautions are recommended. Before mounting the wheel, "ring" it to make sure it is not cracked. Suspend the wheel on a finger, wood dowel, or the like. Gently tap the suspended wheel with a screwdriver handle or small wooden hammer. Tap the wheel 45° off the suspended centerline on both sides approximately halfway from the center to the rim (see Fig. 3.54). A good, undamaged, sound wheel gives a clear metallic "ring"; a damaged or unsound wheel sounds "dead." This test is valid for wheels with vitrified or metal bond, but rubber or organic bonded wheels do not ring with the same clear sound. Do not use a wheel that sounds cracked. Use clean, fresh blotters on each side of the wheel under the mounting flanges. Tighten the spindle nut only enough to hold the wheel firmly in position. If the mounting hole is too large or small, do not use the wheel. Especially do not force a wheel on the spindle. Check the manufacturer's maximum allowable wheel speed, and be sure not to exceed it when grinding.

Before starting to grind, let the wheel turn for at least 1 min. Stand aside during that time. True up the wheel if it is not running true. Dress the wheel if the cutting action is not effective. Both truing and dressing are done with a diamond; see Fig. 3.55. Truing the wheel makes the wheel diameter concentric with the spindle center by removing the nonconcentric portions. Dressing the wheel removes the "loaded" grinding surface and exposes new sharp grains for cutting. Wheel truing may accomplish both results; if it does not, dressing should follow truing.

When using diamond wheels, always make a special effort to mount the wheel as concentric as possible. Diamond wheels are very expensive, so the least truing required makes the wheel last longer.

When grinding, if possible use cutting fluids or coolants (as with most machining operations). Water-soluble oils and chemical fluids and straight oils are all good for this purpose.

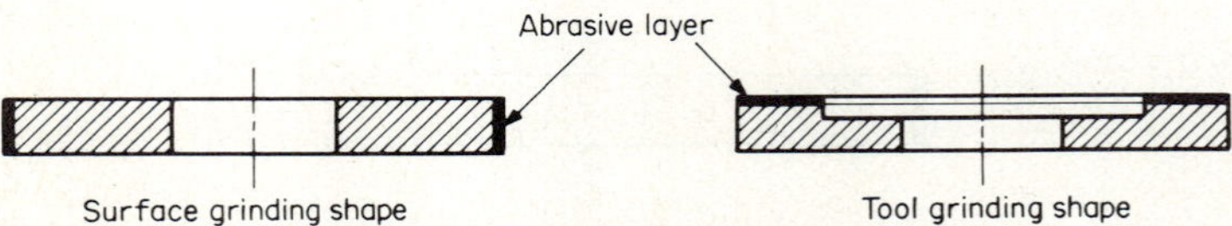

FIG. 3.53 Diamond-wheel shapes for two different applications.

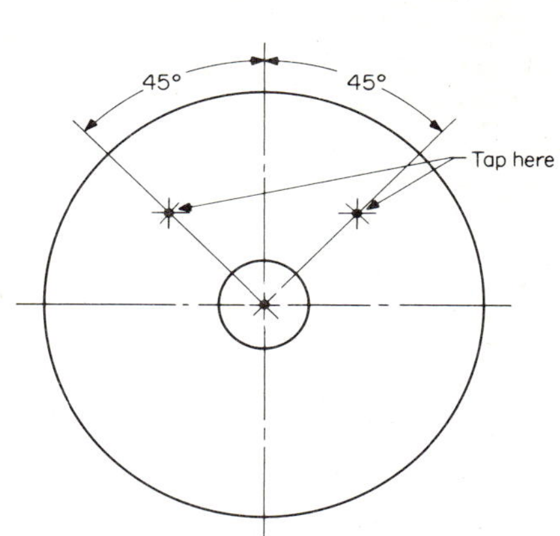

FIG. 3.54 Ring test of grinding wheel.

FIG. 3.55 Diamond-wheel dressing tool.

3.8 LAPPING

Lapping is an abrading process in which extremely accurate, smooth, flat, and/or close-fitting mating surfaces are obtained by rubbing the surface to be lapped against a matching form called a *lap*. Loose abrasive is used in a vehicle of oil or grease base, or the lap is "charged" with a fine abrasive and moistened with an oil-like lard or olive oil. Lapping is a final finishing process and not practical for removing stock. Usually, less than 0.0005 in of metal is removed by the process, although in the special cases of flat lapping by machine it is feasible to remove as much as 0.005 in of material.

Abrasives most commonly used in lapping are silicon-carbide, fused alumina, boron-carbide, and diamond; the latter two abrasives are much more expensive than the former two. These abrasives are available in various grit sizes, from No. 50 to 1000 and finer. The compound and size selected depend upon the planned use of the compound.

Silicon-carbide is extremely hard, and its grit is sharp and brittle. It is recommended for lapping cast iron and hardened steel. Fused alumina, not quite as hard but tougher, is recommended for soft steels and nonferrous metals.

Abrasives such as emery, garnet, and unfused alumina are also used for softer metal and for final lapping parts that require highly reflective surfaces. Generally, the finer the finish required on a piece, the finer the grit size used.

Laps

Laps are made of close-grain soft cast iron, copper, brass, or lead. The material of the lap has to be softer than the material being lapped or the workpiece becomes "charged" and the roles of the lap and workpiece are reversed: Metal is removed from the lap.

Laps for producing flat surfaces are made of cast iron. The surface is grooved or scored if the lap is to be used for roughing. Grooves approximately $\frac{1}{16}$ in deep and $\frac{1}{2}$ in apart are spaced uniformly across and along the length of the lap, forming a grid of ½-in squares. See Fig. 3.56. The roughing lap is charged with No. 100 or 120 grit abrasive and lard oil. An unscored lap is used for finer work and is charged with finer abrasive. Wash off all loose abrasives with gasoline after charging. (*Note:* Use gasoline with caution.) When the lapping process is actually carried out, keep the surface moist with kerosene.

Laps for holes are commonly made of lead cast around a tapered steel arbor. The arbor has a groove along its length, into which the lead flows and acts as a key to prevent the lead from rotating on the arbor. Lead is easily charged with the cutting abrasive. More sophisticated laps for holes have steel arbors with copper or cast iron shells, split so they can be expanded to fit the hole being lapped. See Fig. 3.56.

Lapping Procedure

Lapping is done by either hand or machine. Most engineering technicians do not have a lapping machine in their shops because most types are designed for specific jobs and not adaptable to all lapping operations.

For lapping outer cylindrical surfaces, use a "ring lap," which is cylindrical, with an inside diameter slightly larger than the outside diameter of the piece to be lapped, and is slit axially. See Fig. 3.56. The ring lap is held in a holder (similar to a thread die holder) by a set screw, and the holder has a screw adjustment for slight diameter changes.

When charging a flat lap, spread a very thin layer of abrasive and oil or grease over the surface of the lap and press it into the lap with a hardened steel block. There should be no side or slide motion between the block and lap. After charging the entire surface, clean it with gasoline. The whole surface should have a uniform gray appearance. If not, repeat the process until the surface is uniform in appearance.

To charge a ring lap, roll the abrasive into the inner surface with a hard steel roller slightly smaller in diameter than the inner diameter of the lap. To charge the surface of a lap for holes, spread the prepared abrasive on a hard steel block and roll the lap over the block with pressure firm enough to imbed the abrasive particles in the lap.

To lap a flat surface on a "charged" lap, move the workpiece over the lap in a figure eight motion. Keep the lap moist with kerosene. Do not apply excess pressure or "crowd" the lap. Lapping a flat surface this way is known as the *dry method,* even though the surface of the lap is moistened. An alternative is the *wet method:* The lap is made the same way as in the dry method, except that it

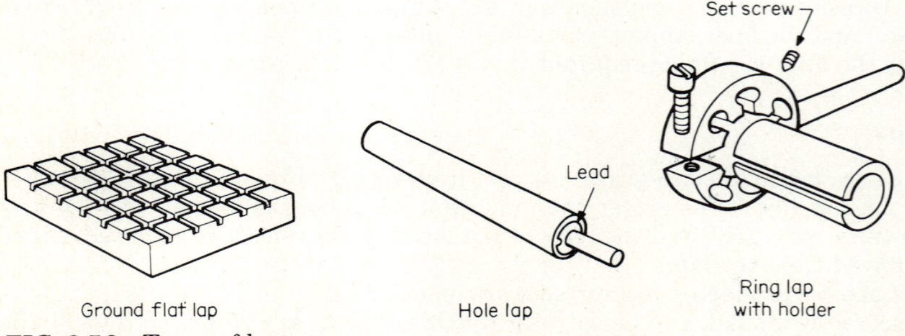

FIG. 3.56 Types of laps.

TABLE 3.35 Lapping Compound, Grit Size, and Uses

Abrasive	Grit size	Use
Silicon-carbide	600–1000	Rough lapping of hardened steel, cast iron
Alumina:		
Fused	500–900	Roughing steel, stainless steel, and chromium plate
Unfused	1–2 μm	Polishing
Garnet	600–800	Finishing brass, bronze
	10 μm	Polishing brass
Chromium oxide	1 μm	Polishing stainless steel
Diamond	No. 0–6 powder	Very hard metals, ceramics

is not charged. Instead, spread an excess amount of oil and abrasive on the lap surface; the oil and abrasive are retained in the grooves. Move the work over the lap in the same manner as described for the dry method. Table 3.35 lists types, grit sizes, and uses of assorted lapping compounds.

When lapping a hole, be sure the lap fills the hole at all times. If the lap is held in a drill press, be sure not to force the lap into the hole or this will load or strip the lap. The lap can be chucked in a lathe and the workpiece run back and forth over the lap. Manually stroke a ring lap back and forth over the rotating cylindrical workpiece held in a chuck or collet of a lathe.

3.9 THREADING AND TAPPING

Threading is a machining process of cutting external threads in both uniform and tapered cylindrical surfaces. Threading is done by several methods: threading dies, thread rolling, thread grinding, thread milling, and single-point turning. (We discussed single-point turning in Sec. 3.4.) Milling, rolling, and grinding threads are processes not likely encountered by the engineering technician; they are used in production work rather than in job shop or model-making work. The technician generally uses die threading to produce an externally threaded workpiece.

Tapping is also a machining or cutting process, to produce internal threads. A "tap" is used, which is a cylindrical or conical cutting tool with many cutting points arranged on its periphery, shaped and spaced to cut the desired thread size.

The technician usually performs threading and tapping operations by hand, in a drill press or lathe, or by turning the tool or workpiece by hand when it is held in the lathe or drill press. A drill press or lathe enables more accurate and true threads to be cut because both tool and workpiece can be held more securely and located more precisely. Where accuracy and alignment requirements are not as stringent, hand operations utilizing a tap wrench and hand die holder can certainly produce very acceptable threads.

Dies

Solid threading dies are made as one piece, with internal cutting edges or replaceable cutting edges called *chasers*. One-piece dies are either adjustable or nonadjustable. Solid dies with chasers are adjustable. One-piece dies are not usu-

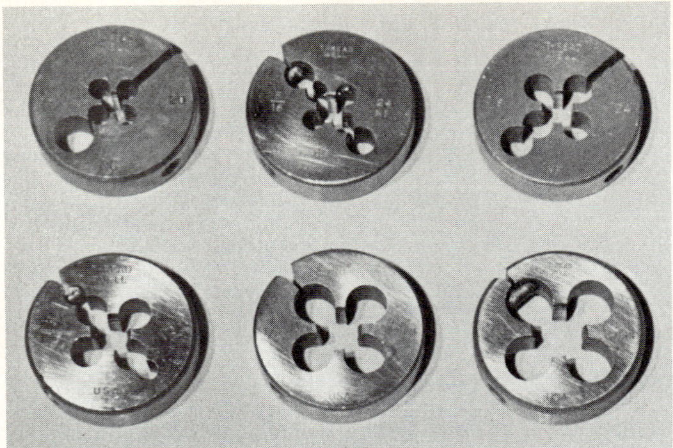

FIG. 3.57 Solid adjusting-screw dies.

ally resharpened but scrapped when worn, whereas inserted chasers can be resharpened.

The most common solid threading die in most technicians' shops is the round, adjusting-screw or button type shown in Fig. 3.57. This die has fixed, hardened cutting edges, clearance holes for chips, a slit and adjusting screw, a spring hole to facilitate the adjustment, and a recess or recesses for set screw points on the outside diameter to prevent the die from rotating in the die holder. A long chamfer is usually on the front face and a short chamfer on the rear face. The short chamfer is for threading close to a shoulder; the long chamfer is for where a couple of incomplete threads are not objectionable. Figure 3.58 details the parts of a typical adjusting-screw solid die. The amount of adjustment is limited, only enough to compensate for slight wear. Too much adjustment may displace the bands off center because of the unequal squeeze at the slit. A solid die must be backed off the thread just cut, or the work must pass through completely.

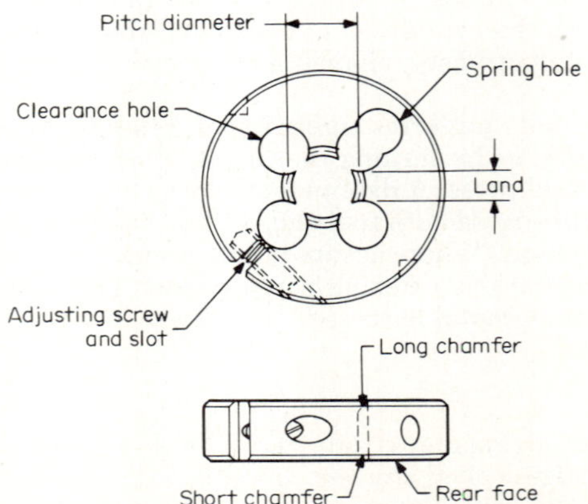

FIG. 3.58 Adjusting-screw solid-die nomenclature.

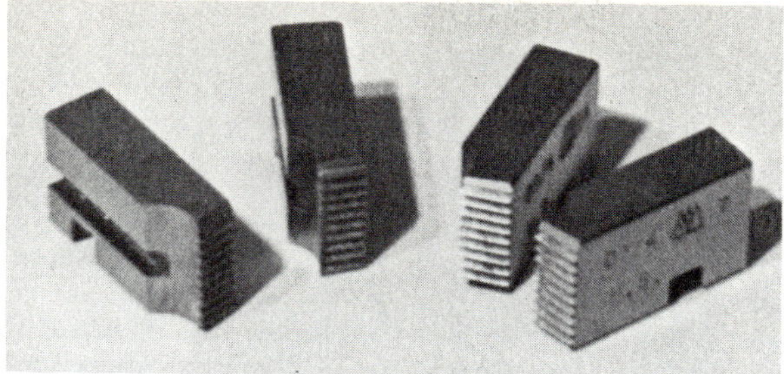

FIG. 3.59 Chasers of self-opening die head.

FIG. 3.60 Self-opening die head.

These dies come in almost all common thread sizes and series, including metric thread sizes. For thread sizes $\frac{3}{16}$ in or less, the dies have three lands. Dies for larger thread sizes have four or five lands.

Another type of die that is less common in the technician's shop area than the adjusting-screw type is the self-opening die head shown in Figs. 3.59 and 3.60. Figure 3.59 is the die "chasers," and Fig. 3.60 is the assembled head and chasers. These dies operate so that when the workpiece advances into the die head a preset distance, the die automatically "opens"; that is, the chasers move out radially and no longer contact the work. The dies can be used on lathes, drill presses, hand screw machines, and turret lathes. If they are available for use in the technician's shop, their advantages over the button-type die are that the four chasers are adjustable concentrically and the work or die does not have to be stopped and reversed to back the die off the work.

Also available are hexagonal rather than round solid dies. The hexagonal dies are intended mainly for reconditioning or reworking damaged threads. Because of their six-sided shape, they can be turned with a wrench. Their sizes cover the common range of bolt diameters used in industry.

Taps

Taps can be classified as

1. Solid
2. Shell

3. Expansion
4. Inserted chaser
5. Sectional
6. Adjustable
7. Collapsible

Of all seven types, the most common in the technician's shop and the job-type machine shop is the solid form.

Solid taps are either straight-thread or taper-thread. Straight-thread taps produce threads having no variation in pitch diameter; taper-thread taps make threads with a pitch diameter that uniformly reduces from thread to thread (pipe threads). Figure 3.61 is the standard nomenclature for solid taps.

Solid taps can have three types of chamfers: (1) taper, (2) plug, and (3) bottoming. Sometimes the tap is designated by its taper. Taper chamfer is chamfered over 7 to 9 threads, plug 3 to 5 threads, and bottoming 1 to 1½ threads; see Fig. 3.62. All cutting by a tap is done by the chamfered threads and the first full thread. As a result, the taper tap distributes the load over the most threads and is easiest to start into the workpiece. However, taper taps are not suitable for blind holes. It is not uncommon to use all three types of chamfers when tapping a blind hole, to take advantage of the best feature of each type.

There are three types of solid tap: (1) regular hand, (2) spiral-point, and (3) spiral-flute. Typical taps are shown in Fig. 3.63. The hand tap is the most common, and although it is used as much in machines as by hand, it still retains the designation "hand." Hand taps are available with straight or spiral flutes. Most have four flutes but come in three- and two-flute design for tapping metal that produces soft stringy chips.

The spiral-point tap has a combination of straight flutes with short spiral flutes near the point. They are best used for tapping through holes because they push the chips ahead of the tap. The flutes are usually shallower than those on a hand tap because the spiral section, where the taper (hence cutting threads) is located, cuts with a shearing action.

Spiral-flute taps are particularly useful for tapping holes with interruptions,

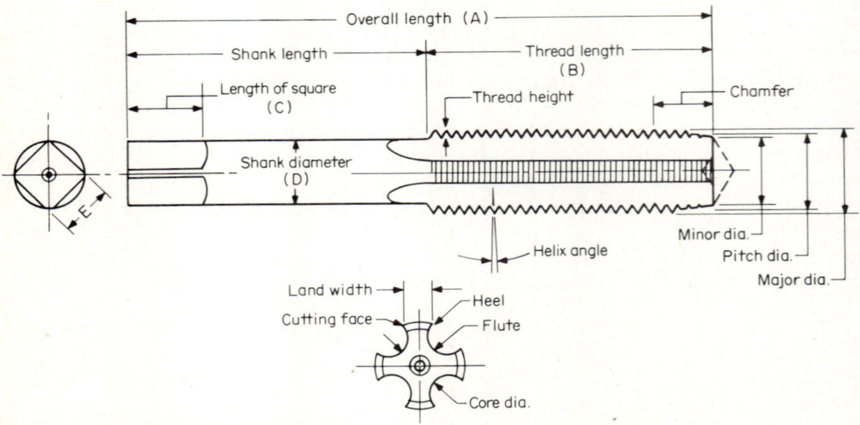

FIG. 3.61 Solid tap nomenclature.

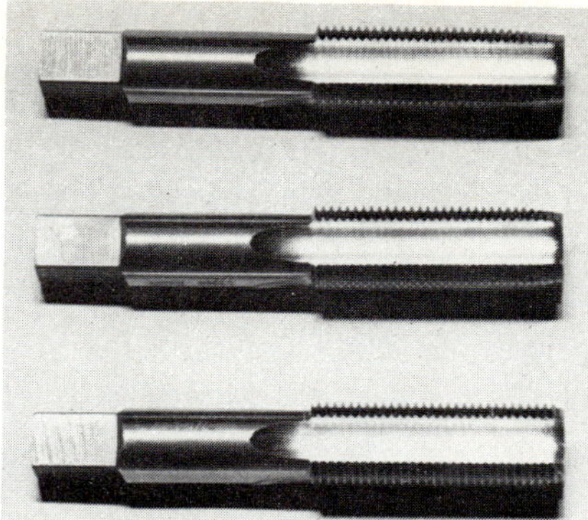

FIG. 3.62 Three tapers of taps.

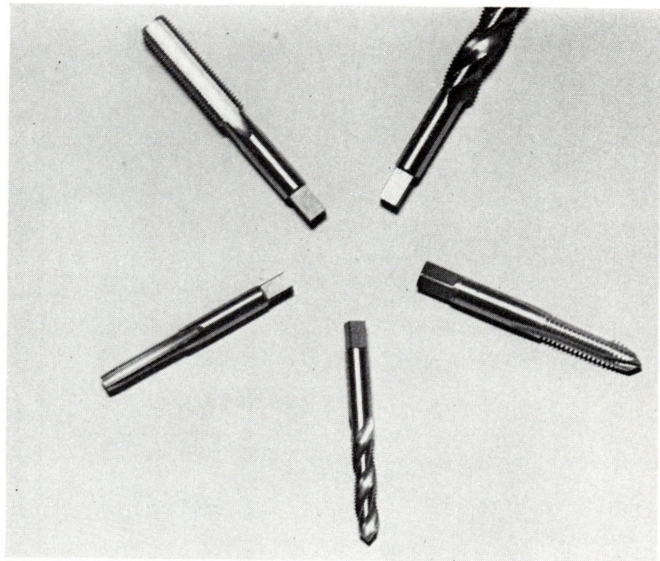

FIG. 3.63 Typical hand and spiral-flute taps.

like keyways. The cutting edges do not all cut into the interruptions, and at the same time they provide less shock load on the tool. This type of tap can have right- or left-hand flutes, and the spiral can be regular or fast. A fast spiral has a larger helix angle (approximately 1½ to 2 times that of regular) for removing chips faster. Most of the many modifications of the three types of solid taps are found in production work, so the engineering technician is not likely to use them.

The general dimensions for hand, machine screw, spiral-fluted, spiral-point, and screw thread insert (STI) taps are given in ANSI B94.9–1971 and shown in Tables 3.36 to 3.46.

Taps are marked with standard symbols for identifying the threads: the nom-

TABLE 3.36 Regular Hand Taps—High-Speed Steel—Ground Thread

All dimensions in inches

Diam. of tap	Threads/in					No. of flutes	Dimensions				
	Carbon steel, cut thread			High-speed steel, cut thread							
	NC UNC	NF UNF	NS UNS	NC UNC	NF UNF		Length overall A	Length of thread B	Length of square C	Diam. of shank D	Size of square E
⅛	...	...	40	...	...	3	1 15/16	⅝	3/16	0.141	0.110
5/32	...	...	32	...	...	4	2⅛	¾	¼	0.168	0.131
3/16	...	...	24, 32	...	...	4	2⅜	⅞	¼	0.194	0.152
7/32	...	...	24	...	...	4	2⅜	15/16	9/32	0.220	0.165
¼	20	28	...	20	28	4	2½	1	5/16	0.255	0.191
5/16	18	24	...	18	24	4	2 23/32	1⅛	⅜	0.318	0.238
⅜	16	24	...	16	24	4	2 15/16	1¼	7/16	0.381	0.286
7/16	14	20	...	14	20	4	3 5/32	1 7/16	13/32	0.323	0.242
½	13	20	...	13	20	4	3⅜	1 21/32	7/16	0.367	0.275
9/16	12	18	...	12	...	4	3 19/32	1 21/32	½	0.429	0.322
⅝	11	18	...	11	18	4	3 13/16	1 13/16	9/16	0.480	0.360
¾	10	16	...	10	16	4	4¼	2	11/16	0.590	0.442
⅞	9	14	...	9	14	4	4 11/16	2 7/32	¾	0.697	0.523
1	8	...	14†	8	...	4	5⅛	2½	13/16	0.800	0.600
1⅛	7	12	...	...	...	4	5 7/16	2 9/16	⅞	0.896	0.672
1¼	7	12*	...	...	...	4	5¾	2 9/16	1	1.021	0.766
1⅜	6†	12*†	...	...	...	4	6 1/16	3	1 1/16	1.108	0.831
1½	6	12*†	...	...	...	4	6⅜	3	1⅛	1.233	0.925
1¾	5†	...	...	...	...	6	7	3 3/16	1¼	1.430	1.072
2	4½†	...	...	...	...	6	7⅝	3 9/16	1⅜	1.644	1.233

These taps are standard, with taper, plug or bottoming chamfer. Cut thread taps, sizes ⅜ in and smaller, have an optional-style center on thread and shank ends; sizes larger than ⅜ in have internal centers in thread and shank ends.

*NF-UNF thread taps have six flutes in these sizes.

†Standard in plug chamfer only.

See Fig. 3.61 for letter-dimension correlation.

SOURCE: "American National Standard Taps—Cut and Ground Threads," ANSI B94.9–1971. By permission of The American Society of Mechanical Engineers.

TABLE 3.37 Two- and Three-Fluted and Spiral-Pointed Hand Taps* —High-Speed Steel—Ground Thread

All dimensions in inches

Diam. of tap	Thread/in NC UNC	Thread/in NF UNF	No. of flutes STD	No. of flutes OPT	Pitch diam. limits and chamfers† H1	H2	H3	H4	H5	Dimensions: Length overall A	Length of thread B	Length of square C	Diam. of shank D	Size of square E
Spiral pointed														
¼	20	...	2		...	...	PB	...	...	2½	1	5/16	0.255	0.191
¼	20	...	3		P	P	PB	...	PB	2½	1	5/16	0.255	0.191
¼	...	28	2		...	...	PB	...	...	2½	1	5/16	0.255	0.191
¼	...	28	3		...	...	PB	...	...	2½	1	5/16	0.255	0.191
5/16	18	...	2		...	...	PB	...	...	2 23/32	1⅛	⅜	0.318	0.238
5/16	18	...	3		...	P	PB	...	PB	2 23/32	1⅛	⅜	0.318	0.238
5/16	...	24	3		...	...	PB	...	...	2 23/32	1⅛	⅜	0.318	0.238
⅜	16	...	3		P	...	PB	...	PB	2 15/16	1¼	7/16	0.381	0.286
⅜	...	24	3		...	...	PB	...	...	2 15/16	1¼	7/16	0.381	0.286
7/16	14	...	3		...	...	P	...	...	3 5/32	1 7/16	13/32	0.323	0.242
7/16	...	20	3		...	...	P	...	...	3 5/32	1 7/16	13/32	0.323	0.242
½	13	...	3		...	...	PB	...	...	3⅜	1 21/32	7/16	0.367	0.275
½	...	20	3		...	...	P	...	...	3⅜	1 21/32	7/16	0.367	0.275
Fluted														
¼	20	...	2	...	P	P	PB	...	P	2½	1	5/16	0.255	0.191
¼	20	...	...	3	...	...	P	...	P	2½	1	5/16	0.255	0.191
¼	...	28	2	...	P	P	PB	P	...	2½	1	5/16	0.255	0.191
¼	...	28	...	3	...	P	...	P	...	2½	1	5/16	0.255	0.191
5/16	18	...	2	...	P	P	PB	...	P	2 23/32	1⅛	⅜	0.318	0.238
5/16	18	...	...	3	...	...	P	...	P	2 23/32	1⅛	⅜	0.318	0.238
5/16	...	24	2	...	P	P	PB	P	...	2 23/32	1⅛	⅜	0.318	0.238
5/16	...	24	...	3	...	P	...	P	...	2 23/32	1⅛	⅜	0.318	0.238
⅜	16	...	3	...	P	P	P	...	P	2 15/16	1¼	7/16	0.381	0.286
⅜	...	24	3	...	P	P	P	P	...	2 15/16	1¼	7/16	0.381	0.286
7/16	14	20	3	...	...	P	P	...	P	3 5/32	1 7/16	13/32	0.323	0.242
½	13	20	3	...	P	P	P	...	P	3⅜	1 21/32	7/16	0.367	0.275
⅝	11	...	3	...	...	...	P	...	P	3 13/16	1 13/16	9/16	0.480	0.360
¾	10	...	3	...	...	...	P	...	P	4¼	2	11/16	0.590	0.442

*See Fig. 3-61 for letter-dimension correlation.

†Chamfer designation: P = plug, B = bottoming.

SOURCE: "American National Standard Taps—Cut and Ground Threads," ANSI B94.9–1971. By permission of the American Society of Mechanical Engineers.

TABLE 3.38 Fast Spiral-Fluted Hand Taps* —High-Speed Steel—Ground Thread

All dimensions in inches

Diam. of tap	Threads/in NC UNC	Threads/in NF UNF	No. of flutes	Dimensions: Length overall *A*	Length of thread *B*	Length of square *C*	Diam. of shank *D*	Size of square *E*
¼	20	28	3	2½	1	5/16	0.255	0.191
5/16	18	24	3	2 23/32	1⅛	⅜	0.318	0.238
⅜	16	24	3	2 15/16	1¼	7/16	0.381	0.286
7/16	14	20	3	2 5/32	1 7/16	13/32	0.323	0.242
½	13	20	3	3⅜	1 21/32	7/16	0.367	0.275

*These taps are standard with plug or bottoming chamfer in H3 limits and have right-hand spiral flutes with a helix angle from 45 to 60°. See Fig. 3.61 for letter-dimension correlation.

SOURCE: "American National Standard Taps—Cut and Ground Threads," ANSI B94.9–1971. By permission of the American Society of Mechanical Engineers.

inal size, number of threads per inch, and a symbol that identifies the thread form. Left-hand taps are marked with an additional L.H. or "lefthand." Table 1.1 lists several such symbols.

Taps are not marked with a U for unified; the symbol for the corresponding American Standard thread form is used. As an example, a ¼-in, 20 threads per inch American National (or Unified) coarse-thread Series tap is marked ¼″-20NC; if left hand, it is ¼″-20NC L.H.

Ground-thread taps made under- or oversized by grinding are used to produce classes of thread that cannot be achieved by cut-thread taps. The manufacturer supplies the taps with standard established tolerances. All are now designated as ground thread instead of the previous classification of commercial ground, commercial ground high, and precision ground. The standard pitch-diameter tolerance for a ground-thread tap in sizes from No. 0 to 1-in diameter inclusive is 0.0005 in; from 1 to 1½ in inclusive, it is 0.001 in.

Ground taps are ground over or under the basic pitch diameter of the tap. If made oversized, they are designated high taps, if undersized, low. Identifying letters are H or L. The tolerance range is denoted by a numeral after the H or L that indicates the number of half-thousandths (0.0005) inches larger or smaller the tap is than basic size. Pitch-diameter-limit numbers indicate the following for taps to 1 in in diameter, inclusive (basic denotes basic tap pitch diameter):

GL1 = basic to basic minus 0.0005 in

GH1 = basic to basic plus 0.0005 in

GH2 = basic, plus 0.005 to 0.001 in

GH3 = basic, plus 0.001 to 0.0015 in

GH4 = basic, plus 0.0015 to 0.002 in

GH5 = basic, plus 0.002 to 0.0025 in

GH6 = basic, plus 0.0025 to 0.003 in

For taps 1 to 1½ in in diameter, inclusive:

GH2 = basic to 0.001 in

GH4 = basic plus 0.001 to 0.002 in

TABLE 3.39 Regular Machine Screw Taps with Standard Number of Flutes—High-Speed Steel—Ground Thread

All dimensions in inches

		Threads/in				Pitch diam. limits and chamfers*				Dimensions				
Size	Basic major diam.	NC UNC	NF UNF	NS	No. of flutes	H1	H2	H3	H7†	Length overall *A*	Length of thread *B*	Length of square *C*	Diam. of shank *D*	Size of square *E*
0	0.060	...	80	...	2	TPB	PB	...	...	1⅝	5/16	3/16	0.141	0.110
1	0.073	64	...	...	2	TPB	P	...	...	1 11/16	⅜	3/16	0.141	0.110
1	0.073	...	72	...	2	TPB	PB	...	...	1 11/16	⅜	3/16	0.141	0.110
2	0.086	56	...	...	3	TPB	TPB	...	...	1¾	7/16	3/16	0.141	0.110
2	0.086	...	64	...	3	...	TPB	...	...	1¾	7/16	3/16	0.141	0.110
3	0.099	48	...	...	3	P	TPB	...	...	1 13/16	½	3/16	0.141	0.110
3	0.099	...	56	...	3	...	TPB	...	...	1 13/16	½	3/16	0.141	0.110
4	0.112	...	...	36	3	...	TPB	...	...	1⅞	9/16	3/16	0.141	0.110
4	0.112	40	...	...	3	TPB	TPB	...	...	1⅞	9/16	3/16	0.141	0.110
4	0.112	...	48	...	3	P	TPB	...	...	1⅞	9/16	3/16	0.141	0.110
5	0.125	40	...	...	3	PB	TPB	...	...	1 15/16	⅝	3/16	0.141	0.110
5	0.125	...	44	...	3	...	TPB	...	...	1 15/16	⅝	3/16	0.141	0.110
6	0.138	32	...	...	3	TPB	TPB	TPB	PB	2	11/16	3/16	0.141	0.110
6	0.138	...	40	...	3	P	TPB	...	...	2	11/16	3/16	0.141	0.110
8	0.164	32	...	...	4	TPB	TPB	TPB	PB	2⅛	¾	¼	0.168	0.131
8	0.164	...	36	...	4	P	TPB	...	...	2⅛	¾	¼	0.168	0.131
10	0.190	24	32	...	4	TPB	TPB	TPB	PB	2⅜	⅞	¼	0.194	0.152
12	0.216	24	...	...	4	...	...	TPB	...	2⅜	15/16	9/32	0.220	0.165
12	0.216	...	28	...	4	P	...	TPB	...	2⅜	15/16	9/32	0.220	0.165

*Chamfer designations: T = taper, P = plug, B = bottoming.

†Formerly designated G o/s.

See Fig. 3.61 for letter-dimension correlation.

SOURCE: "American National Standard Taps—Cut and Ground Threads," ANSI B94.9–1971. By permission of the American Society of Mechanical Engineers.

TABLE 3.40 Spiral-Pointed Machine Screw Taps—High-Speed Steel—Ground Thread

All dimensions in inches

		Threads/in				Pitch diam. limits and chamfers*				Dimensions				
Size	Basic major diam.	NC UNC	NF UNF	NS	No. of flutes	H1	H2	H3	H7†	Length overall *A*	Length of thread *B*	Length of square *C*	Diam. of shank *D*	Size of square *E*
0	0.060	. . .	80	. . .	2	PB	PB	. . .	. . .	1⅝	5/16	3/16	0.141	0.110
1	0.073	64	72	. . .	2	P	P	. . .	. . .	1 11/16	⅜	3/16	0.141	0.110
2	0.086	56	. . .	. . .	2	PB	PB	. . .	. . .	1¾	7/16	3/16	0.141	0.110
2	0.086	. . .	64	. . .	2	P	P	. . .	. . .	1¾	7/16	3/16	0.141	0.110
3	0.099	48	. . .	. . .	2	P	PB	. . .	. . .	1 13/16	½	3/16	0.141	0.110
3	0.099	. . .	56	. . .	2	P	P	. . .	. . .	1 13/16	½	3/16	0.141	0.110
4	0.112	. . .	. . .	36	2	. . .	P	. . .	. . .	1⅞	9/16	3/16	0.141	0.110
4	0.112	40	. . .	. . .	2	PB	PB	. . .	. . .	1⅞	9/16	3/16	0.141	0.110
4	0.112	. . .	48	. . .	2	P	PB	. . .	. . .	1⅞	9/16	3/16	0.141	0.110
5	0.125	40	. . .	. . .	2	P	PB	. . .	. . .	1 15/16	⅝	3/16	0.141	0.110
5	0.125	. . .	44	. . .	2	. . .	P	. . .	. . .	1 15/16	⅝	3/16	0.141	0.110
6	0.138	32	. . .	. . .	2	PB	PB	PB	PB	2	11/16	3/16	0.141	0.110
6	0.138	. . .	40	. . .	2	P	PB	. . .	. . .	2	11/16	3/16	0.141	0.110
8	0.164	32	. . .	. . .	2	PB	PB	PB	PB	2⅛	¾	¼	0.168	0.131
8	0.164	. . .	36	. . .	2	P	P	. . .	. . .	2⅛	¾	¼	0.168	0.131
10	0.190	24	. . .	. . .	2	P	PB	PB	P	2⅜	⅞	¼	0.194	0.152
10	0.190	. . .	32	. . .	2	PB	PB	PB	P	2⅜	⅞	¼	0.194	0.152
12	0.216	24	. . .	. . .	2	P	. . .	PB	. . .	2⅜	15/16	9/32	0.220	0.165
12	0.216	. . .	28	. . .	2	. . .	. . .	P	. . .	2⅜	15/16	9/32	0.220	0.165

*Chamfer designation: P = plug, B = bottoming.

†H7 taps formerly designated G o/s.

See Fig. 3.61 for letter-dimension correlation.

SOURCE: "American National Standard Taps—Cut and Ground Threads," ANSI B94.9–1971. By permission of the American Society of Mechanical Engineers.

TABLE 3.41 Spiral-Pointed Machine Screw Taps*

All dimensions in inches

Size	Basic major diam.	Threads/in				No. of flutes	Dimensions				
		Carbon steel, cut thread		High-speed steel, cut thread			Length overall *A*	Length of thread *B*	Length of square *C*	Diam. of shank *D*	Size of square *E*
		NC UNC	NF UNF	NC UNC	NF UNF						
4	0.112	. . .	. . .	40	. . .	2	1 7/8	9/16	3/16	0.141	0.110
5	0.125	. . .	. . .	40	. . .	2	1 15/16	5/8	3/16	0.141	0.110
6	0.138	32	. . .	32	. . .	2	2	11/16	3/16	0.141	0.110
8	0.164	32	. . .	32	. . .	2	2 1/8	3/4	1/4	0.168	0.131
10	0.190	24	32	24	32	2	2 3/8	7/8	1/4	0.194	0.152
12	0.216	. . .	. . .	24	. . .	2	2 3/8	15/16	9/32	0.220	0.165

*These taps are standard with plug chamfer only. See Fig. 3.61 for letter-dimension correlation.

SOURCE: "American National Standard Taps—Cut and Ground Threads," ANSI B94.9–1971. By permission of the American Society of Mechanical Engineers.

TABLE 3.42 Spiral-Pointed Machine Screw Taps* with Short Flutes† —High-Speed Steel—Ground Thread

All dimensions in inches

		Threads/in				Dimensions				
Size	Basic major diam.	NC UNC	NF UNF	No. of flutes	Pitch diam. limits	Length overall *A*	Length of thread *B*	Length of square *C*	Diam. of shank *D*	Size of square *E*
4	0.112	40	. . .	2	H2	1⅞	9/16	3/16	0.141	0.110
5	0.125	40	. . .	2	H2	1 15/16	⅝	3/16	0.141	0.110
6	0.138	32	. . .	2	H3	2	11/16	3/16	0.141	0.110
8	0.164	32	. . .	2	H3	2⅛	¾	¼	0.168	0.131
10	0.190	24	32	2	H3	2⅜	⅞	¼	0.194	0.152
12	0.216	24	. . .	2	H3	2⅜	15/16	9/32	0.220	0.165

*These taps are standard with plug chamfer only.

†Short-flute spiral-pointed machine screw taps are provided with point only. The balance of the threaded section is left unfluted.

See Fig. 3.61 for letter-dimension correlation.

SOURCE: "American National Standard Taps—Cut and Ground Threads," ANSI B94.9–1971. By permission of the American Society of Mechanical Engineers.

TABLE 3.43 Regular Hand Taps* for Helical Coil Wire Screw Thread Inserts (STI)—High Speed Steel—Ground Thread

All dimensions in inches

	Threads/in		Major diam.			Pitch diam. limits and chamfers†		Dimensions				
Nominal size (STI)	NC UNC	NF UNF	Min.	Max.	No. of flutes	H2	H3	Length overall *A*	Length of thread *B*	Length of square *C*	Diam. of shank *D*	Size of square *E*
1/4	20	. . .	0.3177	0.3187	3	. . .	PB	2 23/32	1 1/8	3/8	0.318	0.238
1/4	. . .	28	0.2985	0.2995	3	PB	. . .	2 23/32	1 1/8	3/8	0.318	0.238
5/16	18	. . .	0.3874	0.3884	4	. . .	PB	2 15/16	1 1/4	7/16	0.381	0.286
5/16	. . .	24	0.3690	0.3700	4	PB	. . .	2 15/16	1 1/4	7/16	0.381	0.286
3/8	16	. . .	0.4592	0.4602	4	. . .	PB	3 3/8	1 21/32	7/16	0.367	0.275
7/16	14	. . .	0.5333	0.5343	4	. . .	P	3 19/32	1 21/32	1/2	0.429	0.322
7/16	. . .	20	0.5052	0.5062	4	. . .	P	3 3/8	1 21/32	7/16	0.367	0.275
1/2	13	. . .	0.6032	0.6042	4	. . .	P	3 13/16	1 13/16	9/16	0.480	0.360
1/2	. . .	20	0.5677	0.5687	4	. . .	P	3 19/32	1 21/32	1/2	0.429	0.322

*These taps are oversize to the extent that the internal thread they produce will accommodate a helical coil wire screw thread insert, which at final assembly will accept a screw thread of the nominal size and pitch.

†Chamfer designation: P = plug, B = bottoming.

See Fig. 3.61 for letter-dimension correlation.

SOURCE: "American National Standard Taps—Cut and Ground Threads," ANSI B94.9–1971. By permission of the American Society of Mechanical Engineers.

TABLE 3.44 Spiral-Pointed Hand Taps for Helical Coil Wire Screw Thread Inserts (STI)—High-Speed Steel—Ground Thread

All dimensions in inches

Nominal size* (STI)	Threads/in		Major diam.		No. of flutes	Pitch diam. limits	Dimensions				
	NC UNC	NF UNF	Min.	Max.			Length overall *A*	Length of thread *B*	Length of square *C*	Diam. of shank *D*	Size of square *E*
¼	20	...	0.3177	0.3187	2	H2, H3	2²³⁄₃₂	1⅛	⅜	0.318	0.238
¼	...	28	0.2985	0.2995	2	H2, H3	2²³⁄₃₂	1⅛	⅜	0.318	0.238

*These taps are oversize to the extent that the internal thread they produce will accommodate a helical coil wire screw thread insert, which at final assembly will accept a screw thread of the nominal size and pitch.

Taps are standard with plug chamfer only.

See Fig. 3.61 for letter-dimension correlation.

SOURCE: "American National Standard Taps—Cut and Ground Threads," ANSI B94.9–1971. By permission of the American Society of Mechanical Engineers.

TABLE 3.45 Regular Machine Screw Taps for Helical Coil Wire Screw Thread Inserts (STI) —High-Speed Steel—Ground Thread

All dimensions in inches

Nominal size* (STI)	Threads/in		Major diam.		No. of flutes	Pitch diam. limits†		Dimensions				
	NC UNC	NF UNF	Min.	Max.		H2	H3	Length overall *A*	Length of thread *B*	Length of square *C*	Diam. of shank *D*	Size of square *E*
4	40	...	0.1463	0.1473	3	PB	...	2	11/16	3/16	0.141	0.110
6	32	...	0.1807	0.1817	3	...	PB	2⅜	⅞	¼	0.194	0.152
8	32	...	0.2067	0.2077	3	...	PB	2⅜	15/16	9/32	0.220	0.165
10	24	...	0.2465	0.2475	3	PB	...	2½	1	5/16	0.255	0.191
10	...	32	0.2327	0.2337	3	PB	PB	2½	1	5/16	0.255	0.191

*These taps are oversize to the extent that the internal thread they produce will accommodate a helical coil wire screw thread insert, which at final assembly will accept a screw thread of the nominal size and pitch.

†Chamfer designation: P = plug, B = bottoming.

See Fig. 3.61 for letter-dimension correlation.

SOURCE: "American National Standard Taps—Cut and Ground Threads," ANSI B94.9–1971. By permission of the American Society of Mechanical Engineers.

TABLE 3.46 Spiral-Pointed Machine Screw Taps for Helical Coil Wire Screw Thread Inserts (STI)—High-Speed Steel—Ground Thread

All dimensions in inches

Nominal size* (STI)	Threads/in		Major diam.		No. of flutes	Pitch diam. limits†	Dimensions‡				
	NC UNC	NF UNF	Min.	Max.			Length overall *A*	Length of thread *B*	Length of square *C*	Diam. of shank *D*	Size of square *E*
4	40	...	0.1463	0.1473	2	H2	2	11/16	3/16	0.141	0.110
6	32	...	0.1807	0.1817	2	H2, H3	2⅜	⅞	¼	0.194	0.152
8	32	...	0.2067	0.2077	2	H2, H3	2⅜	15/16	9/32	0.220	0.165
10	...	32	0.2327	0.2337	2	H2	2½	1	5/16	0.255	0.191

*These taps are oversize to the extent that the internal thread they produce will accommodate a helical coil wire screw thread insert, which at final assembly will accept a screw thread of the nominal size and pitch.

†These taps are standard with plug chamfer only.

‡See Fig. 3.61 for letter-dimension correlation.

SOURCE: "American National Standard Taps—Cut and Ground Threads," ANSI B94.9–1971. By permission of the American Society of Mechanical Engineers.

TABLE 3.47 Fractionall Size Taps—Unified and American Form—Cut Thread
All dimensions in inches

				Standard Thread Limits					
	Threads/in			Major diam.			Pitch diam.		
Size	NC UNC	NF UNF	NS UN	Basic	Min.	Max.	Basic	Min.	Max.
1/16	...	...	64	0.0625	0.0635	0.0650	0.0524	0.0526	0.0536
3/32	...	...	48	0.0938	0.0951	0.0966	0.0803	0.0805	0.0815
1/8	...	...	40	0.1250	0.1266	0.1286	0.1088	0.1090	0.1105
5/32	...	...	32	0.1563	0.1585	0.1605	0.1360	0.1365	0.1380
5/32	...	...	36	0.1563	0.1580	0.1600	0.1382	0.1384	0.1399
3/16	...	...	24	0.1875	0.1903	0.1923	0.1604	0.1609	0.1624
3/16	...	...	32	0.1875	0.1897	0.1917	0.1672	0.1677	0.1692
7/32	...	...	24	0.2188	0.2216	0.2236	0.1917	0.1922	0.1937
7/32	...	...	32	0.2188	0.2210	0.2230	0.1985	0.1990	0.2005
1/4	20	...	...	0.2500	0.2532	0.2557	0.2175	0.2180	0.2200
1/4	...	...	24	0.2500	0.2528	0.2553	0.2229	0.2234	0.2254
1/4	...	28	...	0.2500	0.2524	0.2549	0.2268	0.2273	0.2288
1/4	...	...	32*	0.2500	0.2522	0.2547	0.2297	0.2302	0.2317
5/16	18	...	...	0.3125	0.3160	0.3185	0.2764	0.2769	0.2789
5/16	...	24	...	0.3125	0.3153	0.3178	0.2854	0.2859	0.2874
3/8	16	...	...	0.3750	0.3789	0.3814	0.3344	0.3349	0.3369
3/8	...	24	...	0.3750	0.3778	0.3803	0.3479	0.3484	0.3499
7/16	14	...	...	0.4375	0.4419	0.4449	0.3911	0.3916	0.3941
7/16	...	20	...	0.4375	0.4407	0.4437	0.4050	0.4055	0.4075
1/2	13	...	...	0.5000	0.5047	0.5077	0.4500	0.4505	0.4530
1/2	...	20	...	0.5000	0.5032	0.5062	0.4675	0.4680	0.4700
9/16	12	...	...	0.5625	0.5675	0.5705	0.5084	0.5089	0.5114
9/16	...	18	...	0.5625	0.5660	0.5690	0.5264	0.5269	0.5289
5/8	11	...	...	0.6250	0.6304	0.6334	0.5660	0.5665	0.5690
5/8	...	18	...	0.6250	0.6285	0.6315	0.5889	0.5894	0.5914
11/16	...	...	11	0.6875	0.6929	0.6969	0.6285	0.6290	0.6320
11/16	...	...	16	0.6875	0.6914	0.6954	0.6469	0.6474	0.6499
3/4	10	...	...	0.7500	0.7559	0.7599	0.6850	0.6855	0.6885
3/4	...	16	...	0.7500	0.7539	0.7579	0.7094	0.7099	0.7124
7/8	9	...	...	0.8750	0.8820	0.8860	0.8028	0.8038	0.8068
7/8	...	14	...	0.8750	0.8799	0.8839	0.8286	0.8296	0.8321
1	8	...	...	1.0000	1.0078	1.0118	0.9188	0.9198	0.9228
1	...	12	...	1.0000	1.0055	1.0095	0.9459	0.9469	0.9499
1	...	...	14	1.0000	1.0049	1.0089	0.9536	0.9546	0.9571
1 1/8	7	...	...	1.1250	1.1337	1.1382	1.0322	1.0332	1.0367
1 1/8	...	12	...	1.1250	1.1305	1.1350	1.0709	1.0719	1.0749
1 1/4	7	...	...	1.2500	1.2587	1.2632	1.1572	1.1582	1.1617
1 1/4	...	12	...	1.2500	1.2555	1.2600	1.1959	1.1969	1.1999
1 3/8	6	...	...	1.3750	1.3850	1.3895	1.2667	1.2677	1.2712
1 3/8	...	12	...	1.3750	1.3805	1.3850	1.3209	1.3219	1.3249
1 1/2	6	...	...	1.5000	1.5100	1.5145	1.3917	1.3927	1.3962
1 1/2	...	12	...	1.5000	1.5055	1.5100	1.4459	1.4469	1.4499
1 3/4	5	...	...	1.7500	1.7602	1.7657	1.6201	1.6216	1.6256
2	4 1/2	...	...	2.0000	2.0111	2.0166	1.8557	1.8572	1.8612

SOURCE: "American National Standard Taps—Cut and Ground Threads," ANSI B94.9–1971. By permission of the American Society of Mechanical Engineers.

TABLE 3.48 Recommended Taps for Classes 2, 3, 2B*, and 3B of Unified and American Standard Screw Threads—Numbered Sizes

Size	Threads/in		Recommended tap			
	NC UNC	NF UNF	Class 2	Class 3	Class 2B*	Class 3B
0	...	80	G H1	G H1	G H2	G H1
1	64	...	G H1	G H1	G H2	G H1
1	...	72	G H1	G H1	G H2	G H1
2	56	...	G H1	G H1	G H2	G H1
2	...	64	G H1	G H1	G H2	G H1
3	48	...	G H1	G H1	G H2	G H1
3	...	56	G H1	G H1	G H2	G H1
4	40	...	G H2	G H1	G H2	G H2
4	...	48	G H1	G H1	G H2	G H1
5	40	...	G H2	G H1	G H2	G H2
5	...	44	G H1	G H1	G H2	G H1
6	32	...	G H2	G H1	G H3	G H2
6	...	40	G H2	G H1	G H2	G H2
8	32	...	G H2	G H1	G H3	G H2
8	...	36	G H2	G H1	G H2	G H2
10	24	...	G H3	G H1	G H3	G H3
10	...	32	G H2	G H1	G H3	G H2
12	24	...	G H3	G H1	G H3	G H3
12	...	28	G H3	G H1	G H3	G H3

*Cut-thread taps in sizes 3 to 12 NC and NF inclusive may be used under normal conditions and in average materials for producing tapped holes to this classification.

SOURCE: Erik Oberg, Franklin D. Jones, and Holbrook L. Horton (eds.), *Machinery's Handbook,* 20th ed., Industrial Press, New York, 1975.

GH6 = basic plus 0.002 to 0.003 in

GH8 = basic plus 0.003 to 0.004 in

For example, if the ¼″-20 NC tap cited is a ground-thread tap, it is further marked ¼″-20NC GH3, meaning it was ground with tolerance limit on the pitch diameter 0.001 to 0.0015 in above the basic pitch diameter for that size tap. (See Table 3.47 for basic fractional cut-thread tap dimensions.) If it is marked ¼″-20NC GLI, the pitch diameter is 0.0000 to 0.0005 in undersize.

Class 1B tapped holes can be made with cut-thread taps. Recommended ground taps for other classes of fits are in Tables 3.48 and 3.49.

Threading Procedure

Cutting fluids, or lubricants, are very important during tapping and threading operations; however, no one lubricant serves all situations. A rule of thumb is to use light lubricant for light material, progressively heavier to heavy lubricant for hard materials. Get the lubricant down to the cutting edges of the tap or die by whatever means available. The lubricant should be there at the start of operating as well as during the removal of the tool. Use sulfur-base oil for most steels (including stainless) and monel metal. Use kerosene and lard oil on aluminum alloys and castings. Brass, bronzes, and copper call for a light-base oil lubricant. Use no lubricant with cast iron.

TABLE 3.49 Recommended Taps for Classes 2, 3, 2B, and 3B Unified and American Standards Screw Threads—Fractional Sizes

Size	Threads		Recommended tap			
	NC UNC	NF UNF	Class 2	Class 3	Class 2B*	Class 3B
¼	20	...	G H3	G H2	G H5	G H3
¼	...	28	G H2	G H1	G H4	G H2
5/16	18	...	G H3	G H2	G H5	G H3
5/16	...	24	G H3	G H1	G H4	G H3
⅜	16	...	G H3	G H2	G H5	G H3
⅜	...	24	G H3	G H1	G H4	G H3
7/16	14	...	G H5	G H3	G H5	G H3
7/16	...	20	G H3	G H1	G H5	G H3
½	13	...	G H5	G H3	G H5	G H3
½	...	20	G H3	G H1	G H5	G H3
9/16	12	...	G H5	G H3	G H5	G H3
9/16	...	18	G H3	G H2	G H5	G H3
⅝	11	...	G H5	G H3	G H5	G H3
⅝	...	18	G H3	G H2	G H5	G H3
¾	10	...	G H5	G H3	G H5	G H5
¾	...	16	G H3	G H2	G H5	G H3
⅞	9	...	G H6	G H4	G H6	G H4
⅞	...	14	G H4	G H2	G H6	G H4
1	8	...	G H6	G H4	G H6	G H4
1	...	12	G H4	G H2	G H6	G H4
1	14 NS		G H4	G H2	G H6	G H4
1⅛	7	...	G H8	G H4	G H8	G H4
1⅛	...	12	G H4	G H4	G H6	G H4
1¼	7	...	G H8	G H4	G H8	G H4
1¼	...	12	G H4	G H4	G H6	G H4
1⅜	6	...	G H8	G H4	G H8	G H4
1⅜	...	12	G H4	G H4	G H6	G H4
1½	6	...	G H8	G H4	G H8	G H4
1½	...	12	G H4	G H4	G H6	G H4

*Cut-thread taps may be used under normal conditions and in average materials for producing tapped holes to this classification.

SOURCE: Erik Oberg, Franklin D. Jones, and Holbrook L. Horton (eds.), *Machinery's Handbook*, 20th ed., Industrial Press, New York, 1975.

The hole drilled previous to tapping the thread is important. The drill diameter should be such that the depth of thread produced is approximately 75 percent of full depth. For holes whose thread engagement is more than 1½ times the nominal diameter, a 60 percent thread is usually satisfactory. Recommended drill sizes are in Tables 3.50 to 3.52.

Threading and tapping pipe threads is performed in much the same way as for straight threads. Taps and dies are of the same types, except they have a taper or straight pipe thread form. The same machines and cutting fluid can be used.

Recommended tap-drill sizes for pipe taps are in Table 3.53. For tapered pipe thread, also ream the hole before tapping with a taper reamer that has ¾ in to the foot taper (prescribed taper in pipe threads). This lessens the amount of work required to do the job and helps keep the threads aligned properly.

TABLE 3.50 Tap-Drill Sizes for American National Threads

Screw thread		Commercial tap drills*		Screw thread		Commercial tap drills*	
Outside diam.–pitch	Root diam.	Size or no.	Decimal equiv.	Outside diam.–pitch	Root diam.	Size or no.	Decimal equiv.
1/16–64	0.0422	3/64	0.0469	–27	0.4519	15/32	0.4687
–72	0.0445	3/64	0.0469	9/16–12	0.4542	31/64	0.4844
5/64–60	0.0563	1/16	0.0625	–18	0.4903	33/64	0.5156
–72	0.0601	52	0.0635	–27	0.5144	17/32	0.5312
3/32–48	0.0667	49	0.0730	5/8–11	0.5069	17/32	0.5312
–50	0.0678	49	0.0730	–12	0.5168	35/64	0.5469
7/64–48	0.0823	43	0.0890	–18	0.5528	37/64	0.5781
1/8–32	0.0844	3/32	0.0937	–27	0.5769	19/32	0.5937
–40	0.0925	38	0.1015	11/16–11	0.5694	19/32	0.5937
9/64–40	0.1081	32	0.1160	–16	0.6063	5/8	0.6250
5/32–32	0.1157	1/8	0.1250	3/4–10	0.6201	21/32	0.6562
–36	0.1202	30	0.1285	–12	0.6418	43/64	0.6719
11/64–32	0.1313	9/64	0.1406	–16	0.6688	11/16	0.6875
3/16–24	0.1334	26	0.1470	–27	0.7019	23/32	0.7187
–32	0.1469	22	0.1570	13/16–10	0.6826	23/32	0.7187
13/64–24	0.1490	20	0.1610	7/8–9	0.7307	49/64	0.7656
7/32–24	0.1646	16	0.1770	–12	0.7668	51/64	0.7969
–32	0.1782	12	0.1890	–14	0.7822	13/16	0.8125
15/64–24	0.1806	10	0.1935	–18	0.8028	53/64	0.8281
1/4–20	0.1850	7	0.2010	–27	0.8269	27/32	0.8437
–24	0.1959	4	0.2090	15/16–9	0.7932	53/64	0.8281
–27	0.2019	3	0.2130	1–8	0.8376	7/8	0.8750
–28	0.2036	3	0.2130	–12	0.8918	59/64	0.9219
–32	0.2094	7/32	0.2187	–14	0.9072	15/16	0.9375
5/16–18	0.2403	F	0.2570	–27	0.9519	31/32	0.9687
–20	0.2476	17/64	0.2656	1 1/8–7	0.9394	63/64	0.9844
–24	0.2584	I	0.2720	–12	1.0168	1 3/64	1.0469
–27	0.2644	J	0.2770	1 1/4–7	1.0644	1 7/64	1.1094
–32	0.2719	9/32	0.2812	–12	1.1418	1 11/64	1.1719
3/8–16	0.2938	5/16	0.3125	1 3/8–6	1.1585	1 7/32	1.2187
–20	0.3100	21/64	0.3281	–12	1.2668	1 19/64	1.2969
–24	0.3209	Q	0.3320	1 1/2–6	1.2835	1 11/32	1.3437
–27	0.3269	R	0.3390	–12	1.3918	1 27/64	1.4219
7/16–14	0.3447	U	0.3680	1 5/8–5 1/2	1.3888	1 29/64	1.4531
–20	0.3726	25/64	0.3906	1 3/4–5	1.4902	1 9/16	1.5625
–24	0.3834	X	0.3970	1 7/8–5	1.6152	1 11/16	1.6875
–27	0.3894	Y	0.4040	2–4 1/2	1.7113	1 25/32	1.7812
1/2–12	0.3918	27/64	0.4219	2 1/8–4 1/2	1.8363	1 29/32	1.9062
–13	0.4001	27/64	0.4219	2 1/4–4 1/2	1.9613	2 1/32	2.0312
–20	0.4351	29/64	0.4531	2 3/8–4	2.0502	2 1/8	2.1250
–24	0.4459	29/64	0.4531	2 1/2–4	2.1752	2 1/4	2.2500

*These tap-drill diameters allow approximately 75 percent of a full thread. For small thread sizes, the larger drills will reduce tap breakage.

SOURCE: Erik Oberg, Franklin D. Jones, and Holbrook L. Horton (eds.), *Machinery's Handbook,* 20th ed., Industrial Press, New York, 1975.

TABLE 3.51 Tap and Clearance Drills for American National Thread Form Machine Screws

Size of screw			Tap drills		Clearance-hole drills			
					Close fit		Free fit	
No. or diam.	Decimal equiv.	Threads/in	Drill size	Decimal equiv.	Drill size	Decimal equiv.	Drill size	Decimal equiv.
0	0.060	80	3/64	0.0469	52	0.0635	50	0.0700
1	0.073	64	53	0.0595	48	0.0760	46	0.0810
		72	53	0.0595	″	″		
2	0.086	56	50	0.0700	43	0.0890	41	0.0960
		64	50	0.0700	″	″		
3	0.099	48	47	0.0785	37	0.1040	35	0.1100
		56	45	0.0820	″	″		
4	0.112	36*	44	0.0860	32	0.1160	30	0.1285
		40	43	0.0890	″	″		
		48	42	0.0935	″	″		
5	0.125	40	38	0.1015	30	0.1285	29	0.1360
		40	37	0.1040	″	″		
6	0.138	32	36	0.1065	27	0.1440	25	0.1495
		40	33	0.1130	″	″		
8	0.164	32	29	0.1360	18	0.1695	16	0.1770
		36	29	0.1360	″	″		
10	0.190	24	25	0.1495	9	0.1960	7	0.2010
		32	21	0.1590	″	″		
12	0.216	24	16	0.1770	2	0.2210	1	0.2280
		28	14	0.1820	″	″		
14	0.242	20*	10	0.1935	D	0.2460	F	0.2570
		24*	7	0.2010	″	″		
1/4	0.250	20	7	0.2010	F	0.2570	H	0.2660
		28	3	0.2130	″	″		
5/16	0.3125	18	F	0.2570	P	0.3230	Q	0.3320
		24	I	0.2720	″	″	X	0.3970
3/8	0.375	16	5/16	0.3125	W	0.3860		
		24	Q	0.3320	″	″		
7/16	0.4375	14	U	0.3680	29/64	0.4531	15/32	0.4687
		20	25/64	0.3906	″	″		
1/2	0.500	13	27/64	0.4219	33/64	0.5156	17/32	0.5312
		20	29/64	0.4531	″	″		

*Asterisk (*) denotes screws not in the American Standard; they are from the former ASME Standard.

SOURCE: Erik Oberg, Franklin D. Jones, and Holbrook L. Horton (eds.), *Machinery's Handbook*, 20th ed., Industrial Press, New York, 1975.

3.10 BRAZING

Brazing is a process for joining like or dissimilar metals. A nonferrous filler metal is used, at temperatures above 800°F and below the melting point of the base metals. The filler metal or brazing alloys, when molten, readily flow by capillary action into and between the closely fitted surfaces of the mating parts being joined. The American Welding Society (AWS) divides the commonly used brazing alloys into seven classifications: (1) aluminum-silicon, (2) copper-phospho-

TABLE 3.52 Recommended Tap-Drill Sizes for ISO Thread Form (Metric)

Thread designation		Thread designation	
Diam. × pitch, mm	Tap drill, mm	Diam. × pitch, mm	Tap drill, mm
M 1 × 0.25	0.75	M 9 × 1.25	7.8
M 1.2 × 0.25	0.95	M 10 × 1.5	8.5
M 1.4 × 0.3	1.1	M 11 × 1.5	9.5
M 1.5 × 0.3	1.2	M 12 × 1.75	10.2
M 1.6 × 0.35	1.25	M 13 × 1.75	11.25
M 1.7 × 0.35	1.35	M 14 × 2	12
M 1.8 × 0.35	1.45	M 15 × 2	13
M 2 × 0.4	1.6	M 16 × 2	14
M 2 × 0.45	1.55	M 17 × 2	15
M 2.2 × 0.45	1.75	M 18 × 2.5	15.5
M 2.3 × 0.4	1.9	M 19 × 2.5	16.5
M 2.5 × 0.45	2.05	M 20 × 2.5	17.5
M 2.6 × 0.45	2.15	M 22 × 2.5	19.5
M 3 × 0.5	2.5	M 23 × 2.5	20.5
M 3 × 0.6	2.4	M 24 × 3	21
M 3.5 × 0.6	2.9	M 27 × 3	24
M 4 × 0.7	3.3	M 30 × 3.5	26.5
M 4 × 0.75	3.2	M 33 × 3.5	29.5
M 4.5 × 0.75	3.7	M 36 × 4	32
M 5 × 0.8	4.2	M 39 × 4	35
M 5 × 0.9	4.0	M 42 × 4.5	37.5
M 5.5 × 0.9	4.5	M 45 × 4.5	40.5
M 6 × 1	5.0	M 48 × 5	43
M 7 × 1	6.0	M 52 × 5	47
M 8 × 1.25	6.8		

TABLE 3.53 Pipe-Thread Tap-Drill Sizes, In

Nominal pipe size	Taper thread	Straight thread
1/16	0.250	0.250
1/8	21/64	11/32
1/4	27/64	7/16
3/8	9/16	37/64
1/2	11/16	23/32
3/4	57/64	59/64
1	1 1/8	1 5/32
1 1/4	1 15/32	1 1/2
1 1/2	1 23/32	1 3/4
2	2 3/16	2 7/32
2 1/2	2 19/32	2 21/32

SOURCE: Theodore Baumeister (ed.), *Standard Handbook for Mechanical Engineers,* McGraw-Hill, New York, 1967.

TABLE 3.54 Brazing Filler Metals

AWS classification†	Nominal composition, %*									Temperature, °F			Standard form‡	Uses
	Ag	Cu	Zn	Al	Ni	Cr	B	Si	Ot	Solid	Liquid	Brazing range		
BAlSi-2	...	...	...	92.5	...	...	...	7.5	...	1070	1135	1110–1150	5	For joining aluminum alloys 1060, EC, 1100, 3003, 3004, 5005, 5050, 6053, 6061, 6062, 6063, 6951 and cast alloys A612 and C612. All these filler metals are suitable for furnace and dip brazing. BAlSi-3, -4, and -5 are suitable for torch brazing. Used with lap and tee joints rather than butt joints. Joint clearances run from 0.006 to 0.025 in.
BAlSi-3	...	4	...	86	...	...	...	10	...	970	1085	1060–1120	2, 3, 5	
BAlSi-4	...	...	...	88	...	...	...	12	...	1070	1080	1080–1120	2, 3, 4, 5	
BAlSi-5	...	...	...	90	...	...	...	10	...	1070	1095	1090–1120	5	
BCuP-1	...	95	...	...	...	...	...	...	P, 5	1310	1695	1450–1700	1	For joining copper and its alloys, with limited use on silver, tungsten, and molybdenum. Not for use on ferrous or nickel-base alloys. Are used for cupro-nickels, but exercise caution when nickel content is greater than 30 percent. Suitable for all brazing processes. Lap joints recommended, but butt joints may be used. Clearances used range from 0.001 to 0.005 in.
BCuP-2	...	93	...	...	...	...	...	...	P, 7	1310	1460	1350–1550	2, 3, 4	
BCuP-3	5	89	...	...	...	...	...	...	P, 6	1190	1485	1300–1500	2, 3, 4	
BCuP-4	6	87	...	...	...	...	...	...	P, 7	1190	1335	1300–1450	2, 3, 4	
BCuP-5	15	80	...	...	...	...	...	...	P, 5	1190	1475	1300–1500	1, 2, 3, 4	
BAg-1	45	15	16	...	...	...	...	...	Cd, 24	1125	1145	1145–1400	1, 2, 4	For joining most ferrous and nonferrous metals except aluminum and magnesium. These filler metals have good brazing properties and are
BAg-1a	50	15.5	16.5	...	...	...	...	...	Cd, 18	1160	1175	1175–1400	1, 2, 4	
BAg-2	35	26	21	...	...	...	...	...	Cd, 18	1125	1295	1295–1550	1, 2, 4	
BAg-2a	30	27	23	...	...	...	...	...	Cd, 20	1125	1310	1310–1550	1, 2, 4	
BAg-3	50	15.5	15.5	...	3	...	...	...	Cd, 16	1170	1270	1270–1500	1, 2, 4	

TABLE 3.54 (*Continued*)

AWS classification†	Nominal composition, %*									Temperature, °F			Standard form‡	Uses
	Ag	Cu	Zn	Al	Ni	Cr	B	Si	Ot	Solid	Liquid	Brazing range		
BAg-4	40	30	28	...	2	...	...	...	...	1240	1435	1435–1650	1, 2	suitable for preplacement in the joint or for manual feeding into the joint. All methods of heating may be used. Lap joints are generally used, however, butt joints may be used. Joint clearances of 0.002 to 0.005 in are recommended. Flux is generally required.
BAg-5	45	30	25	...	...	...	...	...	...	1250	1370	1370–1550	1, 2	
BAg-6	50	34	16	...	...	...	...	...	...	1270	1425	1425–1600	1, 2	
BAg-7	56	22	17	...	...	...	...	...	Sn, 5	1145	1205	1205–1400	1, 2	
BAg-8	72	28	...	...	...	...	...	...	...	1435	1435	1435–1650	1, 2, 4	
BAg-8a	72	27.8	...	...	...	...	...	...	Li, 2	1410	1410	1410–1600	1, 2	
BAg-13	54	40	5	...	1	...	...	...	...	1325	1575	1575–1775	1, 2	
BAg-13a	56	42	...	...	2	...	...	...	...	1420	1640	1600–1800	1, 2	
BAg-18	60	30	...	...	...	...	...	...	Sn, 10	1115	1325	1325–1550	1, 2	
BAg-19	92.5	7.3	...	...	...	...	...	...	Li, 0.2	1435	1635	1610–1800	1, 2	
BNi-1	...	...	...	...	74	14	3.5	4	Fe, 4.5	1790	1900	1950–2200	1, 2, 3, 4	For brazing AISI 300 and 400 series stainless steel and nickel- and cobalt-base alloys. Particularly suited to vacuum systems and vacuum-tube applications because of their very low vapor pressure. The limiting element is chromium in those alloys in which it is employed.
BNi-2	...	...	...	...	82.5	7	3	4.5	Fe, 3	1780	1830	1850–2150	1, 2, 3, 4	
NBi-3	...	...	...	...	91	...	3	4.5	Fe, 1.5	1800	1900	1850–2150	1, 2, 3, 4	
BNi-4	...	...	...	...	93.5	...	1.5	3.5	Fe, 1.5	1800	1950	1850–2150	1, 2, 3, 4	
BNi-5	...	...	...	...	71	19	...	10	...	1975	2075	2100–2200	1, 2, 3, 4	
BNi-6	...	...	...	...	89	...	...	...	P, 11	1610	1610	1700–1875	1, 2, 3, 4	
BNi-7	...	...	...	...	77	13	...	...	P, 10	1630	1630	1700–1900	1, 2, 3, 4	
BCu-1	...	100	...	...	...	...	...	...	...	1980	1980	2000–2100	1, 2, 3	For joining various ferrous and nonferrous metals. They can also be used with various brazing processes. Avoid overheating the Cu-Zn alloys. Lap and butt joints are commonly used.
BCu-1a	...	99	...	...	...	...	...	...	O, 1	1980	1980	2000–2100	4	
BCu-2	...	86.5	...	...	...	...	...	...	O, 13.5	1980	1980	2000–2100	6	
RBCu Zn-A	...	59	41	...	...	...	...	...		1630	1650	1670–1750	1, 2, 3	
RBCu Zn-D		48	42	...	10	...	...	...		1690	1715	1720–1800	1, 2, 3	
BMg-1	...	...	2	9	...	...	...	...	Mg89	830	1110	1120–1160	2, 3	BMg-1 and BMg-2a are used for joining AZ10A, K1A, and M1A magnesium-base metals. BMg-2a is also used for joining AZ31B
BMg-2a	...	...	5	12	...	...	...	...	Mg83	770	1050	1080–1130	1, 2, 3, 4	

														and ZE10A metals. Joint clearances used run from 0.004 to 0.010 in.
BAu-1	...	63	...	...	...	...	...	...	Au, 37	1815	1860	1860–2000	2, 4, 5	For brazing iron, nickel, and cobalt-base metals where resistance to oxidation or corrosion is required. Low rate of interaction with base metal facilitates use on thin base metals. Used with induction, furnace, or resistance heating in a reducing atmosphere or in a vacuum and with no flux. For other applications, a borax-boric acid flux is used.
BAu-2	...	20.5	...	...	...	...	...	...	Au, 79.5	1635	1635	1635–1850	2, 4, 5	
BAu-3	...	62.5	...	...	3	...	...	...	Au, 34.5	1785	1885	1885–1995	2, 4, 5	
BAu-4	...	...	...	...	18.5	...	...	...	Au, 81.5	1740	1740	1740–1840	2, 4, 5	

*Trace elements may be present in small amounts in these nominal compositions and are not shown. Abbreviations: Ag, silver; Cu, copper; Zn, zinc; Al, aluminum; Ni, nickel; Cr, chromium; B, boron; Si, silicon; Ot, other, which includes P, phosphorus; Cd, cadmium; Sn, tin; Li, lithium; Fe, iron; O, oxygen; Mg, magnesium; and Au, gold.

‡Numbers specify standard forms as follows: 1, strip; 2, wire; 3, rod; 4, powder; 5, sheet; and 6, paste.

†These classifications contain chemical symbols preceded by B, which denotes brazing filler metal.

rus, (3) silver, (4) nickel, (5) copper and copper zinc, (6) magnesium, and (7) precious metals. The technician most likely will use the silver brazing alloys from which the term *silver soldering,* sometimes used synonymously with brazing, emanated. There are a number of such alloys, most of which contain silver, copper, zinc, and cadmium. Table 3.54 lists all the various brazing filler alloys as classified by the AWS, their composition, the temperature ranges over which they are used, the forms in which they are available, and their uses.

Brazing Methods

There are several methods for joining metals by brazing; the manner in which the heat is applied in the operation designates the method. *Torch brazing,* probably the most common method, is used by the engineering technician. Use oxyacetylene, oxyfuel gas, air-acetylene, air-fuel gas, and/or oxyhydrogen in a handheld torch, with the flame adjusted for a slightly reducing effect. Clean and flux the parts to be joined and either assemble them or hold them in a fixture. Keep the flame in motion at all times during the heating, to prevent localized overheating or burning. When the joint area is at the right temperature as judged by eye and experience, apply the brazing filler metal, which then melts and flows into the joint until the joint is filled and filleted. This method of brazing is particularly useful when the parts are of unequal mass.

In induction brazing, the heat used to join the parts is generated in the parts themselves. A high-frequency current is passed through an appropriately shaped water-cooled coil (usually in a shape similar to the parts being joined) near or surrounding the parts but not in contact with them and induces a current in the pieces. The pieces' resistance to the flow of the induced current heats them to the desired temperature.

Insert preformed shapes of brazing alloy between the fitted pieces, and apply the flux before starting the heating. When the eye sees that the brazing alloy is melted and has been drawn into the joint, turn off the current and let the pieces cool. This brazing method is becoming increasingly popular as smaller and better high-frequency induction-heating generators are becoming available. Because the voltages used are high, take adequate precautions to prevent serious or fatal injury to personnel operating or near the equipment. Keep metal tables, stands, and frames well-grounded. Where multiturn coils are used, be careful in humid atmospheres that condensation does not drip from turn to turn on the coils and cause breakdown. Covering the coils with fiberglass sleeving or the like may be helpful.

Furnace brazing with a gas-tight furnace is another popular method. A reducing atmosphere such as hydrogen or an inert gas at a pressure slightly above atmosphere prevents air and water vapor from entering the furnace. It is necessary to also remove water vapor from the gas used as the atmosphere, to prevent or at least greatly reduce oxide formation. The furnace is either electrically heated or gas-fired. Most furnaces in the technician's area are electric and equipped with temperature, atmosphere, and pressure controls. Furnace brazing is a relatively simple operation. Hold parts in place by either gravity or simple fixtures, place the brazing alloy at the joint or place preformed shapes in or at the joint, and bring up the whole assembly to the melting temperature of the alloy used. Flux is used except when an atmosphere of hydrogen, which performs the same function, is present.

Dip brazing utilizes a bath of molten alloy or molten salt. Parts must be properly jigged and fixtured. The inconvenience of handling molten alloy or salt in

TABLE 3.55 Guide to Selecting Brazing Filler Metals and Fluxes

Base metals being brazed	Filler metals recommended*	Flux				
		AWS brazing flux type no.	Effective temperature range, °F	Ingredients contained therein	Form supplied in	Methods of use†
All brazeable aluminum alloys	BAlSi	1	700–1190	Chlorides, fluorides	Powder	1, 2 3, 4
All brazeable magnesium alloys	BMg	2	900–1200	Chlorides, fluorides	Powder	3, 4
Alloys such as aluminum-bronze; aluminum-brass containing additions of aluminum of 0.5% or more	BCuZn BCuP	4‡	1050–1800	Chlorides, fluorides, borates, wetting agent	Paste or powder	1, 2, 3
Titanium and zirconium in base alloys	BAg	6	700–1600	Chlorides, fluorides, wetting agent	Paste or powder	1, 2, 3
Any other brazeable alloys not listed above	All brazing filler metals except BAlSi and BMg	3	700–2000	Boric acid, borates, fluorides, fluoborates (must contain fluorine compound), wetting agent	Paste, powder, or liquid	1, 2, 3
	All brazing filler metals except BAlSi, BAg–I through BAg–7	5	1000–2200	Borax, boric acid, borates, wetting agent, no fluorine in any form	Paste, powder, or liquid	1, 2, 3

*Abbreviations: B, brazing filler metal; Al, aluminum; Si, silicon; Mg, magnesium; Cu, copper; Zn, zinc; P, phosphorus; and Ag, silver.

†Numbering system: 1, dry powder sprinkled in joint region; 2, heated metal filler rod dipped into powder or paste; 3, flux mixed with such things as alcohol, water, monochlorobenzene to form a paste or slurry; 4, flux used molten in a bath.

‡Types 1 and 3 fluxes, alone or in combination, may be used with some of these base metals also.

SOURCE: Erik Oberg, Franklin D. Jones, and Holbrook L. Horton (eds.), *Machinery's Handbook,* 20th ed., Industrial Press, New York, 1975.

quantity is the main disadvantage of this brazing method, which is probably the least used. Technicians are not likely to encounter this method.

Resistance brazing uses current-carrying carbon electrodes that produce heat which in turn is conducted to the parts to be joined. Alternating current is passed through the parts themselves, which are heated by their own resistance to current flow. The disadvantages of this brazing method are the inaccuracy of the control of temperature and the difficulties of holding the parts together tightly during and after heating.

Brazing Procedure

In all brazing methods the joint area of the parts to be joined must be cleaned of all oil, cutting fluids, dirt, oxides, or any other foreign matter. Cleaning can be done chemically—e.g., degreasing compounds and solvents—or mechanically—e.g., filing and grinding—or by a combination of both methods.

Fluxes are used to prevent oxides from forming on the joint surfaces. Fluxes help remove oxides that may still be present on the base metal after proper cleaning, but do not rely upon them solely for this purpose. The flux must remain active at the temperatures used in the brazing operations and must remain in contact with the joint surfaces to prevent further formations of oxides. Fluxes are available as powder, paste or liquid; paste is the most commonly used form, especially in torch brazing. The flux used depends upon the brazing method, the metals to be joined, and the filler alloy. Table 3.55 is a guide to selecting brazing alloys and fluxes.

3.11 SOLDERING

Soldering is joining metals by using a nonferrous metal whose melting point is below 800°F (700 K) and also below the melting point of the base metal. Tin or lead-base alloys called soft solders or binary tin-lead solders are used. Soldering is for joints not requiring much mechanical strength. If mechanical strength is needed, brazing is recommended (see Sec. 3.10). Soldering is used extensively in electrical work to ensure good electric contact between such parts as terminals and wires and junctions. Soldering is often used with a mechanical means of fastening to seal against leakage.

There are a number of soft solder alloys; Table 3.56 lists their composition and uses. Note that the higher the tin content, the lower the temperature at which the solder reaches the liquid condition. (Above the eutectic composition of approximately 63 percent tin, 37 percent lead, the reverse occurs.) Figure 3.64 is a phase diagram for lead-tin upon which the information is based. Those alloy combinations having a partially molten state over a large range of temperature (solid plus liquid on the phase diagram) are preferred for wiping applications.

Heat is applied by a soldering iron, solder bath, torch, induction heating, or resistance heating. Hand soldering by iron or torch is the method the technician usually uses. In all cases, clean surfaces help to ensure a good efficient joint. Problems from soldering can usually be traced to inadequate or improper preparation or treatment of the surfaces before or after the actual soldering is done, particularly when joining metals with heavy oxide coatings. Fluxes help remove any oxide coatings present on the base metal, prevent oxide films from forming, and increase the wetting properties of the solder. Zinc chloride and ammonium chloride, separately or in combination, and fluxes containing muriatic acid or orthophosphoric acid remove oxide but must be neutralized or cleansed away after soldering is complete or they may cause corrosion. Thoroughly washing

TABLE 3.56 Properties of Solder Alloys

Nominal composition, %				Specific gravity	Melting ranges				Uses
					Solid		Liquid		
Tin	Lead	Antimony	Silver		°C	°F	°C	°F	
									Tin-lead alloys
70	30	...	...	8.32	183	361	192	378	For coating metals.
63	37	...	...	8.40	183	361	183	361	As lowest melting (eutectic) solder for both dip and hand soldering methods.
60	40	...	...	8.65	183	361	190	374	Fine solder. For general purposes, but particularly where the temperature requirements are critical.
50	50	...	...	8.85	183	361	216	421	For general purposes. Most popular of all.
45	55	...	...	8.97	183	361	227	441	For automobile radiator cores and roofing seams.
40	60	...	...	9.30	183	361	238	460	Wiping solder for joining lead pipes and cable sheaths. For automobile radiator cores and heating units.
35	65	...	...	9.50	183	361	247	477	General-purpose and wiping solder.
30	70	...	...	9.70	183	361	255	491	For machine and torch soldering.
25	75	...	...	10.00	183	361	266	511	For machine and torch soldering.
20	80	...	...	10.20	183	361	277	531	For coating and joining metals. For filling dents or seams in automobile bodies.
15	85	...	...	10.50	227	440	288	550	For coating and joining metals.
10	90	...	...	10.80	268	514	299	570	For coating and joining metals.
5	95	...	...	11.30	270	518	312	594	For coating and joining metals.
									Tin-lead-antimony alloys
40	58	2	...	9.23	185	365	231	448	Same uses as (50-50) tin-lead but not recommended for use on galvanized iron.
35	63.2	1.8	...	9.44	185	365	243	470	For wiping and all uses except on galvanized iron.
30	68.4	1.6	...	9.65	185	365	250	482	For torch or machine soldering, except on galvanized iron.
25	73.7	1.3	...	9.96	184	364	263	504	For torch and machine soldering, except on galvanized iron.
20	79	1	...	10.17	184	363	270	517	For machine soldering and coating metals and for tipping and like uses, but not recommended for use on galvanized iron.
									Tin-antimony alloys
95	...	5	...	7.25	234	452	240	464	For joints on copper: electrical, plumbing, and heating.

TABLE 3.56 (*Continued*)

Nominal composition, %					Melting ranges				
					Solid		Liquid		
Tin	Lead	Antimony	Silver	Specific gravity	°C	°F	°C	°F	Uses
									Silver-lead alloys
0	97.5	...	2.5	11.35	304	579	304	579	For use on copper, brass, and similar metals with torch heating. Not recommended in humid environments because of its known susceptibility to corrosion.
1	97.5	...	1.5	11.28	309	588	309	588	For use on copper, brass, and similar metals with torch heating.

SOURCE: *1974 Annual Book of ASTM Standard,* part 8 (B32-70), table 1, pp. 30–31. By permission of the American Society for Testing and Materials.

with commercial water-soluble detergent and water, or just hot water in some cases, removes flux residue. Rosin flux is used for electrical applications because it is not corrosive, but it does not effectively remove oxide film.

3.12 FINISHES

The surface finish for an item is often stated on a drawing (see Sec. 1.7), and the manufacturing technique must provide the finish. Alternatively, the technician, proceeding without a formal drawing, must judge the finish required. In either case, the manufacturing process (including drilling, turning, and polishing) has to be selected so it is inherently possible to meet the specific needs. Table 3.57 shows typical end results of various manufacturing processes.

With any particular process there are many possible finishes. Finish depends

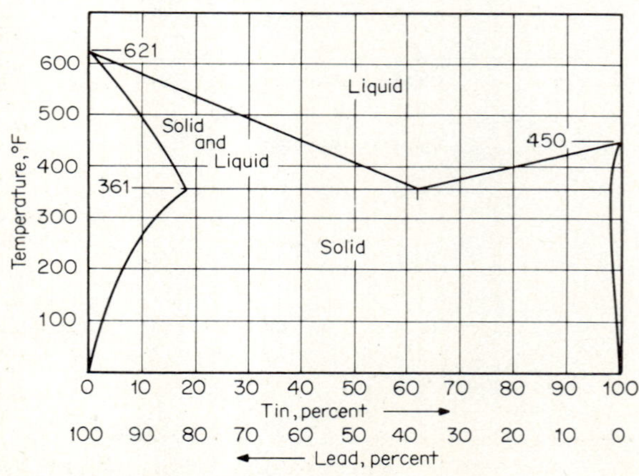

FIG. 3.64 Phase diagram for lead-tin.

TABLE 3.57 Surface Roughness Produced by Common Manufacturing Processes

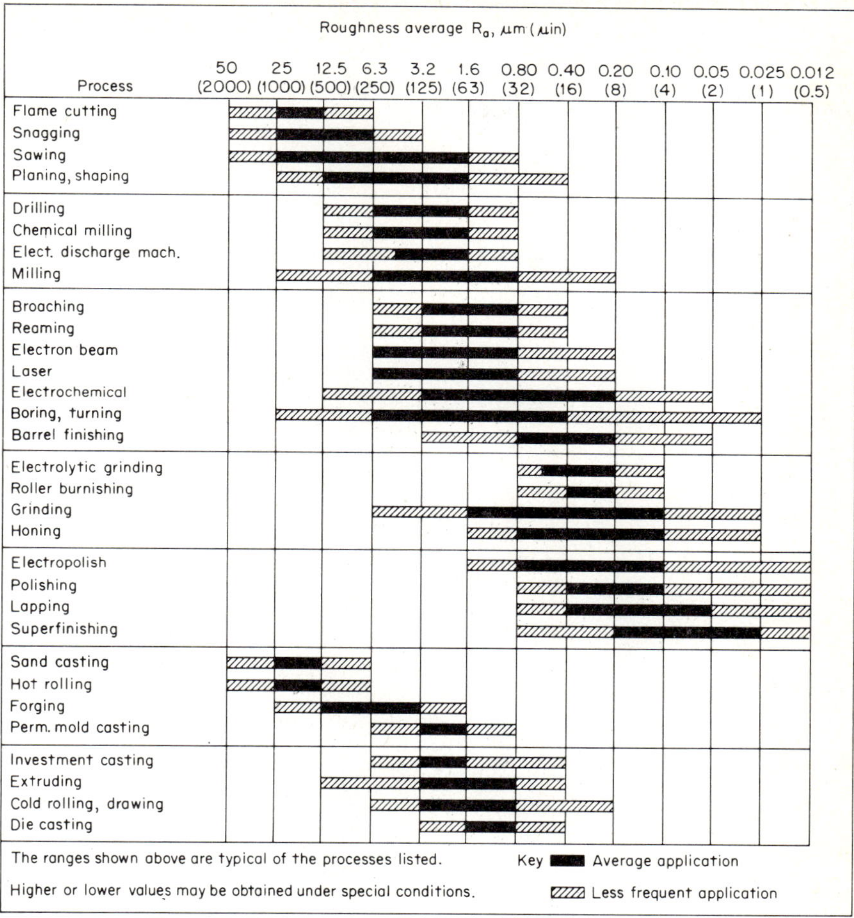

SOURCE: American Society of Mechanical Engineers, ANSI B46.1–1978.

upon the condition of the cutting tool; the condition of the equipment; the cutting speed, tool lubricant, or coolant; the material being processed; and the skill of the operator. The technician should have no problem achieving the average results in Table 3.57. To do better often requires slower speeds and feeds, frequent attention to tool condition, and maintaining the equipment in good condition. It is uneconomical to produce finishes better than required.

Surface finish can be measured with a profilometer, but usually a visual comparative gage is used. Such a gage has examples of various finishes (roughness in microinches) simulating different manufacturing processes or lay (see Fig. 1.27); thus visual comparison can be readily made. A surface-finish comparator is a necessary part of every mechanical technician's tool kit.

chapter 4

Selecting Components and Equipment

Maurice J. Webb

4.1 GENERAL CONSIDERATIONS

The variety of components and equipment that a technician could use in a mechanical laboratory or in pilot plant work is so great that some screening was done to bring the subject into manageable proportions. This was not as difficult as it might first appear and does not mean that significant topics are omitted. Usually the mechanical technician assembles and tests mechanical systems that predominantly include fluid-flow management, with more than the usual provision for instrumentation to provide the engineer or scientist with detailed information regarding the basic phenomena taking place. Emphasis is given to components and equipment available on the market because it is usually quicker to use or adapt something readily obtainable than it is to build completely. When several available items ostensibly fill the same requirements, emphasis is more on the differences and limits of applicability of each. Undoubtedly, new items will come on the market, but it should be possible to make valuable assessments regarding their suitability by referring to the attributes and limitations of those already available, thus minimizing errors in selection. Such errors are crucial elements in meeting performance and schedule requirements in research and development work. The technician is strongly advised to select only those items which provide, with high confidence, the performance sought.

4.2 THREADED FASTENERS

While the types of bolts and screws proliferate and their applications for fastening frequently overlap, there are some general rules that the technician can follow when there is no engineering drawing or sketch to specify the kind to be used. When repairing or reassembling components, it is important to use bolting of exactly the kind used on the original equipment because sometimes the type of fastener was selected with more thought than is readily apparent. If confronted with a situation where a threaded hole already exists but not the corresponding screw, merely using any screw of the right thread size is not sufficient. Consider the application in exactly the same way as though it were an original application.

In choosing a bolt or screw there are four primary considerations: the material, the size, the type of thread, and the type of head. By far the most common screw material is steel, carbon and stainless. There are many grades of steel screw offered, but specific grades should be left to the engineer to specify. The material grades are often stamped on the heads of bolts and, if replacing lost ones, then make sure the same grade is obtained by referring to the ones remaining. More generally the technician can consider the steels in three categories: ordinary or regular steel screws, high-strength steel, and stainless steel. While this is a simplification it will permit guidelines for general good practice. For structural work (e.g., erecting steel frames, displays, test fixtures) use regular steel bolts or screws. For corrosive and high-temperature (above 850°F, or 730 K) work, then stainless steel is the preferred choice. Stainless steels, particularly the 300 series, have a tendency for the threads to gall and become completely jammed or seized. This tendency can be reduced by visually inspecting the threads to make sure there are no nicks or burrs on them and by applying antigalling compounds. There are many proprietary antigalling compounds on the market, but those with a lead or silver base are generally satisfactory.

Brass is the next most common material for small screws. Its primary advantage is its nonrusting and nongalling characteristics. Brass screws are ideal for small assemblies requiring low screw loads and for low-temperature applications (below 500°F, or 530 K).

The size of screw or bolt used is determined primarily from load-magnitude and load-distribution considerations. While a simple calculation may show that a single bolt can provide the load, it is frequently obvious that a larger number of smaller bolts will distribute the load more uniformly. This is especially true for bolting components that are to be sealed against fluid pressure but also for other applications. A single or few load-bearing bolts can readily cause distortion of components being held together. The technician is cautioned against using the full load-carrying capacity of a screw or bolt as published by manufacturers. Use no more than one-quarter of the published yield load when selecting the number of bolts from tensile strength considerations. This is because other loads are present, and the bolt fails from a combination of stresses applied to it. Even an engineer has difficulty in designing a bolt to be used to its maximum combined stresses because the values of frictional force between nut face and bolted surface are so much a function of the material finishes and the lubrication applied that they are difficult or impossible to estimate.

Threads used for screws and bolts are most frequently UNC or UNF (see Tables 1.1, 3.21, and 3.22). For the smaller sizes, less than ¼-in (6-mm) diameter, the threads are designated by number (e.g., size 10). It is customary to identify the thread by both its size and pitch (for example, 10–32). The pitch determines whether it is a fine (UNF) or coarse (UNC) thread. Similarly, for screws above ¼ in (6 mm), the description includes the diameter and pitch (for example, ½–12 for a ½-in UNC thread).

The fine threads provide greater strength than the coarse ones because the material diameter at the root of the thread is larger and therefore the core of the thread has a greater area. Coarse threads are better in softer materials such as plastics, aluminum, and brass. They are also preferred for cast and brittle materials where fine threads may chip when cut. Fine threads have a greater tendency to gall, especially in aluminum and stainless steel. Overall, coarse threads are less likely to give problems.

There are many standard head designs for bolts and screws. Bolts are used for fastening parts together with the use of a nut on the threaded end which is rotated while the bolt is held stationary. Screws are turned into threaded holes. Since bolts are intended for use with nuts, there are a few head configurations specifically intended for bolts alone, and these are shown in Fig. 4.1. Head configurations used predominantly for screws are shown in Fig. 4.2. Included in both these figures are typical applications for each type head.

A special type of screw is the lag screw (Fig. 4.3). It is available with either a square or a hexagonal head. It is intended to be screwed into wood or an expansion shield. (An expansion shield is a soft metal, wood, or plastic tubular insert used for securing screws in masonry floors and walls.)

When using a nut with a bolt it should preferably be of the same material as the bolt. Most applications will use a regular nut, but if vibration or periodic loading exists it is better to use a locknut. There are many proprietary locknuts

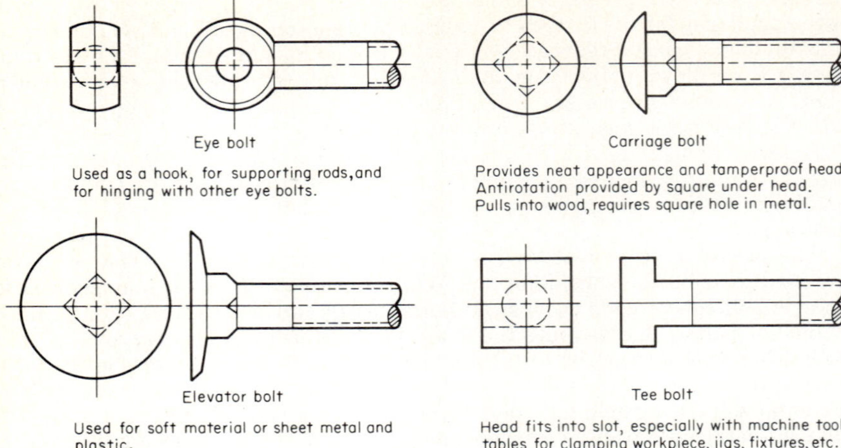

FIG. 4.1 Bolt head designs.

on the market. If a locknut is not readily available, then good locking can be achieved with either epoxy or anaerobic adhesives. Alternatively, if the bolt is long enough, then two nuts may be employed, one jammed or tightened against the other. Frequently the jam nut is a thin nut (Fig. 4.4), but it does not have to be. Two full-length nuts may be used. In any case it is the top nut that takes the load and should be full-length.

For situations where locking must be positive (i.e., no possible chance for loosening permitted), it is best to use a castellated or slotted nut. With these nuts a hole is drilled through the bolt and a cotter pin passed through both the bolt and a convenient slot in the nut so that the nut cannot possibly rotate. The ends of the pin are bent to prevent it from falling out. Figure 4.5 shows the arrangement.

Washers are used under the heads of bolts, screws, and nuts to distribute the load over a larger area so that the material under the head is not crushed. They are especially used when fastening wood and soft metals (aluminum, brass, etc.). Some washers have locking capability, too, and may serve the dual role of locking and load distribution.

Setscrews are special screws used chiefly to hold components to shafts, such as knobs, handles, pulleys, or gears. They are normally headless so that they do not protrude beyond the hub of the component (Fig. 4.6). This makes a neat installation and for rotating machinery eliminates the possibility for anything getting caught on the screw. Typically setscrews are hardened to enable them to "kite" into the shaft. Either a screwdriver slot or a socket head for an Allen (hexagonal) wrench is provided for tightening (Fig. 4.7).

There are five common setscrew tips in use, as shown in Fig. 4.6. Each tip is intended for a specific holding situation, as explained in Fig. 4.7.

4.3 PIPING

Conduits that convey fluids and contain pressures are referred to as *piping*, but piping is generally considered as being one of two types, pipes and tubes. Now

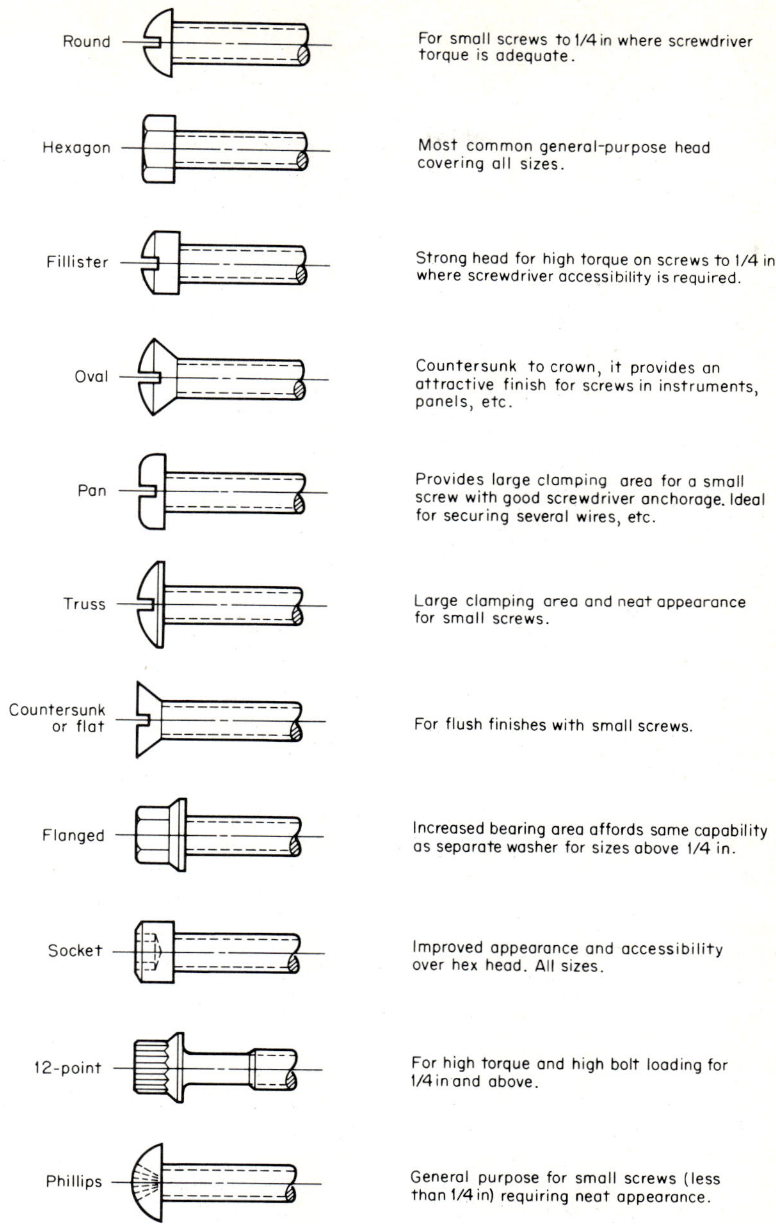

FIG. 4.2 Common screw heads and applications.

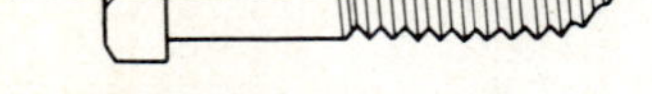

FIG. 4.3 Lag screw.

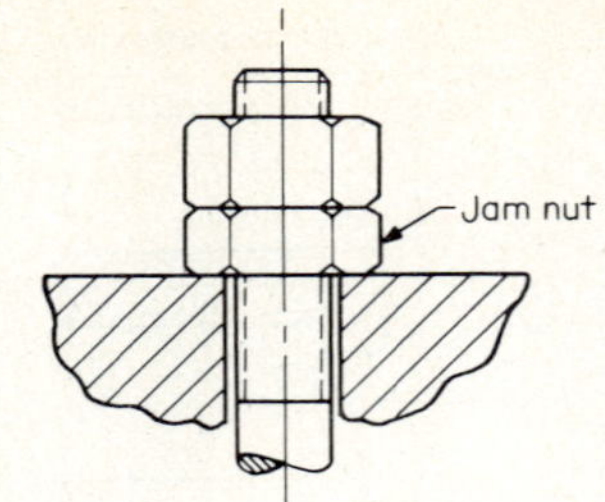

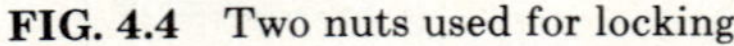

FIG. 4.4 Two nuts used for locking.

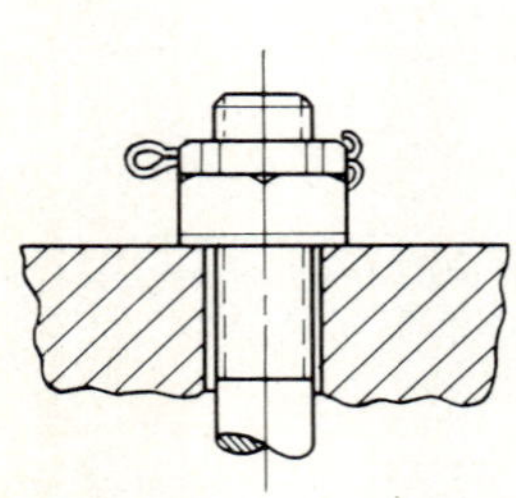

FIG. 4.5 Castellated nut used with pin.

FIG. 4.6 Setscrew used for fastening component to a shaft.

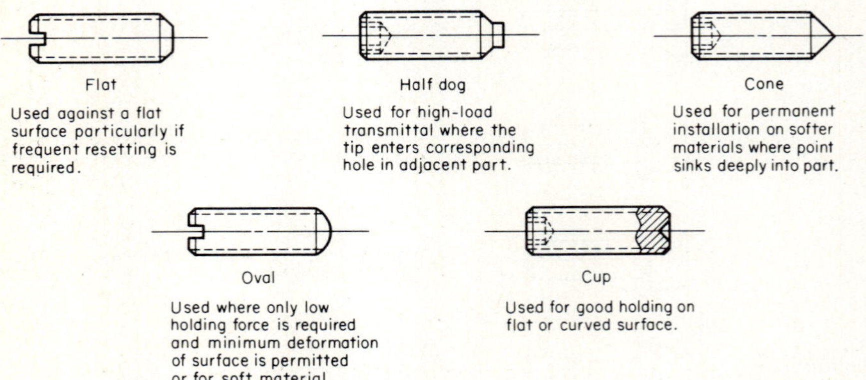

FIG. 4.7 Setscrew tips and their applications.

pipe is typically a heavy-wall conduit usually larger than 1 in (25 mm) in diameter, and tube is a thin-wall conduit usually smaller than 1 in (25 mm) in diameter. Each is specified differently. Pipe is specified by a nominal diameter or size (for example, 4-in pipe) and a wall thickness or *Schedule* (e.g., Schedule 40). The nominal pipe diameter and the actual diameter are different except for the 14-in size and larger, when they are dimensionally equal (Table 4.1). For a given Schedule the wall thickness increases with increasing size (Table 4.1) so that 2-in Schedule 40 pipe has a thinner wall than 6-in Schedule 40 pipe. However, the outside diameter of a specific nominal pipe size is fixed regardless of the Sched-

TABLE 4.1 Diameter and Wall Thickness of Standard Pipe

Nominal pipe size, in	Actual pipe OD, in (mm)	OD tolerance, in (mm) (ASTM A-53 and A-120)		Schedule 40 wall thickness, in (mm)
⅛	0.405 (10.28)	+0.0156 (0.396)	−0.0312 (0.792)	0.068 (1.73)
¼	0.540 (13.72)	+0.0156 (0.396)	−0.0312 (0.792)	0.088 (2.24)
⅜	0.675 (17.14)	+0.0156 (0.396)	−0.0312 (0.792)	0.091 (2.31)
½	0.840 (21.34)	+0.0156 (0.396)	−0.0312 (0.792)	0.109 (2.77)
¾	1.050 (26.67)	+0.0156 (0.396)	−0.0312 (0.792)	0.113 (2.87)
1	1.315 (33.40)	+0.0156 (0.396)	−0.0312 (0.792)	0.133 (3.38)
1½	1.900 (48.26)	+0.0156 (0.396)	−0.0312 (0.792)	0.145 (3.68)
2	2.375 (60.33)	±0.024 (0.610)		0.154 (3.91)
3	3.500 (88.90)	±0.035 (0.889)		0.216 (5.49)
4	4.500 (114.3)	±0.045 (1.143)		0.237 (6.02)
6	6.625 (168.3)	±0.066 (1.676)		0.280 (7.11)
8	8.625 (219.1)	±0.086 (2.184)		0.322 (8.18)
10	10.750 (273.1)	±0.108 (2.743)		0.365 (9.27)
12	12.750 (323.9)	±0.128 (3.251)		0.406 (10.3)
14	14.000 (355.6)	±0.140 (3.556)		0.438 (11.1)
16	16.000 (406.4)	±0.160 (4.06)		0.500 (12.7)
18	18.000 (457.2)	±0.180 (4.57)		0.562 (14.3)
20	20.000 (508.0)	±0.200 (5.08)		0.594 (15.1)

ule, and the internal diameter varies. This constant outside diameter for each pipe size permits the use of specific pipe thread sizes (for example, 2-in NPT) and other standard joining techniques and devices.

The common and standard Schedules are 40, 80, and 160. An extra heavy wall pipe is available in many sizes and is referred to as *Double Extra Strong* or *XX Strong*. Standard schedules alone do not always permit economical piping systems, and other thinner wall Schedules have been adopted from 5 to 30. However, pipe wall below Schedule 40 is not recommended for threading because the wall remaining below the thread root is too small.

Tubing is specified by its actual outside diameter (OD) and wall thickness. For example, a common tube size is ½-in (12-mm) OD by 0.049-in (1.2-mm) wall. Tubing is available in many sizes (for example, 24 in, or 600 mm). Another distinguishing characteristic between tube and pipe is the precision to which it is manufactured. Tubing is usually much more precise than pipe in both outside diameter and wall thickness. For example, a typical 1-in pipe (OD 1.315 in, or 33.4 mm) may have a total diametral tolerance of 0.047 in (1.2 mm), whereas a 1-in (25.4 mm) tube has a tolerance of only 0.012 in (0.3 mm). In keeping with this closer tolerance, tubing has a superior surface finish also.

Pipe is commonly available in carbon steel, stainless steel, aluminum alloy, and plastic. Tubing is readily available in the same materials plus copper. Steel pipe and tubing are manufactured by one of two processes, welded seam, referred to simply as *welded*, and *seamless*. The welded material is typically less expensive and usually has a different pressure capability.

The first consideration in selecting piping is compatibility with the fluid that will be passing through it. The second consideration is the size (internal diameter) of the pipe to carry the fluid, which is usually determined from the acceptable pressure drop for flow (see Sec. 8.8). Even if there is essentially no fluid flow, as for many instrumentation lines, it is unusual to use tubing below ¼-in (6-mm) OD. The third consideration is the piping wall thickness or Schedule. This is determined primarily from the working pressure and temperature of the fluid. Tables 4.2 through 4.17 give maximum working pressures at various fluid (piping) temperatures for pipes and tubes of common materials. These tables are for commonly available grades of material, and the values given use the safety factors and material strengths recognized by national codes and standards.

TABLE 4.2 Maximum Recommended Working Pressure, psi (MPa), for Nylon Tubing and Polyethylene Tubing at Room Temperature

Tubing OD, in (mm)	Nylon tubing					Polyethylene tubing			
	Tube wall thickness, in (mm)								
	0.030 (0.762)	0.045 (1.14)	0.062 (1.57)	0.080 (2.03)	0.125 (3.18)	0.045 (1.14)	0.062 (1.57)	0.090 (2.29)	0.125 (3.18)
3⁄16 (4.76)	530 (3.65)	790 (5.45)				190 (1.31)			
¼ (6.35)	400 (2.76)	590 (4.07)	810 (5.58)			140 (0.966)	195 (1.34)		
⅜ (9.53)	260 (1.79)	390 (2.69)	540 (3.72)	700 (4.83)		95 (0.655)	130 (0.896)	190 (1.31)	
½ (12.7)		300 (2.07)	410 (2.83)	530 (3.65)	820 (5.65)	70 (0.483)	100 (0.689)	140 (0.966)	200 (1.38)
¾ (19.05)			270 (1.86)	350 (2.41)	550 (3.79)	45 (0.310)	65 (0.448)	95 (0.655)	130 (0.986)
1 (25.4)			200 (1.38)	260 (1.79)	410 (2.83)		50 (0.345)	70 (0.483)	100 (0.689)

TABLE 4.3 Maximum Recommended Working Pressure, psi (MPa), for Laboratory-Grade Tygon Tubing at Room Temperature

Tubing OD, in (mm)	Tube wall thickness, in (mm)				
	0.062 (1.57)	0.094 (2.39)	0.125 (3.18)	0.188 (4.78)	0.250 (6.35)
3/16 (4.76)	75 (0.517)				
1/4 (6.35)	65 (0.448)				
3/8 (9.53)	55 (0.379)	65 (0.448)			
1/2 (12.7)	45 (0.310)	60 (0.414)	70 (0.483)		
3/4 (19.05)	...	40 (0.276)	45 (0.310)		
1 (25.4)	...	...	35 (0.241)		
1½ (38.1)	...	...	20 (0.138)	35 (0.241)	45 (0.310)

These working-pressure values assume that no stresses other than those caused by the fluid pressure are present in the piping, but this is seldom true in typical applications. Additional stresses are imposed upon the piping from supports, from components (valves, etc.), from thermal expansion, and from mechanical use or handling loads, all of which reduce the piping working pressure. It is not unusual to find from pressure considerations alone that very thin wall tubing or pipe is acceptable but that such piping is so easily damaged (dented, kinked, bent, etc.) as to render it unsuitable. As a guide for general-purpose work it is best not to use pipe below Schedule 10 to 6-in (150-mm) nominal size and Schedule 30 above 6-in (150-mm) size, or tubing to ½-in (12-mm) OD with a wall less than 0.035 in (0.9 mm) and from ½- to 1-in (12- to 25-mm) OD with a wall less than 0.049 in (1.2 mm).

4.4 PIPE AND TUBE COUPLINGS AND FITTINGS

Fluid systems and associated instrumentation require extensive joining of pipes and tubes to one another as well as to components and instruments. To facilitate this joining a variety of devices, sometimes quite ingenious, are marketed. These joining devices are termed *fittings* or *couplings* and should always be used in piping installations. They provide the most reliable and the safest way to make piping connections.

Because there is a marked difference in the surface finish and dimensional tolerances between pipes and tubes, two entirely different systems of fittings and couplings have evolved and are considered separately here. In many instances, however, it is preferred to change from piping to tubing in different sections of

TABLE 4.4 Maximum Working Pressure, psi (MPa), for Seamless Pipe with Threaded Connections, Carbon Steel SA-106 Grade C

	Temperature											
	−20 to 650°F (244 to 617 K)				700°F (644 K)				800°F (700 K)			
	Schedule											
Pipe size, in	40	80	160	XXS	40	80	160	XXS	40	80	160	XXS
⅛	3640 (25.1)	6562 (45.2)			3453 (23.8)	6225 (42.9)			2496 (17.2)	4500 (31.0)		
¼	3045 (21.0)	5460 (37.6)			2888 (19.9)	5179 (35.7)			2088 (14.4)	3744 (25.8)		
½	2275 (15.7)	4095 (28.2)	6177 (42.6)	12,757 (88.0)	2158 (14.9)	3884 (26.8)	5860 (40.4)	12,101 (83.4)	1560 (10.8)	2808 (19.4)	4236 (29.2)	8748 (60.3)
¾	1942 (13.4)	3482 (24.0)	6107 (42.1)	10,342 (71.3)	1843 (12.7)	3303 (22.8)	5793 (39.9)	9811 (67.6)	1332 (9.18)	2388 (16.5)	4188 (28.9)	7092 (48.9)
1	1767 (12.2)	3132 (21.6)	5407 (37.3)	9292 (64.1)	1676 (11.6)	2971 (20.5)	5129 (35.4)	8815 (60.8)	1212 (8.36)	2148 (14.8)	3708 (25.6)	6372 (43.9)
1¼	1540 (10.6)	2730 (18.8)	4182 (28.8)	7770 (53.6)	1461 (10.1)	2590 (17.9)	3967 (27.4)	7370 (50.8)	1056 (7.28)	1872 (12.9)	2868 (19.8)	5328 (36.7)
1½	1452 (10.0)	2555 (17.6)	4287 (29.6)	7052 (48.6)	1378 (9.50)	2424 (16.7)	4067 (28.0)	6690 (46.1)	996 (6.87)	1752 (12.1)	2940 (20.3)	4836 (33.3)
2	1295 (8.93)	2310 (15.9)	4445 (30.6)	6142 (42.3)	1228 (8.47)	2191 (15.1)	4216 (29.1)	5827 (40.1)	888 (6.12)	1584 (10.9)	3048 (21.0)	4212 (29.0)

TABLE 4.5 Recommended Maximum Working Pressure, psi (MPa), for Seamless Pipe with Welded Connections, Carbon Steel SA-135 Grade B

	Temperature											
	−20 to 650°F (244 to 617 K)				700°F (644 K)				800°F (700 K)			
	Schedule											
Pipe size, in	40	80	160	XXS	40	80	160	XXS	40	80	160	XXS
⅛	4928 (34.0)	7328 (50.5)			4734 (32.6)	7039 (48.5)			3570 (24.6)	5308 (36.6)		
¼	4762 (32.8)	6794 (46.8)			4575 (31.5)	6527 (45.0)			3450 (23.8)	4922 (33.9)		
½	3683 (25.4)	5169 (35.6)	6883 (47.5)	12,344 (85.1)	3538 (24.4)	4965 (34.2)	6612 (45.6)	8894 (61.3)	2668 (18.4)	3744 (25.8)	4986 (34.4)	6707 (46.2)
¾	2997 (20.7)	4216 (29.1)	6325 (43.6)	9728 (67.1)	2879 (19.8)	4050 (27.9)	6076 (41.9)	7210 (49.7)	2171 (15.0)	3054 (21.1)	4582 (31.6)	5437 (37.5)
1	2794 (19.3)	3873 (26.7)	5690 (39.2)	8839 (60.9)	2684 (18.5)	3721 (25.7)	5466 (37.7)	6478 (44.7)	2024 (14.0)	2806 (19.3)	4122 (28.4)	4885 (33.7)
1¼	2299 (15.9)	3213 (22.2)	4343 (29.9)	7163 (49.4)	2208 (15.2)	3087 (21.3)	4172 (28.8)	5417 (37.3)	1665 (11.5)	2328 (16.1)	3146 (21.7)	4085 (28.2)
1½	2070 (14.3)	2921 (20.1)	4254 (29.3)	6426 (44.3)	1989 (13.7)	2806 (19.3)	4087 (28.2)	4917 (33.9)	1499 (10.3)	2116 (14.6)	3082 (21.2)	3708 (25.6)
2	1740 (12.0)	2515 (17.3)	4153 (28.6)	5461 (37.7)	1671 (11.5)	2416 (16.7)	3989 (27.5)	4282 (29.5)	1260 (8.69)	1822 (12.6)	3008 (20.7)	3229 (22.3)

TABLE 4.6 Maximum Allowable Working Pressure, psi (MPa), for Seamless Pipe with Welded Connections, Aluminum Alloy 6061-T6 ASTM-SB241

	Temperature					
	100 to 200°F (311 to 367 K)			400°F (478 K)		
	Schedule					
Pipe size, in	40	80	160	40	80	160
⅛	2328 (16.1)	3462 (23.9)		1358 (9.36)	2019 (13.9)	
¼	2250 (15.5)	3210 (22.1)		1312 (9.05)	1872 (12.9)	
½	1740 (12.0)	2442 (16.8)	3252 (22.4)	1015 (7.00)	1424 (9.82)	1897 (13.1)
¾	1416 (9.76)	1992 (13.7)	2988 (20.6)	826 (5.70)	1162 (8.01)	1743 (12.0)
1	1320 (9.10)	1830 (12.6)	2688 (18.5)	770 (5.31)	1067 (7.36)	1568 (10.8)
1¼	1086 (7.49)	1518 (10.9)	2052 (14.4)	633 (4.36)	855 (5.89)	1197 (8.25)
1½	978 (6.74)	1380 (9.51)	2010 (13.9)	570 (3.93)	805 (5.55)	1172 (8.08)
2	822 (5.67)	1188 (8.19)	1962 (13.5)	479 (3.30)	693 (4.78)	1144 (7.89)

TABLE 4.7 Maximum Allowable Working Pressure, psi (MPa), for Seamless Pipe with Threaded Connections, Aluminum Alloy 6061-T6 ASTM-SB241

	Temperature					
	100 to 200°F (311 to 367 K)			400°F (478 K)		
	Schedule					
Nominal pipe size, in	40	80	160	40	80	160
⅛	2184 (15.1)	3937 (27.1)		915 (6.31)	1650 (11.4)	
¼	1827 (12.6)	3276 (22.6)		766 (5.28)	1373 (9.47)	
½	1365 (9.41)	2457 (16.9)	3706 (25.6)	572 (3.94)	1030 (7.10)	1553 (10.7)
¾	1165 (8.03)	2089 (14.4)	3664 (25.3)	488 (3.36)	876 (6.04)	1536 (10.6)
1	1060 (7.31)	1879 (13.0)	3244 (22.4)	444 (3.06)	788 (5.43)	1360 (9.38)
1¼	924 (6.37)	1638 (11.3)	2509 (17.3)	387 (2.67)	686 (4.73)	1052 (7.25)
1½	871 (6.01)	1533 (10.6)	2572 (17.7)	365 (2.52)	642 (4.43)	1078 (7.43)
2	777 (5.34)	1386 (9.56)	2667 (18.4)	326 (2.25)	581 (4.01)	1118 (7.71)

TABLE 4.8 Recommended Maximum Working Pressure, psi (MPa), for 304 Stainless-Steel Welded Tubing, ASTM A-249 or Equivalent

Tubing OD, in (mm)	Temp., °F (K)	Wall thickness, in (mm)								
		0.020 (0.508)	0.028 (0.711)	0.035 (0.889)	0.049 (1.245)	0.065 (1.651)	0.083 (2.108)	0.095 (2.413)	0.109 (2.769)	0.120 (3.048)
⅛ (3.175)	−20 to 100 (244 to 311)	5835 (40.2)	8475 (58.4)	10,748 (74.0)						
	400 (478)	5028 (34.6)	7302 (50.3)	9261 (63.8)						
	600 (589)	4955 (34.1)	7196 (49.6)	9126 (62.9)						
	800 (700)	4734 (32.6)	6876 (47.4)	8720 (60.1)						
3/16 (4.763)	−20 to 100 (244 to 311)	3705 (25.6)	5390 (37.1)	6932 (47.7)		13,165 (90.7)				
	400 (478)	3192 (22.0)	4644 (32.0)	5973 (41.1)	8617 (59.4)	11,344 (78.1)				
	600 (589)	3146 (21.7)	4577 (31.6)	5886 (40.6)	8492 (58.5)	11,178 (77.0)				
	800 (700)	3006 (20.7)	4373 (30.1)	5624 (38.7)	8114 (55.9)	10,681 (73.6)				
¼ (6.35)	−20 to 100 (244 to 311)	2719 (18.7)	3911 (26.9)	5009 (34.5)	7314 (50.4)	9938 (68.5)				
	400 (478)	2343 (16.1)	3370 (23.2)	4315 (29.8)	6302 (43.4)	8562 (59.0)				
	600 (589)	2309 (15.9)	3321 (22.8)	4253 (29.3)	6210 (42.8)	8438 (58.2)				

	800 (700)	2206 (15.2)	3173 (21.8)	4063 (28.0)	5934 (40.9)	8062 (55.5)				
⁵⁄₁₆ (7.938)	−20 to 100 (244 to 311)	2147 (14.8)	3069 (21.1)	3911 (26.9)	5708 (39.3)	7807 (53.8)				
	400 (478)	1849 (12.7)	2644 (18.2)	3370 (23.2)	4918 (33.9)	6727 (46.4)				
	600 (589)	1823 (12.5)	2606 (18.0)	3321 (22.9)	4847 (33.4)	6629 (45.7)				
	800 (700)	1741 (12.0)	2490 (17.2)	3173 (21.8)	4631 (31.9)	6334 (43.6)				
⅜ (9.525)	−20 to 100 (244 to 311)		2528 (17.4)	3212 (22.1)	4643 (32.0)	6392 (44.0)	8364 (57.6)			
	400 (478)		2178 (15.0)	2767 (19.1)	4000 (27.6)	5507 (37.9)	7206 (49.6)			
	600 (589)		2147 (14.8)	2727 (18.8)	3942 (27.1)	5427 (37.4)	7101 (48.9)			
	800 (700)		2051 (14.1)	2606 (18.0)	3767 (26.0)	5186 (35.7)	6785 (46.8)			
½ (12.7)	−20 to 100 (244 to 311)		1860 (12.8)	2353 (16.2)	3380 (23.3)	4611 (31.8)	6090 (42.0)	7576 (52.2)		
	400 (478)		1603 (11.0)	2028 (14.0)	2913 (20.0)	3073 (21.1)	5247 (36.2)	6096 (42.0)		
	600 (589)		1580 (10.9)	1998 (13.8)	2870 (19.8)	3915 (27.0)	5171 (35.6)	6008 (41.4)		
	800 (700)		1509 (10.4)	1909 (13.2)	2742 (18.9)	3741 (25.8)	4941 (34.0)	5740 (39.5)		
⅝ (15.88)	−20 to 100 (244 to 311)			1860 (12.8)	2655 (18.3)	3609 (24.9)	4722 (32.5)	5501 (37.9)	6424 (44.2)	

TABLE 4.8 (*Continued*)

Tubing OD, in (mm)	Temp., °F (K)	Wall thickness, in (mm)								
		0.020 (0.508)	0.028 (0.711)	0.035 (0.889)	0.049 (1.245)	0.065 (1.651)	0.083 (2.108)	0.095 (2.413)	0.109 (2.769)	0.120 (3.048)
	400 (478)			1603 (11.0)	2288 (15.8)	3110 (21.4)	4069 (28.0)	4740 (32.7)	5535 (38.1)	
	600 (589)			1580 (10.9)	2255 (15.5)	3065 (21.1)	4010 (27.6)	4671 (32.2)	5454 (37.6)	
	800 (700)			1509 (10.4)	2154 (14.8)	2928 (20.2)	3831 (26.4)	4463 (30.7)	5212 (35.9)	
¾ (19.05)	−20 to 100 (244 to 311)			1542 (10.6)	2194 (15.1)	2957 (20.4)	3864 (26.6)	4484 (30.9)	5231 (36.0)	5835 (40.2)
	400 (478)			1329 (9.2)	1891 (13.0)	2548 (17.6)	3329 (22.9)	3863 (26.6)	4507 (31.1)	5028 (34.6)
	600 (589)			1310 (9.0)	1863 (12.8)	2511 (17.3)	3281 (22.6)	3807 (26.2)	4442 (30.6)	4955 (34.2)
	800 (700)			1251 (8.6)	1780 (12.3)	2399 (16.5)	3135 (21.6)	3638 (25.1)	4244 (29.2)	4734 (32.3)
⅞ (22.23)	−20 to 100 (244 to 311)			1313 (9.0)	1860 (12.8)	2512 (17.3)	3260 (22.5)	3784 (26.0)	4404 (30.3)	4897 (33.7)
	400 (478)			1132 (7.8)	1603 (11.0)	2165 (14.9)	2808 (19.4)	3261 (22.5)	3795 (26.1)	4220 (29.1)
	600 (589)			1115 (7.7)	1580 (10.9)	2133 (14.7)	2768 (19.1)	3213 (22.1)	3740 (25.8)	4158 (28.7)
	800 (700)			1065 (7.3)	1509 (10.4)	2038 (14.1)	2644 (18.2)	3070 (21.2)	3573 (24.6)	3973 (27.4)

1 (25.4)	−20 to 100 (244 to 311)			1145 (7.9)	1622 (11.2)	2178 (15.0)	2830 (19.5)	3275 (22.6)	3800 (26.2)	4214 (29.0)
	400 (478)			986 (6.8)	1397 (9.6)	1877 (12.9)	2439 (16.8)	2822 (19.4)	3274 (22.5)	3630 (25.0)
	600 (589)			972 (6.7)	1377 (9.5)	1850 (12.7)	2403 (16.5)	2781 (19.1)	3227 (22.2)	3578 (24.7)
	800 (700)			929 (6.4)	1316 (9.1)	1767 (12.2)	2296 (15.8)	2657 (18.3)	3083 (21.2)	3418 (23.6)

TABLE 4.9 Allowable Working Pressure, psi (MPa), for 304 Stainless-Steel Annealed Seamless Tubing, ASTM A-213 or Equivalent

Tubing OD, in (mm)	Temp., °F (K)	Wall thickness, in (mm)								
		0.020 (0.508)	0.028 (0.711)	0.035 (0.889)	0.049 (1.245)	0.065 (1.651)	0.083 (2.108)	0.095 (2.413)	0.109 (2.769)	0.120 (3.048)
⅛ (3.175)	−20 to 100 (244 to 311)	6863 (47.3)	9967 (68.7)	12,641 (87.2)						
	400 (478)	5909 (40.7)	8581 (59.2)	10,884 (75.0)						
	600 (589)	5835 (40.2)	8475 (58.4)	10,748 (74.1)						
	800 (700)	5542 (38.2)	8048 (55.5)	10,208 (70.4)						
3/16 (4.763)	−20 to 100 (244 to 311)	4357 (30.0)	6339 (43.7)	8153 (56.2)	11,762 (81.1)	15,484 (106.8)				
	400 (478)	3751 (25.9)	5458 (37.6)	7020 (48.4)	10,127 (69.8)	13,331 (91.9)				
	600 (589)	3705 (25.5)	5390 (37.2)	6932 (47.8)	10,001 (69.0)	13,165 (90.8)				
	800 (700)	3518 (24.3)	5119 (35.3)	6584 (45.4)	9498 (65.5)	12,503 (86.2)				
¼ (6.35)	−20 to 100 (244 to 311)	3198 (22.0)	4600 (31.7)	5891 (40.6)	8602 (59.3)	11,688 (80.6)				
	400 (478)	2753 (19.0)	3961 (27.3)	5071 (35.0)	7406 (51.1)	10,062 (69.4)				
	600 (589)	2719 (18.7)	3911 (27.0)	5009 (34.5)	7314 (50.4)	9938 (68.5)				

	800 (700)	2582 (17.8)	3715 (25.6)	4756 (32.8)	6946 (47.9)	9437 (65.1)				
⁵⁄₁₆ (7.938)	−20 to 100 (244 to 311)	2525 (17.4)	3609 (24.9)	4600 (31.7)	6713 (46.3)	9182 (63.3)				
	400 (478)	2173 (15.0)	3107 (21.4)	3961 (27.3)	5780 (39.9)	7905 (54.5)				
	600 (589)	2147 (14.8)	3069 (21.2)	3911 (27.0)	5708 (39.4)	7807 (53.8)				
	800 (700)	2038 (14.1)	2914 (20.1)	3715 (25.6)	5421 (37.4)	7414 (51.1)				
⅜ (9.525)	−20 to 100 (244 to 311)		2973 (20.5)	3777 (26.0)	5460 (37.6)	7517 (51.8)	9836 (38.3)			
	400 (478)		2560 (17.7)	3252 (22.4)	4701 (32.4)	6472 (44.6)	8469 (33.0)			
	600 (589)		2528 (17.4)	3212 (22.1)	4643 (32.0)	7392 (44.1)	8363 (32.6)			
	800 (700)		2401 (16.6)	3050 (21.0)	4409 (30.4)	6070 (41.9)	7943 (30.9)			
½ (12.7)	−20 to 100 (244 to 311)		2188 (15.1)	2768 (19.1)	3976 (27.4)	5423 (37.4)	7162 (27.9)	8322 (57.4)		
	400 (478)		1884 (13.0)	2383 (16.4)	3423 (23.6)	4669 (32.2)	6166 (24.0)	7164 (49.4)		
	600 (589)		1860 (12.8)	2353 (16.2)	3380 (23.3)	4611 (31.8)	6090 (23.7)	7076 (48.8)		
	800 (700)		1767 (12.2)	2235 (15.4)	3210 (22.1)	4379 (30.2)	5783 (22.5)	6719 (46.3)		
⅝ (15.88)	−20 to 100 (244 to 311)			2188 (15.1)	3123 (21.5)	4245 (29.3)	5554 (21.6)	6470 (44.6)	7555 (52.1)	

TABLE 4.9 (*Continued*)

Tubing OD, in (mm)	Temp., °F (K)	Wall thickness, in (mm)								
		0.020 (0.508)	0.028 (0.711)	0.035 (0.889)	0.049 (1.245)	0.065 (1.651)	0.083 (2.108)	0.095 (2.413)	0.109 (2.769)	0.120 (3.048)
	400 (478)			1884 (13.0)	2689 (18.5)	3655 (25.2)	4782 (18.6)	5571 (38.4)	6504 (44.8)	
	600 (589)			1860 (12.8)	2655 (18.3)	3609 (24.9)	4722 (18.4)	5501 (37.9)	6424 (44.3)	
	800 (700)			1767 (12.2)	2522 (17.4)	3428 (23.6)	4485 (17.5)	5225 (36.0)	6100 (42.1)	
¾ (19.05)	−20 to 100 (244 to 311)			1814 (12.5)	2581 (17.8)	3478 (24.0)	4544 (17.7)	5273 (36.4)	6152 (42.4)	6863 (47.3)
	400 (478)			1562 (10.8)	2222 (15.3)	2995 (20.6)	3912 (15.2)	4540 (31.3)	5297 (36.5)	5909 (40.7)
	600 (589)			1542 (10.6)	2194 (15.1)	2957 (20.4)	3864 (15.0)	4484 (30.9)	5231 (36.1)	5835 (40.2)
	800 (700)			1465 (10.1)	2084 (14.4)	2809 (19.4)	3669 (14.3)	4258 (29.4)	4968 (34.3)	5542 (38.2)
⅞ (22.23)	−20 to 100 (244 to 311)			1545 (10.7)	2188 (15.1)	2955 (20.4)	3834 (14.9)	4451 (30.7)	5180 (35.7)	5760 (39.7)
	400 (478)			1330 (9.2)	1884 (13.0)	2544 (17.5)	3300 (12.9)	3832 (26.4)	4460 (30.8)	4959 (34.2)
	600 (589)			1313 (9.1)	1860 (12.8)	2512 (17.3)	3260 (12.7)	3784 (26.1)	4404 (30.4)	4897 (33.8)
	800 (700)			1247 (8.6)	1767 (12.2)	2386 (16.5)	3095 (12.1)	3594 (24.8)	4183 (28.8)	4651 (32.1)

1 (25.4)	−20 to 100 (244 to 311)			1346 (9.3)	1907 (13.1)	2562 (17.7)	3329 (13.0)	3852 (26.6)	4469 (30.8)	4956 (34.2)
	400 (478)			1159 (8.0)	1642 (11.3)	2206 (15.2)	2866 (11.2)	3317 (22.9)	3848 (26.5)	4266 (29.4)
	600 (589)			1145 (7.9)	1622 (11.2)	2178 (15.0)	2830 (11.0)	3275 (22.6)	3800 (26.2)	4214 (29.1)
	800 (700)			1087 (7.5)	1540 (10.6)	2069 (14.3)	2688 (10.5)	3111 (21.4)	3609 (24.9)	4001 (27.6)

TABLE 4.10 Allowable Working Pressure, psi (MPa), for 316 Stainless-Steel Welded Tubing, ASTM A-249 or Equivalent

Tubing OD, in (mm)	Temp., °F (K)	Wall thickness, in (mm)								
		0.020 (0.508)	0.028 (0.711)	0.035 (0.889)	0.049 (1.245)	0.065 (1.651)	0.083 (2.108)	0.095 (2.413)	0.109 (2.769)	0.120 (3.048)
⅛ (3.175)	−20 to 100 (244 to 311)	5835 (40.2)	8475 (58.4)	10,748 (74.1)						
	400 (478)	5615 (38.7)	8155 (56.2)	10,343 (71.3)						
	600 (589)	5285 (36.4)	7675 (52.9)	9734 (67.1)						
	800 (700)	4918 (33.9)	7142 (49.2)	9058 (62.5)						
3⁄16 (4.763)	−20 to 100 (244 to 311)	3705 (25.5)	5390 (37.2)	6932 (47.8)	10,001 (69.0)	13,165 (90.8)				
	400 (478)	3565 (24.6)	5187 (35.8)	6671 (46.0)	9624 (66.4)	12,668 (87.3)				
	600 (589)	3355 (23.1)	4882 (33.7)	6278 (43.3)	9058 (62.5)	11,923 (82.2)				
	800 (700)	3122 (21.5)	4543 (31.3)	5842 (40.3)	8429 (58.1)	11,095 (76.5)				
¼ (6.35)	−20 to 100 (244 to 311)	2719 (18.7)	3911 (27.0)	5009 (34.5)	7314 (50.4)	9938 (68.5)				
	400 (478)	2616 (18.0)	3764 (26.0)	4819 (33.2)	7038 (48.5)	9562 (65.9)				
	600 (589)	2462 (17.0)	3542 (24.4)	4536 (31.3)	6624 (45.7)	9000 (62.1)				

	800 (700)	2291 (15.8)	3296 (22.7)	4221 (29.1)	6164 (42.5)	8375 (57.7)				
5/16 (7.938)	−20 to 100 (244 to 311)	2147 (14.8)	3069 (21.2)	3911 (27.0)	5708 (39.4)	7807 (53.8)				
	400 (478)	2065 (14.2)	2953 (20.4)	3764 (26.0)	5498 (37.9)	7512 (51.8)				
	600 (589)	1944 (13.4)	2779 (19.2)	3542 (24.4)	5170 (35.6)	7070 (48.7)				
	800 (700)	1809 (12.5)	2586 (17.8)	3296 (22.7)	4811 (33.2)	6579 (45.4)				
3/8 (9.525)	−20 to 100 (244 to 311)		2528 (17.4)	3212 (22.1)	4643 (32.0)	6392 (44.1)	8364 (57.7)			
	400 (478)		2433 (16.8)	3091 (21.3)	4468 (30.8)	6151 (42.4)	8048 (55.5)			
	600 (589)		2290 (15.8)	2909 (20.1)	4205 (29.0)	5789 (39.9)	7574 (52.2)			
	800 (700)		2131 (14.7)	2707 (18.7)	3913 (27.0)	5387 (37.1)	7048 (48.6)			
1/2 (12.7)	−20 to 100 (244 to 311)		1860 (12.8)	2353 (16.2)	3380 (23.3)	4611 (31.8)	6090 (42.0)	7576 (52.2)		
	400 (478)		1790 (12.3)	2065 (14.2)	2264 (15.6)	3253 (22.4)	4437 (230.6)	5860 (40.4)		
	600 (589)		1685 (11.6)	2131 (14.7)	3061 (21.1)	4176 (28.8)	5515 (38.0)	6408 (44.2)		
	800 (700)		1568 (10.8)	1983 (13.7)	2849 (19.6)	3886 (26.8)	5132 (35.4)	5963 (41.1)		
5/8 (15.88)	−20 to 100 (244 to 311)			1860 (12.8)	2655 (18.3)	3609 (24.9)	4722 (32.6)	5501 (37.9)	6424 (44.3)	

TABLE 4.10 *(Continued)*

Tubing OD, in (mm)	Temp., °F (K)	Wall thickness, in (mm)								
		0.020 (0.508)	0.028 (0.711)	0.035 (0.889)	0.049 (1.245)	0.065 (1.651)	0.083 (2.108)	0.095 (2.413)	0.109 (2.769)	0.120 (3.048)
	400 (478)			1790 (12.3)	2555 (17.6)	3473 (23.9)	4544 (31.3)	5294 (36.5)	6181 (42.6)	
	600 (589)			1685 (11.6)	2405 (16.6)	3269 (22.5)	4277 (29.5)	4982 (34.3)	5818 (40.1)	
	800 (700)			1568 (10.8)	2238 (15.4)	3042 (21.0)	3980 (27.4)	4636 (32.0)	5414 (37.3)	
¾ (19.05)	−20 to 100 (244 to 311)			1542 (10.6)	2194 (15.1)	2957 (20.4)	3864 (26.6)	4484 (30.9)	5231 (36.1)	5835 (40.2)
	400 (478)			1484 (10.2)	2111 (14.6)	2846 (19.6)	3718 (25.6)	4315 (29.8)	5034 (34.7)	5615 (38.7)
	600 (589)			1397 (9.6)	1987 (13.7)	2678 (18.5)	3499 (24.1)	4061 (28.0)	4738 (32.7)	5285 (36.4)
	800 (700)			1300 (9.0)	1849 (12.7)	2492 (17.2)	3256 (22.4)	3779 (26.1)	4409 (30.4)	4918 (33.9)
⅞ (22.23)	−20 to 100 (244 to 311)			1313 (9.1)	1860 (12.8)	2512 (17.3)	3260 (22.5)	3784 (26.1)	4404 (30.4)	4897 (33.8)
	400 (478)			1264 (8.7)	1790 (12.3)	2417 (16.7)	3136 (21.6)	3641 (25.1)	4238 (29.2)	4712 (32.5)
	600 (589)			1189 (8.2)	1685 (11.6)	2275 (15.7)	2952 (20.4)	3427 (23.6)	3989 (27.5)	4435 (30.6)
	800 (700)			1107 (7.6)	1568 (10.8)	2117 (14.6)	2747 (18.9)	3189 (22.0)	3712 (25.6)	4127 (28.5)

1 (25.4)	−20 to 100 (244 to 311)			1145 (7.9)	1622 (11.2)	2178 (15.0)	2830 (19.5)	3275 (22.6)	3800 (26.2)	4214 (29.1)
	400 (478)			1102 (7.6)	1561 (10.8)	2096 (14.5)	2723 (18.8)	3152 (21.7)	3657 (25.2)	4054 (28.0)
	600 (589)			1037 (7.1)	1469 (10.1)	1973 (13.6)	2563 (17.7)	2966 (20.4)	3442 (23.7)	3816 (26.3)
	800 (700)			965 (6.7)	1367 (9.4)	1836 (12.7)	2385 (16.4)	2760 (19.0)	3203 (22.1)	3551 (24.5)

TABLE 4.11 Allowable Working Pressure, psi (MPa), for 316 Stainless-Steel Annealed Seamless Tubing, ASTM A-213 or Equivalent

Tubing OD, in (mm)	Temp., °F (K)	Wall thickness, in (mm)								
		0.020 (0.508)	0.028 (0.711)	0.035 (0.889)	0.049 (1.245)	0.065 (1.651)	0.083 (2.108)	0.095 (2.413)	0.109 (2.769)	0.120 (3.048)
⅛ (3.175)	−20 to 100 (244 to 311)	6863 (47.3)	9967 (68.7)	12,641 (87.2)						
	400 (478)	6606 (45.5)	9594 (66.1)	12,168 (83.9)						
	600 (589)	6239 (43.0)	9061 (62.5)	11,492 (79.2)						
	800 (700)	5799 (40.0)	8421 (58.1)	10,681 (73.6)						
3/16 (4.763)	−20 to 100 (244 to 311)	4357 (30.0)	6339 (43.7)	8153 (56.2)	11,762 (81.1)	15,484 (107)				
	400 (478)	4194 (28.9)	6102 (42.1)	7848 (54.1)	11,322 (78.1)	14,904 (103)				
	600 (589)	3961 (27.3)	5763 (39.7)	7412 (51.1)	10,693 (73.7)	14,076 (97.0)				
	800 (700)	3681 (25.4)	5356 (36.9)	6889 (45.5)	9938 (68.5)	13,082 (90.1)				
¼ (6.35)	−20 to 100 (244 to 311)	3198 (22.0)	4600 (31.7)	5891 (40.6)	8602 (59.3)	11,688 (80.6)				
	400 (478)	3078 (21.2)	4428 (30.5)	5670 (39.1)	8280 (57.1)	11,250 (77.6)				
	600 (589)	2907 (20.0)	4182 (28.8)	5355 (36.9)	7820 (53.9)	10,625 (73.3)				

	800 (700)	2702 (18.7)	3887 (26.8)	4977 (34.3)	7268 (50.1)	9875 (68.1)				
5⁄16 (7.938)	−20 to 100 (244 to 311)	2525 (17.4)	3609 (24.9)	4600 (31.7)	6713 (46.3)	9182 (63.3)				
	400 (478)	2430 (16.8)	3474 (24.0)	4428 (30.5)	6462 (44.6)	8838 (60.9)				
	600 (589)	2295 (15.8)	3281 (22.6)	4182 (28.8)	6103 (42.1)	8347 (57.6)				
	800 (700)	2133 (14.7)	3049 (21.0)	3887 (26.8)	5672 (39.1)	7758 (53.5)				
3⁄8 (9.525)	−20 to 100 (244 to 311)		2973 (20.5)	3777 (26.0)	5460 (37.6)	7517 (51.8)	9836 (67.8)			
	400 (478)		2862 (19.7)	3636 (25.1)	5256 (36.2)	7236 (49.9)	9468 (65.3)			
	600 (589)		2703 (18.6)	3434 (23.7)	4964 (34.2)	6834 (47.1)	8942 (61.7)			
	800 (700)		2512 (17.3)	3192 (22.0)	4614 (31.8)	6352 (43.8)	8311 (57.3)			
1⁄2 (12.7)	−20 to 100 (244 to 311)		2188 (15.1)	2768 (19.1)	3976 (27.4)	5423 (37.4)	7162 (49.4)	8322 (57.4)		
	400 (478)		2106 (14.5)	2664 (18.4)	3827 (26.4)	5220 (36.0)	6894 (47.5)	8010 (55.2)		
	600 (589)		1989 (13.7)	2516 (17.3)	3614 (24.9)	4930 (34.0)	6511 (44.9)	7565 (52.1)		
	800 (700)		1849 (12.7)	2338 (16.1)	3359 (23.2)	4582 (31.6)	6051 (41.7)	7031 (48.5)		
5⁄8 (15.88)	−20 to 100 (244 to 311)			2188 (15.1)	3123 (21.5)	4245 (29.3)	5554 (38.3)	6470 (44.6)	7555 (52.1)	

TABLE 4.11 *(Continued)*

Tubing OD, in (mm)	Temp., °F (K)	Wall thickness, in (mm)								
		0.020 (0.508)	0.028 (0.711)	0.035 (0.889)	0.049 (1.245)	0.065 (1.651)	0.083 (2.108)	0.095 (2.413)	0.109 (2.769)	0.120 (3.048)
	400 (478)			2106 (14.5)	3006 (20.7)	4086 (28.2)	5346 (36.9)	6228 (42.9)	7272 (50.1)	
	600 (589)			1989 (13.7)	2839 (19.6)	3859 (26.6)	5049 (34.8)	5882 (40.6)	6868 (47.4)	
	800 (700)			1849 (12.7)	2639 (18.2)	3587 (24.7)	4693 (32.4)	5467 (37.7)	6383 (44.0)	
¾ (19.05)	−20 to 100 (244 to 311)			1814 (12.5)	2581 (17.8)	3478 (24.0)	4544 (31.3)	5273 (36.4)	6152 (42.4)	6863 (47.3)
	400 (478)			1746 (12.0)	2484 (17.1)	3348 (23.1)	4374 (30.1)	5076 (35.0)	5922 (40.8)	6606 (45.5)
	600 (589)			1649 (11.4)	2346 (16.2)	3162 (21.8)	4131 (28.5)	4794 (33.1)	5593 (38.6)	6239 (43.0)
	800 (700)			1433 (9.9)	2180 (15.0)	2939 (20.3)	3839 (26.5)	4456 (30.7)	5198 (35.8)	5799 (40.0)
⅞ (22.23)	−20 to 100 (244 to 311)			1545 (10.7)	2188 (15.1)	2955 (20.4)	3834 (26.4)	4451 (30.7)	5180 (35.7)	5760 (39.7)
	400 (478)			1487 (10.3)	2106 (14.5)	2844 (19.6)	3690 (25.4)	4284 (29.5)	4986 (34.4)	5544 (38.2)
	600 (589)			1404 (9.8)	1989 (13.7)	2686 (18.5)	3485 (24.0)	4046 (27.9)	4709 (32.5)	5236 (36.1)
	800 (700)			1305 (9.0)	1849 (12.7)	2496 (17.2)	3239 (22.3)	3760 (25.9)	4377 (30.2)	4866 (33.5)

1 (25.4)	−20 to 100 (244 to 311)			1346 (9.3)	1907 (13.1)	2562 (17.7)	3329 (23.0)	3852 (26.6)	4469 (30.8)	4956 (34.2)
	400 (478)			1296 (8.9)	1836 (12.7)	2466 (17.0)	3204 (22.1)	3708 (25.6)	4302 (29.7)	4770 (32.9)
	600 (589)			1224 (8.4)	1734 (12.0)	2329 (16.1)	3026 (20.9)	3502 (24.1)	4063 (28.0)	4505 (31.1)
	800 (700)			1138 (7.8)	1612 (11.1)	2165 (14.9)	2812 (19.4)	3255 (22.4)	3776 (26.0)	4187 (28.9)

TABLE 4.12 Allowable Working Pressure, psi (MPa), for Carbon-Steel Welded Tubing, SA-556 or Equivalent

Tubing OD, in (mm)	Temp., °F (K)	Wall thickness, in (mm)								
		0.020 (0.508)	0.028 (0.711)	0.035 (0.889)	0.049 (1.245)	0.065 (1.651)	0.083 (2.108)	0.095 (2.413)	0.109 (2.769)	0.120 (3.048)
⅛ (3.175)	−20 to 600 (244 to 589)	6422 (44.2)	9327 (64.3)	11,830 (81.6)						
	800 (700)	4404 (30.3)	6396 (44.1)	8112 (55.9)						
3⁄16 (4.763)	−20 to 600 (244 to 589)	4077 (28.1)	5932 (40.9)	7630 (52.6)	11,007 (75.8)	14,490 (99.8)				
	800 (700)	2796 (19.3)	4068 (28.0)	5232 (36.0)	7548 (52.0)	9936 (68.5)				
¼ (6.35)	−20 to 600 (244 to 589)	2992 (20.6)	4305 (29.6)	5512 (38.0)	8050 (55.5)	10,937 (75.4)				
	800 (700)	2052 (14.1)	2952 (20.3)	3780 (26.0)	5520 (38.0)	7500 (51.7)				
5⁄16 (7.938)	−20 to 600 (244 to 589)	2362 (16.3)	3377 (23.3)	4305 (29.6)	6282 (43.3)	8592 (59.2)				
	800 (700)	1620 (11.2)	2316 (15.9)	2952 (20.3)	4308 (29.6)	5892 (40.6)				
⅜ (9.525)	−20 to 600 (244 to 589)		2782 (19.2)	3535 (24.4)	5110 (35.2)	7035 (48.5)	9205 (63.4)			

	800 (700)		1908 (13.1)	2424 (16.7)	3504 (24.1)	4824 (33.2)	6312 (43.4)			
½ (12.7)	−20 to 600 (244 to 589)		2047 (14.1)	2590 (17.8)	3720 (25.6)	5075 (35.0)	6702 (46.2)	7787 (53.7)		
	800 (700)		1404 (9.7)	1776 (12.3)	2551 (17.6)	3480 (24.0)	4596 (31.7)	5340 (36.8)		
⅝ (15.88)	−20 to 600 (244 to 589)			2047 (14.1)	2922 (20.1)	3972 (27.4)	5197 (35.8)	6055 (41.8)	7070 (48.7)	
	800 (700)			1404 (9.6)	2004 (13.8)	2724 (18.7)	3564 (24.5)	4152 (28.6)	4848 (33.4)	
¾ (19.05)	−20 to 600 (244 to 589)			1697 (11.7)	2415 (16.6)	3255 (22.7)	4252 (29.3)	4935 (34.0)	5757 (39.7)	6422 (44.2)
	800 (700)			1164 (8.0)	1656 (11.4)	2232 (15.4)	2916 (20.1)	3384 (23.3)	3948 (27.2)	4404 (30.3)
⅞ (22.23)	−20 to 600 (244 to 589)			1445 (10.0)	2047 (14.1)	2765 (19.1)	3587 (24.7)	4165 (28.7)	4847 (33.4)	5390 (37.1)
	800 (700)			991 (6.8)	1404 (9.6)	1896 (13.1)	2460 (16.9)	2856 (19.7)	3324 (22.9)	3696 (25.5)
1 (25.4)	−20 to 600 (244 to 589)			1260 (8.7)	1785 (12.3)	2397 (16.5)	3115 (21.5)	3605 (24.8)	4182 (28.8)	4637 (32.0)
	800 (700)			864 (5.9)	1224 (8.4)	1644 (11.3)	2136 (14.7)	2472 (17.0)	2868 (19.8)	3180 (21.9)

TABLE 4.13 Allowable Working Pressure, psi (MPa), for Carbon-Steel Welded Tubing, SA-557 or Equivalent

Tubing OD, in (mm)	Temp., °F (K)	Wall thickness, in (mm)								
		0.020 (0.508)	0.028 (0.711)	0.035 (0.884)	0.049 (1.245)	0.065 (1.651)	0.083 (2.108)	0.095 (2.413)	0.109 (2.769)	0.120 (3.048)
⅛ (3.175)	−20 to 650 (244 to 617)	5505 (38.0)	7995 (55.1)	10,140 (69.6)						
	800 (700)	3743 (25.8)	5437 (37.5)	6895 (47.5)						
3/16 (4.763)	−20 to 650 (244 to 617)	3495 (24.1)	5085 (35.0)	6540 (45.1)	9435 (65.0)	12,420 (85.6)				
	800 (700)	2377 (16.4)	3458 (23.8)	4447 (30.6)	6416 (44.2)	8446 (58.2)				
¼ (6.35)	−20 to 650 (244 to 617)	2565 (17.7)	3690 (25.4)	4725 (32.6)	6900 (47.5)	9375 (64.7)				
	800 (700)	1744 (12.0)	2509 (17.3)	3213 (22.1)	4692 (32.3)	6375 (43.9)				
5/16 (7.938)	−20 to 650 (244 to 617)	2025 (14.0)	2895 (20.0)	3690 (25.4)	5385 (37.1)	7385 (50.9)				
	800 (700)	1377 (9.5)	1969 (13.6)	2509 (17.3)	3662 (25.2)	5008 (34.5)				
⅜ (9.525)	−20 to 650 (244 to 617)		2385 (16.4)	3030 (20.9)	4380 (30.2)	6030 (41.5)	7890 (54.4)			

	800 (700)		1622 (11.2)	2060 (14.2)	2978 (20.5)	4100 (28.2)	5365 (37.0)			
½ (12.7)	−20 to 650 (244 to 617)		1755 (12.1)	2220 (15.3)	3189 (22.0)	4350 (30.0)	5745 (39.6)	6675 (46.0)		
	800 (700)		1193 (8.2)	1510 (10.4)	2169 (15.0)	2958 (20.4)	3907 (26.9)	4539 (31.3)		
⅝ (15.88)	−20 to 650 (244 to 617)			1755 (12.1)	2505 (17.3)	3405 (23.5)	4455 (30.7)	5190 (35.8)	6060 (41.8)	
	800 (700)			1193 (8.2)	1703 (11.7)	2315 (16.0)	3029 (20.9)	3529 (24.3)	4121 (28.4)	
¾ (19.05)	−20 to 650 (244 to 617)			1455 (10.0)	2070 (14.3)	2790 (19.2)	3645 (25.1)	4230 (29.1)	4935 (34.0)	5505 (37.9)
	800 (700)			989 (6.8)	1408 (9.7)	1897 (13.1)	2479 (17.1)	2876 (19.8)	3356 (23.1)	3743 (25.8)
⅞ (22.23)	−20 to 650 (244 to 617)			1239 (8.5)	1755 (12.1)	2370 (16.3)	3075 (21.2)	3570 (24.6)	4155 (28.6)	4620 (31.8)
	800 (700)			843 (5.8)	1193 (8.2)	1612 (11.1)	2091 (14.4)	2428 (16.7)	2825 (19.5)	3142 (21.6)
1 (25.4)	−20 to 650 (244 to 617)			1080 (7.4)	1530 (10.5)	2055 (14.2)	2670 (18.4)	3090 (21.3)	3585 (24.7)	3975 (27.4)
	800 (700)			734 (5.0)	1040 (7.2)	1397 (9.6)	1816 (12.5)	2101 (14.5)	2438 (16.8)	2703 (18.6)

TABLE 4.14 Allowable Working Pressure, psi (MPa), for Carbon-Steel Annealed Seamless Tubing, ASTM A-179 or Equivalent

		Wall thickness, in (mm)								
Tubing OD, in (mm)	Temp., °F (K)	0.020 (0.508)	0.028 (0.711)	0.035 (0.884)	0.049 (1.245)	0.065 (1.651)	0.083 (2.108)	0.095 (2.415)	0.109 (2.769)	0.120 (3.048)
⅛ (3.175)	−20 to 650 (244 to 617)	4294 (29.6)	6236 (43.0)	7909 (54.5)						
	800 (700)	3376 (23.3)	4904 (33.8)	6219 (42.9)						
3/16 (4.763)	−20 to 650 (244 to 617)	2726 (18.8)	3966 (27.4)	5101 (35.1)	7359 (50.7)	9688 (66.8)				
	800 (700)	2144 (14.7)	3119 (21.5)	4011 (27.6)	5787 (39.9)	7618 (52.5)				
¼ (6.35)	−20 to 650 (244 to 617)	2001 (13.8)	2878 (19.8)	3685 (25.4)	5382 (37.1)	7312 (50.4)				
	800 (700)	1573 (10.8)	2263 (15.6)	2898 (20.0)	4232 (29.1)	5750 (39.6)				
5/16 (7.938)	−20 to 650 (244 to 617)	1579 (10.9)	2258 (15.6)	2878 (19.8)	4200 (28.9)	5745 (39.6)				
	800 (700)	1242 (8.5)	1776 (12.2)	2263 (15.6)	3303 (22.7)	4517 (31.1)				
⅜ (9.525)	−20 to 650 (244 to 617)		1860 (12.8)	2363 (16.3)	3416 (23.6)	4703 (32.4)	6154 (42.4)			

	800 (700)		1463 (10.1)	1858 (12.8)	2686 (18.5)	3698 (25.5)	4839 (33.3)			
½ (12.7)	−20 to 650 (244 to 617)		1369 (9.4)	1732 (11.9)	2487 (17.1)	3393 (23.4)	4481 (30.9)	5206 (35.9)		
	800 (700)		1076 (7.4)	1362 (9.4)	1956 (13.5)	2668 (18.4)	3524 (24.3)	4094 (28.2)		
⅝ (15.88)	−20 to 650 (244 to 617)			1369 (9.4)	1954 (13.4)	2656 (18.3)	3475 (23.9)	4048 (27.9)	4727 (32.6)	
	800 (700)			1076 (7.4)	1536 (10.6)	2088 (14.4)	2732 (18.8)	3183 (21.9)	3717 (25.6)	
¾ (19.05)	−20 to 650 (244 to 617)			1135 (7.8)	1615 (11.2)	2176 (15.0)	2843 (19.6)	3299 (22.7)	3849 (26.5)	4294 (29.6)
	800 (700)			892 (6.1)	1270 (8.8)	1711 (11.8)	2236 (15.4)	2594 (17.8)	3027 (20.9)	3376 (23.2)
⅞ (22.23)	−20 to 650 (244 to 617)			966 (6.7)	1369 (9.4)	1849 (12.7)	2398 (16.5)	2785 (19.2)	3241 (22.3)	3604 (24.8)
	800 (700)			760 (5.2)	1076 (7.4)	1454 (10.0)	1886 (13.0)	2190 (15.1)	2548 (17.6)	2834 (19.5)
1 (25.4)	−20 to 650 (244 to 617)			842 (5.8)	1193 (8.2)	1603 (11.0)	2083 (14.3)	2410 (16.6)	2796 (19.3)	3100 (21.4)
	800 (700)			662 (4.5)	938 (6.5)	1260 (8.7)	1638 (11.3)	1895 (13.1)	2199 (15.2)	2438 (16.8)

TABLE 4.15 Allowable Working Pressure, psi (MPa), for Aluminum Alloy 6061-T4&T6 Welded Tubing, ASTM–SB-210, for Temperatures to 100°F (311K)

Tubing OD, in (mm)	Wall thickness, in (mm)								
	0.020 (0.508)	0.028 (0.711)	0.035 (0.889)	0.049 (1.245)	0.065 (1.651)	0.083 (2.108)	0.095 (2.413)	0.109 (2.769)	0.120 (3.048)
⅛ (3.175)	1284 (8.85)	1865 (12.9)	2366 (16.3)						
3/16 (4.763)	815 (5.62)	1186 (8.18)	1526 (10.5)	2201 (15.2)	2898 (20.0)				
¼ (6.35)	598 (4.12)	861 (5.94)	1102 (7.60)	1610 (11.1)	2187 (15.1)				
5/16 (7.938)	472 (3.25)	675 (4.65)	861 (5.94)	1256 (8.66)	1718 (11.8)				
⅜ (9.525)		556 (3.83)	707 (4.87)	1022 (7.05)	1407 (9.70)	1841 (12.7)			
½ (12.7)		409 (2.82)	518 (3.57)	744 (5.13)	1015 (7.00)	1340 (9.24)	1557 (10.7)		
⅝ (15.88)			409 (2.82)	584 (4.03)	794 (5.47)	1039 (7.16)	1211 (8.35)	1414 (9.75)	
¾ (19.05)			339 (2.34)	483 (3.33)	651 (4.49)	850 (5.86)	987 (6.81)	1151 (7.94)	1284 (8.85)
⅞ (22.23)			289 (1.99)	409 (2.82)	553 (3.81)	717 (4.94)	833 (5.74)	969 (6.68)	1078 (7.43)
1 (25.4)			252 (1.74)	357 (2.46)	479 (3.30)	623 (4.30)	721 (4.97)	836 (5.76)	927 (6.39)

TABLE 4.16 Allowable Working Pressure, psi (MPa), for Aluminum Alloy 6061-T6 Annealed Seamless Tubing, ASTM B-210 or Equivalent

Tubing OD, in (mm)	Temp., °F (K)	Wall thickness, in (mm)								
		0.020 (0.508)	0.028 (0.711)	0.035 (0.884)	0.049 (1.245)	0.065 (1.651)	0.083 (2.108)	0.095 (2.415)	0.109 (2.769)	0.120 (3.048)
⅛ (3.175)	−20 to 100 (244 to 311)	3854 (26.5)	5597 (38.6)							
	400 (478)	1615 (11.2)	2345 (16.2)	2974 (20.5)						
3/16 (4.763)	−20 to 100 (244 to 311)	2447 (16.9)	3560 (24.5)	4578 (31.6)	6605 (45.5)					
	400 (478)	1025 (7.1)	1492 (10.3)	1918 (13.2)	2768 (19.1)	3643 (25.1)				
¼ (6.35)	−20 to 100 (244 to 311)	1796 (12.4)	2583 (17.8)	3308 (22.8)	4830 (33.3)	6563 (45.2)				
	400 (478)	752 (5.2)	1082 (7.4)	1386 (9.6)	2024 (13.9)	2750 (18.9)				
5/16 (7.938)	−20 to 100 (244 to 311)	1418 (9.8)	2027 (14.0)	2583 (17.8)	3770 (26.0)	5156 (35.5)				
	400 (478)	594 (4.1)	849 (5.9)	1082 (7.4)	1580 (10.9)	2160 (14.9)				
⅜ (9.525)	−20 to 100 (244 to 311)		1670 (11.5)	2121 (14.6)	3066 (21.1)	4221 (29.1)	5523 (38.0)			
	400 (478)		700 (4.8)	889 (6.1)	1285 (8.9)	1769 (12.2)	2314 (15.9)			
½ (12.7)	−20 to 100 (244 to 311)		1229 (8.5)	1554 (10.7)	2232 (15.4)	3045 (21.0)	4022 (27.7)	4673 (32.2)		

TABLE 4.16 *(Continued)*

Tubing OD, in (mm)	Temp., °F (K)	Wall thickness, in (mm)								
		0.020 (0.508)	0.028 (0.711)	0.035 (0.884)	0.049 (1.245)	0.065 (1.651)	0.083 (2.108)	0.095 (2.415)	0.109 (2.769)	0.120 (3.048)
	400 (478)		515 (3.5)	651 (4.5)	935 (6.5)	1276 (8.8)	1685 (11.6)	1958 (13.5)		
⅝ (15.88)	−20 to 100 (244 to 311)			1229 (8.5)	1754 (12.1)	2384 (16.4)	3119 (21.5)	3633 (25.0)		
	400 (478)			515 (3.5)	735 (5.1)	999 (6.9)	1307 (9.0)	1522 (10.5)	1778 (12.3)	
¾ (19.05)	−20 to 100 (244 to 311)			1019 (7.0)	1449 (10.0)	1953 (13.4)	2552 (17.6)	2961 (20.4)	3455 (23.8)	3854 (26.5)
	400 (478)			427 (3.0)	607 (4.2)	818 (5.6)	1069 (7.4)	1241 (8.5)	1448 (10.0)	1615 (11.1)
⅞ (22.23)	−20 to 100 (244 to 311)			867 (6.0)	1229 (8.5)	1659 (11.4)	2153 (14.8)	2499 (17.2)	2909 (20.0)	3234 (22.3)
	400 (478)			363 (2.5)	515 (3.5)	695 (4.8)	902 (6.2)	1047 (7.2)	1219 (8.4)	1355 (9.4)
1 (25.4)	−20 to 100 (244 to 311)			756 (5.2)	1071 (7.4)	1439 (9.9)	1869 (12.9)	2163 (14.9)	2510 (17.3)	2783 (19.2)
	400 (478)			317 (2.1)	449 (3.1)	603 (4.2)	783 (5.4)	906 (6.2)	1052 (7.2)	1166 (8.0)

TABLE 4.17 Allowable Working Pressure, psi (MPa), for Copper Annealed Seamless Tubing, ASTM B-75 or Equivalent, for Temperatures to 100°F (311K)

Tubing OD, in (mm)	Wall thickness, in (mm)								
	0.028 (0.711)	0.032 (0.813)	0.035 (0.889)	0.049 (1.245)	0.065 (1.651)	0.083 (2.108)	0.095 (2.413)	0.109 (2.769)	0.120 (3.048)
1/8 (3.175)	3198 (22.0)	3690 (25.4)	4056 (28.0)						
3/16 (4.763)	2034 (14.0)	2370 (16.3)	2616 (18.0)	3774 (26.0)					
1/4 (6.35)	1476 (10.2)	1710 (11.8)	1890 (13.0)	2760 (19.0)	3750 (25.9)				
5/16 (7.938)		1338 (9.2)	1476 (10.2)	2154 (14.9)	2946 (20.3)				
3/8 (9.525)		1098 (7.57)	1212 (8.36)	1752 (12.1)	2412 (16.6)	3156 (21.8)			
1/2 (12.7)			888 (6.12)	1276 (8.80)	1740 (12.0)	2298 (15.8)			
5/8 (15.88)			702 (4.84)	1002 (6.91)	1362 (9.39)	1782 (12.3)	2076 (14.3)		
3/4 (19.05)			582 (4.01)	828 (5.71)	1116 (7.69)	1458 (10.1)	1692 (11.7)	1974 (13.6)	
7/8 (22.23)			496 (3.42)	702 (4.84)	948 (6.54)	1230 (8.48)	1428 (9.85)	1662 (11.5)	
1 (25.4)			432 (2.98)	612 (4.22)	822 (5.67)	1068 (7.36)	1236 (8.52)	1434 (9.89)	1590 (11.0)

one installation for reasons of economy and convenience. Consequently many tube fittings are available that permit the transition from tube to pipe, and vice versa.

Pipe Couplings and Fittings

Three types of all-metal fittings are commonly used with pipe: screwed fittings, socket-weld fittings, and flanged fittings. There are also a number of proprietary pipe couplings on the market that provide enormous savings in installation time. They should be checked first for pressure, temperature, and chemical compatibility before use. Pipe fittings are usually pressure-rated, as is piping itself. Therefore, having selected the correct piping Schedule for the application, fittings of the same Schedule or pressure capability should be used.

Screwed Pipe Fittings

Pipe fittings already prepared for screwed connections are readily available and are usually machined from forgings and castings. They are made in many configurations to facilitate installations. Figure 4.8 shows some of the configurations frequently used. Both externally and internally threaded connections are used, with the latter being most common since they accept threaded pipe.

These fittings are provided with tapered threads of the NPT Series (see Table 1.1). As the threaded parts are screwed together, they become increasingly tight because the increasing size of one taper and the decreasing size of the other cause interference. Sealing is partially provided by this interference, but for complete sealing a thread sealant must be applied. The average amount of thread engagement required to make up a pipe joint is shown in Table 4.18 for various sizes of pipe. The actual amount of engagement will vary slightly according to the thread tolerances.

TABLE 4.18 Allowance for Thread Makeup for Standard Taper Pipe Threads (NPT)

Nominal pipe size, in	Thread makeup engagement, in (mm)
⅛	0.26 (6.6)
¼	0.40 (10)
⅜	0.41 (10)
½	0.53 (13)
¾	0.55 (14)
1	0.68 (17)
1½	0.72 (18)
2	0.75 (19)

The sealant selected must be chemically compatible with the fluid and adequate for the temperature and pressure service. For most applications sealing with Teflon tape wrapped on the male thread (1½ to 2 wraps) is best. Since the action with taper threads is one of forcing and sliding the metal parts together, the tendency for thread galling is high. Metals such as aluminum, stainless steel, and monel require special care, and it is good practice to apply a liquid Teflon dispersion to the female threads as well as Teflon tape to the male thread.

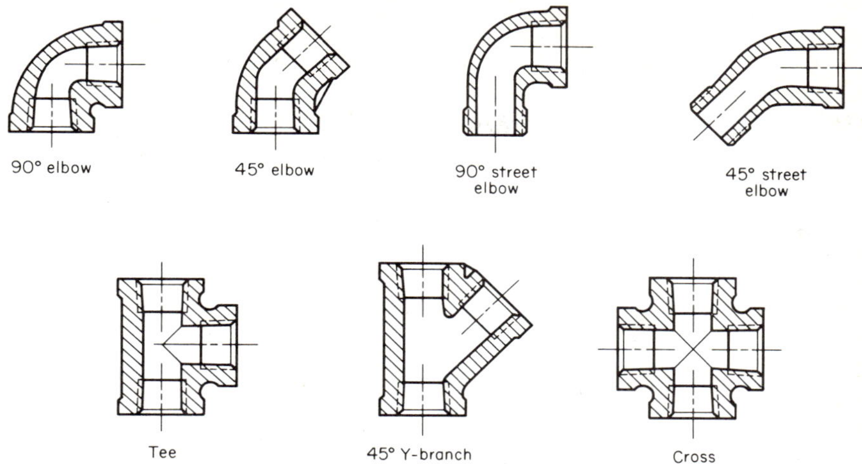

FIG. 4.8 Commonly used pipe-fitting configurations.

A particular form of the taper pipe thread, the Dryseal thread, is used without sealant. The thread form is slightly different from the standard pipe thread, and it is designated "NPTF" on drawings, taps, dies, etc. Its tendency to gall is somewhat greater than for standard NPT threads. The NPTF thread is used where a sealant is objectionable for reasons of contamination of the piping system, chemical incompatibility between the sealant and fluid to be used in the system, or where the service temperature exceeds that for available sealants. If silver plating of the male thread is acceptable from chemical considerations, it is very helpful in reducing galling of Dryseal pipe threads and may be used to temperatures of 1500°F (1100 K).

When assembling fluid systems using screwed pipe fittings, cut the pipe to length, making an allowance at each end for the thread makeup engagement shown in Table 4.18. Thread each end of the pipe, apply sealant to the threads, and then screw the pipe tightly into a fitting using the engagement guide of Table 4.18. Then screw the appropriate fitting or component to the other end of the pipe. Where component orientation is important, as is frequently the case, the piping or component is typically tightened to slightly greater values than those given in the table rather than slightly less to ensure a leaktight joint. When orienting a component rotate it carefully so as not to pass the desired position—advancing another full turn or nearly so may not be possible because of the tapered thread and to go back a part of a turn usually results in a leaky joint. Frequently check for joint leak tightness as system assembly proceeds. Check by capping or plugging the last installation and pressurizing the system. Screwed piping systems are especially difficult to go back to and correct for a leaking joint because rotating a component to effect a tighter seal may be impossible (e.g., elbows, tees). Rotating the piping tightens one end while loosening the other.

Fluid systems using NPT joints may be performed with Schedule 40 pipe or heavier. Pipe sizes from ⅛ to 2 in (3.2 to 50 mm) may be used. The high torque required for pipe thread cutting and subsequent assembly limits the practical maximum size using typical laboratory equipment to 2 in NPT.

Flange Fittings

Most fluid components for use with pipe size 1 in (25 mm) or greater are available with flange connections. Figure 4.9 is a typical flanged-valve installation. Flanges on the piping correspond to like flanges on the valve (or other components), and the flange pairs are held together with bolts. There is always a fluid seal between the mating flange surfaces. Flanges are standardized with respect to size (OD and thickness) and number and size of bolt holes.

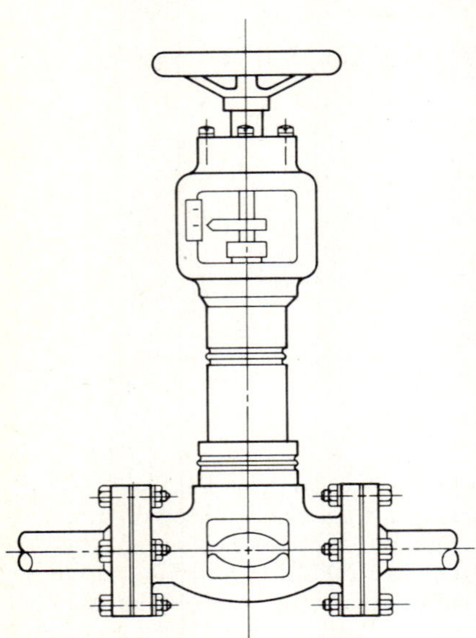

FIG. 4.9 Fluid component (valve) installed with flanges.

Seals for flanges take various forms. The simplest is a flat ring of plastic or rubber (Fig. 4.10*a*). The seal is squeezed between the flange faces as the bolts are tightened. This method is fine for low pressures (200 psi, or 1400 kPa) and modest temperatures (200°F, or 370 K). For higher pressures an elastomeric O ring (Fig. 4.10*b*) is good. It is simple, inexpensive, and effective. For high-temperature work a metal O ring can be used, as in Fig. 4.10*c*. For both high temperature and high pressure, composite rings of stainless steel and asbestos are very effective (Fig. 4.10*d*).

A particular attribute of flanged connections is the convenience they provide when removing a single component or piece of piping without having to disturb the adjacent piping or fixtures. Once the bolts have been removed from the flanges, the part of interest is removed in any direction parallel to the flanges. When reinstalling parts it is advisable to renew the flange seals.

Pipe is attached to flanges by one of several means. For small pipe the flanges are often threaded, but for larger ones it is more usual for them to be welded (Fig. 4.11*a*). Welding is sometimes done with the piping in position to maintain correct mating of the flange bolt hole and pipe alignment.

Welding Fittings

There are two types of welding fitting: socket weld and butt weld. Figure 4.11*b* and *c* shows a typical fitting for each, with piping attached to one connection by way of example. Welded fittings are normally used for permanent installations although they can be used in conjunction with flanges (Fig. 4.11*a*) to permit the removal of components for servicing, etc.

The socket weld is more common and easier to use because the pipe is positioned in the fitting and the weld is a simple fillet weld. The butt weld requires careful positioning and fixturing to maintain alignment of both parts of the joint, and the full-penetration weld is more difficult to perform. The butt weld is used where crevices are not acceptable. For example, with some fluids, impurities tend to concentrate in the crevices and lead to excessive corrosion. Another situation

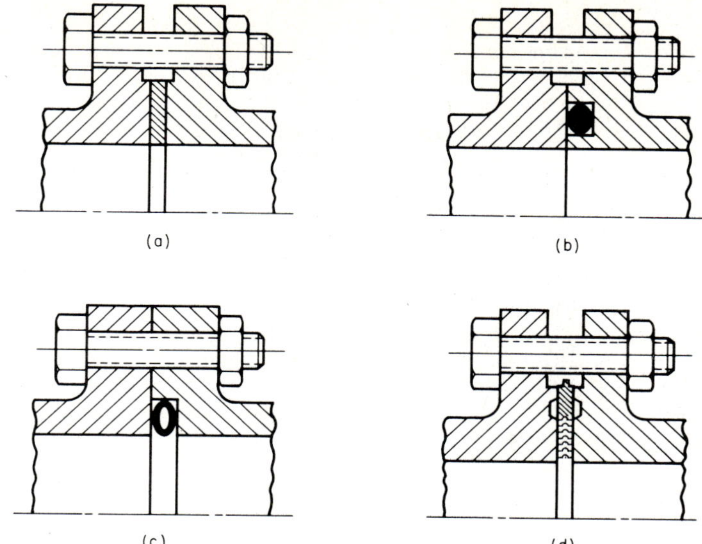

FIG. 4.10 Four common flange-sealing systems.

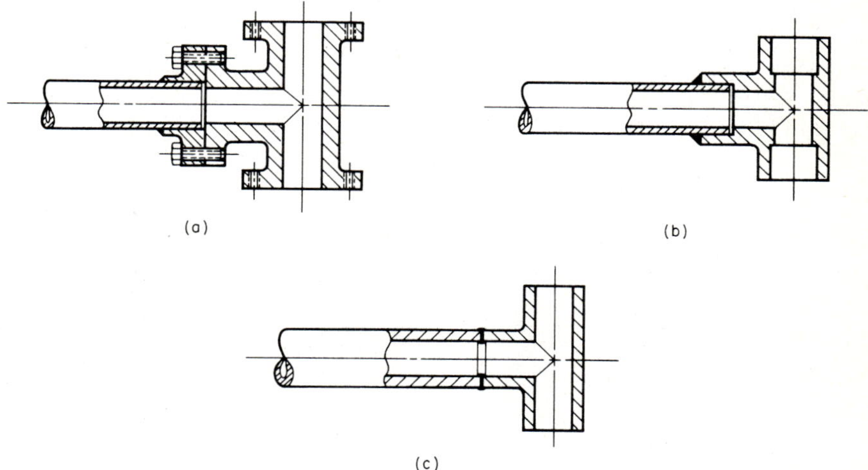

FIG. 4.11 Fittings with welded connections.

could be where cleanliness is important, and the crevices of the socket welds are virtually impossible to clean out in an assembled piping system.

Grooved Pipe Couplings and Fittings

A highly successful system of pipe joining is one where the pipe ends are grooved to provide positive mechanical attachment to couplings and fittings and have specially designed elastomeric gaskets to effect a seal. A partially sectioned coupling showing the components is illustrated in Fig. 4.12. Two semicircular housings are bolted together so that keys along the inside edges of the housings engage grooves in the pipe ends. A circular gasket is contained within the housing halves and seals on the outside diameter of both pipe ends. Initial sealing of the gasket is provided by gasket compression as the housing halves are drawn together with bolts. The pressure-responsive design of the gasket is such that the seal is improved as the internal fluid pressure increases.

Apart from the groove, no special pipe-end preparation is required. Grooving tools are readily available which facilitate correct sizing and positioning of the groove at the pipe end. Providing the pipe end is clean, free from loose scale and deep scratches, no other work is required.

Assembly is simple and quick. Lubricate the gasket, using only manufacturer's recommended lubricant. Then slide the gasket onto one pipe end. Bring the other pipe end into position and center the gasket between the two grooves on each pipe end. Place the two housing halves over the gasket and position so that the keys engage the pipe grooves. Insert and tighten the bolts to bring the housing halves into firm contact.

Many elastomeric gasket materials are available to provide compatibility for a wide range of fluids. The temperature capability is determined by the gasket and extends from −30 to 350°F (240 to 450 K), with each particular elastomer having its own service range. Couplings and fittings are available for pipe sizes from ¾ to 24 in (20 to 600 mm) and for pressure to 2500 psi (170 MPa).

Couplings are available for piping systems requiring either rigid or flexible joints. The flexible permits axial movement to accommodate thermal expansion or contraction and angular positioning or movement.

Sections of piping or components can be removed without disturbing adjacent piping. Grooved fittings having all the regular configurations are available, as are special flanges to permit changing from flanged to grooved systems.

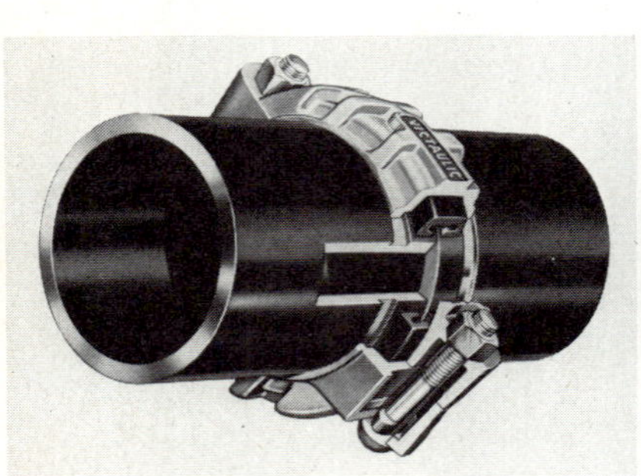

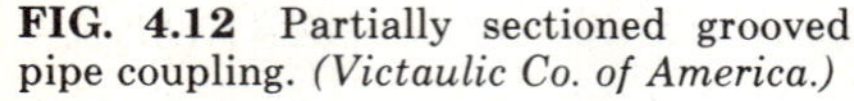

FIG. 4.12 Partially sectioned grooved pipe coupling. *(Victaulic Co. of America.)*

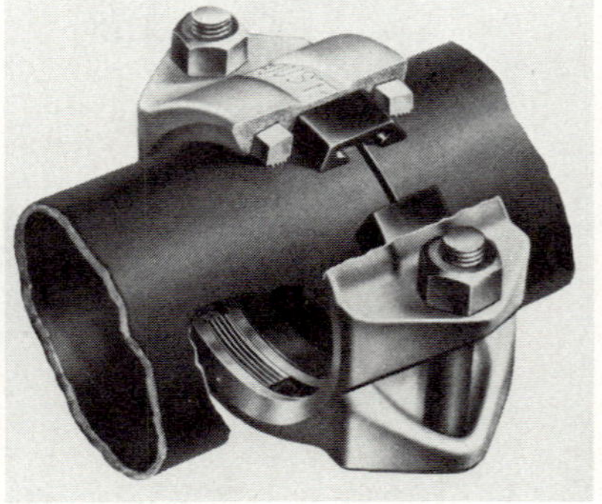

FIG. 4.13 Partially sectioned plain-end pipe coupling. *(Victaulic Co. of America.)*

Plain-End Pipe Couplings

Pipe couplings requiring no special pipe preparation, such as grooving or threading, are available. These couplings clamp right onto the pipe and use a pressure-sensitive rubber gasket seal similar to those used in grooved couplings. The pipe is held rigidly in the coupling by specially designed hardened steel grippers, and they restrain the pipe against axial pressure loads. Figure 4.13 shows a coupling of this type. Special plain-end fittings (e.g., elbows, tees, reducers) are also available for use with these plain-end couplings.

The manufacturer recommends a specific torque for tightening the bolts for these couplings. It is important that these recommendations be followed to provide proper joint integrity. Always use a torque wrench to be sure the torque requirement is met. While these couplings are intended primarily for use on Schedule 40 steel pipe, they may be used with lighter schedules and with other materials (e.g., stainless steel and aluminum), but the manufacturer's data for these applications should be closely followed. Do not use this type of coupling on pipe with hard surfaces (above Rb 80) because the grippers cannot penetrate the pipe surface sufficiently to provide adequate restraint.

While primarily intended for original installations, these couplings are ideal for making alterations to existing piping systems. It becomes a simple matter to cut into a length of pipe and add a branch connection, valve, or other component.

Tube Fittings

Tube fittings requiring very simple installation procedures have reached a state of development that provides high reliability and ease of installation with zero leakage under the most severe service conditions. Their performance has been demonstrated in many applications including extremes of temperature and pressure, thermal cycling, vibration, and nuclear radiation. There are several general types of fittings from which to choose, each offering specific characteristics that may make it the preferred choice for an application.

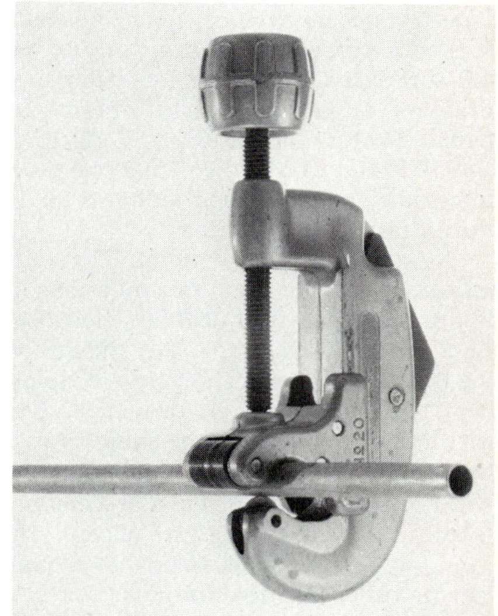

FIG. 4.14 Tube cutter used for cutting thin-wall tubing.

Compression Fittings

Since tubing is used extensively for instrumentation lines and small process lines, fittings requiring the minimum of time and effort for installation have obvious advantages. The all-metal compression fitting provides the simplest plumbing and covers the broadest service conditions. No special preparation of the tube end is required, and either a tube cutter (Fig. 4.14) or a hacksaw for heavy-wall tubing and a wrench are all that are required. Insert the tubing into the fitting and tighten the nut. Only 1¼ turns of the nut are required on most brands to effect a proper seal. Of the various brands available, two-ferrule systems (Fig. 4.15) provide a more reliable and effective seal. In these systems one ferrule, the front ferrule, provides the sealing, while the rear ferrule holds the tubing securely and prevents push-out even with fluids pressurized to the burst pressure of the tubing. This rear-ferrule action is

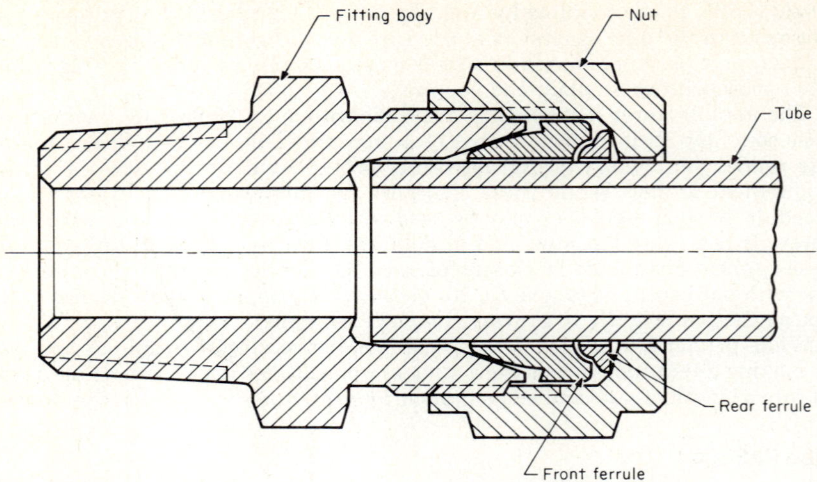

FIG. 4.15 A high-performance, two-ferrule compression tube fitting. (Hoke, Inc.)

particularly advantageous for heavy-wall tubing where other type fittings are less suitable.

Compression fittings may be used to the maximum temperatures and pressures for the corresponding tubing with which they are assembled (see Tables 4.8 through 4.17). Therefore they are suitable for very wide temperature and pressure ranges. Successful operation has been demonstrated at temperatures from -400 to 1500°F (30 to 1100 K) and pressures from 10^{-5} torr to 10,000 psi (0.001 Pa to 70 MPa) for 316 stainless-steel fittings. Because fitting and tubing materials are similar throughout, no differential expansion problems result from fluid thermal transients.

Numerous fitting configurations are commonly available to suit almost any required combination of connections. Figure 4.16 shows the typical range of configurations. For most of them their use is evident, but three deserve special mention: the heat-exchanger tee, the adaptors, and the reducer. The heat-exchanger tee is used for making a simple coaxial-tube heat exchanger, or condenser, or for heat tracing of lines. It provides a ready means for placing one tube coaxially within another and at the same time provides fluid access to the annular space between the tubes as well as to the center tube (Fig. 4.17).

Reducers with tube ends are most convenient to have on hand since they can readily convert fittings to accept different-size tubing (Fig. 4.18), including changing a regular tee to a heat-exchanger tee (Fig. 4.19).

The tube-ended adaptors are used to change the compression fitting to accept a male or female pipe thread (Fig. 4.20). Together with the tube-end reducer, these adaptors extend enormously the many connections that can be accommodated with a few standard fitting configurations; they are particularly useful for small laboratories where it is not economical to keep a large range of fittings in inventory.

Fittings are commonly available in 316 stainless steel, carbon steel, brass (for copper tubing), aluminum, and plastics. Most manufacturers will also provide

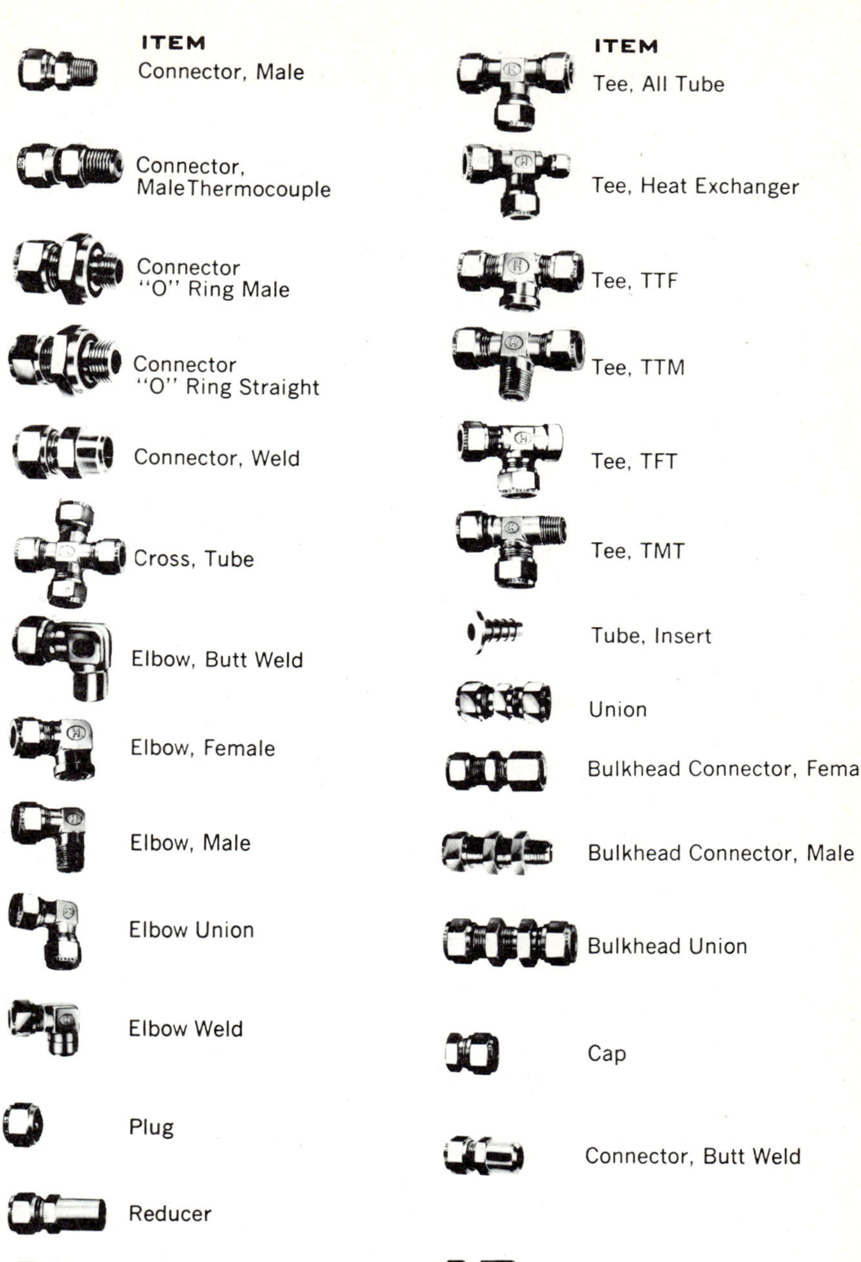

FIG. 4.16 Typical tube-fitting configurations. *(Hoke, Inc.)*

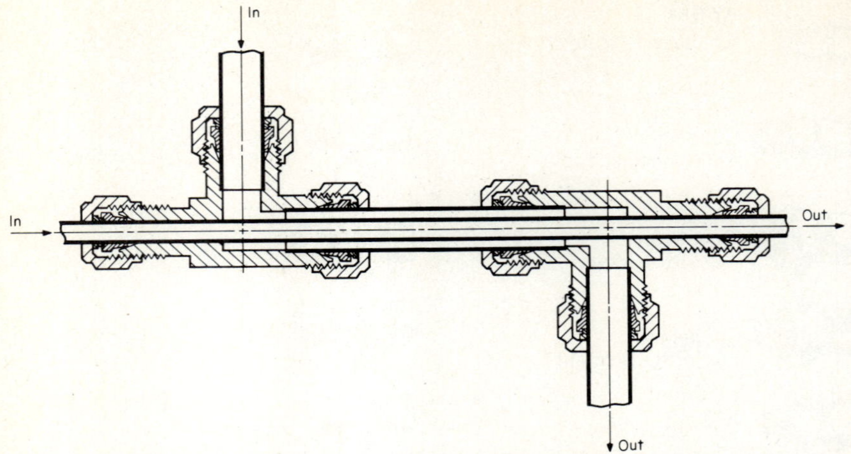

FIG. 4.17 Heat exchanger made using two heat-exchanger tee compression fittings.

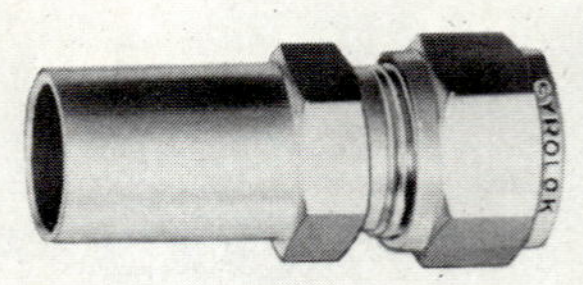

FIG. 4.18 Reducer used to change fitting to accept smaller-size tubing. *(Hoke, Inc.)*

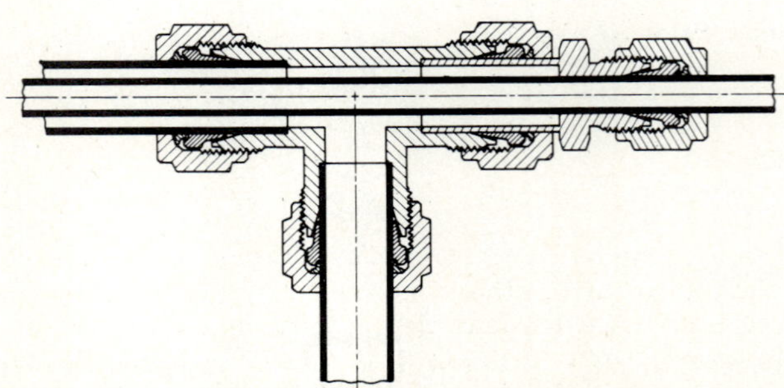

FIG. 4.19 Reducer used to convert regular tube tee to heat-exchanger tee. *(Hoke, Inc.)*

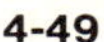

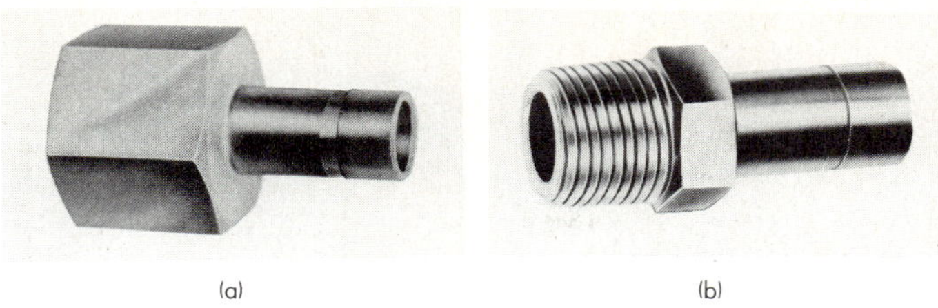

FIG. 4.20 Tube-end adaptors for converting tube-fitting connection to screwed pipe connections. (*a*) Female pipe thread to tube; (*b*) male pipe thread to tube.

them in Nickel, Monel, Inconel, and Hastelloy because these materials are required to provide the requisite corrosion resistance to various chemicals and environments (e.g., saltwater spray). Metal fittings may be used with plastic tubing without difficulty. However, for soft plastic tubing (e.g., Tygon or soft PVC) place an insert in the end of the tubing (Fig. 4.21).

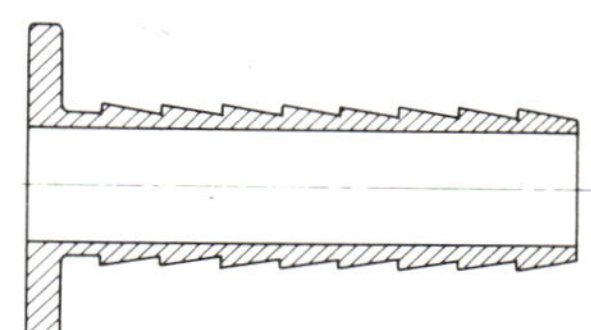

FIG. 4.21 Tube insert for plastic tubing.

Compression fittings are generally available for tubing from 1/16 to 1 in (1.5 to 25 mm) in diameter. The wrench torque required to make up the larger fittings (above 3/4 in, or 20 mm), especially if the tube wall is heavy, often becomes too great for convenient *in situ* assembly. The torque difficulty may be overcome by presetting the ferrules on the tube at a bench. For this purpose special presetting bodies are available so that the actual fitting body can remain with plumbing assemblies already made. After presetting, the makeup torque is much less and is comparable to that required to remake an assembly. Remakes (i.e., disassembly and reassembly of tubing to fitting) may be made extensively regardless of fitting size without degrading functional performance simply by "snugging" the nut at each reassembly.

Flare Fittings

The flare fitting owes its name to the geometry of the tube end, which is "flared" to make an assembly to the fitting (Fig. 4.22). The nut and the ferrule must be placed on the tubing before the tube is flared. Special tools are used to flare the end of the tube, which must first be cut squarely. For small-diameter tubing (to 3/8-in, or 10-mm, OD) of thin wall, a hand flaring tool may be used (Fig. 4.23). For large-size tubing and heavier wall, especially for steel tubing, a flaring machine must be used to provide the necessary force to form the flare at the tube end.

Some flare fittings do not use a separate ferrule. Instead the flared tube end is clamped between the fitting body and the nut. With this arrangement sometimes the tube is rotated as the nut is tightened, which may impart a torque to the tube with attendant torsional stresses. The tube rotation may also upset alignment of the tubing run or other components assembled at the tube end.

Since copper easily work-hardens, flare fittings used with copper tubing are normally restricted to working pressures appreciably below the normal tube working pressure because of potential cracking of the hardened tubing caused by flaring. Typical applications are for 100 psi (700-kPa) air lines.

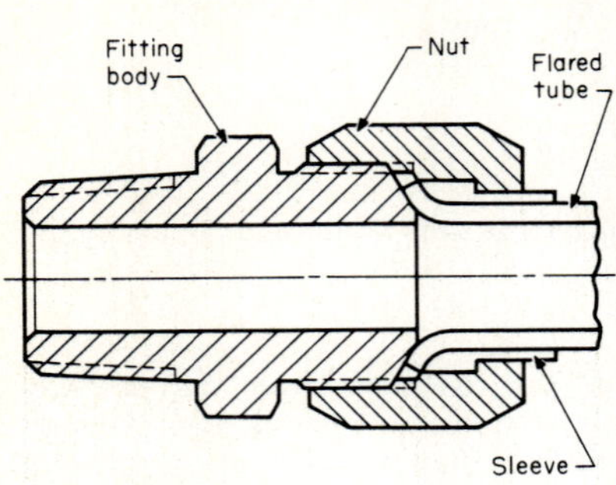

FIG. 4.22 Flare fitting in cross section.

FIG. 4.23 Hand flaring tool for flaring thin-wall tubing.

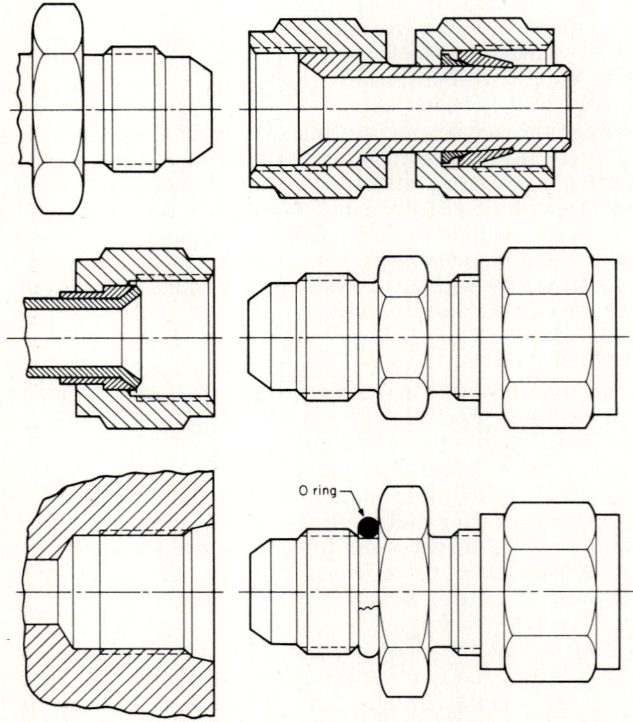

FIG. 4.24 Fittings for changing from flare fitting to compression fitting.

Flare-ended tubing cannot be used with compression fittings, and vice versa. Today the flare-type fitting is used primarily for replacement, for reuse of components designed for its use, and where lack of interchangeability presents logistic problems. Flare fittings are available for all the configurations shown for compression fittings in Fig. 4.16.

Fittings are available for converting from compression to flare, or vice versa. Figure 4.24 shows three of the most commonly used fittings for changing from one system to another.

O-Ring Seal Fittings

Flat-Face Seal • A disadvantage of the compression and flare fittings is that the tubing must be inserted and withdrawn from the fitting body for assembly and disassembly, respectively. While this is unimportant in many applications because the tubing affords sufficient flexibility to accommodate the required movement, there are installations where it can be a problem. For example, in situations where components are closely and intricately connected or where the tubing is inflexible because of large diameter or heavy wall, the required tubing displacement may not be available for the insertion and removal of components (valves, etc.) or to make piping connections in tight spaces. The flat-face O-ring seal fittings permit such assemblies with ease because no movement of the adjacent piping is needed, as shown in Figs. 4.25 and 4.26.

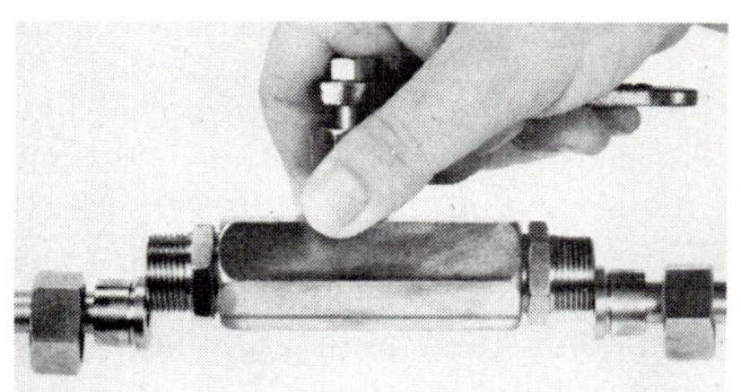

FIG. 4.25 Flat-faced O-ring tube fittings provide slip-in–slip-out accessibility of components. *(Combination Pump Valve Co.)*

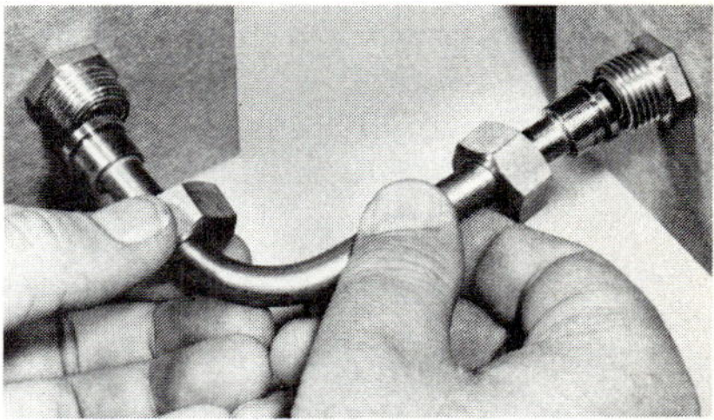

FIG. 4.26 Flat-faced O-ring tube fitting for makeup in close quarters. *(Combination Pump Valve Co.)*

The O-ring seal fitting comprises four components: the body, the union nut, the O ring, and the tailpiece (Fig. 4.27). Bodies and tailpieces are available for either pipe or tubing. Pipe may be screwed, brazed, or socket-welded to the fitting tailpiece, while tubing is normally brazed.

These fittings are suitable for pressures to 3000 psi (20 MPa) and temperatures from −20 to 225°F (240 to 380 K). The temperature range can be extended with other than standard O rings. Consult the manufacturer for these alternate O rings before making substitutions.

This fitting can be remade indefinitely. It assembles with a low torque so there is little stress on the components. The only component likely to wear is the O ring, and that is easily replaced.

Fittings are available in sizes for tubing or pipe up to 2 in (50 mm). Because of the low torque required for assembly, fittings this size can easily be assembled with a wrench.

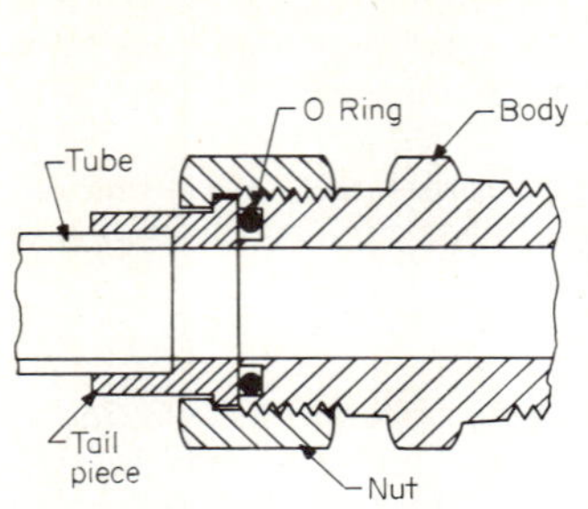

FIG. 4.27 Section of flat-faced O-ring tube fitting. *(Combination Pump Valve Co.)*

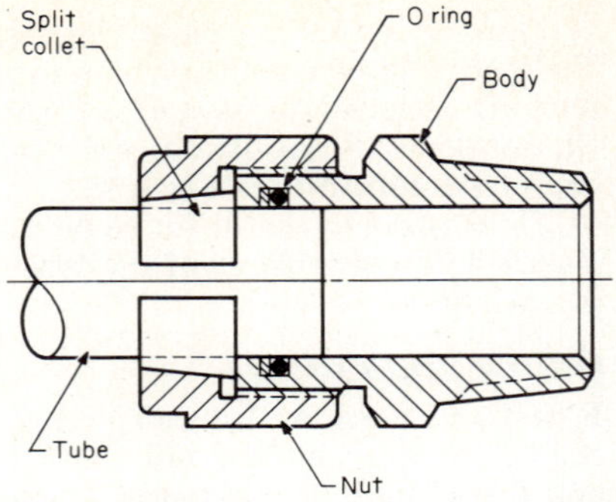

FIG. 4.28 OD tube O-ring seal tube fitting.

OD Tube Seal • This fitting provides a seal with an O ring on the outside diameter of the tubing. The O ring is located in a groove in the fitting body (Fig. 4.28). A Teflon backup ring improves the performance of the seal. To prevent the tubing from being pushed out under pressure, the frictional grip of a split collet is employed. The outer surface of the collet is tapered and matches a corresponding taper in the nut. As the nut is tightened its forward movement squeezes the collet firmly against the tube. One turn of the nut from handtight secures the assembly.

Prior to assembly remove all burrs from the outer edge of the tube end so that the O ring is not damaged when the tube is inserted into the fitting body. In addition, lubricate the O ring so that it does not chafe as it moves under pressure variations.

No precise length of tube is necessary in making an installation. Excess length is accommodated simply by sliding the tubing farther into the fitting body. This feature facilitates assemblies for close-quarter runs (for example, between two fixed bulkheads) and for short bends.

With this fitting the nut rotation and tightening torque tend to be transmitted to the tubing. It is good practice, then, on a length of tubing to hold securely with one wrench the first nut tightened while tightening the second nut with another wrench so that the first nut is not loosened.

Fittings are available in common configurations for tubing up to 2 in (50 mm) in diameter in carbon and stainless steels.

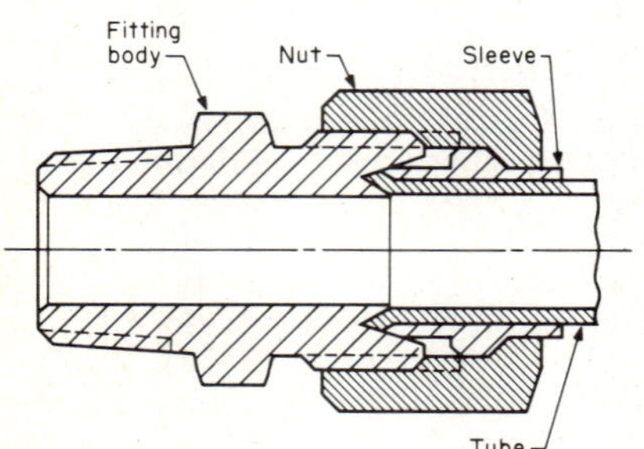

FIG. 4.29 Self-flare tube fitting.

Self-Flare Fittings

The disadvantage of the conventional flare fitting is that special tools and a tube-flaring operation are required prior to fitting makeup. The self-flare fitting is made up by using a specially designed sleeve and body. Simply push the tubing into the fitting (Fig. 4.29) and rotate the nut until a significant increase in torque is obtained. The fitting is then ready for use. It may be disassembled and reassembled repeatedly.

Prepare tubing ends with more than usual care. Cut the tube squarely and carefully remove all burrs, especially at the ID; otherwise leakage will occur.

The tubing should be thin wall and fully annealed (soft). If heavy-wall and harder tubing are to be used it is necessary to preset the sleeve on the tubing, and hydraulically operated presetting equipment is available for this purpose. If the tube wall is extra heavy, a special fitting having a different configuration is also required.

Fittings for tubing up to 2-in (50-mm) OD are available in many materials for service up to the normal working pressure of the tubing.

Instant Fittings

For plastic tubing fittings are available that provide the ultimate in simplicity and essentially zero assembly time. All that is required is to push the tubing into the fitting, whereupon it automatically locks in place and seals. An internal collet performs the locking and a Buna-N O ring provides the seal (Fig. 4.30). To disconnect the tubing all that is required is to push the collet toward the fitting and pull out the tubing. Any semirigid tubing with closely held tolerances on the OD may be used. Tubing made from nylon, polypropylene, polyethylene, and Teflon are well-suited.

Since there is no nut to be tightened, as is typical of other fittings, instant fittings can be located in compact arrays (Fig. 4.31) because it is unnecessary to leave the usual space between adjacent fittings for a nut wrench. In addition, fittings provided with a male threaded connector on one end may be obtained with an internal hexagonal hole (Fig. 4.31) for the purpose of tightening with an Allen wrench, which allows a fitting to be removed or replaced from the center of an array without difficulty.

Fittings are available for plastic tubing to ½-in (12-mm) OD and are suitable for the normal working pressure of the tubing or a maximum of 250 psi (1500 kPa) and temperatures from 5 to 150°F (260 to 340 K).

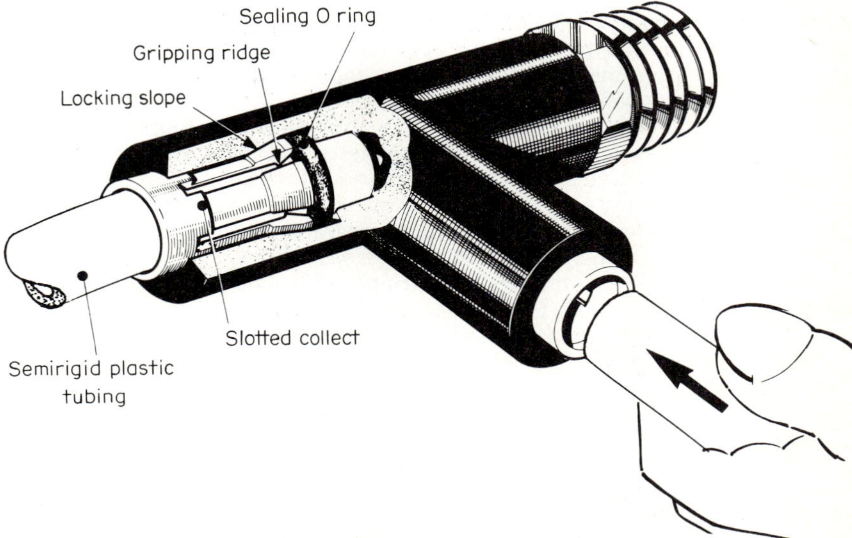

FIG. 4.30 Instant fitting for plastic tubing. *(Legris, Inc.)*

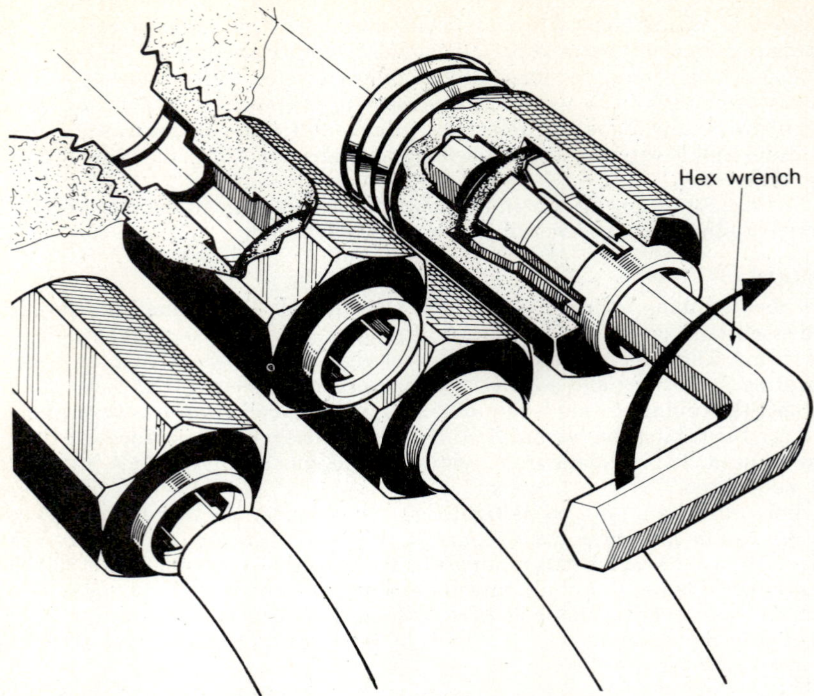

FIG. 4.31 Close-packed array of instant fittings. *(Legris, Inc.)*

Plastic Fittings

Fittings made entirely of plastic for plastic tubing and plastic fittings with a metal tube retainer for metal tubing are available. They offer two advantages: they are inexpensive, and they can provide complete similarity of material (tubing and fitting) throughout.

The all-plastic fitting comes either with a gripping and sealing sleeve integral with the nut (Fig. 4.32*a*) or with a separate plastic ferrule for gripping and sealing (Fig. 4.32*b*). The integral sleeve is for lower-pressure applications. For low pressure (25 psi, or 172 Pa) it is often adequate to tighten the nut with fingers alone, no wrench being required. The fitting for the separate plastic ferrules may also be finger-tightened, but it is best to use a wrench.

Plastic fittings with a stainless-steel gripper (Fig. 4.32*c*) provide the highest pressure capability because the tube is held more securely and is not so easily pushed out by the fluid pressure. Using metal tubing, working pressures as high as 500 psi (3 MPa) are claimed. However, since plastic materials cold-flow under stress, it is advisable to use much lower pressures (100 psi max., or 700 kPa). Long-term exposure to pressure causes the plastic bodies to expand slowly from cold-flow, and the effectiveness of the gripping device is reduced, resulting eventually in tube push-out.

Plastic fittings are available in Teflon, nylon, polyethylene, acetal copolymer, and polypropylene.

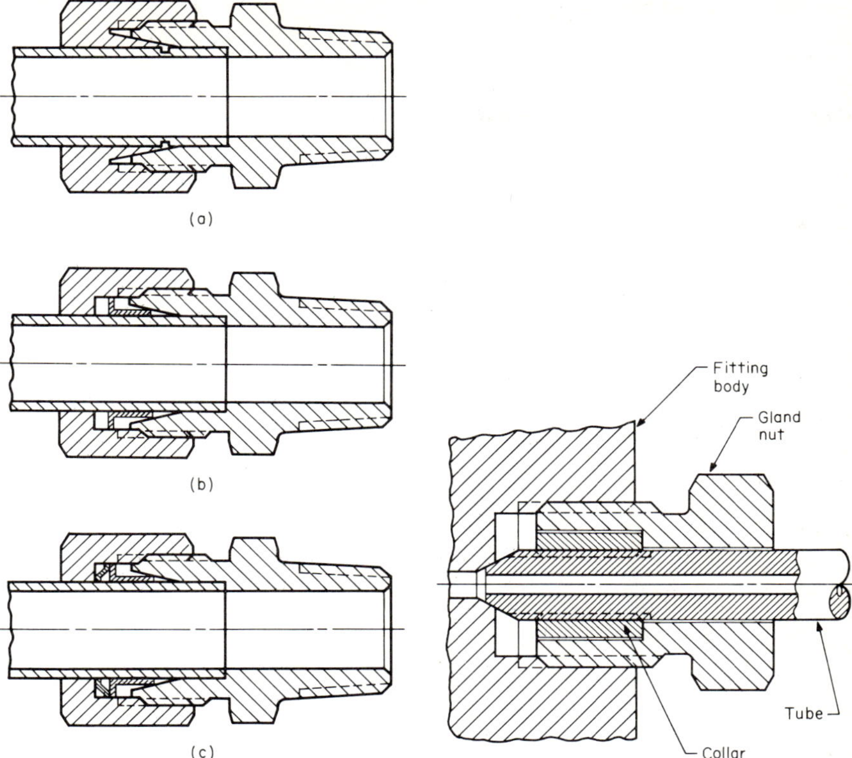

FIG. 4.32 Plastic fittings. (*a*) Integral ferrule and nut; (*b*) separate plastic ferrule; (*c*) metal tube gripper and plastic ferrule.

FIG. 4.33 High-pressure fitting.

High-Pressure Fittings

For very high pressure service, to 200,000 psi (1400 MPa), specially designed fittings are available. These all-metal fittings, which are naturally used with very heavy wall tubing, employ a specially shaped tube end to effect a seal (Fig. 4.33). The tube end is shaped to a cone and the tube exterior threaded. A special collar threads onto the tube. A gland nut screws into the body and forces the tube into a cone of slightly different angle at the bottom of the fitting cavity or socket. A line contact metal-to-metal seal is formed between the end of the tube and the body.

Special tools are available for coning and threading the tube ends. First cut the tubing to length with a hacksaw, then cone, and finally thread. Cutting oil is recommended for use with these tools, which gives better tool life and better cut finishes.

Fittings are available for tubing with ¼-, 5⁄16-, ⅜-, and 9⁄16-in (6-, 8-, 10-, and 14-mm) OD.

4.5 VALVES AND REGULATORS

Fluids are controlled primarily by valves and regulators and secondarily by pumps (Sec. 4.8). The types of valves and their manufacturers proliferate. Their applications frequently overlap, and often more than one type of valve will perform adequately. The technician can easily become deluged with manufacturers' literature. Nevertheless, this literature should be read carefully, using the material below as a checklist in the selection process.

Valve Types

In selecting a valve for a fluid system, there are a series of questions which the user should first answer. The answers will automatically narrow the selection of the valve types from which to choose. This is important because some preliminary screening is paramount, and ultimately all the questions must be considered anyway. The questions are:

1. What size valve?
2. What is the nature of the fluid?
3. What is the pressure and temperature of the fluid?
4. What type of control is required (e.g., on-off or throttling)?
5. What type of operator?
6. What is the life expectancy?
7. What type of end connections?
8. What space or weight limitations exist?
9. What special environmental conditions exist?
10. What special features are required?

Each of these areas is discussed below, and the principal valve types and their description and typical applications are listed in Table 4.19.

Valve Size

Valves are most often selected so that the end connections suit the piping system into which they are to fit. However, this consideration alone is not sufficient. The flow capacity of the valve must also be considered, otherwise the fluid pressure drop through one or more valves may be too high for the system to operate properly. The flow capacity for a valve is invariably given in the manufacturer's literature in terms of the valve flow coefficient, or C_v value. Unless otherwise stated, the C_v is always given for the valve fully open.

The flow coefficient is defined as the quantity of water (sp gr = 1.0) in U.S. gallons per minute that will pass through the valve when the fluid pressure drop is 1 psi. The C_v takes into account all the factors leading to pressure drop through the valve from the inlet plane to the outlet plane, not just the drop for the valve port or orifice alone. To determine the pressure drop for any liquid, the formula is

$$\text{Pressure drop (psi)} = \frac{Q^2}{C_v} \times \text{sp gr}$$

where Q = flow of liquid, gal/min

TABLE 4.19 Valve Types and Their Applications

Valve type	Description	Application
Globe	Generally a globe or spherical body with inlet and outlet connections in-line. A stem raises and lowers disk (or plug) in and out of an orifice. A common configuration for valves over ½ in size. Frequently all-metal construction. Wide size ranges available.	A most versatile valve for low to high pressure; cryogenic to liquid-metal temperature ranges. Excellent sealing capabilities in clean fluids. Long lasting and minimal maintenance.
Needle	Name applied to small valves where stem is needlelike with point fitting into the valve orifice. Different stem points provide a variety of flow characteristics.	For instrumentation lines and flow control in small pipelines and pilot plants. Provides very tight shutoff even to high pressures and high vacuums. Best service where no particulate matter in fluid.
Ball	Principal control element is a ball with a hole through it. A 90° turn of the ball takes valve from fully closed to fully open.	For quick action, full flow, and minimum pressure drop. Not well-suited to throttling. Wide pressure-range capability. Wide size range. Can tolerate some particulate matter in fluid.
Metering	For use with clean fluids where fine control or metering is required. Vacuum to high pressure.	Typically needle-type valve with small orifices and gradually contoured stem point to provide wide-range flow control. For clean or filtered fluids.
Gate	Orifice plate or gate slides across full flow port. Wedge-shaped gates provide tight seals.	Full flow capability with slow action. Some throttle capability. Vacuum to moderate pressures. For use with all types of fluid including slurries.
Butterfly	Rotating disk (circular or nearly so) operating in a circular duct provides quick-action full flow and low-pressure drop with 90° turn action. Wide size ranges available.	Quick on-off action and some throttle capability. For low-pressure systems. Can tolerate some particulate matter in fluid. A very versatile valve.
Toggle	Usually a small globe-type valve type provided with a 90° cam-operated handle.	Rapid action (flick of finger), full open from full closed. Moderate pressure and good sealing capability. Used mostly for instrumentation.
Relief	Designed to pass or relieve at a predetermined level, closing again when pressure decays below that value. Often capable of being set to relieve over wide pressure ranges.	For maintaining back-pressure on upstream pressure that is anticipated to fluctuate routinely. Not intended as a primary pressure safety valve where pressure increases above normal operating level are unintentional.
Check	Provides for flow in only one direction. Gravity-held or lightly spring-loaded ball or poppet easily opens under influence of fluid pressure in flow direction and is held tightly closed by fluid pressure with reverse flow.	For capturing fluid in a system in the event of failure or disconnect of fluid source. Prevents reverse fluid flow if one portion of system could produce a higher-than-normal pressure.

TABLE 4.19 (*Continued*)

Valve type	Description	Application
Safety relief	Provides rapid venting of system to prevent damage and accidents. Similar in operations to a relief valve except typically provides greater flow capacity at selected relief pressure and designed to function in the rare event that system pressure exceeds design value. Relieving pressure level often preset and adjustment device tamperproof.	For protecting system against pressure levels that would be dangerous as, for example, when system supplied from pressure source higher-than-normal operating pressure or when pressure may elevate through faster-than-normal chemical reaction rates, or increasing pressure from temperature, etc.
Diaphragm	A diaphragm, usually of elastomeric material, is forced against a weir or orifice by the valve stem to control flow.	For low pressures and tight shutoff. Used for viscous fluid, slurries, etc., as well as clean fluids.
Hermetically sealed	Hermetic sealing of stem and other valve joints is provided. Normally of two types: diaphragm-sealed or bellows-sealed. In either case the flexible member (diaphragm or bellows) is welded to stem and body to provide hermetic seal and positive containment. Mostly globe and needle type.	For high-vacuum service or handling hazardous materials, such as radioactive fluids, pyrophoric fluids, and highly toxic fluids. Also used for very high temperature service such as combustion bleed gases and liquid metals where normal stem seals have poor service life and for cryogenic liquids. For very high vacuum to high pressures.
Plug	Cylindrical or conical stem or plug closely fitted to body of corresponding shape. Hole through plug aligns with one or more holes in body. Small angular rotation of stem opens and closes ports to fluid.	Suited for quick action and selector valves (two-, three-, four-, or five-way). For clean fluid service and moderate pressures.
Control	Usually a globe valve with longer stem and plug travel. Plug contoured to give close control of flow area with stem travel. Often used with pneumatic or electric operators that position the stem and plug to regulate flow for a specific-magnitude signal to the operator.	For service where frequent changes in valve position (flow area) are required to maintain specific fluid flow rate. May be zero to full flow (i.e., on-off control) or fluid metering.
Solenoid	Valve control force provided by electromagnetism from an integral solenoid. Needle-type valve with solenoid surrounding stem. Two-, three-, and four-way valves common.	For primary service providing on-off operation. Remote operation to control operator fluid to pneumatic or hydraulic operators for actuating larger valves.
Selector	One port of valve can be selectively opened to two or more ports, usually one at a time.	For applying a single fluid source sequentially to a number of individual components or systems or for taking individual samples from many sources and directing them to a single instrument.

For example, for a valve with a C_v of 3.5 and a desired flow rate of 12 gal/min of alcohol at a temperature of 70°F (sp gr = 0.72), the pressure drop would be

$$\frac{(12)^2}{3.5} \times 0.72 = 29.6 \text{ psi (204 kPa)}$$

Of less significance for flow capacity are valves used for instrumentation purpose, where the flow is essentially zero and the valve is used so that the instrument can be isolated for calibration, servicing, etc.

Nature of the Fluid

The nature of the fluid—its chemical composition, viscosity, cleanliness, reactivity with the atmosphere, and toxicity—is of prime importance in selecting the type of valve and materials of construction.

Valves are widely available in stainless steel, carbon steel, and brass. However, these are the construction materials used for the main components (body, bonnet, stem, and plug), and other materials will be used for stem and seat seals, bonnet seals, etc. Therefore in selecting any fluid-control item it is important to look beyond the main materials of construction to all the components exposed to the fluid or environment, using the data of Chap. 2 to ensure complete suitability.

Do not use either brass components or soldered joints in systems where mercury is to be found. This applies not only to liquid mercury but also to its vapor. The vapor easily travels from the liquid along tubes and pipes to permeate an entire system.

Stainless-steel components, particularly where they are highly stressed (as with high-pressure service) and at elevated temperatures (400°F, or 500 K, and above), corrode with water containing chlorides. Since most common water supplies do contain chlorides, use demineralized or distilled water to avoid problems.

Pressure and Temperature of Fluid

The valve must be capable of withstanding the fluid pressure at the operating temperature. Besides the normal working pressure, there are other pressures the valve may be required to withstand. There may be a system hydrostatic test to be performed that typically will exceed the maximum operating pressure by a factor of 1.2 to 1.5. Perhaps pressure pulses or shocks are anticipated in the system, and their maximum value must be ascertained. Similarly, not only the normal operating temperature is to be considered but thermal shocks as well, such as those a valve encounters when first opened to high-temperature fluid.

Manufacturers' literature gives the pressure and temperature ratings of their valves, and most often it is stamped on the valve body or nameplate also. However, neither the nameplate nor the body are normally large enough to contain all the information required and the maximum working conditions that are typically used. It should not be assumed that both maximums can be used simultaneously, and usually a pressure-versus-temperature curve is available in the literature. Figure 4.34 is typical.

There are designs where the valve trim, stem or seat seals, is the temperature-limiting item of a valve. Frequently stem seals operate at lower temperatures than the fluid because they are removed from the mainstream. The pressure and temperature conditions given for valves refer to the mainstream fluid, not to the ambient temperature. Normally a valve will operate well in ambient temperatures to 120°F (320 K), and if higher temperatures are to be encountered, consult the manufacturer regarding the suitability of the product. Where a valve is specifically designed for high ambient temperature, such as for bake-out applications, the manufacturers' data will be specific in this regard.

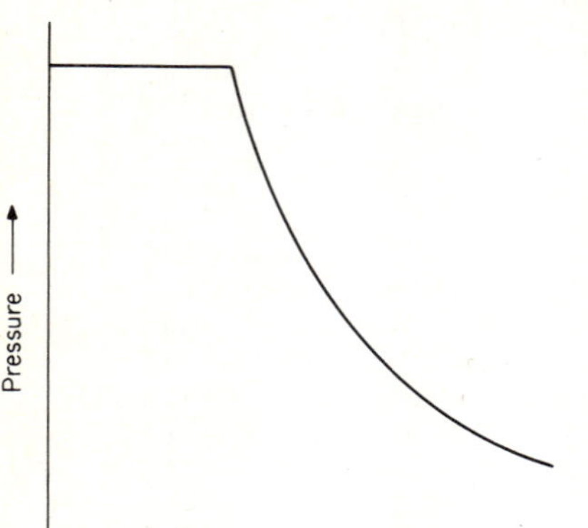

FIG. 4.34 Typical temperature-versus-pressure operating curve for a valve.

Type of Valve Control

The inherent capability of the valve to control the fluid passing through it is another important consideration. Valves provide on-off and throttling service. Those intended for on-off service are referred to as *shut-off valves*, whereas those intended to meter the flow over wide ranges are called *control valves* or *metering valves*.

While some essentially shut-off valves may provide a degree of fluid throttling by partial opening, control valves are specially designed to provide close control or metering of the fluid. Different flow characteristics may be designed into a control valve to meet specific flow requirements. Two types of characteristics are commonly used: *linear flow characteristics*, in which equal movements of the valve stem produce equal changes in flow, and *proportional flow characteristics*, in which either stem position or operator signal produces a specific flow (i.e., flow is proportional to a given signal).

Another factor in flow control is the allowable leakage when the valve is closed. Three leakage standards are in common use. One is determined by the amount of water (cubic inches per minute) permitted to pass when the pressure drop across the valve is specified. Another is a bubble test where an amount of gas (usually air) is permitted to pass the valve with a given pressure drop. Both these standards may be from zero indication over a specified time to some specific rate of leakage, such as five bubbles per minute. The last, and most demanding leakage standard, is a helium leak test (HLT), where the leakage rate of helium is measured by a mass spectrometer. Industrial mass spectrometers reliably measure to 2×10^{-10} mL/s (milliliters per second).

Type of Operator

The control capability of a valve is exercised through some form of operator, be it manual, pneumatic, hydraulic, or electric. Manual operation with a handwheel or lever is certainly most common. A circular handwheel is almost always used on a valve that requires more than one turn to bring it to its extreme of operation. Where only a quarter- or half-turn is needed, then a lever-type handle is usually employed. The latter is especially useful for indicating valve position, too, and is a distinct safety feature.

When a valve is to be operated remotely, when continuous readjustment is required to maintain desired flow, when special and precise time sequencing of a valve is required, or when especially fast or slow actuation is needed, then some operator other than a manual operator is needed. Most frequently a pneumatic operator is used. Pneumatic operators can provide linear motion (to replace a

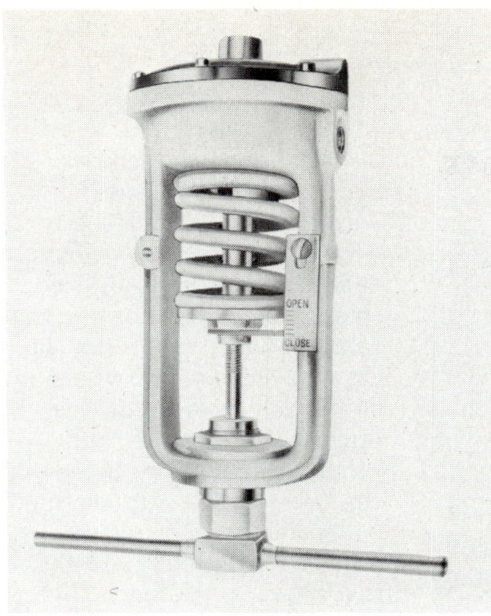

(a)

(b)

FIG. 4.35 Valve pneumatic operators. (*a*) Linear-motion operator with tube-end valve; (*b*) rotary-motion operator with three-way ball valve. *(Hoke, Inc.)*

stem thread and handwheel) or rotary motion (to replace the lever-type handle operation) (Fig. 4.35). They are capable of fast response by using a solenoid valve to control the operator air supply. Quite often the pneumatic operator is fitted with a strong spring (Fig. 4.35*a*) so that motion in one direction of valve stem movement is by spring force alone. The spring simplifies the operator air supply plumbing and valving but is more often used as a fail-safe feature (i.e., in the event of air-pressure failure the valve is moved to a safe condition, fully closed or fully open).

Electric operators are electric motors that drive the valve stem through a reduction gear (Fig. 4.36). Reversing motors are typical so that the valve can be placed in any position and moved from that position in either direction for more or less flow. In the event of an electric power failure, electric operators normally remain in their last position, which sometimes is the safest condition and may be selected for that reason. Often they are provided with a manual override so that the valve may be placed in the desired position in the absence of electric power.

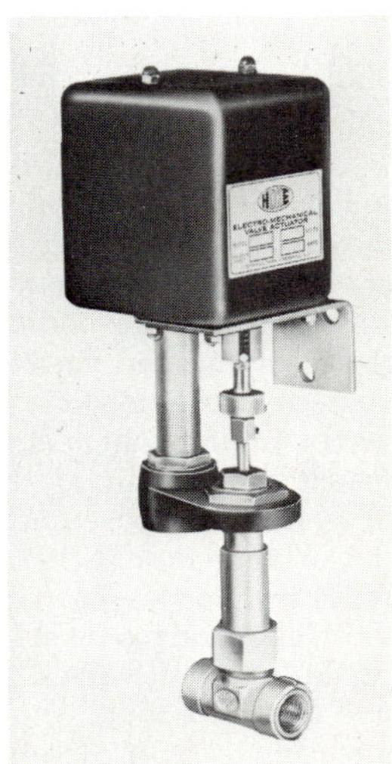

FIG. 4.36 Valve with electric operator. *(Hoke, Inc.)*

Life Expectancy

Valves, like most other devices, vary greatly in their ability to function properly over long periods of sustained use. The technician must decide how long the valve is required to be in the system and how many times it is likely to be actuated during that time. When the operational requirements are particularly arduous or the removal of a valve from a system likely to be costly, then either a valve of superior performance or one easily serviced is required. The most likely causes for limited valve life are seat leakage, stem-seal leakage, and stem-thread wear (where applicable). A number of features, sometimes optional additions and inherent in other designs, may be applied to give the desired life. Among them are specially hardened seats, removable seats, nonrotating plugs or disks, back seating, stem packing below the power threads, and union bonnet designs. The valve in Fig. 4.37 has many of these features.

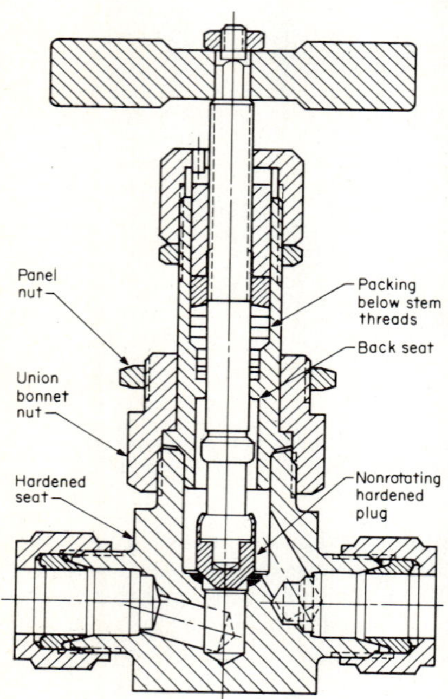

FIG. 4.37 Valve incorporating features for long service life. *(Hoke, Inc.)*

Hardened surfaces on one or both of the seating components reduces deformation and wear of the mating surfaces and also reduces galling. In addition, if a particle of dirt gets trapped between the sealing surfaces, it is less likely to cause damage if the seats are hard, and it may be flushed away by opening the valve slightly. A small particle on softer seats not only may leave a permanent impression and potential leakage path but also may become embedded in the seal surface. Hardened surfaces are provided by using heat-treatable materials (especially for small valves), by depositing special corrosion-resistant alloys (Stellite is most common), or by using hardened inserts.

Seat inserts in valve bodies are typically screwed into place and are made from specially hardened material. When screwed, the seat may be removed from the valve body and a new one inserted. In this case, the valve body remains connected to the piping, which is especially helpful if welded connections have been used because then removing the whole valve is a major undertaking.

Small, general-purpose needle valves often have the stem point or plug integral with the stem so that there is a rotary sliding motion as a needle point seats. Such motion, when the seating surfaces are brought together, obviously induces wear. While large valves invariably allow the stem to rotate independently of the plug or disk, for small valves this feature is not always provided but is required for a valve to have a long, troublefree seat life.

Back seating is the provision of a stem seat below the stem packing (Fig. 4.37). When the valve is fully opened (back-seated), the packing is relieved from the fluid pressure. This is not only an added safety feature should the packing deteriorate but also prolongs the packing life since it affords added protection during normal service.

Small valves often have the stem packing located above the stem threads. While this mechanical arrangement leads to simpler designs and lower-cost valves, it has its disadvantage. Fluid passing through the valve can remove the lubricant and lead to excessive stem-thread wear. When exposure of the stem thread to the fluid is likely to pose problems, then designs having the packing below the threads must be used.

Screwed or bolted bonnet valves provide access to the valve internals for servicing (e.g., seat replacement). Screwed bonnets have a tendency to gall at the threads. On small valves a Union nut can be employed (Fig. 4.37) to secure the bonnet to the body, and the threads are not exposed to the fluid. At the same time ready access to the valve seat is provided.

End Connections

The valve-end connections should be compatible with the piping system. If the valve is to be connected to pipe, then purchase the valve body with appropriate connections, such as NPT male or female threads, flanged, or with welding connections (butt or socket). If the valve is to be connnected with tubing, then provide one of the tube fittings of Sec. 4.4 or otherwise prepare for welded or brazed tube connection. Since a wide range of options is available, it is sensible to select the ones that make assembly easiest. The size of the connection will be dictated by the piping.

Space and Weight

The size of the valve must suit the installation and proximity of other plumbing and components. Valves placed in panels require special attention to space not only for the valve itself but also for sufficient room to assemble the connections behind the panel. Weight is seldom a factor unless the installation requires large valves.

Pneumatic, hydraulic, and electric actuators add significantly to the space requirements and to the weight of valves. Check closely the installation dimensions and weights of these assemblies.

Special Environmental Conditions

A number of environmental conditions that arise quite frequently and must be evaluated when selecting a valve are:

1. External heat: May be caused by radiant heat sources or by a requirement to heat the entire piping system to keep the material liquid (e.g., sodium).
2. Nuclear radiation: When handling radioactive materials or in proximity to nuclear radiation sources, certain materials, mostly seal materials, deteriorate quickly.
3. Seismic loading: In locations where earthquakes are anticipated, the piping and valves may be required to withstand anticipated shock. Valves with operators require especially close attention.
4. External leakage: Of special importance in handling toxic, radioactive, or pyrophoric fluids. Complete hermetic sealing is usually required, and valve selection is limited to ones having these design features.

Special Features

A number of other features are available that add significantly to installation, servicing, or operating convenience. They can often be obtained or added to a standard valve.

Soft-Seated Valves • For low pressures (vacuum to several hundred psi) and moderate temperatures (−60 to 250°F, or 220 to 400 K), valves with soft seats, such as Teflon or Kel-F, can provide excellent service, and, of course, with the mating part metal there is no wear. However, for these valves to last for long periods the fluids must be clean and the technicians well-instructed in their correct use if they are manually operated. Very little pressure and low handwheel torques are characteristics of soft-seated valves, and overtightening quickly deforms the soft material which then requires seat replacement.

Panel Mounting • Small valves that may be used in panels are frequently available with panel-mounting features. Figure 4.37 shows a valve with a panel-mounting nut that holds the valve securely in the panel with the other panel-mounted instruments.

Outside Screw and Yoke (OS&Y) • Valves from size about ½ in (12 mm) and up are often provided with an *outside screw and yoke*. That is, the stem thread is outside the valve bonnet and supported by a yoke (Fig. 4.38). The yoke is big enough to allow the stem packing to be changed while the valve is in service. To do this back-seat the valve, then loosen the packing gland bolts and slide the gland up the stem to provide access to the packing material. Remove the old packing and replace with new. Tighten the gland into place and the valve is ready.

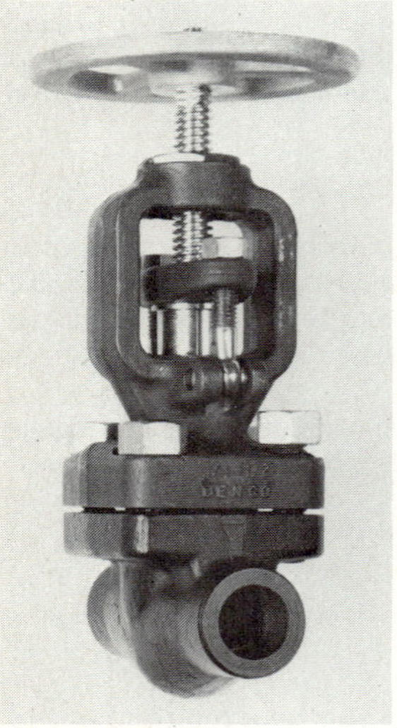

FIG. 4.38 Valve with outside screw and yoke (OS&Y). *(Hoke, Inc.)*

Regulator Types

Regulators fall into three types: (1) regulation or control of the *downstream* fluid pressure, (2) regulation of the *upstream* fluid pressure, and (3) regulation of vacuum. The first is by far the most common and is understood to be of that type when just the word "regulator" is used. The second type of regulator is usually termed a *back-pressure* regulator and operates similar to a relief valve in maintaining the upstream pressure constant, and the third is a vacuum regulator. Either the second or third type of regulator can be manual or remotely operated by a suitable fluid-pressure source (pilot pressure) or by electric actuation.

A common regulator used for regulating the supply pressure of a gas from a cylinder is shown in Fig. 4.39. This regulator is equipped with two pressure gages, one showing the upstream or cylinder pressure and the other the regulated or

FIG. 4.39 Cylinder pressure regulator with gages. *(Tescom Corp.)*

FIG. 4.40 Precision two-stage pressure regulator. *(Fairchild Industrial Products Div.)*

downstream pressure. The handle is turned to select the desired downstream-regulated pressure level.

Once the pressure level has been selected, it should remain at that value even though the upstream pressure may vary or the flow rate change, providing the correct-size regulator is used. Actually there may be some small change in the regulated pressure value if the upstream pressure change is large, as in the case of a gas supply from a cylinder, or if the flow rate change is large. When fine control of pressure is required under these widely varying conditions of supply pressure and flow, a two-stage regulator should be used (Fig. 4.40).

A two-stage regulator combines in a single unit two single-stage regulators in series. Therefore the pressure variations from the first-stage regulation are the only variations experienced by the second stage. Consequently the second stage operates from an almost invariant supply pressure, and its output or final pressure remains very constant.

When a regulator is to be used with a high-pressure gas supply, it is usual to use the supply gas itself as a servofluid. These regulators are referred to as *ballast regulators* and incorporate a ballast or pressure chamber as a reference pressure for controlling the downstream pressure. To reduce the regulated pressure, some gas must be vented from the ballast, and typically this is achieved using a single control lever or handle. A small movement of the handle in one direction charges the ballast chamber, while movement in the opposite direction discharges or vents it. The venting feature of these regulators precludes their use for toxic or flammable gases unless a vent line is attached for the safe disposal of vented gas.

A disadvantage of the ballast regulator is the variation in ballast pressure that occurs as a result of temperature change. The increase or decrease in ballast pressure in turn changes the regulator control pressure, and periodic readjustment is necessary to maintain constant regulated supply. It is not uncommon for regulators to become quite cold because of the expansion of the gas flowing through them. In fact, one sometimes sees regulators covered with ice that has frozen zen from the water vapor in the air. So temperature change is much more than that due just to ambient fluctuations.

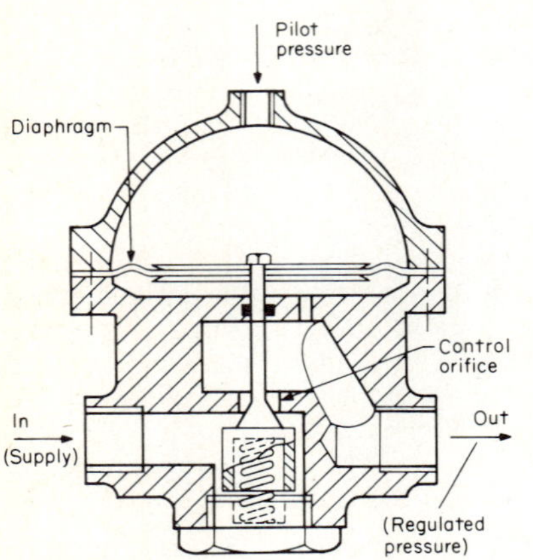

FIG. 4.41 Pilot-pressure-operated regulator.

To control the pressure of toxic or flammable gas or to control fluid pressure remotely, a pilot-operated regulator is used (Fig. 4.41). A normal regulator is used to supply the pilot pressure to the ballast chamber of a pilot-operated regulator. As the pilot pressure is increased or decreased, so the delivery pressure for the pilot-operated regulator increases and decreases. This same technique is also used, regardless of the nature of fluid to be pressure-controlled, when fluid flow rates are large. A small, conveniently located pilot-gas pressure regulator controls a large pilot-operated regulator.

Electrically controlled pilot-operated regulators are also available. Push-button control is a feature of these regulators, which otherwise are used in similar situations as the manual-controlled units described above.

Back-pressure regulators are usually remotely controlled, using a pilot pressure in much the same way as the pilot-pressure regulators. They are especially useful for wide-range selected pressure relief and can be installed without regard to accessibility, unlike the conventional relief valve that requires manual adjustment for each relief pressure selected.

Vacuum regulators are typically used between a vacuum pump and the systems to be evacuated. They are ideal for maintaining the vacuum at the desired level while the pump runs continuously. If a regulator is not used then the system gets pumped down to the full capability of the pump, which may be greater than desired. Laboratory models as well as large industrial units are available, and they are designed to handle almost any gas or vapor.

4.6 PRESSURE SEALS

Seals are considered here from the viewpoint of the technician, not the design engineer. The technician's task is frequently to effect a seal in a plumbing system or piece of apparatus. When repairing or servicing purchased equipment, purchase spare replacement seals from the equipment manufacturer or follow the manufacturer's recommendation for type and brand name replacements. Careful design and extensive experience go into the selection of seals in manufactured equipment, and to use different seals for convenience or cost saving is to invite problems.

Because there is considerable difference in how they are employed, if not in actual design, seals are discussed for their two types of application: static and dynamic. Static seals are used to seal two or more adjacent parts that have no relative motion (e.g., pipe joints). Dynamic seals are used when there is relative motion (e.g., shaft seals).

Static Seals

The most common, versatile, easiest to use, and least-expensive seal is the O ring. It is a circular cross-section ring usually made from an elastomeric material. For an O ring to seal, it must experience some initial squeeze between the components to be sealed and be adequately supported against the fluid pressure forces. It is the application of pressure which "energizes" the O ring to seal tighter (Fig. 4.42).

Even though the adjacent components may be static, the O ring itself actually moves as pressure is applied. For this reason, the component sealing surfaces must have a good finish (63 microinches, or 1.6 μm, finish or better), otherwise the O ring will wear from abrasion as pressure is applied and released. Even if the pressure does not fluctuate, similar finish ought to be provided. Otherwise continuous contact will not be provided and leakage will result.

O rings are available in a number of standard elastomeric compounds that cover a wide range of fluid compatibility. The common materials are Buna-N, Viton, silicone rubber, and polyurethane.

The most frequent use of an O ring is in a groove, as shown in Fig. 4.42. They may be used effectively, however, in other configurations, such as those shown in Fig. 4.43, and it is these alternatives that can often aid the technician most. The squeeze to be made on an O ring is typically 18 percent on the cross-sectional diameter, but since there are several materials readily available, it is better to consult the O-ring manufacturers' data for a specific recommendation.

O-ring seals can effectively seal to pressures of 100,000 psi (700 MPa). Failure attributed to the seal itself is often as much due to the movement of the components under pressure and the resulting changing configuration. The effect of increasing pressure on a simple O-ring installation is illustrated (and exagger-

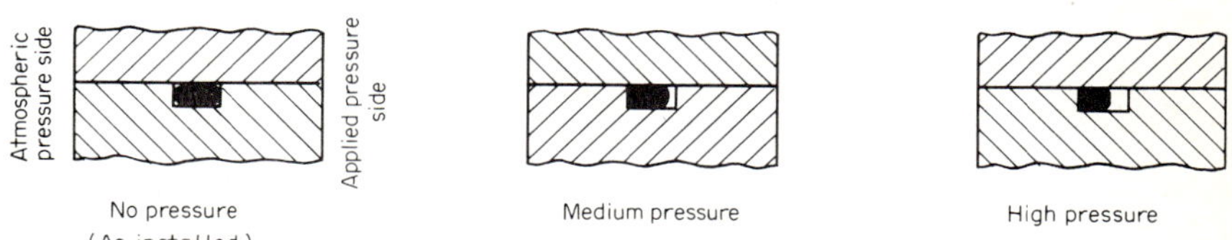

FIG. 4.42 Movement of static O ring under increasing pressure.

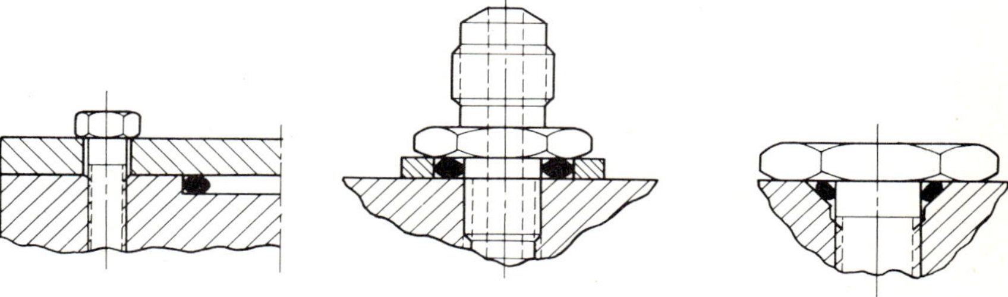

FIG. 4.43 Some O-ring applications.

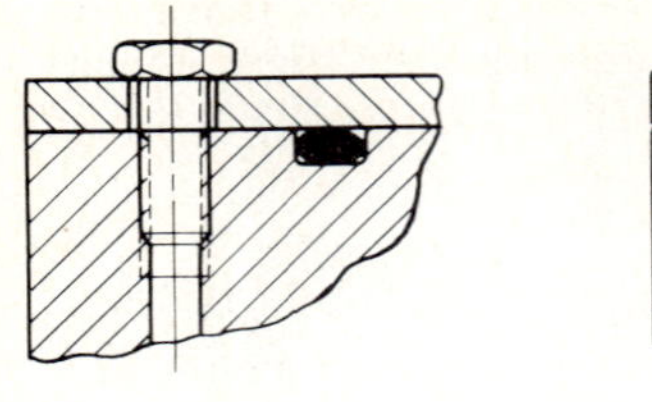

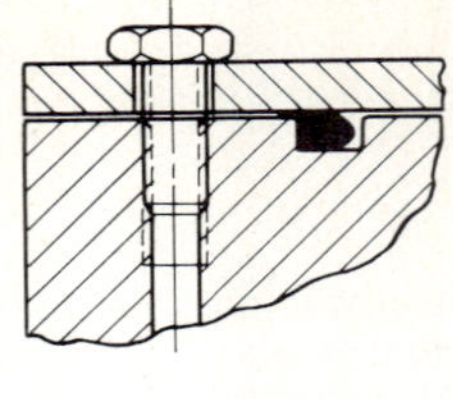

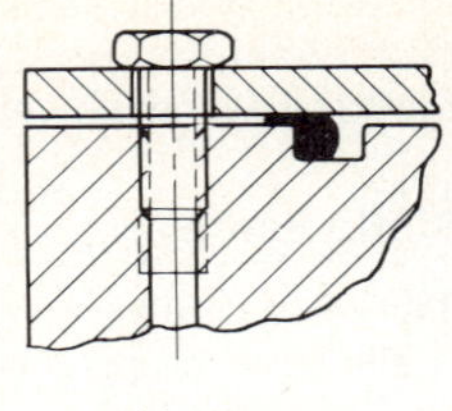

FIG. 4.44 O ring in high-pressure application.

ated) in Fig. 4.44. As the pressure increases the two components separate because the bolts stretch. The O ring starts to extrude into the gap between the components, and eventually a piece gets sheared off. Adequate bolting and component design (thickness) can delay this effect.

Teflon can be used as a very effective seal and is sometimes preferred because of its excellent compatibility with most chemicals. Since it is available in rod and sheet form it is simple to make special seals from it. Teflon is not elastomeric and does cold-flow (deform) under stress. Therefore the Teflon seal has to be compressed at assembly and restrained or encapsulated to prevent subsequent movement. Initial compression of 12 percent is typical.

A Teflon-sealed joint is shown in Fig. 4.45. Surface finishes of 32 microinches (0.8 μm) or better are required for the sealing surfaces. Seals to 25,000 psi (170 MPa) can be made this way for room-temperature service and somewhat lower if elevated temperatures are to be encountered. The maximum service temperature for Teflon is 450°F (500 K) for these applications. Again the seal will fail first from the separation of the components rather than from a failure of the seal material itself.

A solid slug of Teflon compressed with a simple gland screw (Fig. 4.46) makes an effective seal for rods, wires, etc. The higher the pressure the closer the components must fit to eliminate extrusion of the Teflon between the gaps. For service to 5000 psi (35 MPa), the gaps should be no greater than 0.003 in (0.08 mm).

While elastomers and plastics provide convenient sealing systems and are compatible with a very wide range of chemicals, they are limited in their temperature capability. When it is necessary to seal at higher temperatures (above

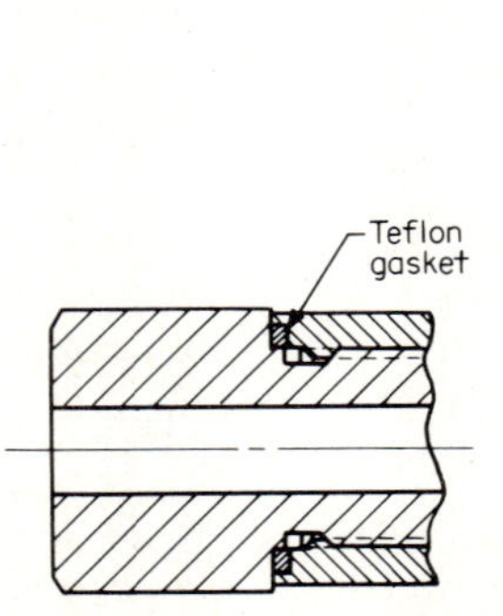

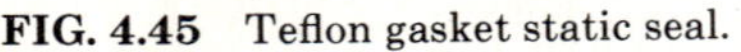
FIG. 4.45 Teflon gasket static seal.

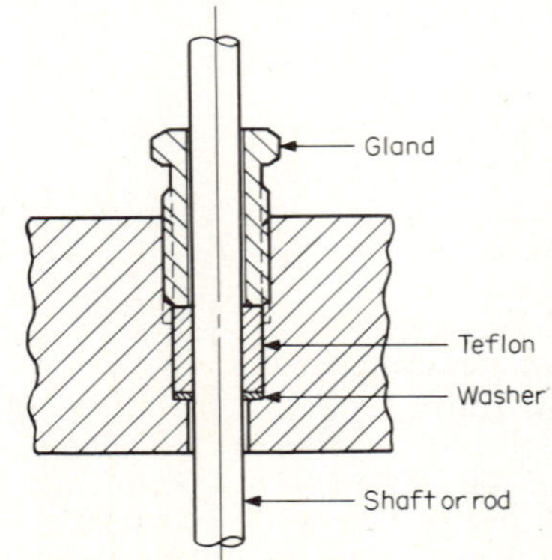

FIG. 4.46 Teflon bushing seal.

450°F, or 500 K), several proprietary metal seal rings are available. These rings require specific groove dimensions and especially good groove surface finishes (8 to 16 microinches, or 0.2 to 0.4 μm). For the technician it is often easier to resort to a metal-to-metal seal of simpler design. A typical metal-to-metal face seal on a plug is shown in Fig. 4.47. Good machine finishes (8 to 16 microinches, or 0.2 to 0.4 μm) are still required on the sealing surfaces, but the geometry is simple. The surface area of the seat must be small enough so that the load (stress) is sufficiently high to take the material beyond its yield point for best results. Sealing to 20,000 psi (140 MPa) and good to 1000–1500°F (800–1100 K) can be obtained in this manner provided the threaded or bolted connection will not yield under the application conditions. It is necessary to maintain the sealing load under the temperature and pressure excursions.

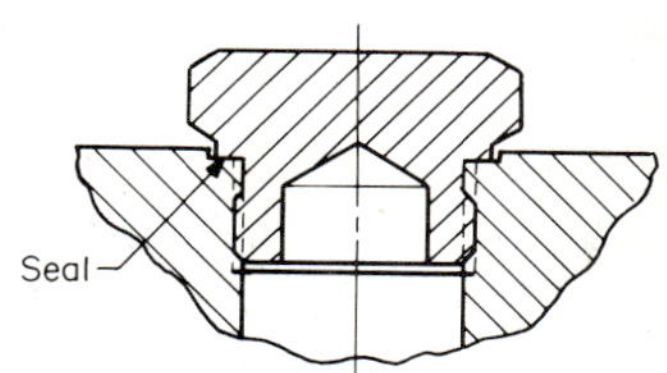

FIG. 4.47 Metal-to-metal static seal.

The surface finishes on simple metal-to-metal seals can be reduced if a soft-metal gasket or washer can be employed. Copper is a good gasket material, but sometimes it cannot be used because of fluid compatibility. Fully annealed stainless-steel gaskets may also be used. In any event, if large temperature fluctuations are to occur, the thermal coefficients of expansion of the gasket and seating components must be matched.

Dynamic Seals

Again, the O ring is a most versatile dynamic seal and can be used by the technician to pressures of 5000 psi (35 MPa) without difficulty. O rings can be used as dynamic seals much above this pressure, but then special design considerations are required. A simple O-ring installation for either oscillatory or rotary shaft motion is improved by using standard backup rings for pressures above 2000 psi (14 MPa). The shape and dimension of the O-ring cavity is more important for a dynamic seal. Excessive squeeze leads to high friction and poor life; too little squeeze causes leaks. Consult the O-ring manufacturer's catalog for groove size.

For good sealing and long seal life the moving component (shaft, rod, etc.) must have a good finish where it is to slide on the seal. Finishes better than 32 microinches (0.8 μm) are typical, and this is true regardless of the type of seal.

Where an O-ring material cannot provide the necessary compatibility with the fluid to be sealed, a Teflon seal can be made, as shown for a static seal in Fig. 4.46. The length of Teflon should be three to four times the diameter of the rod or shaft. Seals made in this way will seal well to 5000 psi (35 MPa) and where the sliding motion is not too fast (a few inches per minute, or about 0.003 m/s). With faster motion, friction-generated heat can raise the Teflon temperature to where it easily extrudes even past close tolerances of the seal housing and shaft.

An improved seal using Teflon can be made for rods and shafts by replacing the solid hollow cylinder of Teflon in Fig. 4.46 with alternate washers of Teflon and metal. The metal washers should provide only a small clearance between the shaft and gland cavity (about 0.002 in, or 0.05 mm, in diameter). When this stack of washers is compressed, the Teflon flows radially to effect a seal. Four or five Teflon washers suffice for pressures to 10,000 psi (70 MPa).

For service temperatures above 450°F (500 K), neither an O ring nor Teflon can be used. There are many compression packing materials on the market in the form of continuous strands of various cross sections. Even these are limited in temperature to about 900°F (750 K), and having a suitable size and shape strand on hand may present a problem. A versatile material of even better temperature capability for these applications is Grafoil (Union Carbide Corporation).

Grafoil is a resilient form of graphite and may be used to temperatures of 1500°F (1100 K). It is available in sheet form from which packing may be cut into rings of appropriate size. In packing Grafoil into the seal cavity, each ring must be compressed before the next one is inserted, otherwise the packing farthest from the gland screw does not get compressed as much as that close to it and a poor seal results. Grafoil packings of this type can seal up to 2500 psi (17 MPa).

Where even higher-temperature service is required, an all-metal labyrinth seal is one of the easiest to construct (Fig. 4.48). It comprises a series of metal washers separated by spacer rings. The torturous path presented to the fluid results in a lower pressure between each successive washer. A small leakage is the result. If this small leakage is objectionable it can be eliminated by using a sacrificial bleed gas. With a longer labyrinth seal a bleed gas is introduced at about the midposition by using a lantern-ring spacer (Fig. 4.49). The pressure of the bleed gas at the lantern-ring location is arranged to be high enough to prevent the process fluid from leaking. Instead, the bleed gas takes its place.

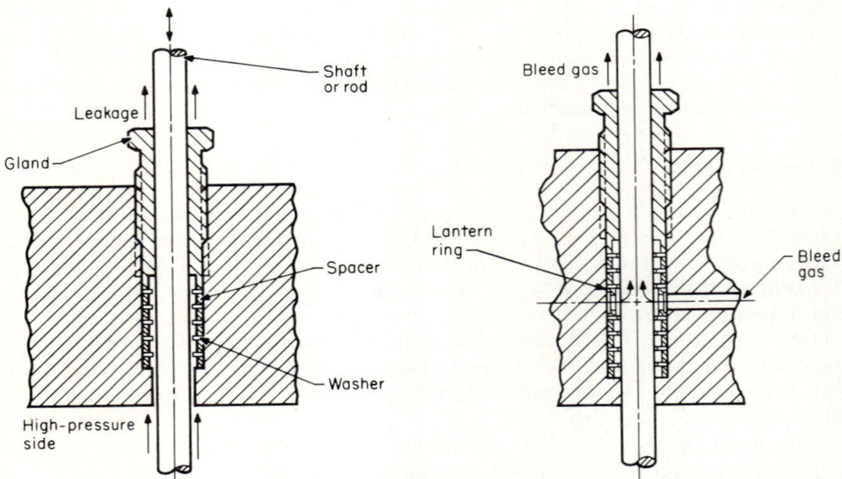

FIG. 4.48 All-metal labyrinth seal.

FIG. 4.49 All-metal labyrinth seal with bleed gas.

4.7 FILTERS

Filters are a key component in any fluid system. It is not sufficient to assume that because clean liquids or gases are introduced into a system that the system will remain clean. First of all, particulate matter often enters the system during assembly. A common source of particles is from tightening threaded joints, especially NPT threads, when sharp thread crests can be broken off and enter the system as small slivers. Also sealants, such as Teflon tape placed on NPT threads, can be cut into small shreds, some of which may not be trapped between the threads, and also enter the system. Other sources of particles are burrs from the cut ends of tube and pipe, loose scale from fluid-wetted surfaces which may build up with use, improperly cleaned components and piping, dust particles, wear particles from components, and sediments formed from fluid chemical changes. Many fluid components are damaged or their life shortened by the presence of particles in the system. Filters trap the particles and materially assist in providing long, troublefree service and should always be considered.

A filter presents some resistance to flow, and there will be a pressure drop across the filter that will generally be greater as the filtration capability improves (smaller particles removed). As the pores of the filter get clogged, the pressure drop increases or the flow is reduced. This increase in pressure drop is sometimes used as an indication of the filter condition. The finer the filter the quicker it becomes clogged. Obviously the selection of a filter is not simple, for if too fine a filter is used, unnecessarily high flow restriction results and frequent filter replacement is required. If too coarse a filter element is used, the system will not be adequately protected.

The pore size in a filter is measured in micrometers (μm).[1] (1 μm $= 1 \times 10^{-3}$ mm, or about 0.00004 in.) Particles larger than the pore size are retained, while those smaller pass through. The pores in a filter vary in size, and the degree of variation depends upon filter material and the technique used to manufacture it. Pore size ratings are either *absolute* or *mean.* The absolute rating is the largest particle that will pass through, whereas the mean rating is the average of the pore sizes. Pore sizes for metal screens start at about 500 μm and for cellulose filters go down to 0.01 μm. The size to choose will depend upon what the manufacturer requires for the best operation of equipment. It is not unusual to have several different filters in one system, each having different filtering capability to suit the component it is protecting.

In addition to removing particulate material from fluids, filters also are used to remove liquids from gases, especially oil and water from compressed air. Small droplets suspended in the gas collect in the filter element. Eventually one or more additional small droplets will impact and coalesce with the first one until a droplet is formed that is bigger than the pore size. The pressure drop across the filter element pushes these larger, coalesced droplets through the filter element where they run down the outer surface and collect in the bottom of the filter (Fig. 4.50). Normally a valve is situated at the bottom of the filter container so that the accumulated liquid may be drained periodically. Liquids do not plug filters, only the solid particles do that.

Filters are usually designed so that the filter element can be changed without disturbing piping connections (Fig. 4.51). More often than not, filter elements must be discarded rather than cleaned and reused. Replacing filter elements only takes a few minutes, but if the flow cannot be interrupted then two filters are piped into the system in parallel. Using two three-way valves, one upstream and one downstream of the filters (Fig. 4.51), the flow is diverted through one filter while the other is serviced.

The filter element itself is limited to the maximum pressure drop it can with-

[1]This unit was formerly known as a *micron* (μ).

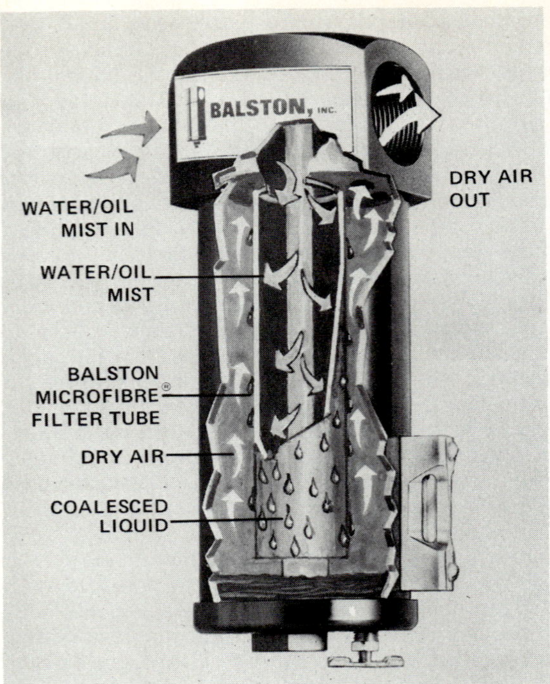

FIG. 4.50 Coalescing filter. *(Balston, Inc.)*

stand without rupture so that pressure protection is often provided with any upstream relief or bypass valve. Whether the exit of the relief valve is connected back into the system or is dumped to a reservoir depends upon how serious it would be to run the system for a time with unfiltered fluid.

Filter elements are available in a wide variety of materials to provide compatibility with all fluids. Common elements are made from sintered metal or ceramic, porous plastic, linen, cellulose, fiberglass, and paper. One filter capable of removing submicrometer-size particles uses a standard 500-sheet roll of two-

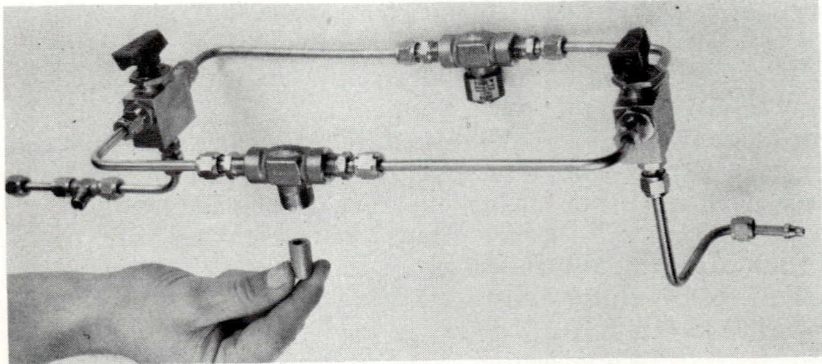

FIG. 4.51 Diverting flow through one filter while the element of the other is changed. *(Hoke, Inc.)*

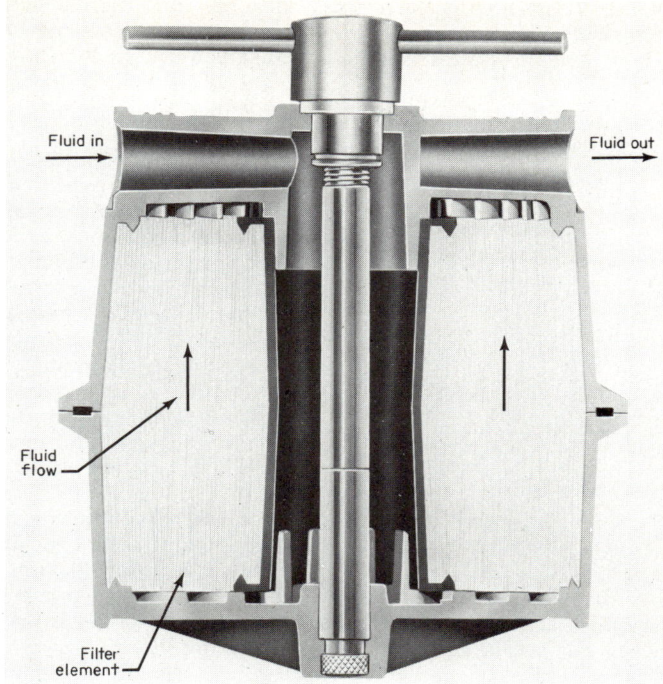

FIG. 4.52 Filter with cylindrical element made from roll of compressed paper. *(Motor Guard Corp.)*

ply bathroom tissue or a roll of nonwoven fabric (Fig. 4.52) and is well-suited for compressed-air systems, removing particles as small as 0.01 μm.

4.8 PUMPS AND COMPRESSORS

The technician is confronted with a wide variety of pumping devices, each with its own special characteristics that make it either suitable or unsuitable for the application. The devices are generally characterized as pumps, compressors, or blowers.

A pump increases the pressure of a fluid, usually liquid, often as a result of a flow restriction downstream of the pump outlet. If there is no downstream restriction, then little pressure is generated and little power is required to drive the pump. A compressor is used for air or other gases, and the pressure of the gas is increased to a higher value regardless of the downstream conditions. Typically, it will compress the gas to a certain limited design value only. If there is no downstream restriction, the gas immediately expands from its compressed value at the compressor outlet to atmospheric pressure. If the downstream restriction is very great, only the design pressure value is reached. Blowers and fans are very low pressure compressors and are used for moving large volumes of air or gas.

Pumps may be placed in one of two categories: positive displacement and nonpositive displacement. The positive-displacement pumps sometimes deliver the fluid in a series of pulses. Where pulsated flow is objectionable it can be smoothed by using a spring- (usually air-) loaded reservoir or accumulator, as in Fig. 4.53.

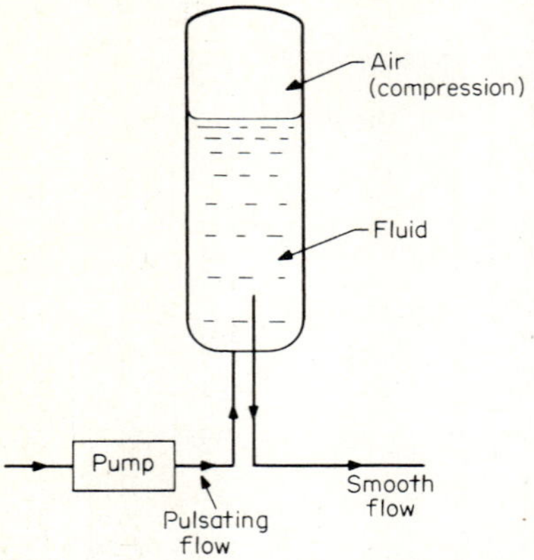

FIG. 4.53 Accumulator used to smooth flow from a pump with pulsating delivery.

If the downstream flow restriction is too great for the specific design of a positive-displacement pump, the liquid pressure continues to increase. If this is permitted the pump can easily be damaged by overpressure. To prevent damage a fluid-relief device, often a pressure-relief valve, is placed in the delivery line of the pump. The relief-valve setting is adjusted to be at or below the pump rated pressure. Usually the relieved fluid or "spill" is returned directly to either the pump inlet (Fig. 4.54*a*) or the fluid reservoir (Fig. 4.54*b*). Sometimes relief devices are an integral part of the pump.

The capacity or delivery capability of a positive-displacement pump is mostly proportional to the pump speed over fairly wide speed ranges (factors of 2 to 25 dependent upon design), which characteristics can be used to advantage (e.g., "Metering Pumps," Sec. 4.8). Where the speed is not easily changed and a varying flow demand is a requirement of the system, the excess pump capacity is again accommodated by using a pressure-relief device employing the system of Fig. 4.54.

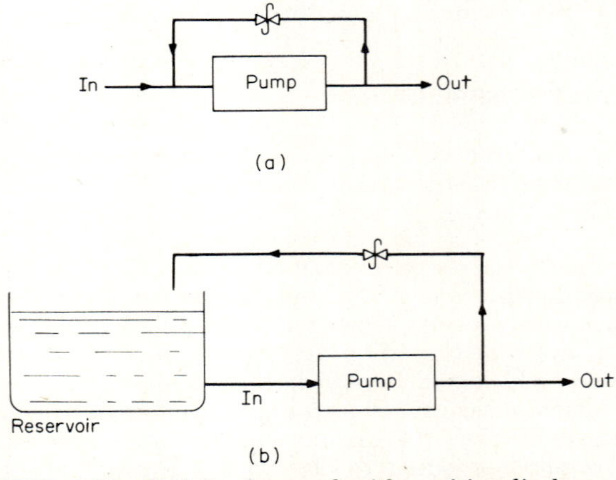

FIG. 4.54 Relief valve used with positive-displacement pump to prevent overpressure. (*a*) Pump "spill" returned to pump inlet; (*b*) "spill" flow returned to reservoir.

Nonpositive-displacement pumps increase the pressure of the fluid by first accelerating the flow and then slowing it down to convert the kinetic energy imparted to the fluid into pressure energy. Restricting the flow at the downstream side of the pump does not cause overpressurization. Most likely the pressure will drop somewhat because the efficiency of the pump drops at low flow. However, running the pump appreciably below its design flow by throttling will cause the fluid and pump to overheat and must be avoided. A temperature-measuring instrument must be attached to the pump if it is likely that this situation can arise from the intended use.

For pressures below atmospheric, a pump is usually referred to as a *vacuum pump.* While some regular pumps can also be used effectively as vacuum pumps, for higher-vacuum requirements (lower than 5 psi absolute, or 35 kPa) special pumps are needed (see "Vacuum Pumps," Sec. 4.8).

Another category of pump is the *metering pump,* which is specially provided with means to regulate the output fluid flow and to maintain the desired flow rate constant (see "Metering Pumps," Sec. 4.8).

Net Positive Suction Head (NPSH)

An important consideration in the installation of a pump is the net positive suction head (NPSH). There are two values of NPSH: that which is available, which depends upon the installation, and that required by the pump itself, which is inherent to the pump design and given in the manufacturer's literature. The available NPSH must always exceed the required NPSH, otherwise cavitation will occur either in the pump suction (supply) pipe or in the pump itself. The result of cavitation will be poor pump efficiency at best and no fluid pumped at all at worst. Operating a pump under cavitating conditions for extended periods, even if adequate pressure and flow are being obtained, can cause wear and damage and should be avoided.

The NPSH may be expressed in terms of *head.* It is the pressure equivalent of a column of water whose height or head in feet equals that pressure, or it may be expressed in pressure (psi). Since there is often interest in pumping liquids other than water, it is necessary to convert from the head of water to the corresponding head of the other liquid. To do this use the following formula:

$$H_{\text{liquid}} = H_{\text{water}} \times \frac{1}{\text{sp gr}_{\text{liquid}}}$$

To convert from head to pressure the formula is

$$P(\text{psi}) = H(\text{ft}) \times \text{density}(\text{lb/ft}^3) \times \frac{1}{144}$$

Always calculate NPSH *before* selecting and installing a pump, and consult an engineer if there are questions about correct vapor pressures, friction pressure drop, etc.

The available NPSH can easily be calculated from the desired installation by the formula

$$\text{Available NPSH} = P_{\text{total}} - P_h - P_f - P_{\text{vap}}$$

where P_{total} = total pressure as measured by an absolute pressure gage connected to a pitot tube in the pump supply line, or atmospheric pressure if an open reservoir

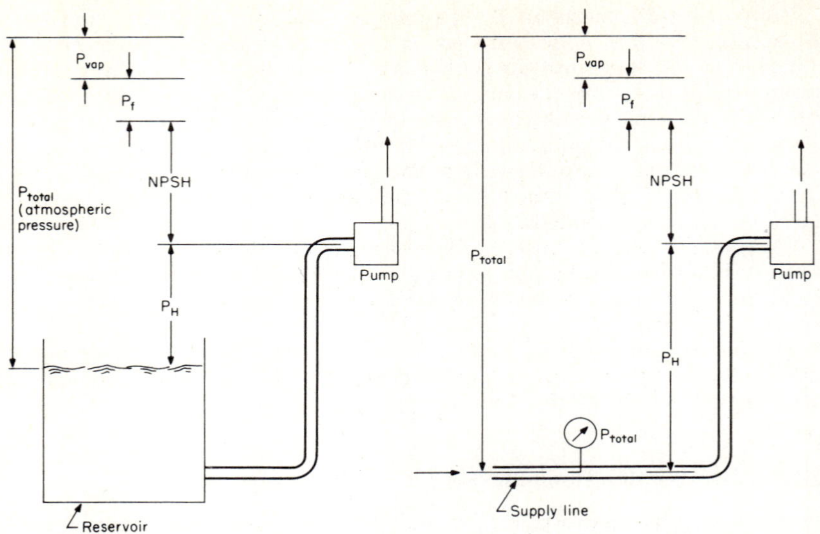

FIG. 4.55 The determination of available net positive suction head (NPSH) for two pump installations.

P_h = equivalent pressure to height or lift of liquid between the pump inlet and the position where P_{total} is taken (see Fig. 4.55)

P_f = pressure loss from pipe friction between the pump inlet and the position where P_{total} is taken

P_{vap} = vapor pressure of the liquid at the pump inlet for the temperature of the liquid. (For most liquids the vapor pressure increases significantly with temperature.)

Positive-Displacement Pumps

Peristaltic Pumps

These are ideal for laboratory bench-top pumps or may be used for sampling systems. Liquid flow rates to 2 gpm (1×10^{-4} m^3/s), liquid pressure to 50 psig (350 kPa), and gas pressure to 15 psig (100 kPa) are obtainable. The pumping principle is one where a fluid is pushed through a flexible (rubber or plastic) tube by successive squeezing actions of a moving cam or roller along a curved length of the tubing (Fig. 4.56). Nothing other than the tubing ever contacts the fluid. Tubing may be selected to provide compatibility with practically all chemicals. Fluid temperatures are limited to about 200°F (370 K) with plastic tubing and 400°F (480 K) with rubber (Viton) tubing. These pumps will handle slurries, gels, and liquids with viscosities to 1000 cP. Suction capability of 4 psia (27 kPa) or 25 ft of water is obtainable. Tubing life varies from 100 to 500 h depending upon type of tubing, temperature of fluid, pressure of fluid, and pump speed.

Peristaltic pumps eliminate fluid contamination and leakage because they are free from seals. Wide flow-range capability is provided by varying pump speed (see "Metering Pumps," Sec. 4.8), and increased flow capacity can be obtained by ganging pump head units (Fig. 4.57).

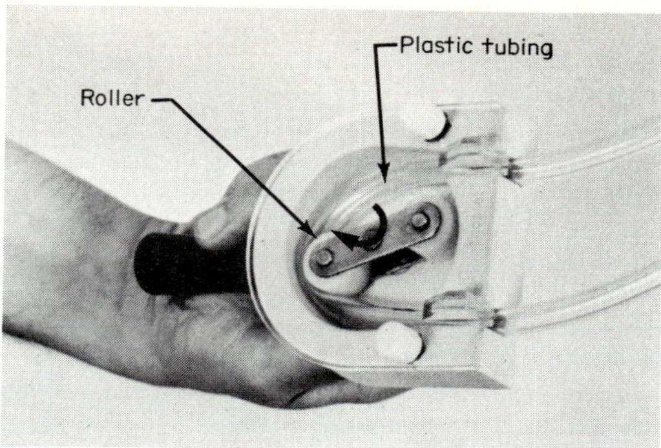

FIG. 4.56 A peristaltic pump with plastic tubing. *(Manostat, Inc.)*

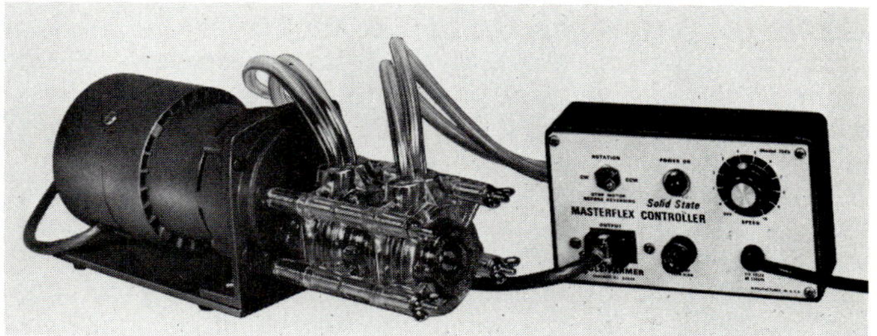

FIG. 4.57 Ganged peristaltic pumps with variable-speed drive. *(Masterflex Inc.)*

Vane Pumps

Vane pumps typically comprise a housing that holds a rotor with radial slots into which the vanes slide (Fig. 4.58). The vanes are spring-loaded so that they are held in contact with the housing. Lengthwise, the vanes are permitted very little

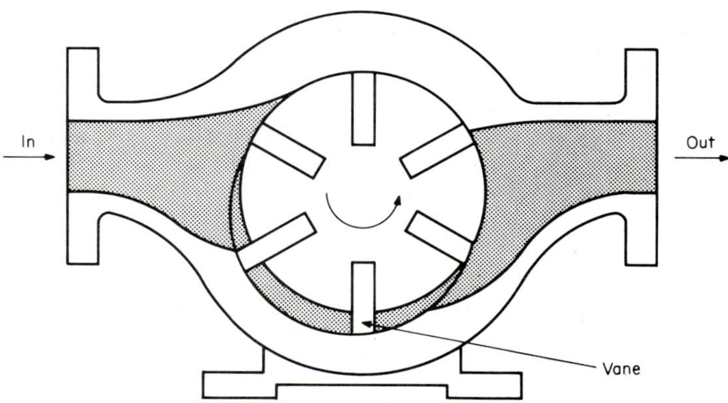

FIG. 4.58 Vane pump.

clearance within the housing. As the rotor rotates, the changing volume between the rotor and housing forces the fluid out at one port and draws fluid in at the other. Variations on the sliding vane include rollers guided by radial slots in the rotor and also nonsliding vanes that flex as the rotor rotates. Pump housings are available in a wide range of metals and plastics. Vanes are made from Teflon, plastic, rubber, and specially compounded materials to provide low friction and long wear life.

These pumps provide pressures to 300 psig (2 MPa) and capacities to 30 gpm (2×10^3 m^3/s), with fluid temperature limited to about 400°F (480 K). It is evident that they are for use with abrasivefree and low-viscosity fluids. They provide good suction capability (5 psia, or 35 kPa), and for their size deliver large flow rates. Similar designs are used for air compressors except that some compression of the gas between vanes is provided.

Gear Pumps

Gear pumps are available in two basic designs, external gear and internal gear. In the external-gear pumps, fluid is trapped between the gear teeth and the pump casing. As the gears rotate, the trapped fluid is transferred from one side of the pump to the other (Fig. 4.59). Meshing of the gears forces the fluid out of the gear teeth. In the internal gear pumps, the two gear wheels are placed one inside the other, and both rotate in the same direction but at different speeds (Fig. 4.60). Fluid is drawn to the gear teeth as they unmesh and is forced out as they mesh. A crescent-shaped stationary web helps to keep fluid between the gear teeth from being forced back toward the inlet as the pressure increases toward the point of gear meshing.

Gear pumps are for handling liquids and are capable of providing pressures to 4000 psig (30 MPa). While the pumps are capable of good suction (about 25 ft water, or 75 kPa), it is good practice to place a filter, typically 25-μm capabil-

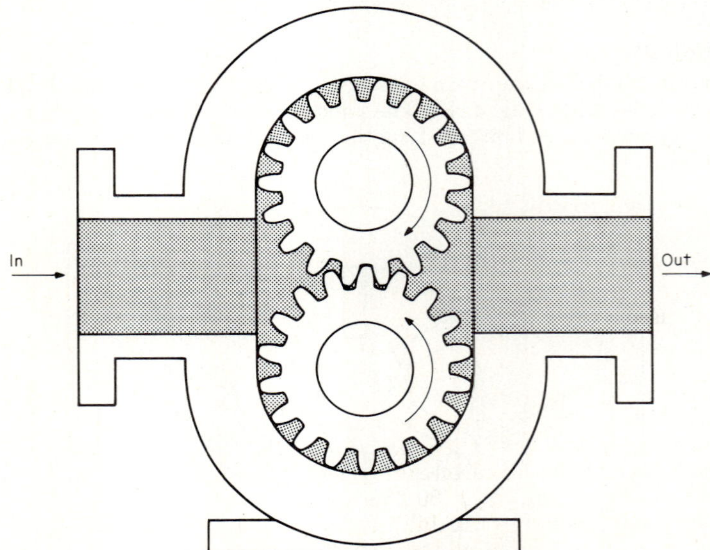

FIG. 4.59 External-gear pump.

ity, at the inlet to reduce the effective suction available. Clean or filtered liquids are necessary for gear pumps to provide long service life because particulate solids in the liquid quickly wear the meshing gears.

Piston Pumps

Piston pumps comprise a piston sliding in a cylinder, with two valves at the head end to control fluid entry and exit. Often these valves are spring-loaded ball-check or reed valves. The reciprocating piston draws in a fresh fluid charge as it moves away from the head and then discharges the fluid on the return stroke. Because of the intermittent nature of the fluid discharge, the flow is smoothed (and pump vibrations reduced) by using multicylinder units (Fig. 4.61). These pumps are constructed from steel alloys employing a wide variety of materials for piston and valve seals. They are well-suited for pumping oils, water, etc., at high pressure. Piston pumps are capable of very high pressure service up to 100,000 psig (700 MPa). While the basic principle is used for either compressors or pumps, the units themselves are completely different. Typical applications are for liquid atomization, industrial cleaning, high-pressure transfer, operating hydraulic equipment, and hydrostatic testing.

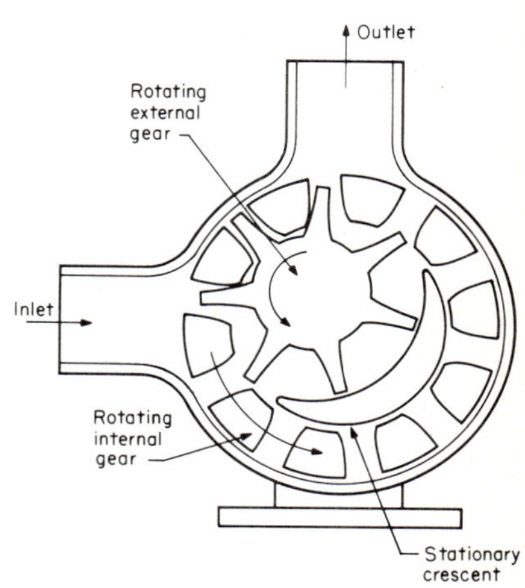

FIG. 4.60 Internal-gear pump.

For the higher-pressure applications, such as hydrostatic testing, piston pumps are available with direct-coupled piston-driven air motors (Fig. 4.62). Because the air motor only requires a standard shop air supply (100 psig, or 700 kPa), these pumps are as convenient for laboratory use as they are for industrial applications. The maximum pressure output from the pump can easily be controlled by regulating the motor air-supply pressure.

Hermetically Sealed Pumps

Of the pumps described thus far only the peristaltic eliminates seals and therefore possible leakage or contamination of the process fluid. But all-metal pumps that eliminate seals and provide the same advantages as the peristaltic pump but with increased life are available. These are bellows-seal and diaphragm-seal pumps.

The bellows-seal pump is similar to a piston pump or compressor except that the piston-to-cylinder seal is a metal bellows (Fig. 4.63). The valves are reed type. The piston head is shaped to accommodate the compressed bellows and provide a small clearance volume for good pumping efficiency. Construction from stainless steel is possible for all surfaces that come into contact with the media to be pumped. Vacuums to 24 inHg (20 kPa) and pressures to 40 psig (275 kPa) are available. The temperature capability is unique because the whole pump can be operated in an environment of 450°F (500 K), using an extended drive for the motor. Contrary to one's intuition, the bellows in these pumps have essentially an infinite life when operated within the recommended conditions. The corrosion

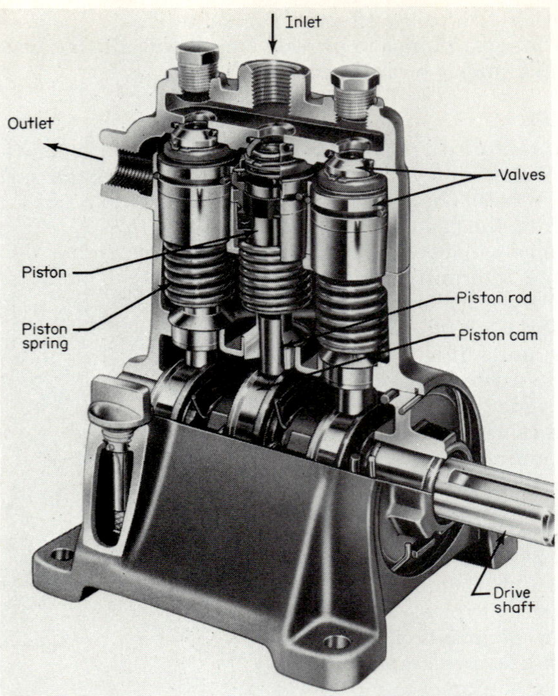

FIG. 4.61 A multicylinder piston pump. *(Hypro Division of Lear Siegler, Inc.)*

resistance, hermetic sealing, and high-temperature service make these pumps ideal for withdrawing and pumping gaseous samples to an analyzer and for pumping radioactive or toxic gases.

In addition to pumps for gases, bellows-seal pumps are also available for liquids (Fig. 4.64). The arrangement of the essential components is similar to the gas pumps. These liquid pumps are capable of delivering pressures to 300 psig (2 MPa) and provide a suction of 10 psig (70 kPa). They can also handle liquids to temperatures as high as 500°F (530 K).

Helical Vane or Rotary Screw Pumps

These pumps comprise two or more engaging helically vaned (screwlike) parallel rotors that entrap fluid between mating components and move it axially in a constant flow through the rotors. Figure 4.65 shows the principal components of a three-rotor pump: one main rotor and two idler rotors. No external gears are required to couple the idlers to the main rotor because the latter does the work while the idlers act as seals. Power is supplied only to the main rotor. No gears are used to drive the idlers.

Screw pumps provide extremely smooth flow, are quiet in operation and free from vibration, and have good suction characteristics. They are capable of delivery pressures to 3000 psi (20 MPa) and because of their high speed (3500 rpm) deliver high flow rates from a small installed volume. To maintain good hydraulic seals between the rotors and casing, fluids with viscosity in excess of 100 SSU at

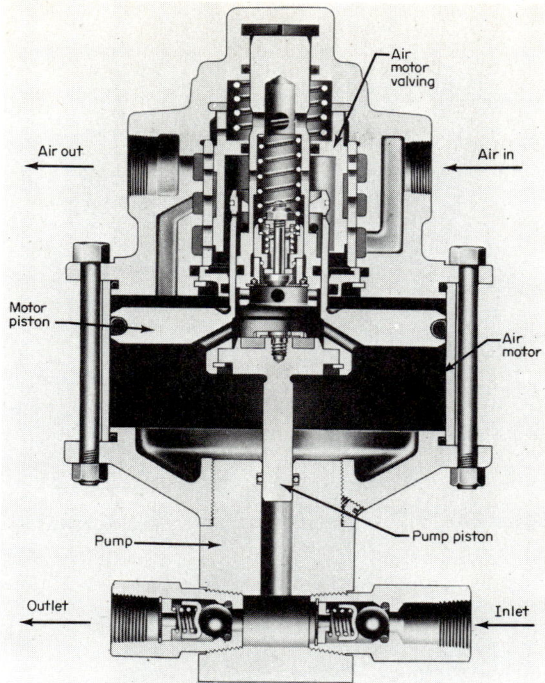

FIG. 4.62 Piston pump with coupled air motor. *(SC Hydraulic Engineering Corp.)*

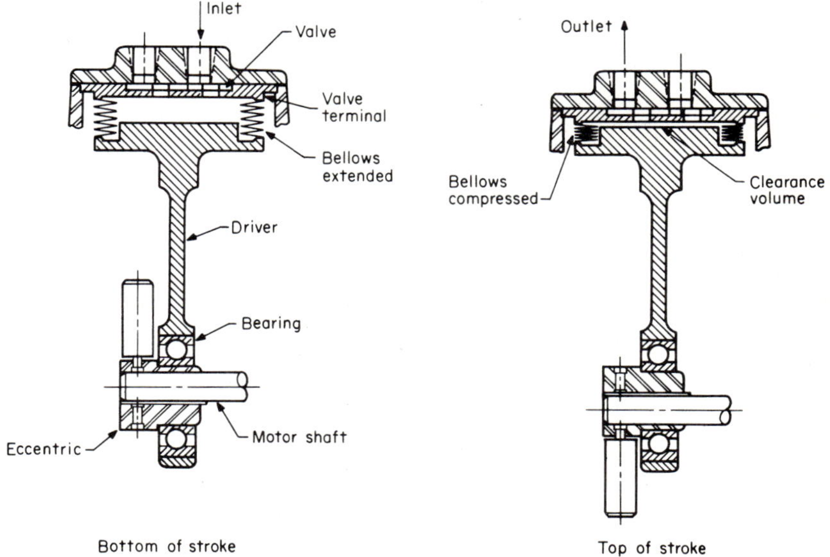

FIG. 4.63 Piston pump with metal bellows seal. *(Metal Bellows Corp.)*

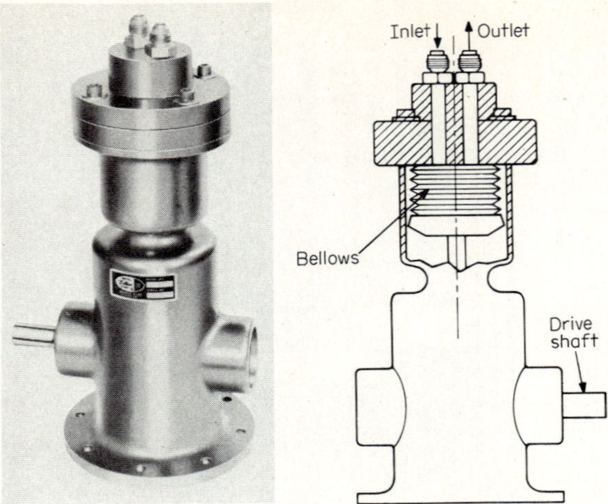

FIG. 4.64 Bellows-seal pump for liquids. *(Metal Bellows Corp.)*

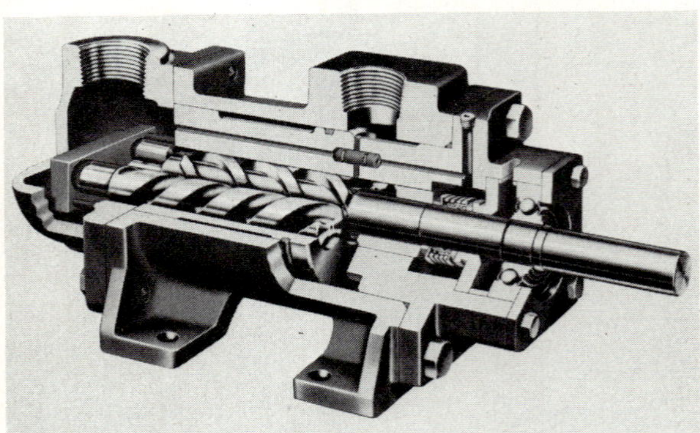

FIG. 4.65 Rotary screw pump with center main drive and two idlers (one visible). *(Imo Pump Div., Transamerica Delaval Inc.)*

pumping temperature are required. This precludes water as the fluid but includes oil-water emulsions, most oils, and fire-resistant liquids. Wide size-range pumps are available starting at 5 gpm (300 $\mu m^3/s$).

Compressors of similar design are also available. These machines are often oil-flooded to eliminate wear between the rotors and improve the seal. The oil is removed by separators and filters prior to use. For dry machines the rotors are kept in accurate mesh and clearance with gears at the end of the rotor shafts.

Nonpositive-Displacement Pumps

Centrifugal Pumps

By far the most common nonpositive-displacement pump is the centrifugal. It is simple in that it comprises only two main components: the impeller and the casing. Fluid is drawn in at the center of the impeller and delivered at the periphery of the casing (Fig. 4.66). Centrifugal pumps are available in a wide range of plastics and metals to suit all kinds of service. Pressures to 300 psig (2 MPa) are available although most of the pumps are designed for pressures of about one-third this value. Typical fluid temperatures are to 400°F (480 K).

These pumps must be primed (liquid-filled) for the pumping action to start. Frequently they are gravity-fed to facilitate starting.

Because there are no close-fitting parts, centrifugal pumps can handle particulate matter in the liquids. However, if the solids are particularly abrasive, then damage from erosion can occur because of the high fluid velocities internal to the pump.

Cavitation and aeration of liquids in centrifugal pumps are not uncommon. They stem from the presence of high local velocities and low local pressures that allows dissolved gases and vapor bubbles to form and to be present at the pump exit and may continue in the fluid for some distance in the downstream piping.

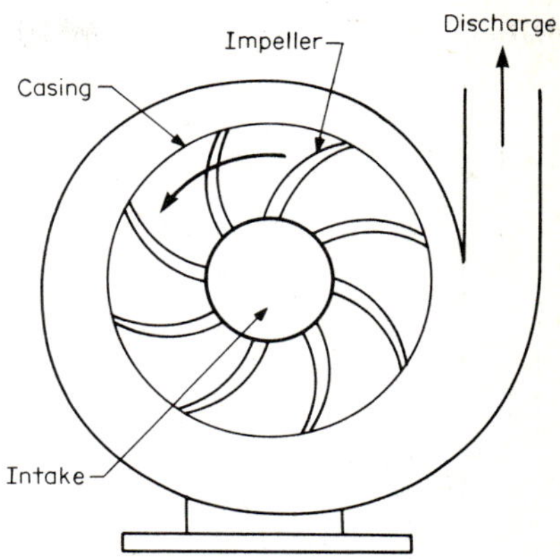

FIG. 4.66 Centrifugal pump.

Axial Pumps

Axial flow pumps are in reality a propeller in a duct. Their applications are for very large flows and small pressures. Since minimum sizes are about 300 gpm (1.9 mm^3/s), they are used in major installations rather than in laboratory or prototype work.

Metering Pumps

Positive-displacement pumps naturally lend themselves to controllable flow by varying pump speed. Some pumps are readily available with variable-speed drives. A peristaltic metering pump and a gear metering pump are shown with their speed controls in Figs. 4.57 and 4.67.

FIG. 4.67 Gear metering pump with pump (left), variable-speed drive (center), and electric motor drive (right). *(Tuthill Pump Co.)*

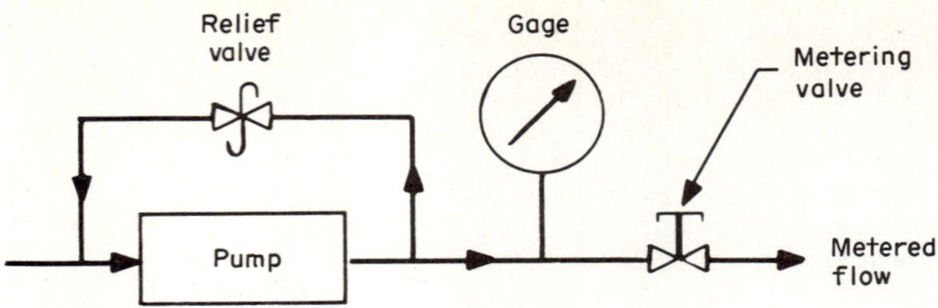

FIG. 4.68 Metering flow with a pump and metering valve.

Alternatively, metering can be done by using a metering valve in a constant-pressure supply system. Constant pressure is provided with an adjustable relief valve (Fig. 4.68). Pump output must exceed the required metered flow so that the relief valve maintains constant pressure to the valve. Excess flow is discharged from the relief valve and returned to the reservoir or pump inlet. The metered flow can be adjusted by adjusting either the metering valve or the relief-valve pressure setting.

Using a metering valve for such applications has some limitations. If the pressure downstream of the metering valve can change, then the pressure drop across the valve changes and so does the flow rate. Therefore a constant flow is only obtained if the downstream pressure is constant, a situation that must be investigated in advance if accurate metering is required.

Vacuum Pumps

Vacuum pumps are selected for a number of reasons. Among the most important are:

1. Ultimate low-pressure capability
2. Ability to maintain a specific vacuum or time to evacuate a system to a specified pressure (i.e., pump capacity or pumping speed)
3. Ability to deal with contaminants (e.g., water vapor or droplets)

Ordinarily the term *vacuum* is applied to any system operating at below atmospheric pressure. Vacuum pressure is measured in a variety of units. Absolute units, where zero represents a perfect vacuum, are most widely used. For low vacuums (i.e., relatively high absolute pressures) the pressure is expressed in inches of mercury absolute (inHgabs), pounds per square inch absolute (psia), or pascals (Pa). A vacuum of 8 inHgabs equals about 4 psia and 28 kPa (Fig. 4.69). For high vacuums it is common to express the pressure in torr or micrometers (μm). Both these units refer to the height of a mercury column (atmospheric pressure is 30 inHg, or 760 mmHg). A torr is equal to one millimeter of mercury pressure (1 mmHg) so that atmospheric pressure is 760 torr (that is, 760 mmHg). The micrometer is one-thousandth of a millimeter of mercury so that 1 torr is equivalent to 1000 μm, or 133 (Pa).

Another method used to express vacuum pressure is by reference to the value below 1 atm. For example, a pressure of 25 inHg vacuum is the same as 5 inHgabs.

Vacuum pumps can attain pressures as low as 1×10^{-12} torr (10^{-10} Pa). Such high vacuums are capable only by using special equipment and techniques

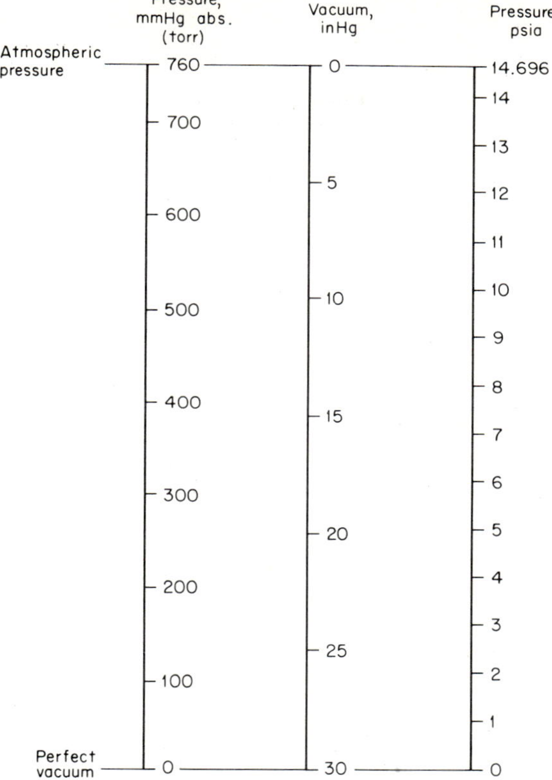

FIG. 4.69 Vacuum pressure scales in common use.

beyond the scope of this book. Standard laboratory pumps provide vacuums to about 1×10^{-3} torr (0.1 Pa), and many applications only require vacuums of this magnitude.

For some applications two pumps, each having different characteristics, are best-suited. For example, one could be a roughing pump, having a high pumping speed but poor ultimate pressure capability, and the other a fine pump, with a low pumping speed but providing high vacuum. The roughing pump quickly evacuates the system to a modest pressure; the fine pump provides a high vacuum. Initial pump cost may be reduced this way and considerable time savings obtained in reaching the desired vacuum.

Where the demand for vacuum is intermittent, then a vacuum reservoir (vacuum tank) may be used with a small pump. Very high effective pumping speeds can be obtained this way, and it is especially useful where high vacuum is not required.

Sliding Vane Vacuum Pumps

Sliding vane or rotary piston pumps are used widely for laboratory and industrial work. The principle of operation is similar to the positive-pressure sliding vane pump described in "Positive-Displacement Pumps," Sec. 4.8 and shown in Fig. 4.58. One important difference between the two is that the vacuum pump runs in oil. The oil provides excellent sealing between the moving parts, and the vacuum capability is provided by attempting to increase the volume of oil between the vanes as they rotate. Since the oil cannot expand, a void is formed between the vanes, and the pressure in this void is no more than the vapor pressure of the oil. The vacuum connection of the pump is ported to the position in the pump casing where this void occurs. Using special oils of very low vapor pressure, good vacuum is obtained. Vacuum pressures of 1×10^{-3} torr (2×10^{-5} psia and 0.1 Pa) are typical for this type of pump. Since the moving parts are well-lubricated, these pumps have a very long service life.

Manufacturers' recommendations for oil replenishment should be strictly followed. The oils are special, and only recommended formulations should be used.

Sometimes these pumps are designed to run quite hot so that contaminants, especially water vapor or droplets, that become entrapped in the oil can be more easily vaporized and removed in the exhaust cycle.

Liquid-Ring Vacuum Pump

The liquid-ring pump operates on a principle not too different from the sliding vane pump. The essential difference is that the volume expansion and contraction is provided not by sliding vanes but by impeller vanes being withdrawn and inserted into a circular liquid surface. Figure 4.70 shows how this is achieved. When the pump is at rest the casing is about half-filled with liquid. During operation the rotation of the impeller distributes the liquid uniformly around the inside of the casing, being maintained there by centrifugal action because the liquid rotates with the impeller. In particular, the inner liquid surface or liquid ring is concentric with the casing. The impeller, on the other hand, is eccentric to the casing. Therefore, as the impeller rotates, a void is formed in the space between the impeller vanes and the liquid ring on the left side (Fig. 4.70), and gases are drawn in via the suction port. The gases are subsequently compressed as the space between the vanes is reduced on the right and are forced out of the exhaust port as the impeller rotates.

Because the vanes are not in contact with the casing, no special lubricating qualities are required of the liquid in these pumps. Often the liquid is water when the vacuum capability is 150 torr (20 kPa). If oil is used as the liquid then

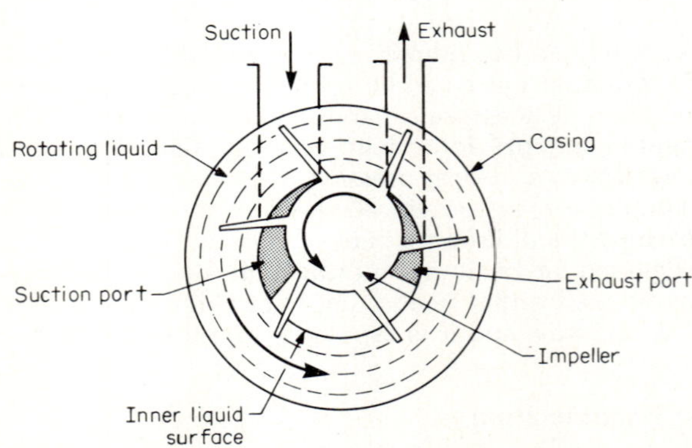

FIG. 4.70 Liquid-ring vacuum pump.

pressure to 10 torr (1 kPa) is possible. The ability to use different liquids in liquid-ring pumps is a distinct advantage especially in dealing with certain vapors when wide-range compatibility can be provided.

Ejector Pumps

Ejector pumps have no moving parts and use a liquid (usually water) or a gas (usually air or steam) as a working fluid. The principle of operation is to entrain gas on the suction side with a high-velocity jet (Fig. 4.71). The jet velocity for liquid ejectors is low compared with air or steam pumps because the latter can be accelerated greatly by using a convergent-divergent nozzle. Liquid ejectors can provide vacuums to 25 torr (3 kPa) and air ejectors to 3 torr (0.4 kPa).

A particular advantage of ejector pumps is that they can handle particulate matter in the vacuum system. They can also handle larger amounts of liquids than other pumps.

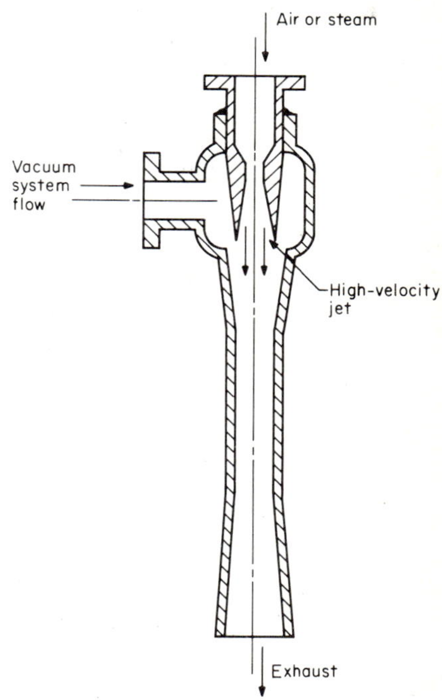

FIG. 4.71 Ejector vacuum pump.

4.9 GEARS

Gears provide a positive rotary drive from one shaft to another. The shafts may have their axes parallel or inclined, and they may rotate at the same or at different speeds. Once selected, shaft speeds and their relative orientations are not easily changed because gears must be positioned accurately and their shafts supported securely for correct functioning so that the installation is quite permanent. Therefore before selecting gear drives one must be sure of the speeds and shaft rotational directions required. In the laboratory, common applications for gears are remote drives through walls and vessels, multipower takeoffs for agitators or for cycle-testing several items simultaneously, and increasing or decreasing speeds of available motor drives.

Where possible, the technician is advised to use "packaged" gear assemblies to satisfy gear-drive requirements. These packaged assemblies eliminate the need for precision alignment of support shafts and bearings. Figure 4.72 shows a number of off-the-shelf packaged gear assemblies.

The starting point in deciding upon a gear assembly is to determine the power (horsepower) to be transmitted. Most likely an electric motor will be the drive source, and its horsepower and shaft speed will be found on the nameplate. This horsepower, however, is that which the motor is capable of producing, but if the load does not require it, then the power will not be produced. If the selected motor is chosen out of convenience it may be that it is overpowered for the job, and there is no need to get a gear drive for more power than the load really needs.

(c) (d)

FIG. 4.72 Packaged gear assemblies. (*a*) Compound helical-gear reducer. *(Renold Inc.)* (*b*) Crossed helical-gear angle drive. *(Tol-O-Matic, Inc.)* (*c*) Bevel- or miter-gear angle drive. *(Johnson and Bassett, Inc.)* (*d*) Worm-gear assembly. *(Eaton Corp., Industrial Drives Div.)*

The size and cost of the gear drive will probably be reduced considerably. Next determine the gear-drive ratio required to change the motor speed to the required speed. For example, a gear drive with a reduction of 3:1 will reduce the motor speed by 3. (Do not assume that a gear reducer can be operated as a gear increaser without the manufacturer's concurrence, because often they are designed for one particular mode of operation only and may not be suitable for the reverse mode except, perhaps, at an entirely different power and speed.) Note that the torque increases as the speed reduces so that a 3:1 gear reducer will increase the torque approximately by a factor of 3. (It would be exactly 3 if it were not for the inefficiency of the gears.) The power and drive ratio are the first two steps for selecting suitable gear drives, and once this selection is narrowed then check the speed capability of gears. Finally, determine the efficiency of the gear drive to verify that the transmission losses are not so high that a larger drive motor is required. If the power required is 2 hp and the drive efficiency is 80 percent, then a drive motor of 2.5 hp (that is, 2/0.8) is needed. Gear-drive efficiencies usually fall in the 75–95 percent range for each pair of mating gears.

The purchaser will find a wide price variation in gears that provide the same basic requirements of speed ratios and power capabilities. This variation arises from the precision and type of gear employed. Simple straight-toothed (spur) gears are the least expensive. However, ones used for high-speed and large load-carrying capacity, long life, and quiet running cost more than slow-speed, low-load, and noisy gears because the former are more precise and the finish on the contacting faces (flanks) is much smoother. Helical gears (Fig. 4.72*a*) have their teeth cut at an angle to the gear axis, are generally quieter, and have superior load-bearing characteristics than spur gears and cost more. In addition, helical gears develop an end thrust that requires the use of a thrust bearing on the shaft. Bevel gears are for angle drives, usually right angle, and the common tooth form is straight. More complex bevel gears (e.g., spiral bevel, hypoid) are much more expensive and used for specific applications for quietness, smoothness, and where the intersection of the shaft axes is objectionable. Crossed helical gears (Fig. 4.72*b*) are also used for right-angle drives and modest ratios up to 2:1. Worm gears (Fig. 4.72*d*) are used for high reduction ratios (for example, 20:1) and provide high-torque advantage. Usually worm gears cannot be driven in reverse (i.e., the larger wheel cannot drive the smaller worm gear). This zero back-drive characteristic of worm gears is sometimes used to advantage when the large wheel holds position unless the worm is rotated, thereby providing a position lock even though there may be torque applied to the large wheel.

If one decides not to use packaged gears, there are some guidelines to be followed that will help prevent service problems. Neither speed reductions of more than 10:1 nor speed increases greater than 5:1 may be made with a single pair of spur gears. If these ratios are to be exceeded, then use a compound gear arrangement. Figure 4.73 illustrates a compound gear arrangement using only two meshing pairs, and Fig. 4.72*a* has three pairs. The overall reduction of a compound gear is the product of the individual ratios (e.g., a 3:1 and a 4:1 ratio yield a 12:1 ratio).

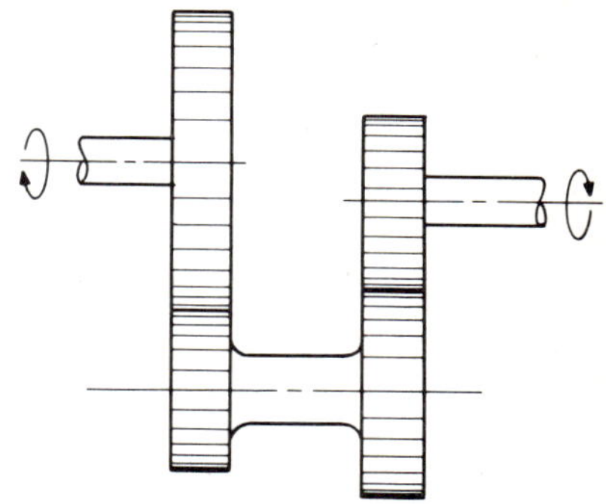

FIG. 4.73 Compound gear arrangement.

Multiple parallel drives should be provided with a common drive shaft and bevel gears located suitably along it, as shown in Fig. 4.74*a*. Long straight-spur gear trains for multiple drives (Fig. 4.74*b*) should be avoided because the power requirements become large due to gear inefficiency. For example, the six-spindle drive of Fig. 4.74*b* has six pairs of meshing gears. If the efficiency of each pair is 85 percent, then the motor torque and power for the end spindle alone must be increased by a factor of 2.65 ($1/0.85^6$) because of gear friction.

Gears are mostly made from steel, and the surfaces of the teeth are often case-

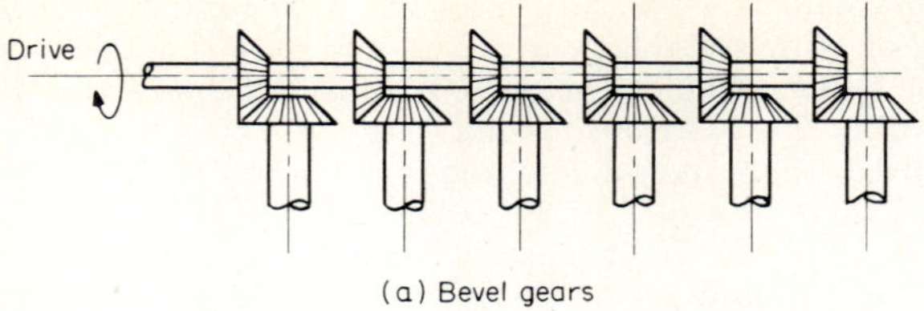

(a) Bevel gears

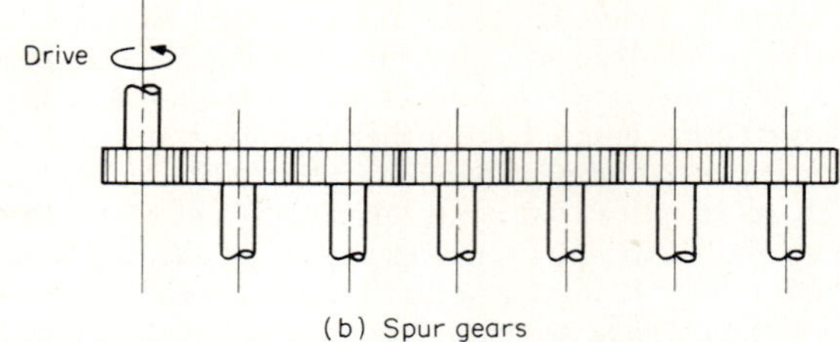

(b) Spur gears

FIG. 4.74 Gearing for multiple parallel drives.

hardened to afford higher load bearing (high torque) and longer life. In all types of gears there is considerable sliding of one tooth against another, requiring adequate lubrication with very clean oil. Small gears for low loading may be made from molded plastic (e.g., nylon and acetal) and are often ideal for laboratory use where the primary reason for gear selection is a positive drive. Plastic gears run very quietly, run mostly without lubrication, and are inexpensive.

4.10 TIMERS

Timers are an essential item for almost every laboratory. They may be used to measure time intervals, initiate and control a sequence of operations, repeat operations at desired intervals, and perform tests and record results in the absence of personnel. Because development work often entails endurance and life-cycle testing, timers are ideally suited for controlling such tests on a continuous basis to minimize the elapsed time. Furthermore, failure-detection devices, for example, pressure and temperature switches, can be used in conjunction with timers to terminate tests and safely deactivate a system without human intervention. Indeed it is sometimes safer and better to have complicated sequences of operations initiated and terminated by these devices rather than to rely upon fallible humans. With a little ingenuity the technician can devise very sophisticated control systems by using timers, relays, electric motor drives, mechanical switches, solenoids, solenoid valves, pressure and temperature switches, and limit, proximity, and vibration switches.

Two types of timers are in use. One measures elapsed time between two events and is typically fitted with a clock face or dial or digital readout. These timers are triggered electrically or mechanically. Ordinarily used to measure the time a piece of equipment has been functional, they may also be arranged to measure the time for a specific event or a selected sequence of events.

The other type of timer does not measure time but initiates and terminates events at predetermined or preset intervals. A very simple but ideal laboratory device is shown in Fig. 4.75. With this device a small synchronous motor drives, through a reduction gear, a cam wheel that actuates a microswitch. Gear ratios are easily changed as is the cam position. This type of timer is continuous or recycling, and the interval for 1 rev of the cam may be as short as $\frac{2}{3}$ s or as long as 72 h. In many laboratory applications, large numbers of components must be sequenced with respect to each other, and to facilitate these installations multicam timers are available. Figure 4.76 is typical of a multicam unit with one electric motor actuating many cams and switches. The microswitches themselves may be used as either *normally open* or *normally closed* by making the appropriate terminal connections. If the electrical capacity of the microswitch is too small for the load that it is to actuate, then a suitable relay must be interposed in the circuit to prevent switch damage. For these cam timers, the intervals are conveniently set by cams with graduated scales and may be subsequently checked with a stopwatch or interval timer for greater precision.

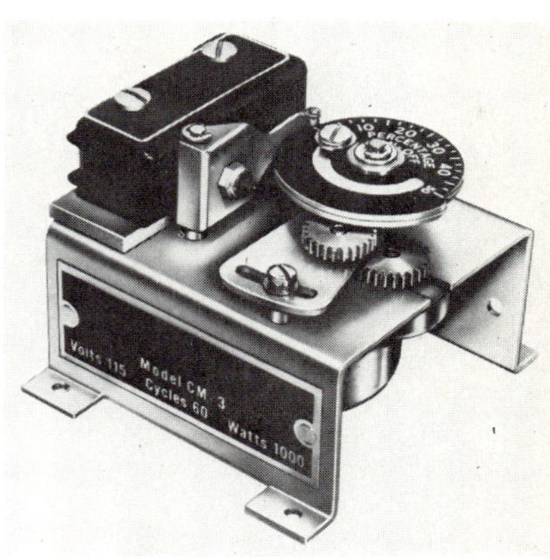

FIG. 4.75 Single-cam recycling timer. *(Meylan Stopwatch Corp.)*

Another type of interval timer is the timing electric relay. In these devices the electric relay contacts do not break (or make) immediately when the coil is energized (or deenergized) but instead do so after a prescribed time lapse. For example, for an on-delay model the time delay starts immediately when the relay coil is energized but the relay contacts are not actuated until the preset time has elapsed. The contacts are usually for both normally open and normally closed wiring in the same unit so that after the prescribed time delay the electric supply is either initiated or terminated as desired. The off-delay models work in the reverse way; that is, the timing sequence starts when the electric supply to the relay coil is terminated. Figure 4.77 shows a typical timing relay.

Timing relays are available covering wide delay times (for example, 0.1 s to 60 min), with individual units having ranges varying by a factor of as much as 100

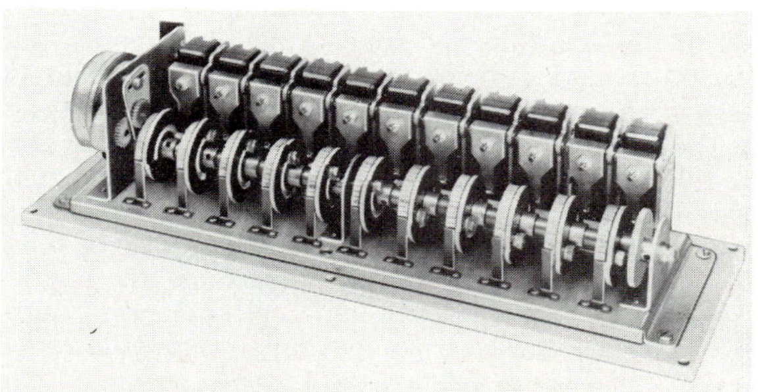

FIG. 4.76 Multicam recycling timer. *(Meylan Stopwatch Corp.)*

(for example, 1 to 100 s). The delay interval is simply set with a dial. Repeatability and accuracy are features of these devices.

Timing relays quickly reset to their original position once the coil is deenergized (or energized if an off-delay model). This is a distinct advantage over cam timers when the unit must completely recycle, a time that depends upon the cam rotational speed. Therefore, following a false start, a system with timing relays is easily readied for a repeat operation without the necessity of having to verify specific cam positions, etc.

FIG. 4.77 Precision timing relay. (*Amerace Corp., Control Products Division.*)

4.11 TEMPERATURE AND PRESSURE SWITCHES

The control of temperature and pressure is important in many laboratory processes and pilot plants for the safe and protective operation of equipment. These switches open or close electric circuits that may provide automatic control (for example, of an electric heating element), signal an alarm, trigger another instrument or apparatus, or actuate any electric device. It is evident then that their response characteristics, accuracy, reliability, and electric load capacity are of prime importance in selection.

Temperature Switches

Temperature switches are generally one of two types. One employs fluid expansion to actuate a switch, and the other employs a bimetallic element to actuate a switch. The fluid-filled devices usually contain liquid because it provides faster response. They may be for either local or remote mounting. For remote mounting the switching-unit liquid capsule is connected with a metallic capillary tube to a temperature-sensing bulb. The bulb, capillary, and capsule are completely filled with liquid. For accurate and repeatable performance these liquid-filled temperature switches need to be compensated for fluctuations in ambient temperature because ambient-temperature changes alone cause the fluid to expand or contract in the capillary and capsule independent of the expansion or contraction of the fluid in the bulb. A typical remote-mounting, ambient-temperature-compensated temperature switch is shown in Fig. 4.78.

Liquid-filled temperature switches may be used over a temperature range from 100 to 600°F (300 to 600 K), and the adjustable range of a specific unit is typically 200°F (370 K). Sensing bulb and capillaries are typically made from stainless steel to provide chemical compatibility to a wide range of chemicals. Copper bulbs are available to provide faster temperature response. The bulbs may be subjected to pressures of 500 psi (3500 kPa) so that they can usually be

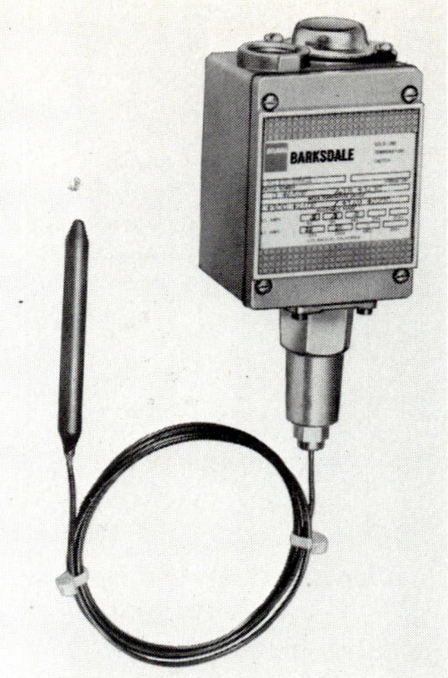

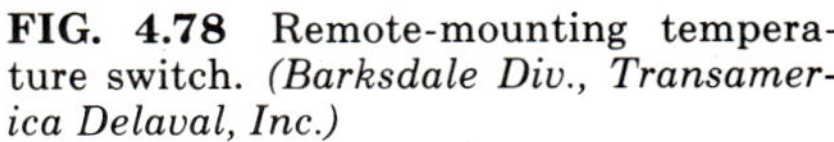
FIG. 4.78 Remote-mounting temperature switch. *(Barksdale Div., Transamerica Delaval, Inc.)*

FIG. 4.79 Local-mounting, all-metal temperature switch. *(Burling Instrument Co.)*

placed directly into the environment to be monitored. If the power capability of the switch is not adequate to actuate the electric equipment to be controlled, then a simple relay installed in the circuit is all that is required to provide the necessary switch capacity.

The response of these devices is typically 10 to 20 s per 100°F (55 K) temperature change. If these response times are not fast enough, then use a small thermocouple with a much lower thermal capacity. However, the electric output from a thermocouple is small, and it is necessary to amplify the signal to actuate switches for reasonable electric power requirements (for example, 20 A, 110 V).

The all-metal, bimetallic types of temperature switch are available for use to temperatures to 2000°F (1400 K) and are of the local-mounting variety (i.e., the switch and temperature-sensing element are located where the temperature is required to be monitored). Typically they are used for monitoring and controlling ovens, dryers, furnaces, engine exhausts, and molten salt baths. A temperature switch of this type is shown in Fig. 4.79.

In these temperature switches the bimetallic element comprises a metal tubular probe, inside of which is a rod of different metal. The rod is attached to the bottom of the tube. The instrument head and mounting plate are secured to the tube. The plate supports a lever arm, and the lever is actuated by the rod. A microswitch is positioned at the free end of the lever.

In use, the tubular probe is immersed in the fluid (liquid or gas) whose temperature is to be monitored. The tube and rod attain the temperature of the fluid, but as the temperature increases the rod expands more than the tube. The differential expansion between the tube and rod is magnified by the lever arm, and the switch is actuated when the temperature reaches the predetermined value.

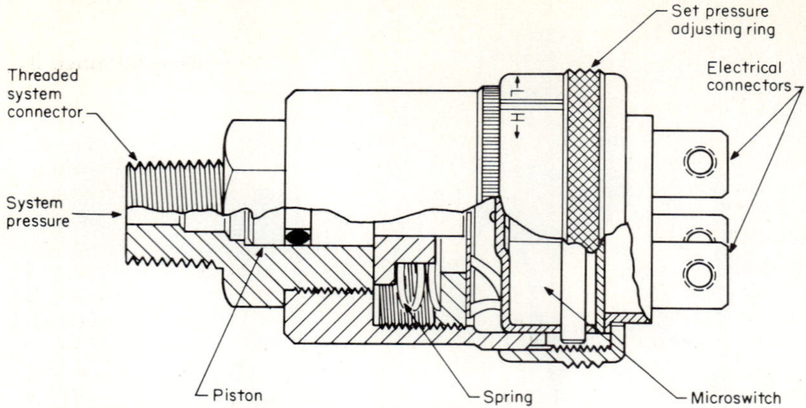

FIG. 4.80 Typical pressure-switch assembly. *(Whitman General Corp.)*

Pressure Switches

Pressure switches comprise a pressure capsule and microswitch combination. Movement of the capsule under increasing or decreasing pressure actuates the switch. The *set point,* that is, the pressure at which the switch is triggered, depends upon the stiffness or spring rate of the capsule and the location of the switch with respect to the capsule. Figure 4.80 illustrates a typical pressure-switch assembly.

Capsules are of four kinds: (1) a piston, either free-floating or spring-loaded (Fig. 4.80), (2) a bellows, (3) a Bourdon tube, and (4) a diaphragm, the most common. Since the modulus of most metals and alloys varies with temperature, so will the spring rate of the capsule; hence the set point will change. However, where good repeatability is required over wide ambient-temperature ranges, diaphragm capsules using a special nickel alloy are employed. The modulus of this alloy is invariant with temperature over the range −65 to 200°F (220 to 370 K). Capsules are normally designed to withstand a pressure 50 percent over the maximum rating for the device to allow some leeway for unexpected system pressure excursions.

Ordinarily pressure switches are designed to sense pressure differential, using atmospheric pressure as a reference. However, for low positive pressures, varia-

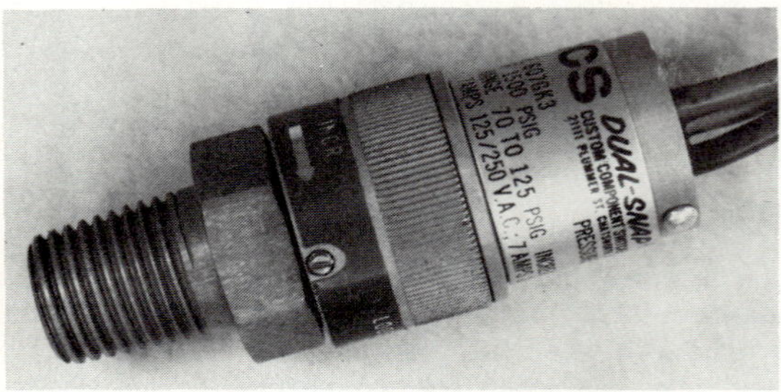

FIG. 4.81 Miniature pressure switch. *(Custom Component Switches, Inc.)*

tions in atmospheric pressure may be significant. For such uses an "absolute" pressure switch is required, and absolute pressure reference is obtained by evacuating and hermetically sealing the interior of the device.

All pressure switches exhibit hysterisis. That is, the pressure value at which the switch is actuated as pressure increases is greater than the value at which the switch is again deactivated as the pressure falls. This pressure band is sometimes referred to as the *actuation value*. For precise monitoring of pressure it is evident that the actuation value should be kept small, but it can never be eliminated. The microswitch itself contributes to the actuation value, with the smaller, lower current-capacity switches contributing less. Therefore a more sensitive pressure switch will have a smaller current rating. If these low current ratings are inadequate for direct control of the electric equipment, then an intermediate relay may be used to accept the electric load. Repeatability of pressure switches (i.e., ability to actuate switch at the same pressure in successive operations) is typically less than 5 percent and in the better units as low as 0.1 percent of the full pressure range.

The life of a pressure switch is usually greater than 100,000 switching cycles and often exceeds 1 million cycles. The fatigue life of the capsule is typically the limiting life-cycle component. Capsule life is dependent upon factors such as the total number of switching cycles, the pressure setting in the adjustable range, the overpressure experienced after actuation, and the frequency and magnitude of pressure fluctuation that are not large enough to actuate the switch.

Pressure switches are constructed from a variety of materials to be compatible with most fluids. They may be factory-set or field-adjustable and are provided with a wide range of electric connectors and explosionproof housings and with flanged or threaded pressure mountings. Figure 4.81 shows a typical miniature unit with a male threaded mounting, adjustable over the pressure range from 2 to 300 psi (15 to 2000 kPa).

chapter 5

Selecting Measuring Instruments

Robert C. Knauer

5.1 GENERAL CONSIDERATIONS

This section will help the technician in industry or science put together a measurement system. In most cases there is sufficient information to assemble components from sensor to indicator and to build a usable system. Actually, this was the way it was done in the "old days." Increasingly, however, the experimenter can and does buy a package that plugs in and is ready to use.

As an example of this trend, once if technicians wanted to measure a temperature, they snipped off 6 ft of thermocouple wire, welded a junction, set up an ice bath reference junction, and hooked up the circuit to a hand-balance potentiometer. Today the technicians select from a catalog an integrated system consisting of a jacketed thermocouple and built-in reference-junction compensation and linearization, and then read the temperature digitally, directly in degrees to a precision of $\pm 0.01°$F or C, at the flip of a switch.

Probably there will always be situations in which no existing equipment will do the job, but most commercial products solve many measurement problems. Let the specialist design your instrumentation.

5.2 TEMPERATURE-MEASURING INSTRUMENTS

Measuring temperature is so common in engineering and everyday life that it is considered easy to do. In reality, there are many possible sources of error in most temperature measurements, and under some conditions accurate measurement is still beyond the state of the art.

The most-used temperature scales are Celsius (C—formerly Centigrade) and Fahrenheit (F). The Celsius scale is based upon the freezing point of water (0.01°C) and boiling point of water (100.00°C), both points taken at normal atmospheric pressure. Fahrenheit is also based upon the properties of water; the freezing point is 32°F and the boiling point 212°F. For use in thermodynamic calculations, two absolute scales are defined:

$$\text{Degrees Kelvin (K)} = {}^\circ\text{C} + 273.16$$
$$\text{Degrees Rankine (R)} = {}^\circ\text{F} + 459.69$$

The international practical temperature scale is identical to the Celsius scale, but it includes additional fixed points both above 100°C and below 0°C. The International System of Units (SI) has adopted the unit kelvin as the standard for temperature, although Celsius and Fahrenheit are still used.

Thermometers

Liquid-Filled Glass Thermometers

These types of thermometers are widely used and familiar to everyone. When mercury is the liquid, the temperature range is from −39°C to 600°F (234 to 590 K). The upper limit is extended to about 1000°F (810 K) by special construction. Various organic liquids such as alcohol or pentane are used to construct thermometers that go to −328°F (73 K). The user of any type of liquid-filled glass thermometer should consider whether the instrument's high and low limits will fit a particular measurement situation.

These glass thermometers are either partial- or full-immersion. A partial-immersion model has been calibrated while the thermometer was immersed in a liquid bath to a partial depth of say 3 in. The remainder of the stem was in air at approximately room temperature. Often there is a mark at the proper immersion depth. A full-immersion thermometer has been calibrated with the entire stem in the bath.

It is possible to use thermometers at different immersions than specified by calculating corrections, but this is tedious and inexact; it is best to select the proper instrument for the application.

When used at elevated temperatures, liquid-filled glass thermometers may undergo permanent shifts in calibration because the glass bulb stretches. These changes are always unpredictable, so the only recourse is to frequently recalibrate thermometers used at temperatures above 150°C (420 K) (see Fig. 5.1).

Filled-System Thermometers

These temperature-measuring thermometers are widely used in industry for indication, recording, and control. Among the advantages are remote indication, no electric-power requirement (except for a chart drive), and sufficient force available to drive a pen. A system has three major components (see Figs. 5.2 and 5.3): a bulb that senses the temperature; a capillary tube that conducts pressure; and a pressure gage that indicates pressure, which is interpreted as temperature. The system is filled with liquid, vapor, or gas and sealed by the manufacturer before calibration. Field disassembly is not practical because loss of some working fluid occurs with the resulting change in calibration. The liquid-filled systems are subject to errors because the liquid in the gage responds to changes in ambient temperature. This problem is taken care of by built-in compensation.

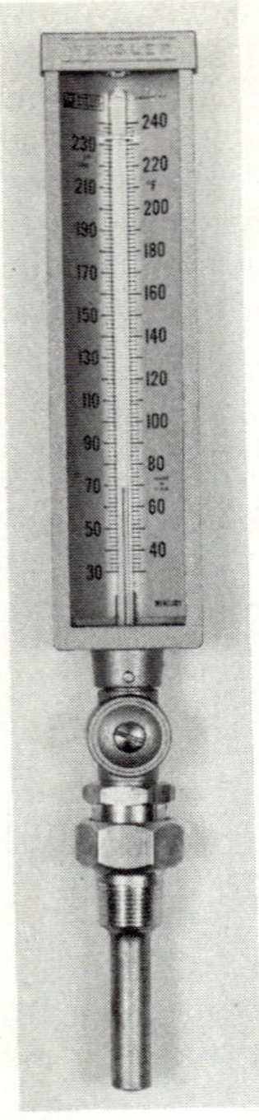

FIG. 5.1 Liquid-in-glass thermometer. *(Weksler Instruments Corp.)*

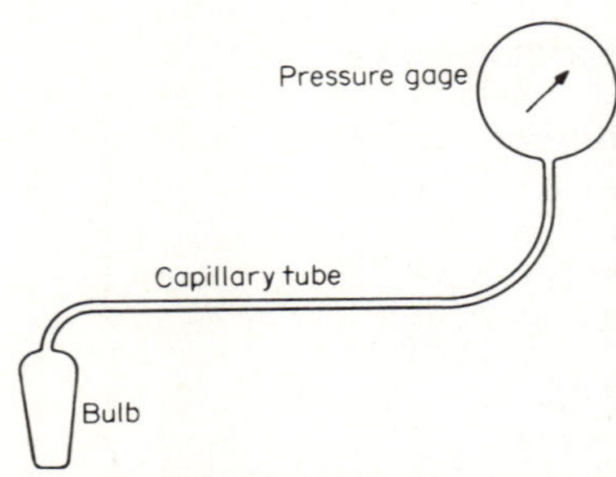

FIG. 5.2 Filled-system thermometer.

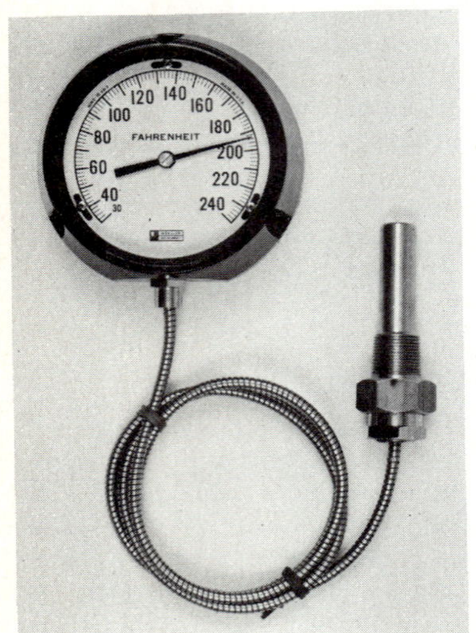

FIG. 5.3 Filled-system thermometer. *(Weksler Instruments Corp.)*

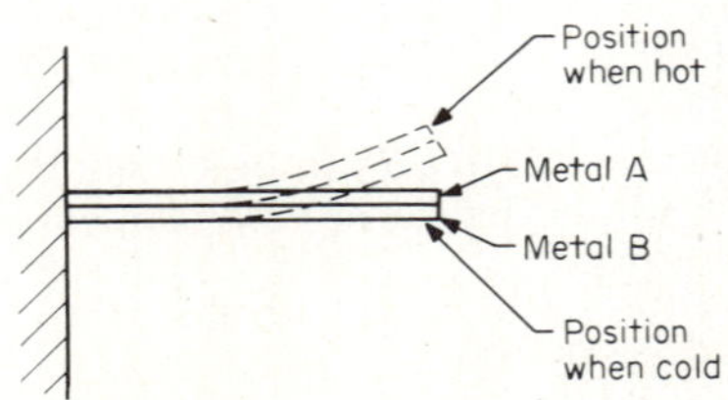

FIG. 5.4 Bimetallic strip.

Liquid- and vapor-filled systems display bulb-elevation error if the bulb is either much higher or much lower in elevation than the indicator. The effect is particularly large for mercury-filled systems. If this condition is anticipated at the time of purchase, the manufacturer can adjust for it by calibration. Gas-filled systems are not subject to this error.

Bimetallic Thermometers

This class of thermometers uses the principle of differential thermal expansion. Refer to Fig. 5.4: Two strips of metal are bonded by welding, riveting, or other means. If metal B has a greater coefficient of expansion than metal A, as the strip is heated it curves upward. Brass and steel are the metals most often used, but other combinations are possible.

In practical instruments, the sensitive element is wound into a helix and inserted into a relatively small bulb. Temperature changes cause the free end of the helix to rotate. This motion is coupled through an extension rod to a pointer that indicates along a circular dial. This is the familiar dial thermometer, which has many applications in industry, the laboratory, and the home. Its advantages include low cost, ease of reading, and durability (Fig. 5.5).

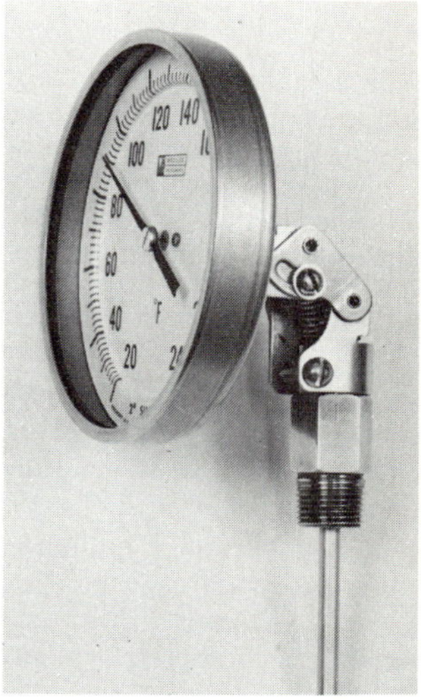

FIG. 5.5 Bimetallic thermometer. *(Weksler Instruments Corp.)*

Thermocouples

Thermocouples are junctions of dissimilar metals that generate small voltages under temperature differences. This was originally known as the thermoelectric Seebeck effect; it was later proved a combination of the Peltier and Thomson effects. The voltage across a junction results from:

1. The contact of two dissimilar metals and the junction temperature
2. Temperature gradients along the conductors

Figure 5.6 is the basic illustration of a practical thermocouple system. Three conditions must be met for a reading to be obtained. (1) Wires A and B must be of different materials, such as copper and constantan. (Constantan is a copper and nickel alloy used in thermocouples and other applications.) (2) T_1 and T_2 must be of different temperatures. (3) No current should flow in the circuit. This last condition is not always met; this is discussed later. The millivolt reading is very nearly proportional to the temperature difference; if the temperature difference is 100°F (311 K), the reading is 2.8 mV. (A millivolt is 1/1000 V.)

A more practical circuit is shown in Fig. 5.7. This circuit uses reference junctions separate from the terminals of the millivoltmeter. In the basic circuit it is necessary to read the junction temperature with a direct-reading thermometer. With the separate reference-junction circuit, the junctions can be made in an ice

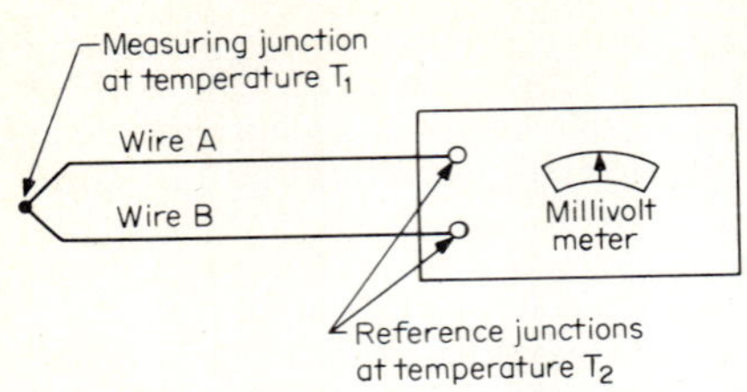

FIG. 5.6 Basic thermocouple circuit.

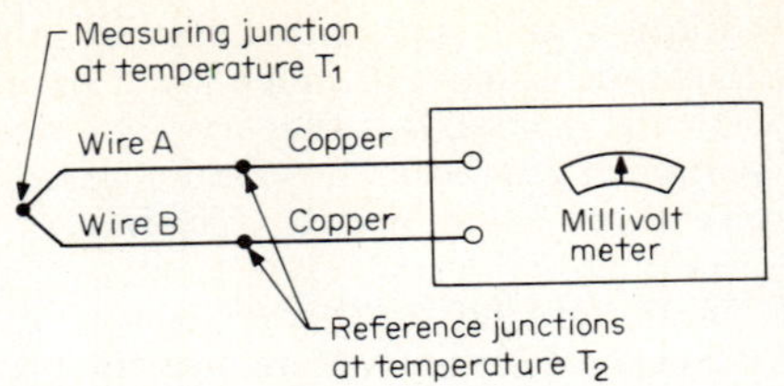

FIG. 5.7 Thermocouple circuit with separate reference junctions.

bath or in a controlled oven. An ice bath holds the reference junctions at +32°F (273 K). Special ovens hold the reference junctions at +150°F (339 K). A typical commercial thermocouple is shown in Fig. 5.8.

In principle, almost any two wires generate thermoelectric signals, but a certain few combinations have been adopted as standard. Here are the more common thermocouple materials and their characteristics.

Copper-constantan (type *T*): A general-purpose thermocouple usable from −300 to +600°F (90 to 590 K). The output is 2.8 mV/100°F (5 mV/100 K).

Iron-constantan (type *J*): This is the most popular thermocouple and can be used in the +32 to +1400°F (273 to 1030 K) range. Oxidizing atmospheres attack the iron and cause a calibration shift unless a jacketed couple is used. The output is 3.2 mV/100°F (5.7 mV/100 K).

Chrome-alumel (type *K*): This is usable from +32 to 2300°F (273 to 1530 K), but it is most often used at the high temperature end. The materials are very resistant to oxidizing atmospheres but subject to damage from reducing atmospheres, unless protected by a jacket. The output is 2.3 mV/100°F (4 mV/100 K).

Platinum-10 percent rhodium platinum (type *S*): This is usable from +1000 to +2600°F (800 to 1700 K). This couple is primarily a standard for the calibration of other thermocouple materials and is relatively expensive. The output is 0.63 mV/°F (1.1 mV/K).

All the listed ranges can be extended at the high end of the temperature span for short times. If higher accuracies are desired, instead of using the millivolts

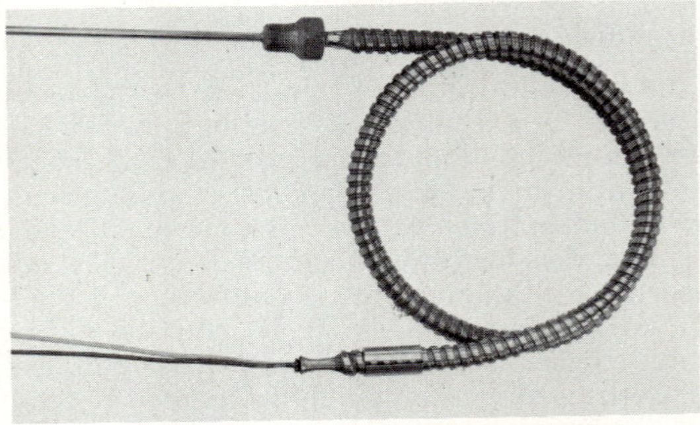

FIG. 5.8 Jacket thermocouple. *(The Foxboro Co.)*

per degree figure, consult a chart of thermocouple output potentials (such charts are available from equipment manufacturers and other sources).

Thermocouple Readout Systems

As useful and as versatile as thermocouples are, they are severely limited by a low-level millivolt output. Detecting these small signals with good accuracy requires special methods and careful layout of the circuitry.

The simplest readout system is shown in Fig. 5.6. The thermocouple leads are connected to a sensitive voltmeter or ammeter whose characteristics are high sensitivity and low resistance. Such a system lets current flow through the junction and the lead wires, and the actual voltage at the meter terminals is not as great as that shown in the tables. Such systems must be calibrated after assembly, and the reference junctions should be held constant. These devices are made commercially and usually contain reference-junction compensation to allow the meter-case temperature to vary over a limited range without having to be corrected or stabilized. The thermocouple itself and its lead wires can be replaced with the same type thermocouple, providing its resistance is that specified on the meter face. The thermocouple and meter systems are simple and convenient but not suited to the most accurate tasks.

For work requiring the greatest accuracy, a potentiometer is a more suitable readout instrument. Figure 5.9 shows the essentials of a thermocouple-potentiometer system. A source of constant voltage is applied to a precision-wound resistor that has an adjustable tap. A calibrated scale is mechanically coupled to the adjustable tap so that the voltage or electromotive force (emf) appearing across points x to y can be read with great accuracy from the calibrated scale. The galvanometer merely detects current. To make a reading, adjust the slide wire until the current, as indicated on the galvanometer, is zero. At this point the output voltage from the slide wire is the same as the output voltage from the measuring and reference junctions, and no current is flowing through the thermocouple leg. This method provides great precision in thermocouple work. Figure 5.10 illustrates a typical hand-balance potentiometer.

Self-balancing potentiometers are often used for thermocouple readout; an unbalanced detector is substituted for the galvanometer. This detector feeds an amplifier that in turn drives a motor. The motor drives the slider so that the system is always in balance; that is, the slide-wire position always corresponds with the thermocouple's output voltage. The motor also has sufficient power to drive a recording pen or large pointer. Such a system is faster than hand balancing but not quite as accurate.

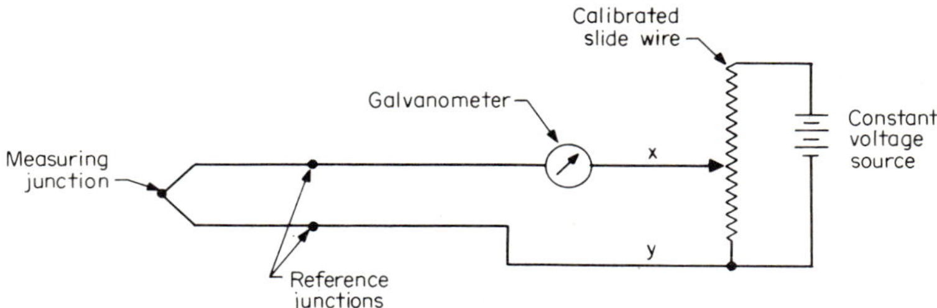

FIG. 5.9 Potentiometer circuit for thermocouple readout.

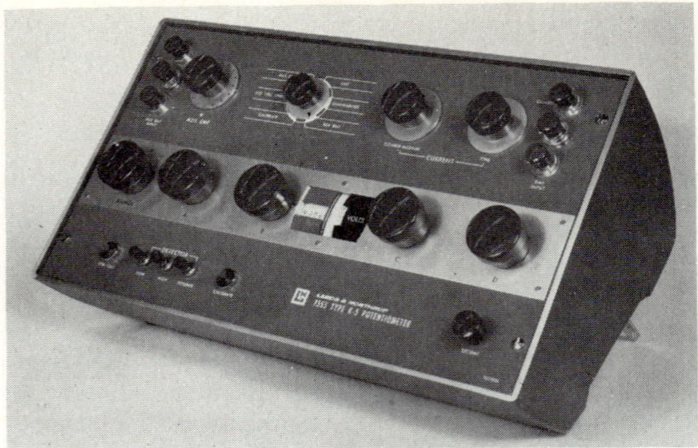

FIG. 5.10 Hand-balance potentiometer. *(Leeds & Northrup Co.)*

Pyrometers

Radiation Pyrometers

Physical Basis for Radiation Pyrometry • When a piece of steel is heated in a flame, it becomes red hot or white hot. The steel object can be seen because it is radiating energy in the visible part of the spectrum. Actually, considerable energy is also being radiated in what is called the *infrared* portion of the spectrum, and to a small extent in the ultraviolet region. All objects are emitting and absorbing radiation at all times unless they are at absolute zero temperature. At ordinary temperatures, this radiation is all in the infrared region. As the temperature increases, the wavelength of the radiation decreases, and some energy falls in the visible range. At no time is the wavelength a discreet value; it is widely distributed over the spectrum. Both the intensity of radiation and the peak wavelength of this radiation indicate the temperature of the object; a measurement of these quantities can be used to infer the temperature of a body. Such a method is called *radiation pyrometry,* which enables measurement at a distance, a great advantage for many applications.

Most bodies do not absorb all radiation falling upon them; they reflect a certain portion of that radiation. The ratio of absorbed radiation to incident radiation is called *absorptivity,* which is a number between 0 and 1. Obviously, if a body reflects all incident radiation, its absorptivity is zero, and if it absorbs all incident radiation, its absorptivity is 1. As examples, a smooth silver plate has an absorptivity of 0.07, and a carbon block has an absorptivity of 0.93. An object with a value of 1 is called a *blackbody,* a useful term theoretically but not easily achieved. The blackbody is the most efficient absorber and most efficient emitter. The emissivity of a body is the ratio of its radiance to that of a blackbody at the same temperature. One fundamental of radiation law is that the absorptivity and emissivity of a body are the same. In many radiation measurements, the emissivity of the surface must be known; there are published tables of the emissivity of many materials for this purpose. True blackbody conditions are approached only when looking into an oven through a small hole. Fortunately, this is a fairly common situation, and the emissivity is taken as 1.

Total Radiation Pyrometers • Such instruments are shown in Figs. 5.11 and 5.12. Radiant energy is collected by a lens or mirror or a combination of both

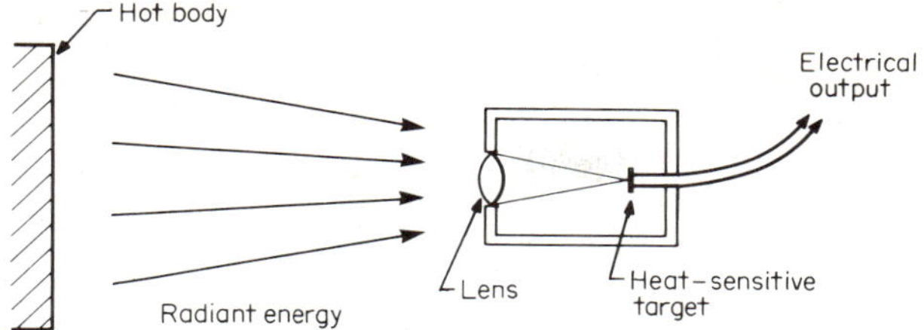

FIG. 5.11 Total radiation pyrometer.

and focused upon a heat-sensitive target, such as a thermopile (many thermocouples in series), a photocell, or other temperature-to-electric detector. The target temperature then reaches some equilibrium temperature that is a function of the hot-body temperature, a relationship established by calibration for a particular application. These pyrometers are usually usable from 1000°F (800 K) upward to 5000°F (3000 K) and even higher, although some operate at temperatures as low as 120°F (320 K).

FIG. 5.12 Total radiation pyrometer. *(Mikron Instrument Co.)*

Optical Pyrometers • Optical pyrometers are sometimes called *brightness pyrometers.* Figure 5.13 shows such an instrument in its simplest form. The observer sights through a telescope at the hot body and sees the hot filament of a lamp contained in the telescope tube. The lamp current is adjusted manually until the filament fades into the background light emitted from the hot body. At this point the lamp-filament temperature is the same as the source. A previous calibration of lamp current versus temperature is furnished with this instrument. Because the source must be emitting energy in the visible region, such pyrometers are usable from only 1400°F (1000 K) upward. In similar but more complex devices, the eye is eliminated in favor of a photomultiplier tube. Electronic circuitry controls the lamp current, and an automatic matching is effected. These systems can work in the infrared region and enable one to read temperatures down to 200°F (360 K).

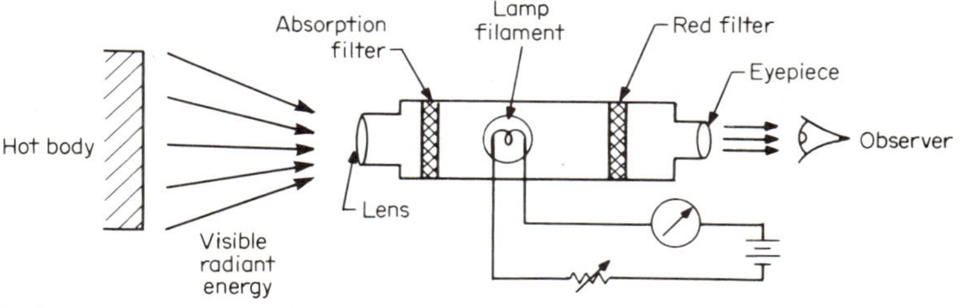

FIG. 5.13 Optical, or brightness, pyrometer.

FIG. 5.14 Optical, or brightness, pyrometer. *(Pyrometer Instrument Co.)*

As with the total radiation pyrometer, the emissivity must be considered when interpreting the results, except for blackbody conditions (Fig. 5.14).

Two-Color Pyrometers • A two-color pyrometer, also known as a *Shawmeter,* eliminates the need for an emissivity correction. It is essentially an automatic brightness pyrometer operating at two wavelengths. The radiation detector in the instrument is rapidly switched from a blue optical path to a red optical path. The ratio of these signals is derived electronically, and the result is a measure of temperature independent of emissivity.

Resistance Thermometers

Most metals exhibit a change in electric resistance with a change in temperature. Not surprisingly, this characteristic is used in temperature measurement, and ready-fabricated sensing devices are widely available. There are only four metals used for most work:

Platinum: Platinum has a high-temperature coefficient of resistance, which in itself recommends its use as a resistance thermometer. It is also very linear and stable. Platinum has been chosen to define the IPTS from −297°F or 90.6 K (liquid oxygen boiling point) to +1168°F or 904.4 K (melting point of antimony).

Nickel: The coefficient of resistance versus temperature of nickel is the highest of all wire materials at about 3.0 Ω/(1000 Ω · °F) [5.4 Ω/(1000 Ω · K)]. However, it is not linear, but this property is easily calibrated out of a system. Because of its relative low cost, nickel wire is the most commonly used for industrial resistance thermometry.

Copper: Copper has a high coefficient and is readily available in controlled purity. Its chief disadvantage is that its resistance is low, so more wire is needed to achieve a useful sensor.

Tungsten: Tungsten has a good coefficient, is extremely linear, has a high melting point, and is compatible with a nuclear environment.

It is impractical for users of a resistance thermometer to fabricate their own sensors. Sensors are available from numerous sources. The general form is a wound spool of fine wire inserted into a protective sleeve, with terminals or leads brought out (Fig. 5.15). All base metals and alloys have a positive temperature coefficient; that is, increasing temperature results in an increasing resistance.

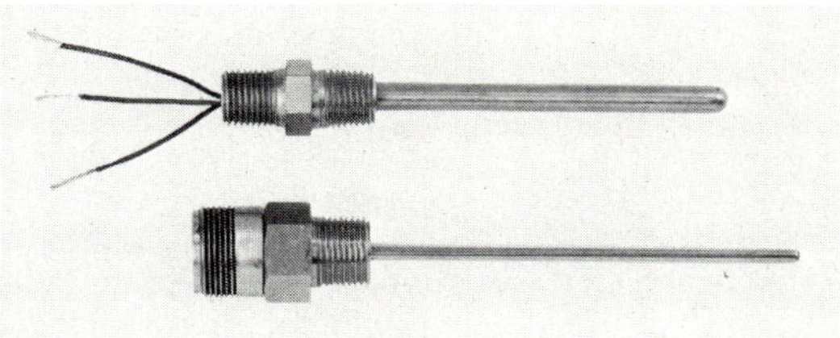

FIG. 5.15 Industrial resistance thermometers, stainless steel sheathed. *(ARI Industries, Inc.)*

The other important class of resistance thermometers is represented by the thermistor, which is a semiconductor material fabricated into a resistor. A thermistor has huge negative temperature coefficients of resistance and many uses in instrumentation, for temperature compensation and various control tasks. As a temperature sensor, the thermistor offers a high-temperature coefficient of resistance, small size (many thermistors are tiny beads the size of a pinhead), and relatively high resistance. The principal disadvantages are nonlinearity and variability of resistance from one sample to another. When a thermistor is used for measuring circuits, it is usually necessary to provide a means of linearization and to match the characteristics of each individual thermistor.

Resistance-Thermometer Readout Systems • Readout or signal conditioning for resistance-thermometer sensing elements is most often accomplished with the Wheatstone bridge, often simply referred to as a bridge. The bridge principle is widely applied in many other electrical measurements; here we discuss general bridge usage.

Why not read out a resistance thermometer by simply hooking an ohmmeter across it? The first objection to this approach is that a temperature range of; say, from 0 to 100°F (256 to 311 K) might represent a resistance range of 84 to 105 Ω, which is not a very convenient scale. Another objection is that a typical ohmmeter scale is not linear, so if one chooses the 0- to 1000-Ω scale, 85 to 105 Ω is a narrow span with poor resolution. The bridge offers a way out of these difficulties.

Figure 5.16 is the simplest bridge circuit. R_1 and R_2 are fixed resistors of the same value, say, 100 Ω. R_3 is an adjustable resistor that covers a range from, say, 50 to 150 Ω. R_4 is called the *active* arm and is a nickel wire with a nominal resistance of 100 Ω at 77°F (298 K). Further suppose that R_4 is held at 77°F (298 K) in a controlled-temperature bath. If a voltage is applied across terminals A and B and R_3 is adjusted to exactly 100 Ω, we have a bridge with equal resistances in all four arms. Under these conditions, the same current flows in the left leg (R_1 and R_3) as flows in the right leg (R_2 and R_4). The bridge is said to be "balanced," and the output detector, which can be a galvanometer or any other type of current- or voltage-sensitive device, reads zero. We have thus eliminated our "zero" reading, and we may use R_3 as a "zero adjust," to set the bridge to read zero output at any temperature from 0 to 100°F (256 to 311 K). We might now set R_3 such that zero bridge output corresponds to 0 V when R_4 is at 0°F (256 K). It is now possible to calibrate bridge output versus the temperature of R_4. If the detector is a typical galvanometer with a scale range of 1 mA, we find that with an excitation of 10 V, there is an output of 56 percent of full scale at 100°F (311 K). Thus the bridge circuit lets us adjust both range and zero for any instrument that has a resistive change as its working parameter.

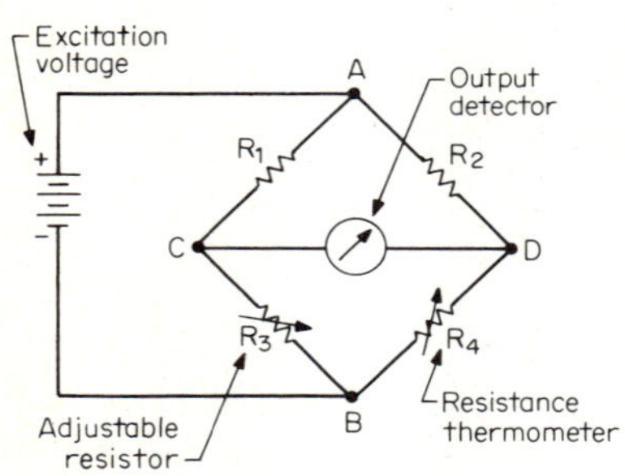

FIG. 5.16 Wheatstone bridge circuit.

A further technique refinement enables us to take measurements that do not depend upon either precise control of the excitation voltage or precise knowledge of the detector's characteristics. This technique involves using the bridge under conditions in which it is always in balance when making measurements.

Resistor R_3 should be a calibrated precision resistor whose ohmic value can be read directly. The bridge is operated as follows: R_4 is held at some fixed temperature to be measured. R_3 is adjusted until the detector reads zero. At this point R_3 and R_4 are equal, and because R_3 is of known calibration, R_4 and hence its temperature are known. This operation mode of a bridge is known as *finding a null.* The user is simply matching the resistors in the bridge; this null is independent within wide limits of the detector's sensitivity and the excitation-voltage value.

A practical consideration involved in selecting the excitation voltage is that it should be low enough so the currents flowing in the bridge do not heat the resistors enough to cause appreciable change. Generally, the inactive arms of the bridge are resistors of low-temperature coefficient wire, but the resistance-thermometer arm, being necessarily a high-temperature coefficient material, must not have sufficient power input to elevate its temperature. This condition is easily met in a circulating-liquid environment, but in gas at rest it is possible to create large errors. The excitation voltage must not be too low, or a sharp null cannot be obtained. Most applications have as an optimum an applied voltage of a few volts.

The null-balanced bridge offers great measurement accuracy because it is essentially a comparison of resistance values rather than a measurement of voltage or current. Resistance standards are much easier to establish than current or voltage standards. There are many variations and modifications of the basic bridge circuit shown here.

Temperature-Sensing Paint and Markers

The change in the physical or chemical state of an applied material can be a convenient way to measure the temperature of an object's surface, such as Seger cones and Tempilstiks (Fig. 5.17). The temperatures at which substances melt or initiate chemical reaction are often known and reproducible characteristics.

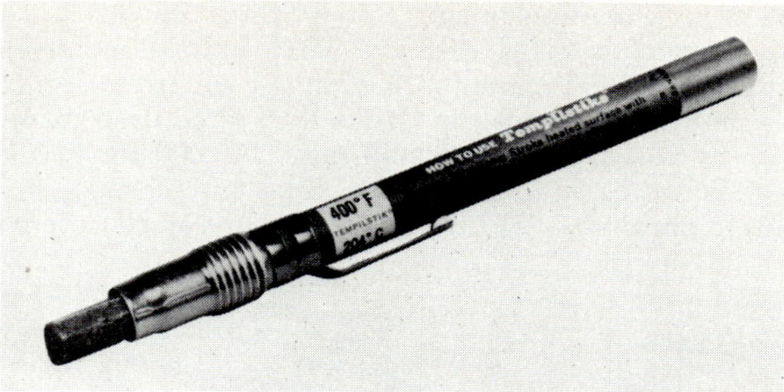

FIG. 5.17 Temperature-sensitive crayons. *(Tempil Division, Big Three Industries, Inc.)*

There are commercial products that cover the temperature range from about 120 to 3600°F (320 to 2260 K), in intervals ranging from 12 to 70°F (6.7 to 38.8 K). The temperature-sensing element can be a solid that softens and changes shape at the critical temperature, or it can be applied as a paint or crayon that changes color or surface appearance.

5.3 PRESSURE-MEASURING INSTRUMENTS

In science and engineering, pressure is force per unit area, and the most often used units are pounds per square inch or psi, or in the SI, the pascal (Pa). This force is called pneumatic pressure when dealing with gases, hydraulic pressure when dealing with liquids. (Pressures do exist in solids, but they are described by more complex relationships, and the subject is more properly referred to as stress analysis.) Here we are concerned with only the measurement of pressure in liquids and gases, for which Pascal's law applies. According to Pascal's law, pressure in a fluid (that is, a liquid or gas) is exerted equally in all directions. This is not true for solids.

Absolute pressure is the pressure in a vessel or container that is a perfect vacuum. *Differential pressure* is the difference in the pressure between two vessels. *Gage pressure* is the difference between the pressure in a vessel and that in the surrounding atmosphere. Gage pressure, the easiest to measure, is the one most commonly meant when we speak of pressure measurement. If we mean absolute or differential pressure, we must say so, but to refer to gage pressure we simply say "The pressure is 10 psi." Pressure expressions are abbreviated as

psi	pounds per square inch
psia	pounds per square inch, absolute
psid	pounds per square inch, differential

There is a definite relationship between psi and psia. Absolute pressure is equal to gage pressure plus atmospheric pressure. The normal atmosphere at sea level is 14.7 psi (101.3 kPa); use this value to convert from one basis to the other for most work, but not for very precise work because the atmospheric pressure varies with time and elevation. If the application is critical, a barometer must be read to determine the true ambient pressure at the time and location of the experiment.

The prevalent pressure unit in the United States today is still psi, but the SI unit is the pascal, which is based upon the metric units of meters, kilograms, and seconds. (Pascal is a special name for newton per square meter; the formula for a newton is $kg \cdot m/s^2$.) The pascal undoubtedly will be used more in the future. Unfortunately, the pascal is a very small unit (100,000 Pa = about 1 atm), so usually pressure is expressed in kilopascals (kPa) and megapascals (MPa). Standard atmospheric pressure is 101.3 kPa; a pressure of 100 psi = 689.3 kPa or 0.6893 MPa.

Bourdon Tube Gages

The Bourdon tube gage is the most common pressure-measuring device. As shown in Fig. 5.18, it is a flattened, spring bronze or steel tube bent into a three-fourths circle. Pressure in the tube tends to straighten the circle. Because one end of the tube is fixed to the pressure inlet, the other end moves outward a distance roughly proportional to the pressure difference between the inside and outside of the tube. This motion is conveyed to the pointer through a pinion-and-sector mechanism. If the case is vented to the atmosphere, as is most often done, the device reads gage pressure. If the case is sealed and evacuated, it becomes an absolute pressure gage; if the sealed case is brought out to a fitting, it becomes a differential pressure gage. Bourdon gages are made in many sizes and qualities for many applications. One important trend in design is the safety gage. Because gages can explode from overpressurization or fatigue of the Bourdon tube, many now have a front plate of heavy sheet material and a soft rubber or plastic back so that any explosion is vented away from personnel.

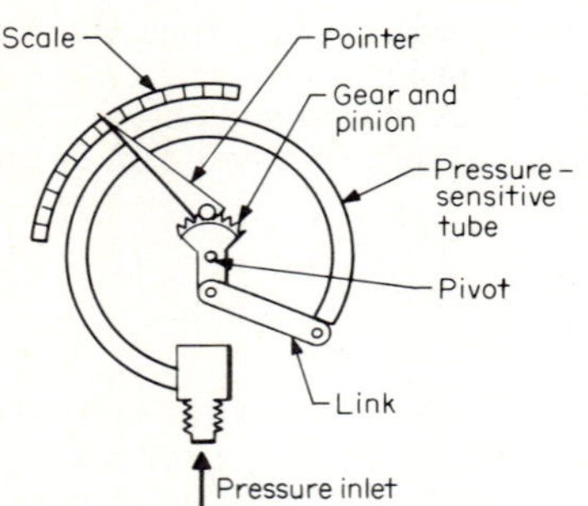

FIG. 5.18 Bourdon tube pressure gage.

Liquid-Filled Manometers

A manometer (Fig. 5.19) measures pressure via the difference in height produced in the working fluid, usually mercury. If two differing pressures P_1 and P_2 are

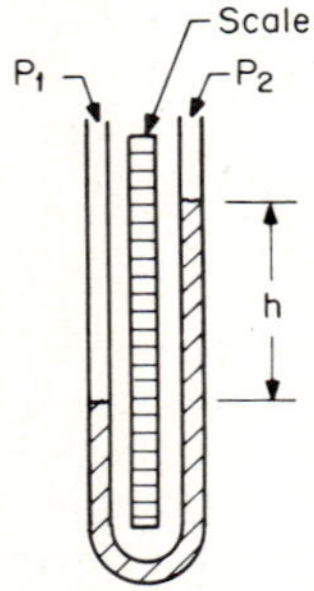

FIG. 5.19 U-tube manometer.

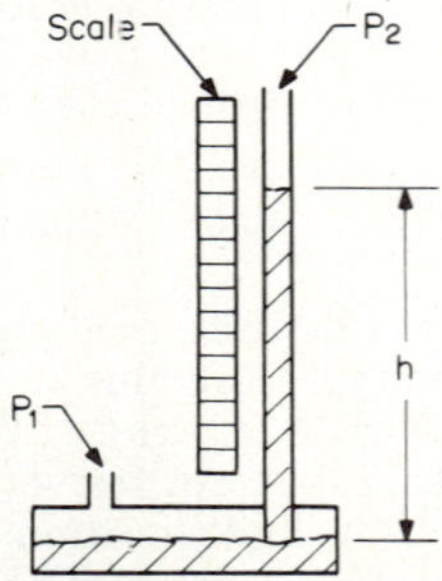

FIG. 5.20 Well-type manometer.

applied to two sides of a U-shaped glass tube, and if a dense liquid such as mercury is placed in the tube, the tube assumes a difference in level proportional to the pressure difference. Such manometers are generally used only with gases because the gas in the column produces a negligible addition to the "head" of the mercury column. Manometers do not require calibration if a fluid with known properties is used. Simply fill the tube with clean mercury, for example, and read the difference in height of the two columns. The pressure difference can be expressed as so many millimeters of mercury or converted to psi by the factor of 51.72 mmHg = 1 psi.

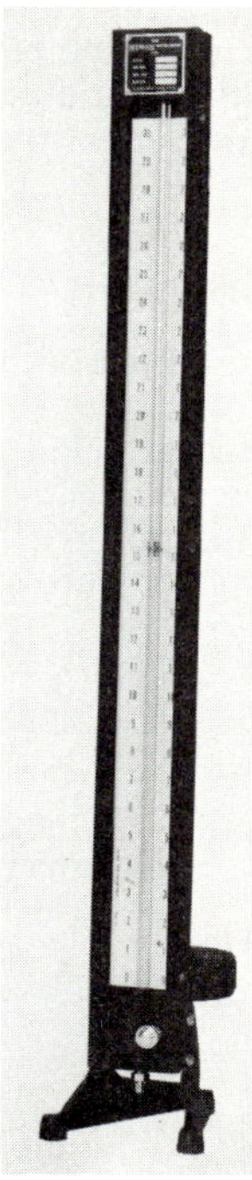

FIG. 5.21 Well-type manometer. *(Meriam Instrument Div., The Scott & Fetzer Co.)*

With the U-tube manometer, two observations of height must be made for each pressure measurement. This can be simplified to one reading by using a reservoir or well, as shown in Figs. 5.20 and 5.21. If the surface of the well is large compared to the cross-sectional area of the column, the well level drops only a small amount compared to the rise of the column. The change of level in the well can be accounted for in two ways. (1) Make the scale adjustable, so that the zero corresponds to the level in the well. (2) Construct a special scale so that the liquid level in the tube as read corresponds to that measured from the true level in the reservoir. In effect, the scale is shrunk by a few percent.

The inclined manometer (Fig. 5.22) tilts the tube for a gain in sensitivity. The pressure difference is still equivalent to the height of the liquid column h, but the scale in effect has been lengthened, so better resolution is possible.

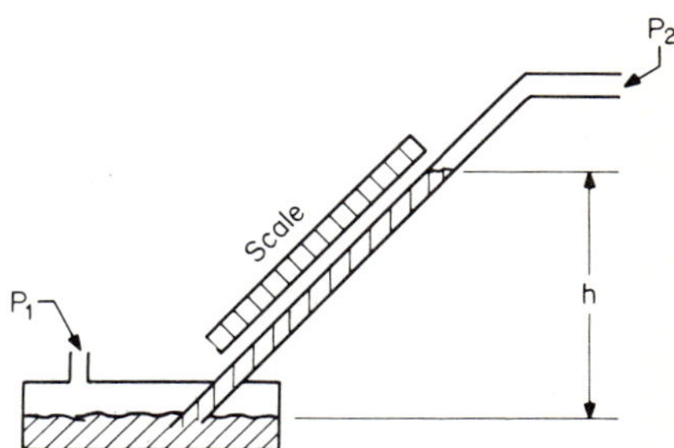

FIG. 5.22 Inclined manometer.

All the manometers described are differential pressure-measuring instruments, but if the space above the fluid column is evacuated, they become absolute pressure gages; if P_1 is exposed to the atmosphere, they become barometers. Such instruments are mercury-filled, with a tube about 32 in long. A normal barometer is 14.7 psia, which corresponds to about 30 inHg (760 mmHg).

In principle, any fluid can be used in a manometer, but the density of such a fluid should be known to good precision and the vapor pressure should be low.

The most common manometer fluids are mercury and water, but special oils are also available. Mercury is usually the preferred fluid because it is readily available in a pure form and its properties are well-known to a high degree of precision. As the pressures involved in a particular task become lower than about 1 psi, the height of the mercury column becomes small, and water or oil are preferred to mercury. One psi is equivalent to 2.036 inHg (51.71 mmHg), but the height of a water column at this pressure is 27.67 in (702.8 mm), so water is the choice of a manometer fluid for such work as furnace drafts and some air-velocity measurements.

Pressure Transducers

All the pressure-measurement methods involve reading an indication visually. Often it is necessary to convert pressure information to another form of energy to transmit it over a long distance or to record it. Such an energy-converting device is called a *transducer*. Strictly speaking, a transducer converts energy into any form, but an electric signal is usually the most useful, so here we describe several types of pressure to electric transducers.

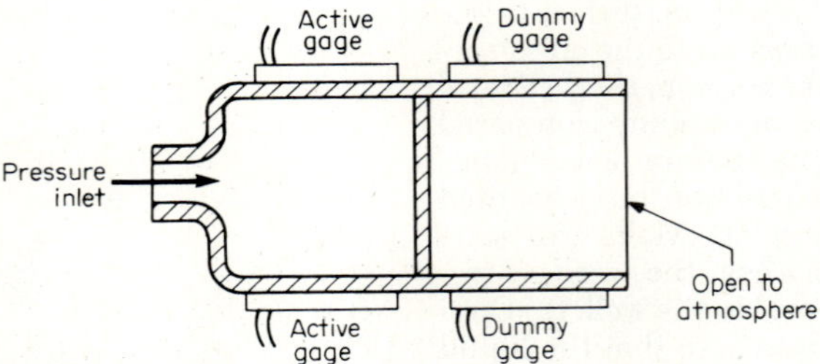

FIG. 5.23 Bonded strain-gage pressure transducer.

Strain-Gage Pressure Transducers

Section 5.5 describes how the bonded-wire resistance strain gage applies to measure strain in a structure. This principle has been used to build pressure transducers that generate millivolt signals proportional to the pressure acting upon them.

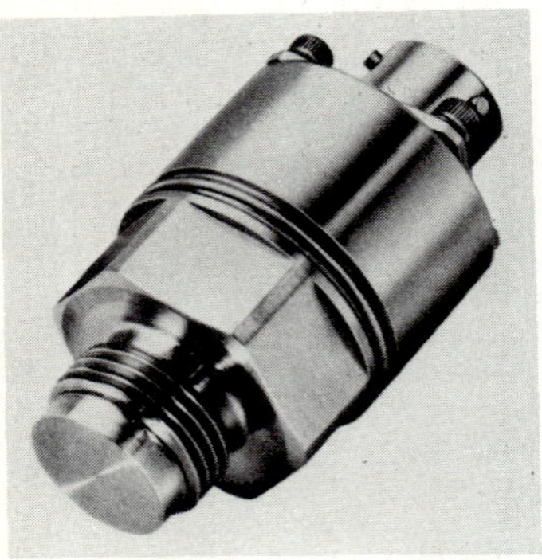

FIG. 5.24 Bonded strain-gage pressure transducer. *(Dynisco.)*

A bonded transducer is shown in both Figs. 5.23 and 5.24; Fig. 5.23 is a section through the strain tube. When pressure is admitted to the tube, the tube undergoes strain; that is, it both lengthens and grows in circumference. Two gages are cemented to the tube and form the active legs of a bridge circuit. Two other gages are cemented to the end of the tube beyond the pressurized region and therefore are not subjected to any strain. They are called dummy gages, and they act as temperature compensators. Because

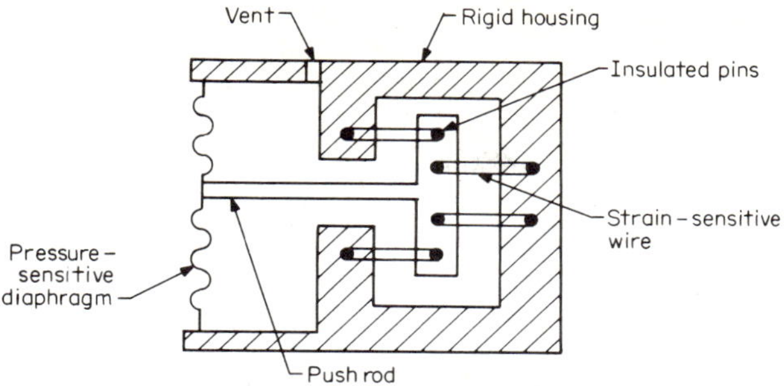

FIG. 5.25 Unbonded strain-gage pressure transducer.

the entire tube tends to be at the same temperature and all four gages are connected in the bridge circuit, temperature changes tend to cancel.

Unbonded transducers are shown in Figs. 5.25 and 5.26. This configuration is used in transducers for the lower-pressure ranges, where a limited amount of effort and strain are available to actuate the gages. With the unbonded arrangement, all four gages can be made active, and full temperature compensation is retained. Because it is possible to achieve increased sensitivity with unbonded gages, why isn't this principle used for all pressure transducers? The answer is that unbonded construction is more expensive than bonded, and the resulting transducer is more sensitive to vibration. The best general rule is to use bonded gages where possible, and to use unbonded gages only where necessary for an adequate output signal.

FIG. 5.26 Unbonded differential strain-gage pressure transducer. *(Gould Inc., Statham Instruments Division.)*

Variable-Reluctance Transducers

Such transducers can be used both for pressure measurement and any quantity that can be converted to motion. A variable-reluctance transducer is often referred to as an LVDT (linear variable differential transformer); the essential elements are shown in Figs. 5.27 and 5.28. An ac signal is required to activate the transformer. This signal may be at 60 Hz (cycles per second) or 1000 Hz—the frequency is not critical.

When the armature is centered in the assembly, the net output signal is zero because the two secondary windings are connected in series-opposed. As the armature (made of a ferromagnetic material) moves away from the center position, a voltage appears at the output terminals. This voltage is proportional to the armature's position, and suitable detector circuits can be built to provide large dc output signals in response to a pressure input. Such systems feature reliability, large signal output, and infinite resolution.

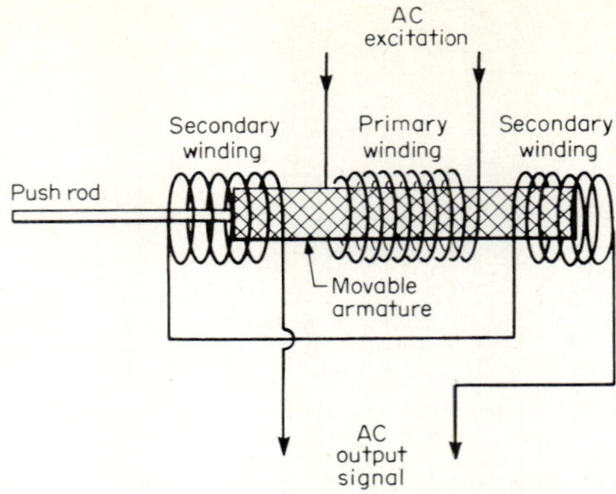

FIG. 5.27 Linear variable differential transformer.

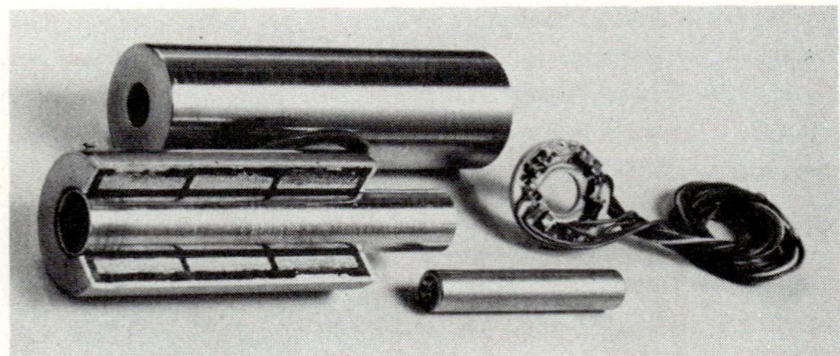

FIG. 5.28 Linear variable differential transformer, cutaway view. *(Schaevitz Engineering.)*

Piezoelectric Transducers

Certain crystalline materials, including quartz, Rochelle salts, and barium titanate, exhibit the phenomenon of piezoelectricity: When the crystal undergoes deflection, an electric charge appears across it. This property has been known for many years and has been widely applied to earphones and phonograph cartridges. The charge and the energy available from the piezoelectric effect are very small, and for a long time it was thought that steady-state (slow-changing) measurements could not be made. But recent advances (the charge amplifier and the field effect transistor) have made the piezoelectric transducer susceptible to ordinary calibration methods, and the device can hold steady-state readings for many minutes. The advantages are small size and rapid response. However, the user must be especially careful with the circuit from transducer to amplifier input or the microscopic charges developed will leak off. Generally, a fully shielded sys-

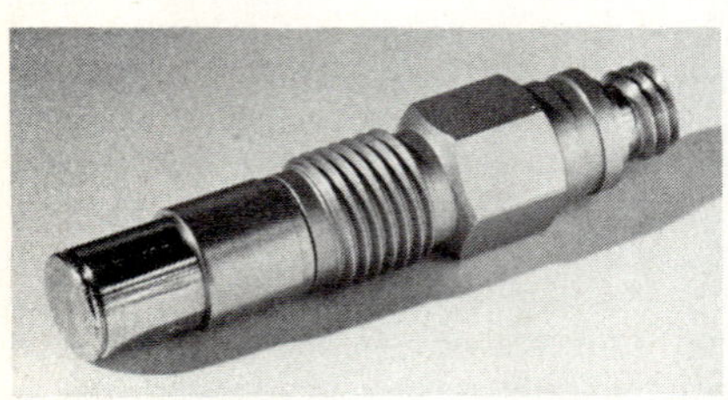

FIG. 5.29 Quartz piezoelectric pressure transducer. *(Sundstrand Data Control, Inc.)*

tem using low-leakage insulation is an important requirement. Also, the entire input system must be kept clean and dry. Many applications are not adaptable to piezoelectric transducers, but for rapid response they have no equal. Refer to Fig. 5.29.

Piezoresistive Transducers

Semiconductor strain gages, sometimes called *piezoresistive strain gages,* are discussed in Sec. 5.5. When applied to pressure measurement, they produce outputs in volts rather than the millivolts obtained from conventional strain gages. At first the transducers presented a large temperature-sensitivity problem, but devised compensation methods now make systems readily available for many applications.

Capacitance Transducers

An electric capacitor is simply two plates in close proximity, isolated electrically. A pressure-sensitive diaphragm is also a plate, supported at its edge and capable of deflection. Capacitance pressure transducers have four advantages. (1) Because the diaphragm is not connected to any other moving member, only very small forces oppose its motion. This means a transducer can be built with great sensitivity. (2) The diaphragm can be made very thin and of light material, meaning the dynamic response can be made fast. (3) The diaphragm is the only temperature-sensitive part of the transducer and can be made of a material virtually immune to temperature effects. (4) The transducer is completely unaffected by magnetic fields.

Figure 5.30 is the common design of a capacitance-type differential pressure transducer. A metal plate or foil is stretched across the inside of the case and supported around the case's periphery. Two other plates are suspended in the

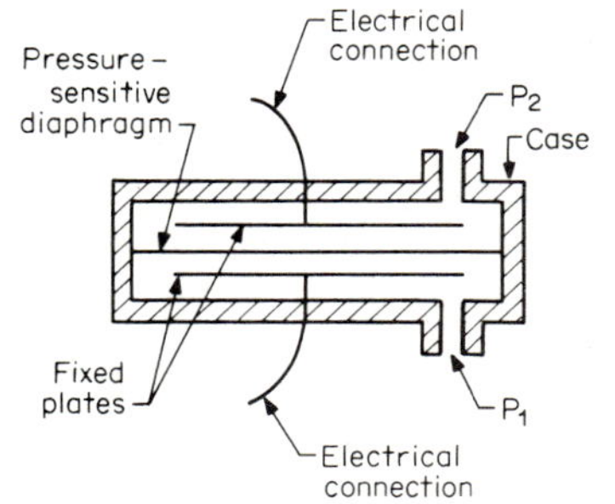

FIG. 5.30 Capacitance pressure transducer.

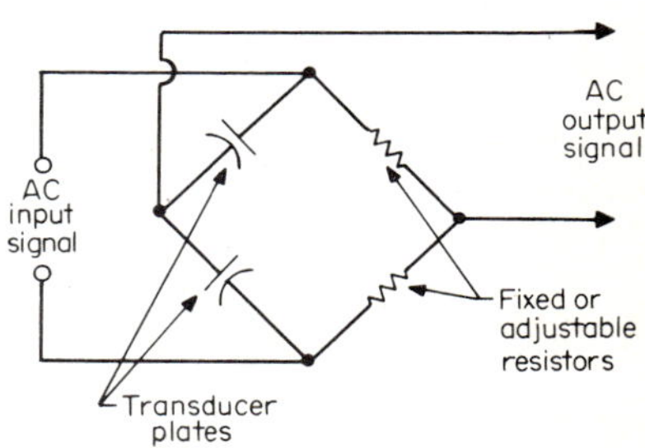

FIG. 5.31 Capacitance bridge.

case close to the diaphragm, but they are insulated from the case. As P_2 becomes greater than P_1, the diaphragm bows downward and the capacitance to the lower plate increases as the capacitance to the upper plate decreases. These elements are connected in a bridge circuit, as shown in Fig. 5.31. The unbalance in the two capacitors is reflected as an unbalance of the ac bridge, and an output signal is generated. As configured in a typical commercial transducer, the entire electronic section is built into one end of the transducer housing: The instrument user merely brings in a source of power and gets out a dc signal proportional to pressure.

Miscellaneous Pressure Transducers

There is almost no limit to the types of pressure transducers that have been conceived, but many are not widely used. One popular type is the simple potentiometric transducer, in which a bellows or Bourdon tube drives an arm along a resistor. These transducers are inexpensive, but their limitations involve resolution, hysteresis, and life. Typical construction is a wire-wound resistor for the resistive element, with a wiper arm traveling along the resistor in response to pressure changes. Although such construction is commonplace in electrical instrumentation, there are certain inherent design problems. If the wire is very fine (many turns), the resolution is good but the windings wear out readily. If the wire is coarse (few turns), resolution becomes poor but the windings last a long time. Any design involves a compromise between these two factors. The wiper arm must bear upon the windings with some finite force to ensure reliable contact, and because friction is never zero, this force amounts to a drag on the arm; this drag is seen as hysteresis.

Pneumatic Pressure Transmitters

Many pressure-measurement applications in the chemical-process industry require nonelectric pressure sensing and transmission over long distances because of safety, cost, and other factors. One cannot just run a pressure line to the control room because process chemicals are often hazardous and can cover many pressures. In a typical installation, all pressures should be in the medium of clean compressed air and cover the same pressure ranges at the point of recording, regardless of what they are in the process. This way all the recorders can cover the same pressure range.

To meet these conditions, the widely used pneumatic pressure transmitter is used; it is shown schematically in Fig. 5.32. The pressure to be measured is admitted to the sensing bellows and acts as a force upon the lever arm, tending to push it down. As the arm moves down, it blocks off the exit of compressed air from the nozzle to the atmosphere and builds up the pressure in the balancing bellows. The system quickly reaches an equilibrium in which the lever arm holds the nozzle in a partially open position. At this balance point there is a fixed relationship between P_1 and P_2, a ratio determined by the point of action of the two bellows and their relative effective areas. For example, say P_1 is the pressure of

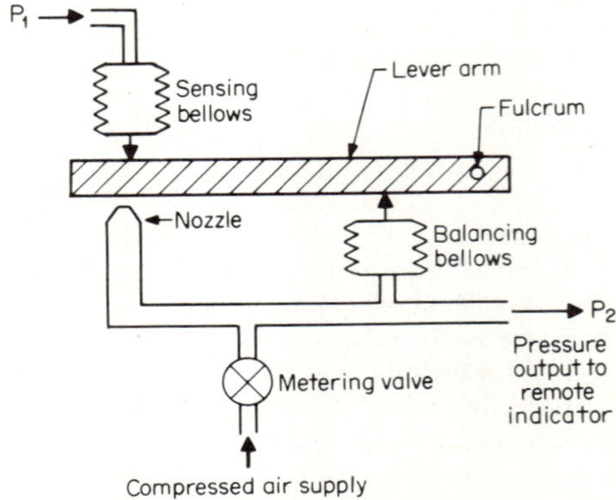

FIG. 5.32 Pneumatic pressure transmitter.

a hot oil line that has a range of 100 to 200 psi. A properly designed pressure transmitter can convert this range to 0 to 15 psi of compressed air, and this 15-psi signal can be transmitted 300 ft to a remote recorder. Pneumatic pressure transmitters are well-developed instrumentation devices, available from many manufacturers. Figure 5.33 is an example of a commercial pneumatic transmitter.

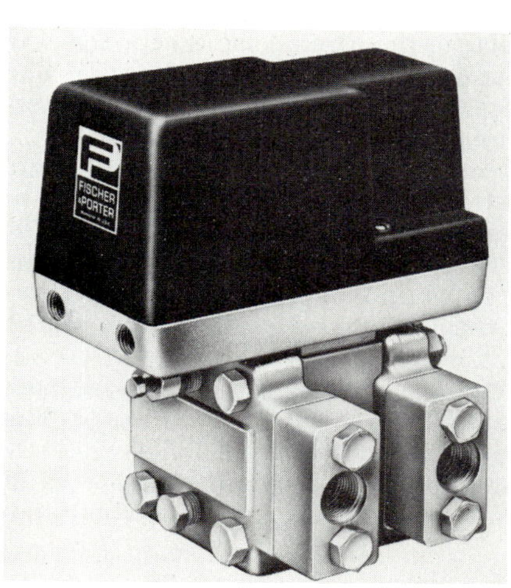

FIG. 5.33 Pneumatic differential pressure transmitter. *(Fischer & Porter.)*

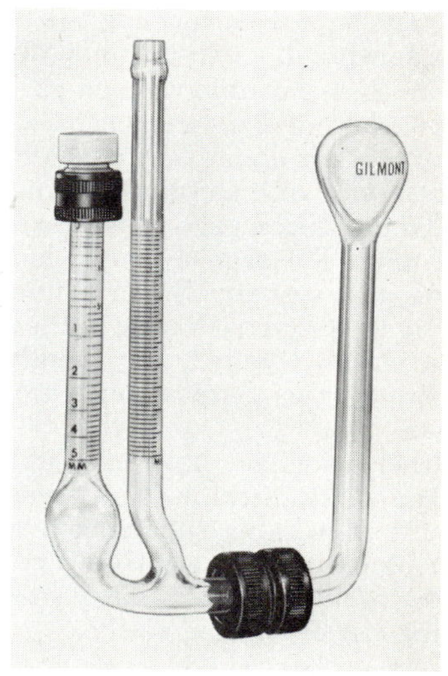

FIG. 5.34 McLeod vacuum gage. *(Roger Gilmont Instruments, Inc.)*

Vacuum Gages

Vacuum measurement involves absolute pressures below 1 atm. Mechanical gages using diaphragms or bellows are suitable for such pressures down to about 1 mmHg, but for lower pressures, other techniques, which depend upon derived properties of a gas, are used to measure other properties, and the pressure is inferred. For a short discussion of vacuum units, see Sec. 4.8.

The McLeod gage is perhaps the oldest and best-known high-vacuum gage. It is often constructed of laboratory glassware by the experimenter and based upon a simple principle: If a volume of gas is compressed isothermally, its pressure rises in proportion to the decrease in volume. Stated mathematically (as Boyle's law),

$$P_1V_1 = P_2V_2$$

In the McLeod gage, the level of mercury in a reservoir is allowed to rise until it traps off a known volume of gas. The mercury is then allowed to continue to rise until the volume is compressed by a known ratio of, for example, 100:1. The elevation of mercury in an adjacent branch tube is read, and the apparent pressure is read as 100 times the actual system pressure. The method has certain limitations; for instance, condensable vapors should not be present, and there will be capillary effects if tubing sizes are too small. Also, it is a sampling method and does not allow for an electric output. McLeod gages cover numerous pressures, being useful from 10 to 10^{-6} torr (1.3 kPa to 13 mPa). See Fig. 5.34.

The Pirani gage is usable in the more limited range of 1 to 10^{-3} torr (130 to 0.1 Pa). However, its continuous electric output makes it preferable to the McLeod gage for many applications. This gage is based upon the heat conductivity of a gas. A wire in the gas is heated electrically, and the heat loss from wire to gas is proportional to the absolute pressure. If the wire is a material with a high thermal coefficient of resistance, the wire resistance depends upon temperature, which in turn depends upon heat loss, which then depends upon gas pressure. The wire is performing the two functions of heat transference and temperature sensing in a two-terminal device.

The thermocouple vacuum gage is similar to the Pirani gage except that the temperature of the wire is sensed by a thermocouple attached to the wire. The useful range is about that of the Pirani gage, but the sensor becomes a four-terminal device, and certain simplifications are attained in the electronics.

The ionization gage greatly extends vacuum measurement down to 10^{-10} torr (10^{-2} μPa). The gage is a multielement electron tube connected pneumatically to the vacuum system. The principle involved is the generation of ions from collisions with electrons flowing in the space between the electrodes. Because the ions are positively charged, they readily collect on a negative plate. The current flowing through this plate is then a measure of the absolute pressure (vacuum) in the system.

Other principles have been used in vacuum gages, but those discussed here are the most often used. All vacuum gages share a common difficulty: The amount of working substance available is small, and instruments of necessity must be extremely sensitive. Typical experimental setups use several types of gages working simultaneously so that errors caused by overranging or malfunctioning can be spotted easily.

5.4 FLUID FLOW

Orifice Flowmeters and Nozzles

These devices are sometimes called *headmeters* or *inferential flowmeters* because there is a pressure drop associated with the flow (a difference in "head") and the flow rate is "inferred" from the pressure drop. Figures 5.35 and 5.36 are two types of such flowmeters. The principle of operation is the same for both, that is, a loss of energy and hence a loss of pressure along the flow stream. The nozzle is somewhat more efficient (lower pressure drop for a particular flow rate). Unfortunately, the differential pressure ($P_1 - P_2$) is not a linear function of flow rate but proportional to the square of the flow rate. Usually the pressure drop is read out differentially rather than by two separate meters. The differential pres-

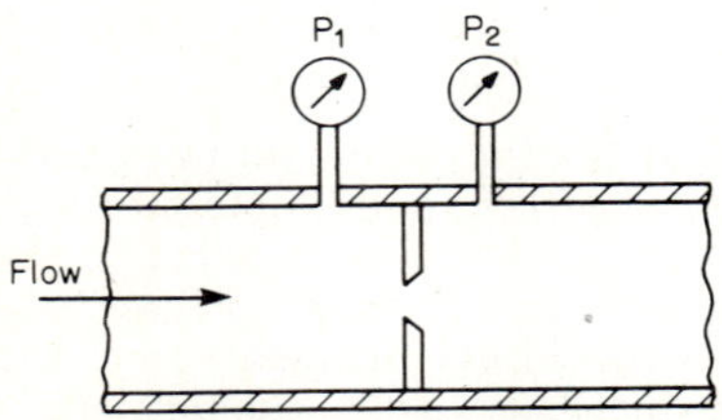

FIG. 5.35 Orifice flowmeter.

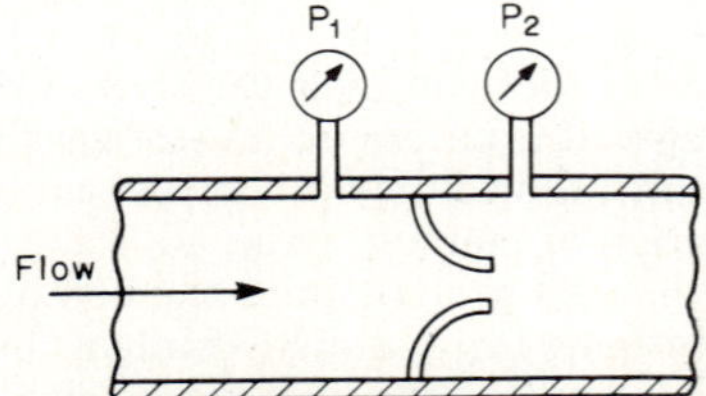

FIG. 5.36 Flow nozzle.

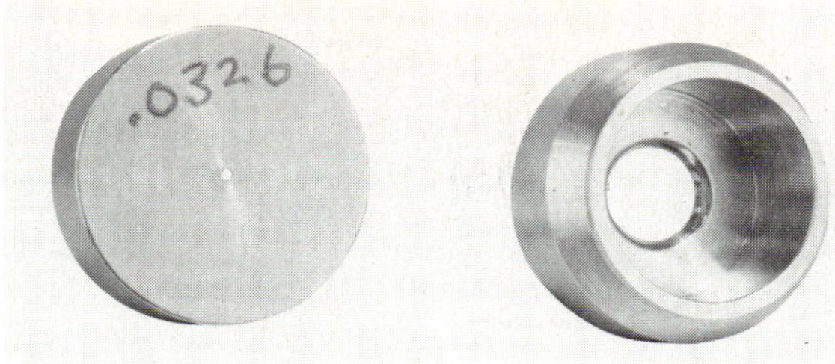

FIG. 5.37 Typical flow orifices. *(The Foxboro Co.)*

sure indicator can be a mercury manometer (for gas flows) or a differential pressure gage.

The approach most common in the design and use of headmeters is the American Society of Mechanical Engineers (ASME) Test Code. This organization has extensively researched and tested the design of such devices, including dimensions, location of pressure taps, and flow coefficients, for both liquids and gases. Two typical orifices are in Fig. 5.37.

Venturis

The venturi (Fig. 5.38) is another form of headmeter based upon the same principle as the orifice. Because of the gradual contraction and expansion of the

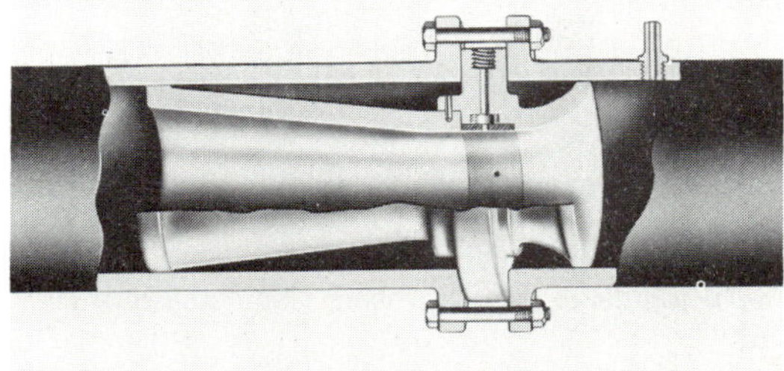

FIG. 5.38 Venturi flow sensor. *(The Foxboro Co.)*

cross-sectional area, it is the most efficient of the three types. Pressure loss is about 30 percent of inlet pressure, as against perhaps 80 to 90 percent for an orifice. Its disadvantages are greater cost and space requirements.

Tapered-Tube Flowmeters

A common example of a variable-area flowmeter is the tapered tube or rotameter, shown schematically in Fig. 5.39 and actually in Fig. 5.40. It essentially consists of only two parts: a conical tube (usually glass) and a plug or ball free to "float" in the upward-moving fluid stream. It must be mounted vertically with flow upward. The float is actually heavier than the fluid, and at zero flow rate it rests at the bottom of the tube. During flow it rises to some equilibrium position where the gravitational force downward is balanced by upward forces of buoyancy,

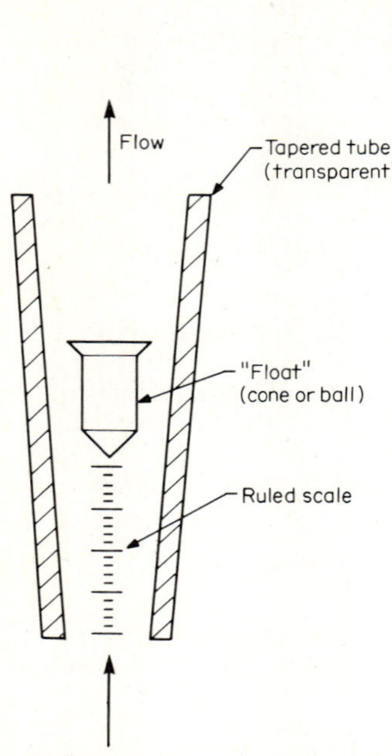

FIG. 5.39 Tapered-tube flowmeter, or rotameter.

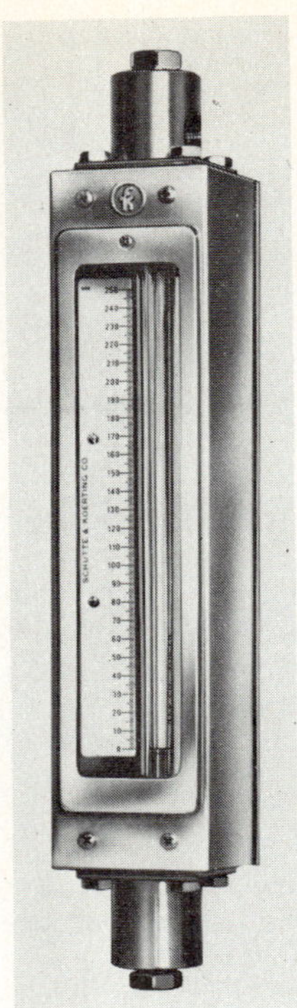

FIG. 5.40 Rotameter. *(Ametek, Scutte & Koerting Div.)*

drag, and pressure differential. Such devices are neither purely volumetric or gravimetric and are sensitive to density and viscosity changes. Careful design of the float can minimize the effect of these variables. The device is best calibrated using the fluid with which it is to be used. Typical accuracies are 2 percent; a typical range is 5:1.

Advantages are low cost and direct visual indication for applications involving moderate pressures. The same principle has been used in high-pressure work; the tube is fabricated of metal. Here some form of electrical or magnetic sensing of float position is required, and the device loses some of its simplicity. Either liquids or gases can be handled. Only flow rates (gallons per minute) can be indicated. Totalizing (gallons, cubic feet) is not possible without an integrating device.

Turbine Flowmeters

Turbine flowmeters are widely used and available from many manufacturers. The principle of operation is simple. A free-turning rotor is inserted into the flow stream and caused to rotate by angled blades (see Figs. 5.41 and 5.42). Rotational speed is proportional to fluid velocity and sensed electromagnetically. The resulting electric signal's most salient characteristic is its frequency (hertz). Such well-designed flow sensors have low friction bearings and small magnetic drag from the pick-off system. Usually a two-pole magnet is embedded in the rotor,

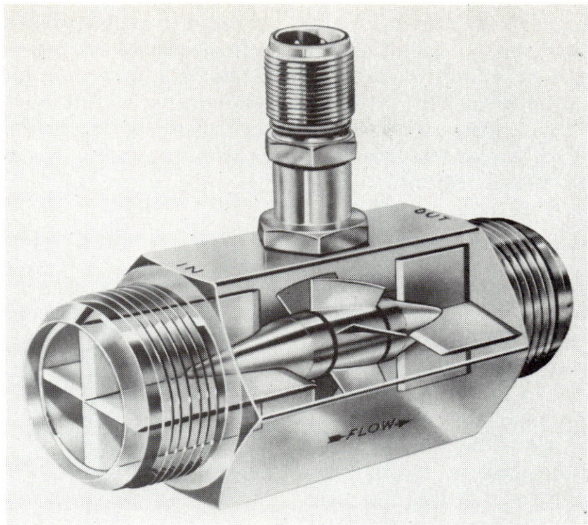

FIG. 5.41 Turbine flow sensor. *(Flow Technology, Inc.)*

although in some sensors each blade tip holds an individual magnet. A coil mounted in the housing receives the impulses of the magnet poles and generates an electric impulse.

These sensors are volumetric devices; the output is an ac signal whose frequency is proportional to, for example, cubic feet per second or gallons per minute. Because a user is often more interested in the actual amount of material passing through the sensor, a more useful result is the weight flow or mass flow. To determine weight flow through a volumetric flowmeter, multiply volume flow rate by the density of the fluid by the density (in pounds per gallon) (for example, lb/h = gph × lb/gal). To measure density, withdraw a sample of the liquid and measure it with a hydrometer. If the fluid's properties are known, the temperature can be measured and the density inferred by referring to an appropriate table.

To automatically correct for density change, some available systems continuously sample the fluid and provide direct readout on a gravimetric (weight) basis.

A typical turbine-flowmeter system consists of a sensor, frequency-to-voltage converter, and chart recorder. Typical system accuracy is ±0.5 percent. There are sensors for full-scale ratings of 0.1 to 12,000 gpm (gallons per minute) (10^{-6} to 10 m^3/s), and there are even larger units.

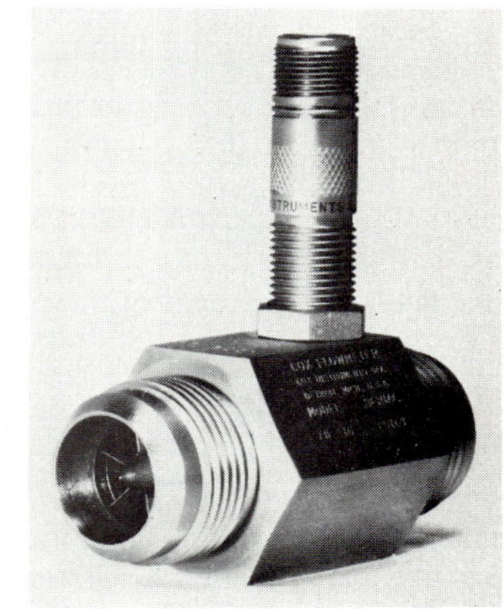

FIG. 5.42 Turbine flow sensor. *(Cox Instrument Div., Lynch Corp.)*

The advantages of turbine flowmeters are rapid response, good accuracy, good repeatability, and easy installation. Disadvantages include relative complexity and maintenance in certain types of service (slurries, for example).

Gas flow can be measured with turbine-flow sensors, but specially designed bearings and blade angles are necessary for such service. Again, the measurement is volumetric. Pressure and temperature of the gas at the sensor must be known for a meaningful measurement.

Vane or Drag-Body Flowmeters

Such flowmeters are simple in concept: To provide a measurement of flow rate, a target is inserted in the flow stream, and the impact forces upon the target are sensed. See Figs. 5.43 and 5.44. This type of flowmeter is most closely related to the headmeter in principle and has a square-law response; that is, force is proportional to the flow rate squared. Readout is customarily by strain gages located external to the fluid stream. These meters have had good commercial acceptance. Advantages include rapid response, a simple and reliable sensing element, and easy installation. The sensing element is electrically compatible with any strain-gage indicating or recording system. The chief disadvantage is nonlinear output, although readout equipment that linearizes the output signal is available.

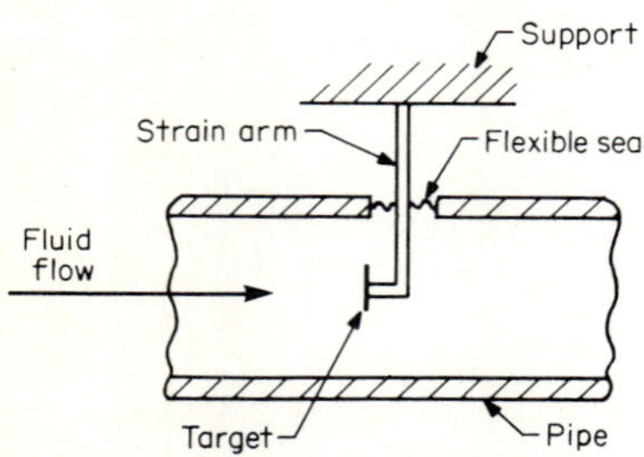

FIG. 5.43 Drag-body flowmeter.

FIG. 5.44 Drag-body flowmeter. *(Ramapo Instrument Co.)*

Sonic Flowmeters

Sonic or ultrasonic flowmeters are based upon the Doppler effect: Sound travels more rapidly downstream than upstream. An everyday example of this phenomenon is the train whose whistle seems to change pitch as it passes us.

In a typical sonic flowmeter (Fig. 5.45), two sound paths in the liquid stream are established, in essentially opposite directions. Electronic circuitry generates pulses of acoustic energy in both paths and measures the difference in speed of propagation. The measurement does not depend upon the velocity of sound in the particular fluid because it is a difference method.

For special applications requiring no obstruction to flow, this meter may be useful. The meter is complex, costly, and not suited for use with gases.

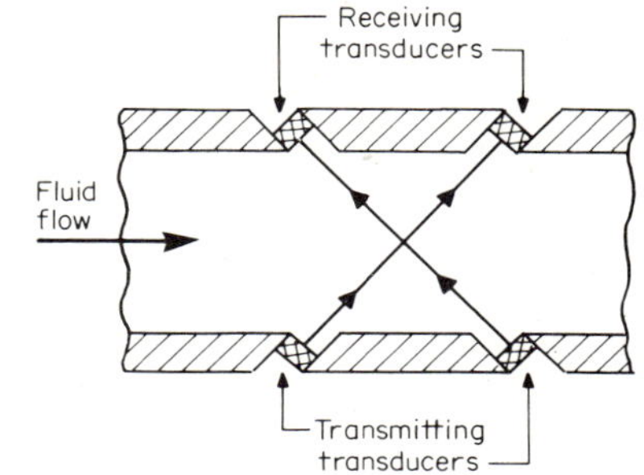

FIG. 5.45 Sonic flowmeter.

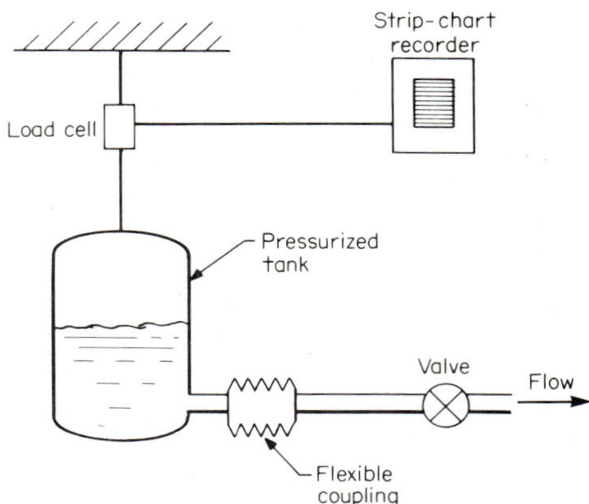

FIG. 5.46 Flow measurement by weight and time.

Weight versus Time

One often-used flow-measurement method does not involve a flowmeter; it simultaneously measures the weight loss or gain of a tank and the time interval. Because both weight and time are readily measurable with great accuracy, this method can be used to calibrate other flowmeters. One such system is shown in Fig. 5.46. The weigh tank is hung from a rigid overhead support by a load cell whose output is recorded against time. The chart drive speed is known, and if the flow rate is held constant, the trace on the chart is a line with uniform slope; that slope in pounds per second is the gravimetric flow rate. Many variations are possible. For example, the tank can rest upon three legs, and the load cells under each can be summed electrically. Alternatively, rather than the source tank being weighed, the fluid may be caught in the weigh tank.

One important consideration is that the flexible coupling exerts a minimum force on the tank, a nonexistent problem if the fluid is dumped into an open vessel. Also, if the pressure is high, and consequently the weight of the gas displacing the liquid is large, a correction should be made for this weight. However, in most practical cases the weight of the pressurizing gas is negligible.

Miscellaneous Flowmeters

There are virtually endless flow-measurement methods; the selection is usually a compromise based upon factors like accuracy, response rate, cost, and fluid-material compatibility. Two rather unique flowmeters utilizing very different principles are electromagnetic flowmeters and vortex-shedding meters.

Electromagnetic Flowmeters

The electromagnetic flowmeter is based upon Faraday's law of induction, which states that if an electric conductor moves perpendicularly to a magnetic field, a voltage is induced in the conductor. This is the operating principle behind electric generators and many other devices. For a flowmeter utilizing the principle, a magnetic field must be established across the flow stream and the fluid must be electrically conductive. This latter requirement is not too stringent because most liquids are suitable, hydrocarbon liquids the chief exception. The flowmeter wall must be nonmagnetic and an electric insulator. Two sensing electrodes are imbedded in the opposite walls on a line at right angles to both the magnetic field and the flow stream. These meters have been used most successfully on problem fluids like slurries and liquid metals. They do not obstruct flow because they are simply a continuation of the connecting pipes (Fig. 5.47).

FIG. 5.47 Electromagnetic flow transmitter. *(The Foxboro Co.)*

Vortex-Shedding Meters

Vortex shedding is the natural effect that can occur when a gas or liquid flows around a blunt or nonstreamlined object (Fig. 5.48). The flow is unable to follow the shape on its downstream side, so it separates from the surface of the object, leaving a highly turbulent wake that can be made to take the form of a continuous series of eddies forming and being swept downstream. Each eddy, or *vortex,* first grows and is then detached or shed from the object—hence the phenome-

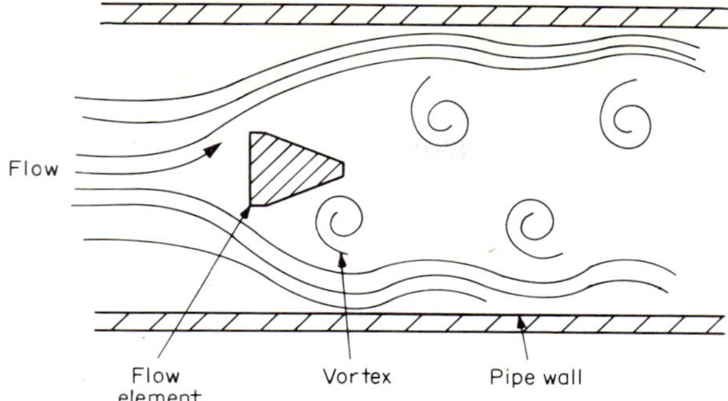

FIG. 5.48 Flow pattern generated by a vortex-shedding flow element. *(Neptune/Eastech, Inc.)*

non's name. The effect is noticeable in the flow around and behind a rock in a stream or in a flag fluttering in the wake generated by the wind and the flagpole.

If the vortex-generating object is correctly shaped and placed in a pipeline of correct relative dimensions, it forms a primary flow element that generates pulse signals over very wide flow ranges at a frequency proportional to the flow rate. The approaching flow separates from the flow element, and the vortices form and shed alternately on either side of the triangular shape. As the flow rate increases, the speed with which the vortex forms or "fills up" increases at the same rate. As a result, the number of vortices generated is directly related to the amount of fluid passing through the device.

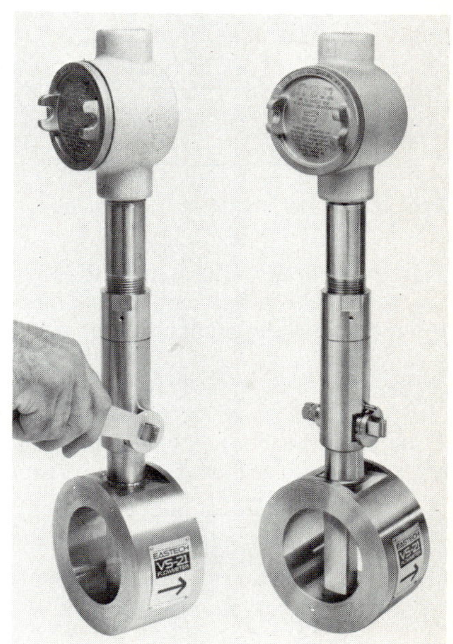

FIG. 5.49 Vortex-shedding flowmeter. *(Neptune/Eastech, Inc.)*

The oscillatory motion, a function developed by the mechanics of the meter, is sensed by a thermistor that counts the vortex rate and converts it to a pulse-train output through an electronic circuit. This motion also can be sensed mechanically by means similar to those used in a turbine meter, and a conventional electromagnet pickup can develop the pulse output.

The calibration factor, or pulses per gallon or cubic foot, is determined only by the dimensions of the flow element and the pipeline. It does not depend upon the specific gravity, viscosity, pressure, or temperature; neither does it depend upon whether a gas or liquid flows through the meter. An example of such a flowmeter is in Fig. 5.49.

5.5 STRAIN MEASUREMENT

In any structure, stress results from the external force applied; strain is the deformation produced by this stress. For simple structures, such as a flat beam or square column, usually it is possible to calculate strain if the stress is known by using Hooke's law. However, for complex structures, and particularly where dynamic stresses (forces variable with time) are present, strain must be experimentally determined. The methods described here specifically cover strain measurement on structures. For example, strain measurement in pressure or force transducers is discussed in specific subsections of Sec. 5.3, "Pressure-Measuring Instruments."

Bonded-Wire Resistance Strain Gages

Strain measurements down to one-millionth of an inch per inch are possible with electric-resistance wire gages. Such gages can be used to measure surface strains on any shape or size of object. Figures 5.50 and 5.51 illustrate schematically the

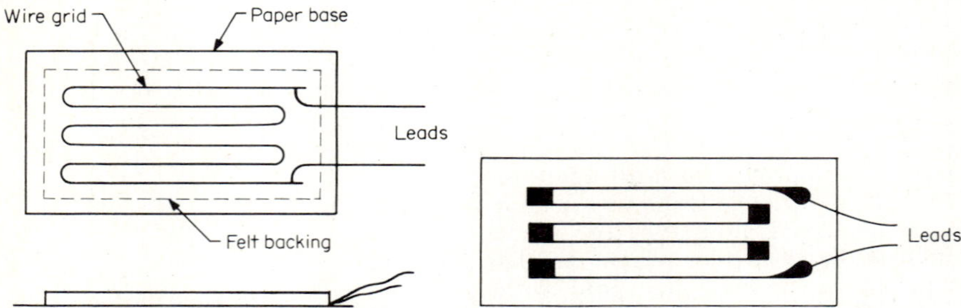

FIG. 5.50 Resistance wire strain gage.

FIG. 5.51 Foil strain gage.

gage construction with a grid of fine alloy wire or thin foil, bonded to paper and protected with a felt pad. In use, the gage is cemented rigidly to the surface of the member to be analyzed. The strain relationship is

$$e = \frac{\Delta R}{R} \times \frac{1}{f} \text{ in/in}$$

Thus, if the resistance R and the gage factor (given by the gage manufacturer) are known and the change in resistance ΔR is measured, the strain that caused the resistance change can be determined and Hooke's law applied to determine the stress.

Gages must be properly selected according to the manufacturer's recommendations. The surface to which the gage is applied must be clean, the proper cement must be used, and the gage assembly must be coated as protection from such environmental conditions as moisture.

A gaging unit, usually a Wheatstone bridge (see Fig. 5.52), is needed to

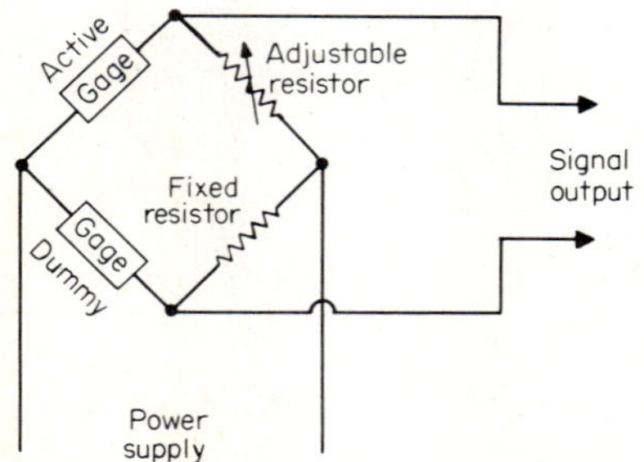

FIG. 5.52 Bridge circuit for strain gages.

detect the signal resulting from the strain gage's change in resistance. The strain and therefore the signal are often too small for direct handling because they are of only a few millivolts, so amplification is required before inputting to recorder or oscilloscope or other readout device.

With the bridge set so that the only unbalance is the change of resistance in the active strain gage, the potential difference between the output terminals becomes a measurement of strain. Because the gage is sensitive to both temperature and strain, it measures the combined effect. Most practical strain-gage wire is selected for low-temperature coefficient of resistivity; temperature sensitivity is a second-order effect, but some temperature-compensation scheme is desirable. If a "dummy" gage, cemented to an unstressed piece of the same metal subjected to the same ambient conditions, is wired into the bridge leg adjacent to the one containing the "active" gage, the temperature effect is cancelled. Thus the active gage reports only what is taking place in the stressed plate. The power supply is most commonly voltage-regulated direct current because the output directly depends upon the input.

It is sometimes useful to make both gages active—for example, mounted on opposite sides of a beam—with one gage subjected to tension and the other to compression. Temperature effects are still compensated, and the output is doubled. In other instances it may be desirable to make all four bridge arms active gages, in which case an adjustable resistor is commonly inserted into the bridge for zero adjustment. The experimenter must determine the most practical arrangement for the problem and remember that the bridge unbalances are proportional to the difference in the strains of gages in adjacent legs and to the sum of strains in gages in opposite legs.

Foil Gages

Foil gages are produced from thin metallic foil by photoetching techniques and are applied, instrumented, read, and evaluated as are wire-grid gages. Being thinner, foil gages can be applied easily to curved surfaces, have lower transverse sensitivity, exhibit negligible hysteresis under cycling loads, creep little under sustained loads, and can be stacked on top of each other. See Fig. 5.53.

Semiconductor Strain Gages

These gages, sometimes called *piezoresistive,* are whiskerlike single crystals of some semiconductor material, usually silicon. Initially they exhibited certain disadvantages, such as high-temperature sensitivity to both resistance and gage factor. They offer greatly increased output. For example, some transducers have direct outputs of 5 V, thus requiring no amplification for most applications. The early problems have been solved by using constant-current excitation and other techniques. A semiconductor strain gage is thus a useful tool for strain measurement.

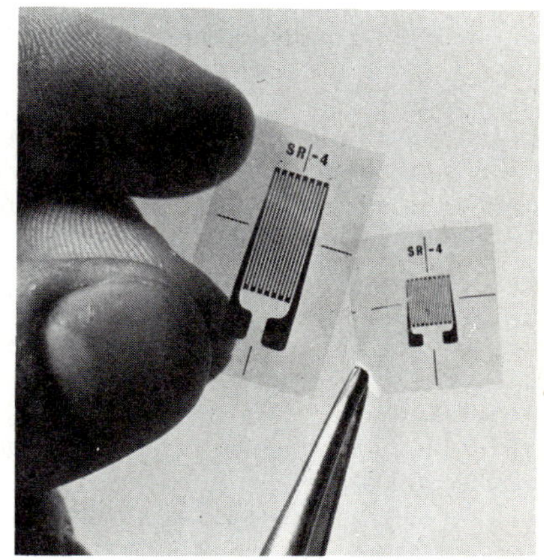

FIG. 5.53 Foil gages ready for application. *(BLH Electronics.)*

Brittle Coatings

Brittle coatings that adhere well to the surface can reveal the strain in the underlying material. The first such coating was millscale, a thin iron oxide that forms on hot-rolled steel stock. Many other coatings, such as whitewash, portland cement, and shellac, also have been tried.

The most popular of today's strain-indicating brittle coatings are the wood-rosin lacquers supplied by the Magnaflux Corporation under the trade name Stresscoat. Several compositions are available; the suitability of a particular lacquer depends upon the prevailing atmospheric conditions. The lacquer is usually sprayed 0.004 to 0.008 in thick (0.1 to 0.2 mm) upon the surface, which must be clean and free of grease and loose particles. Calibration bars are sprayed at the same time. Both the material and bars must be dried at an even temperature for up to 24 h. To facilitate observation of cracks, a bright aluminum undercoating is often applied.

When the cured testpiece is subjected to loads, the lacquer first begins to crack at its threshold sensitivity in the area of the largest principal stress. This information is often sufficient, revealing the critical area and direction of normal stress.

The threshold sensitivity of Stresscoat lacquers is 600 to 800 microinches/in in a uniaxial stress field. Proper selection of lacquer and exact control of thickness, curing, and testing temperatures may reduce the threshold to 400 microinches/in. If desired, the approximate strain (probably within 10 percent) can be established by using the calibration strip that was sprayed with the test part. Place the strip in a loading device and bend it as a cantilever beam by a cam at the free end; this causes the coating to crack on the tension surface. Crack spacing varies with the strain, close at the fixed end and diminishing toward the free end down to the threshold-sensitivity values. Place the strip in a holder containing strain graduations: Visually compare cracks on the testpiece's surface with those on the strip to see the strain magnitude that caused the cracks.

Birefringent Coatings

A birefringent coating becomes double refractive when strained. The principles behind the process are quite old (the same as those for conventional photoelasticity), but recently developed plastics, which adhere to all kinds of materials, have stable optical-strain constants and are sufficiently sensitive to be practical. The trade name of these plastics is Photostress. The plastics are either thin sheets [0.040, 0.080, and 0.120 in thick (1, 2, and 3 mm)] or in liquid form. The sheet material is bonded to a surface with a special adhesive; the liquid is brushed or sprayed on, or the part is dipped into the liquid. The layer should be at least 0.004 in (0.1 mm) thick. It is often necessary to apply several successive coatings, heat curing each layer in turn. Two sheet types and two liquids are available; these differ in stretching ability and magnitude of the optical-strain constant. Each sheet material is metalized on one face, to reflect polarized light even when cemented to a dull surface.

One advantage of Photostress is that the plastic can be applied directly to the part, which can then be subjected to actual operating loads. A special reflecting polariscope must be used. It contains only one polarizer and quarter-wave disk because the light passes back through the same pair after being reflected by the stressed surface-plastic interface. The only limitations depend upon the geometry of the structural component to be examined: It must be possible to apply the plastic to the surface, and the surface must be accessible to light.

5.6 HUMIDITY MEASUREMENT

Humidity is the amount of water in gaseous form present in the atmosphere or in industrial gases. This amount of water vapor is subject to wide variation and is an important variable in many industrial processes and human comfort. Its determination is not usually subject to direct measurement; most often humidity is inferred from related variables.

Humidity Definitions or Limits

The *dew point* is one way to describe the humidity of an air sample. If such a sample is cooled gradually and no moisture is allowed to enter or leave the sample, at some temperature condensation forms, which is the dew point. If that temperature is lower than 32°F (273 K), the condensation is called the *frost point*.

A more familiar way of describing humidity is by *relative humidity* (RH), which is the percentage of the amount of water in the air sample compared to what the sample could hold if saturated. It is not a particularly unique way to describe humidity because it is temperature-sensitive. For example, if an air sample at 80°F (300 K) has a 60% RH, that same sample when heated is then able to hold more water vapor, so even if no water vapor is added or lost, the relative humidity drops with increasing air temperature.

Still another way of specifying humidity is by the wet bulb–dry bulb temperature. Place two thermometers—one dry and the other moistened with water—in an air stream. The wet bulb generally is cooled by evaporation and reads a lower temperature. (In saturated air, both thermometers read the same.) By comparing the wet bulb–dry bulb readings to a psychometric chart, one can infer such factors as dew point, relative humidity, and absolute humidity. The sling psychrometer, shown in Fig. 5.54, is a basic wet bulb–dry bulb instrument. The

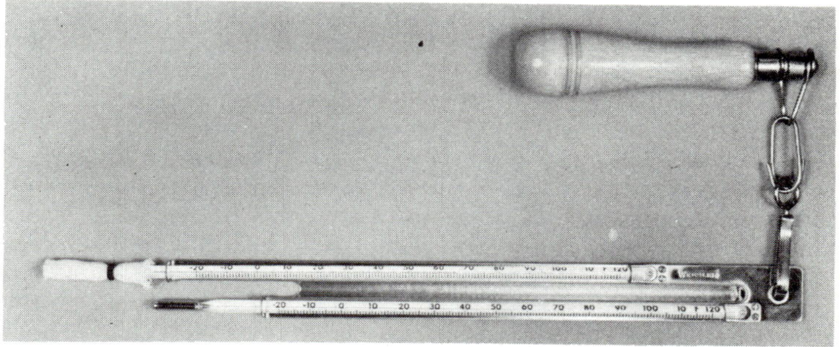

FIG. 5.54 Sling psychrometer. *(Weksler Instruments Corp.)*

wicking at the bulb is saturated with water, and the device is twirled rapidly in the air. After about 1 min both thermometers are read, yielding the so-called wet and dry bulb temperatures.

Absolute humidity is the water content of air, measured in weight of water per pound of dry air. It is also expressed as parts per million (ppm) or even as atmospheres (the partial pressure of the water vapor present in the gas sample in accordance with Dalton's law).

Humidity Sensors

Dew point can be measured directly: Cool a mirror, and when a water film forms, measure the temperature to get the dew point. The formation of condensate can be detected visually for manual instruments; if remote and automatic sensing are required, detect the water film by optical, electrical, or nuclear techniques. Industrial versions of these condensation hygrometers are capable of accuracies of $\pm 1°F$ (0.5 K) over temperature spans of $+200$ to $-100°F$ (370 to 200 K). They are relatively complex and subject to contamination unless used in a very clean environment.

A widely used device for dew-point measurements is the lithium chloride-saturated salt sensor, whose principle is based upon an equilibrium of water content between the heated salt and the surrounding air. Although the temperature at the equilibrium point is not the dew point, it is directly related, so these systems can be calibrated for dew point. Advantages of these devices include low cost and good durability. Disadvantages are slow response (several minutes) and unsuitability for relative humidity lower than 11%. Figure 5.55 shows an instrument that directly indicates dew point and temperature based upon a lithium chloride cell and thermistor probes.

Mechanical hygrometers, often used as indicating instruments, are simple in principle: A material that undergoes dimensional changes when subjected to changes in humidity is used, such as paper, nylon, or hair, with human hair most often chosen. The dimensional change is generally magnified mechanically by levers or gears and displayed directly. This is the common household-humidity indicator. Such mechanical devices can be coupled to electric-ouput devices and thus become transducers, although they are more often merely indicators.

The wet bulb–dry bulb principle can be developed into a transducer. If the sensors are thermocouples or resistance thermometers, an electric signal can be obtained. Generally, the surrounding air must be circulated around the wet sensor by a fan.

Also available commercially are infrared absorption systems, which pass a beam of infrared energy through the gas sample. The water vapor present absorbs a percentage of the energy, and this in turn can be read as the vapor density.

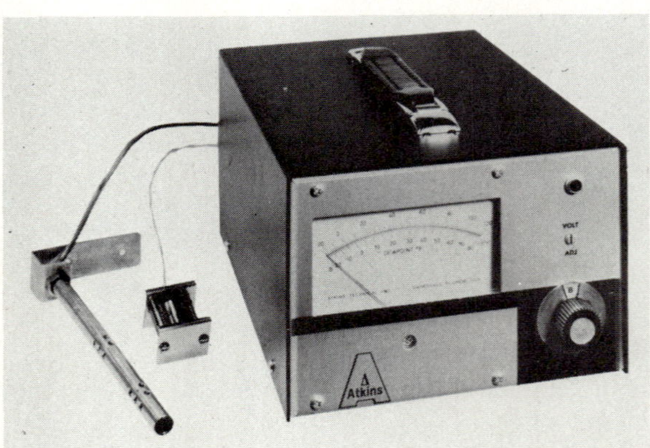

FIG. 5.55 Dew-point and temperature apparatus. *(Atkins Technical Inc.)*

5.7 FORCE MEASUREMENT

Mass, Weight, and Force

For most engineering purposes, 1 lb of mass and 1 lb of force are considered numerically equivalent. Strictly speaking, this is true only at a latitude of 45° and at sea level. At such a location the acceleration of gravity is said to be "standard," which is equal to 32.17 ft/s^2 (9.8 m/s^2). At such a location a "mass" of 1 lb experiences a gravitational attraction of exactly 1 lb. The same mass weighs about 0.0025 percent more at the north or south pole and about 0.0025 percent less at the equator. These are the extremes; for most practical cases the deviation from standard gravity is much less. If working to a precision of better than one-tenth percent, corrections may have to be made when measuring forces derived from mass.

There are two exceptions for which no gravity corrections are needed. For weights on a balance device, gravity acts equally on both sides, so the weights or masses can be compared directly. For forces being measured on a spring-tension device, gravity is not a factor, so any variation in the acceleration of gravity can be ignored.

One other correction that may affect weight measurements is the buoyancy of air. Every object displaces some amount of air and in reality is "floating" in the atmosphere. This buoyant effect is removed in a vacuum. But more practical than weighing in a vacuum is to correct for buoyancy of air (most often the effect is negligible anyway).

The following discussion assumes that gravity is standard and air buoyancy negligibly small. With such assumptions, weights and forces become interchangeable, and dead weights can be used to calibrate spring devices and vice versa. This is realistic for most engineering measurements.

Weight-Balance Systems

The simplest weight-balance system is the familiar laboratory balance shown in Figs. 5.56 and 5.57. It is an equal-arm balance system in which the unknown weight is equal to and balanced by the calibrated or standard weights on the opposite pan. Such balances are suited for weights up to about 1 lb.

For greater weights, systems have unequal arms, such as platform and truck scales. The unequal lever arms achieve force multiplication of 100 : 1 or 1000 : 1, or even greater. Thus it is feasible to weigh a 50-ton railroad car with a 1-lb weight, using a lever system with a factor of 100,000 : 1.

On dial-indicating scales, balance is achieved automatically by the deflection of calibrated pendulum weights from the vertical. The deflection is magnified by the pointer-actuating mechanism, providing a direct-reading weight indication on the dial.

Because the deflection of a spring (within its elastic limit) is directly proportional to the applied force, a calibrated spring is a simple and inexpensive weighing device. Applications include the spring scale and torsion balance. A precision-spring dynamometer is illustrated in Fig. 5.58.

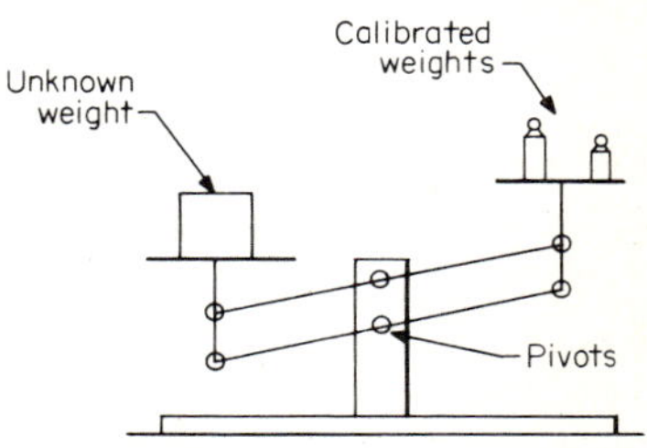

FIG. 5.56 Laboratory balance.

FIG. 5.57 Laboratory balance, Harvard trip style. *(Ohaus Scale Corp.)*

FIG. 5.58 Traction-type dynamometer. *(W. C. Dillon & Co.)*

FIG. 5.59 Strain-gage load cells. *(BLH Electronics.)*

Load Cells

The strain-gage load cell is a versatile device of increasing importance used to sense force and produce an electric signal (Fig. 5.59). It is a rather compact sealed cylinder that contains a four-arm resistance wire strain bridge (refer to Sec. 5.5). Advantages of such load cells include the absence of pivots and the ability to measure forces in any orientation. They may be used to weigh large objects such as tanks or airplanes. Generally, a three-point suspension system is used. The

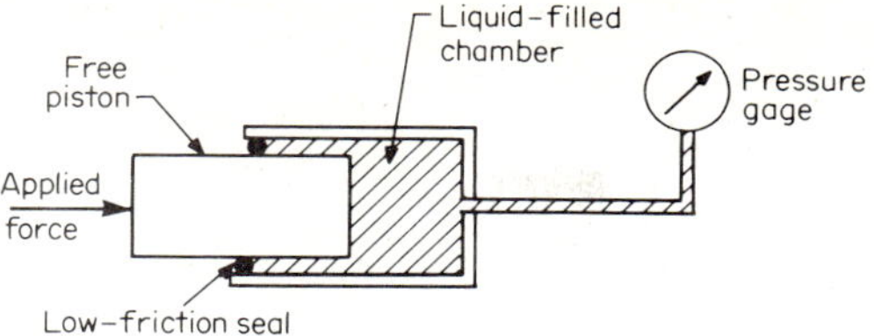

FIG. 5.60 Hydraulic load cell.

weight of the object is the sum of the weights on the three load cells. The output signals are summed electrically and read out by any suitable recorder or indicator. Such electric load cells have good dynamic properties and thus can measure live loads.

The hydraulic load cell in Fig. 5.60 is simple in principle: force applied = piston area × pressure. However, in practice three conditions must be met to obtain accurate results. (1) The liquid-filled volume must be completely filled. Air trapped in the volume causes excessive piston movement, and if the piston bottoms, readings are meaningless. (2) Ideally the seal should not exert any force on the piston. Practically speaking, some friction is present under the best conditions. (3) The effective area of the piston should be slightly greater than the actual piston area. This effective area can be determined by calibrating the system. The frictional forces can also be approximated in such a calibration.

The pneumatic load cell (Fig. 5.61) is actually a null-balance servo system in which the applied force is balanced by pneumatic pressure from a supply line. When the system reaches equilibrium, the applied force is balanced by the pressure behind the piston acting against the effective area of the piston. The load cell operates as described. As a force is applied to the piston, the three-way valve moves inward, letting gas flow pass from the pressure source to the inlet and into the cell body. As the pressure in the cell builds up, the piston reaches equilibrium and the valve closes. If gas is lost through leakage, the piston and valve again move inward and admit more gas. If the applied force is reduced, the piston moves outward and the valve port to the vent is opened, thereby bleeding off pressure until the piston again reaches equilibrium. Properly designed systems

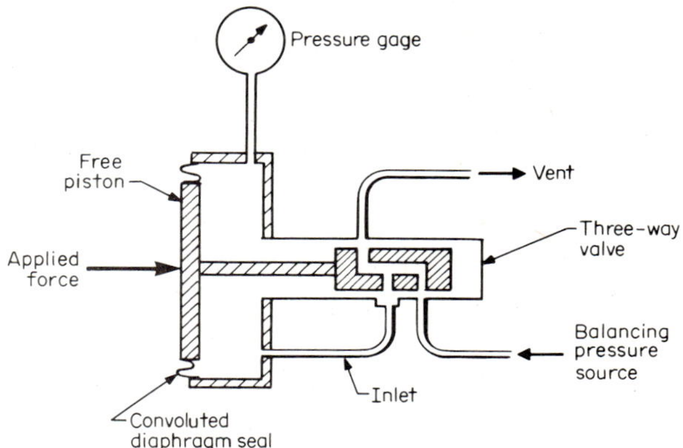

FIG. 5.61 Pneumatic load cell, or force transmitter.

have very little friction and respond rapidly, usually faster than 1 s. For optimum results, the pressure source should be regulated and higher than the maximum working pressure.

5.8 VIBRATION MEASUREMENT

Vibration in an engineering structure or scientific device is oscillatory motion. Thorough analysis of vibration problems can involve extremely complex mathematical procedures, so here we concentrate upon a general description only.

In any vibration problem there are three basic parameters (measurable quantities) that can be observed directly: displacement, velocity, and acceleration. *Displacement* is the distance a point has moved in space. It is easily visualized, but in many practical cases this motion is too small or too rapid to be seen without instruments. Displacement is measured in inches or millimeters or, more often in vibration measurement, in mils (0.001 in or 0.025 mm) or microinches (0.000001 in or 0.000025 mm). *Velocity* is the time rate of change of displacement, the speed of motion of a particle. It is measured in inches or microinches per second (or millimeters per second), or more generally in length per time. *Acceleration* is the time rate of change of velocity and is ordinarily stated in units of inches per second squared. Acceleration is more difficult to visualize than displacement or velocity but is important because it is directly proportional to the force acting upon a body or particle. Often it is the only quantity measured in a vibration-analysis problem.

Sensing Methods

There are accelerometers that transduce vibrational energy into electric energy, including potentiometric, piezoelectric, and strain-gage accelerometers (these transduction methods were described under pressure transducers in Sec. 5.3). Instead of using a pressure-sensitive diaphragm as their primary element, they employ an inertial mass (Fig. 5.62).

FIG. 5.62 Quartz piezoelectric accelerometer. *(Sundstrand Data Control, Inc.)*

Velocity transducers are usually electromechanical; that is, a moving coil is located in the field of a fixed magnet, and the generated electric signal is then proportional to the velocity of motion. Such transducers are more difficult to use because two mechanical attachments are required, one to the device being tested and one to a fixed and rigid support. On the other hand, an accelerometer needs only to be bolted or clamped to the vibrating member.

Optical vibration-sensing methods usually do not require that anything be attached to the equipment being tested. A beam of light is aimed at a reflective area, and then the reflected beam is picked up by a photosensitive electric cell. Such optical methods are best adapted to displacement measurements, but velocity and acceleration can be derived from them.

5.9 RECORDERS

It is possible to conduct experiments and operate plants without recorders. The operator might watch indicating instruments and periodically write down what is read on those instruments. But this approach is generally extremely unsatisfactory because in many experiments the amount of data to be obtained is beyond the capacity of an experimenter to observe, let alone write down, and the observer may make mistakes. Today most data are recorded, if at all possible.

There are a multitude of recorders. Here we discuss only six of the more popular types. The magnetic tape recorder is not covered because it does not provide direct visual output; the operator must replay a magnetic tape and display or record the data on some other device.

Null-Balance Potentiometers

These versatile and popular recorders are also called *self-balancing potentiometers.* An input signal of as little as a few millivolts or even a few microvolts is compared to a generated signal within the recorder. If there is a difference in the value of these two signals, a so-called error signal is developed. The error signal is amplified and used to drive a balancing motor that drives the pen either upscale or downscale until the error signal becomes zero. At this point a null is reached, and the servo system is in balance. There are three main advantages with such a system. (1) The recorder can accept relatively weak signals, and when balance is reached, no current flows in the signal source. This is ideal with thermocouple signals, for example. (2) The recorder may contain its own voltage, current, or resistance standard. (3) Sufficient power is available from the balancing motor to drive the recording pen plus controls or remote indicating devices.

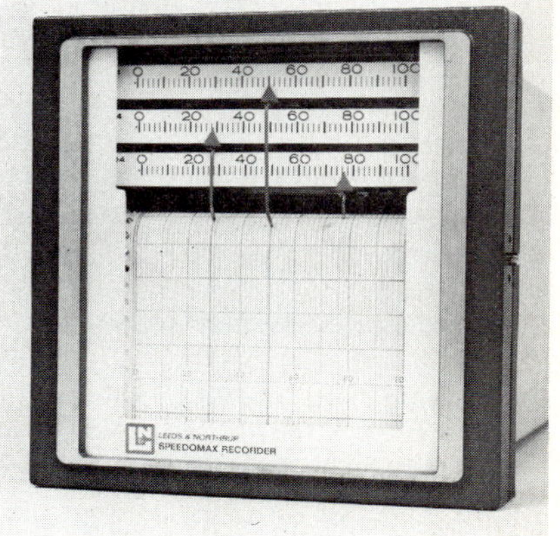

FIG. 5.63 Null-balance potentiometer, three-channel. *(Leeds & Northrup Co.)*

There are three nonsevere limitations to the null-balance potentiometer. (1) Speed of response is limited to 0.025 s for full travel of the pen across the scale. This is adequate for most applications because many sensors have response rates poorer than this. (2) These recorders are relatively complex, although despite their complexity they exhibit excellent reliability. (3) The recorders are expensive, but their long life and great adaptability to most measurement tasks make them a sound investment (Fig. 5.63).

In its most common form, such a recorder uses a paper chart driven by a synchronous motor to provide an accurate time base. Another version—an *x-y* recorder—has another amplifier and balancing motor for the chart drive so that one variable can be plotted against another. Some versions use two pens and two balancing systems plus the synchronous chart drive and can simultaneously record two variables versus time. Still other recorders are so-called multipoint recorders, which have as many as 12 input circuits. The recorder automatically switches to each input channel and prints in succession both the channel number and its value.

Direct-Recording Galvanometers

These recorders were originally developed for the medical field and used for many years for electrocardiograms. They have been adapted within industry and are widely used to record phenomena at rates up to about 100 Hz. The heart of such a recorder is a galvanometer that drives a direct-writing stylus in contact with the chart paper. The writing method is often ink delivered from a fixed well, or sometimes the stylus is a hot wire that contacts a heat-sensitive chart paper. An appreciable amount of power is required to operate these galvanometers; the power is obtained from a special amplifier matched to the characteristics of the galvanometer. The recorders are available in single and multichannel designs. Their chief advantage is speed of response, which is achieved at the expense of a lowered accuracy as compared to the null-balance potentiometer (Fig. 5.64).

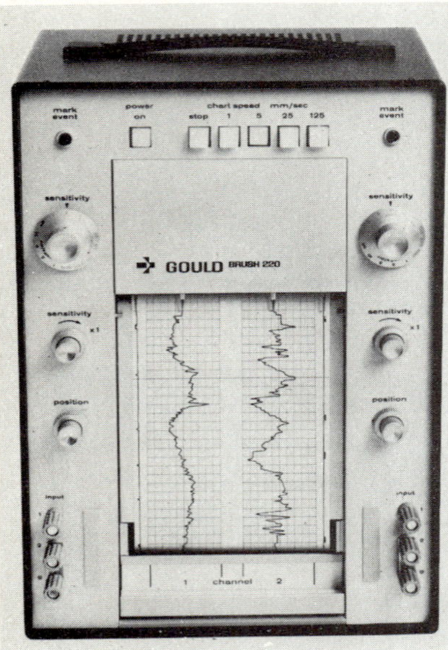

FIG. 5.64 Direct-recording galvanometer, two-channel. *(Gould Inc., Instrument Systems Div.)*

Light-Beam Galvanometers

Recorders using light-beam galvanometers are similar electrically to direct-recording galvanometer systems. The stylus is eliminated and replaced by a small mirror that deflects a beam of light onto a photosensitive paper chart. This redesigned galvanometer accomplishes three things. (1) The mass of the moving parts is greatly reduced, per-

FIG. 5.65 Light-beam galvanometer, six-channel, Visicorder 906c. *(Honeywell, Inc.)*

mitting frequencies of more than 5000 Hz to be recorded. (2) The galvanometer size is reduced, so that as many as 50 channels can be accommodated on a 12-in (300-mm)-wide chart. (3) The effective length of the arm can be increased, and all the channels can occupy the full chart width. Trace identification (the traces will probably cross) is accomplished by momentarily interrupting the optical path of each channel in sequence. Modern photosensitive chart paper requires no development, and the traces become visible in ordinary room light in a few seconds. Records are sensitive to strong ultraviolet and so should be protected from sunlight. Refer to Fig. 5.65.

Direct Nonelectric Recorders

Some instruments have sufficient energy to drive a pen directly. One example is a 100-psi pressure gage. If the Bourdon element is connected with suitable linkage, a direct-inking pen can be made to trace a line on a paper chart. Usually these recorders have a circular clock-driven chart that makes 1 rev in 24 h. The chart-driven motor is ordinarily electrically powered, but it can be a mechanical windup version. Such a recorder might be used in a hazardous area or in a remote area where no power is available. See Fig. 5.66.

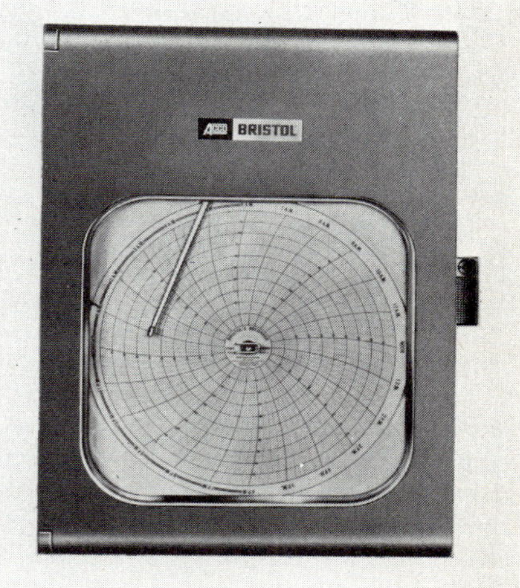

FIG. 5.66 Direct recorder. *(Acco Bristol.)*

Chopper-Bar Recorders

An ordinary d'Arsonval electric meter does not have sufficient power to drive a pen, but there is a scheme whereby an inexpensive recorder can be built around such a meter. The chopper-bar principle involves letting the meter pointer ride free in space, with no drag other than the bearings. Once every 2 s the pointer is held by an actuating bar against the pressure-sensitive chart paper for a fraction of a second. The pointer is released and then has the remainder of the 2-s period to seek the true value. The resulting trace is a series of dots that define a curve. The method seems primitive, but for certain applications it is entirely satisfactory and cannot be equaled for low cost. See Fig. 5.67.

Event Recorders

Event recorders, or operations recorders, provide only two kinds of information about a variable: (1) its state (on or off) and (2) the timing of its change of state (when it turned on or off). Frequently this type of information is sufficient, and often it is really the only information obtainable. For example, an electric switch is either on or off, and many mechanical valves are either open or closed. Other types of recorders could be used for this purpose, but such a degree of sophistication is wasteful. Event recorders are actually simple devices that can be fur-

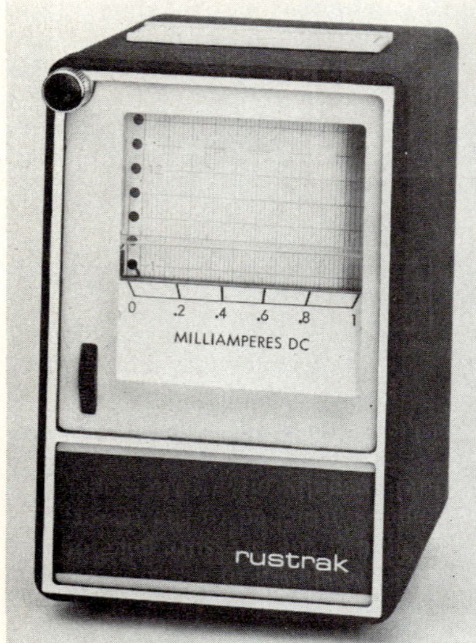

FIG. 5.67 Chopper-bar recorder *(Rustrak Instrument Div., Gulton Industries, Inc.)*

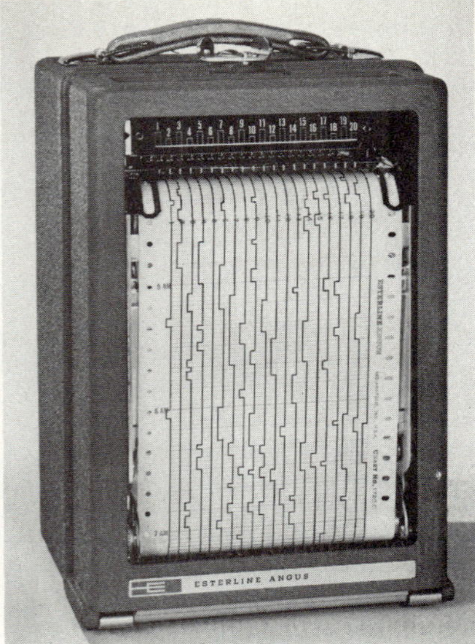

FIG. 5.68 Event recorder, 20-channel. *(Esterline Angus Instrument Corp.)*

nished at a much lower cost per channel than any other recorder. Another advantage is that the chart width required per channel is minimal, so many channels can be accommodated in one recorder. Event recorders are available with up to 100 channels recording on a 12-in (25-mm) chart. Usually chart speeds can be selected by a selector switch or gear changes and can be as slow as inches per day or as fast as inches per second. Figure 5.68 illustrates a 20-channel recorder.

5.10 CATHODE RAY OSCILLOSCOPES

The cathode ray oscilloscope is most familiar as the display element of a television set. The scope, as it is usually called, is also widely used in the engineering and science fields and is perhaps the most versatile instrument in the laboratory. It displays any variable convertible into an electric signal (virtually all variables may be so converted) versus time, at fantastic rates of change. [Scope technology has long passed through the nanosecond range (10^{-9}s).] Generally, these superfast scopes are not needed for mechanical technology because physical objects cannot move that fast, but response and sweep rates in the microsecond area (0.000001 s) are considered "ordinary" and suitable for the mechanical laboratory.

A detailed description of scope circuitry is not of interest to the typical user, so here we discuss only the fundamentals of scope functions and describe scope features from an applications standpoint. A cathode ray tube is an evacuated glass bottle containing an electron gun and having a relatively flat face coated with a phosphorescent material. Where the electron beam strikes the phosphor, a spot of light becomes visible. To provide useful information, the beam is deflected in both the horizontal and vertical directions. Television picture tubes generally use deflection coils external to the glass envelope; this process is called *electromagnetic deflection.* Scopes for laboratories usually employ deflection plates located inside the evacuated bottle—*electrostatic deflection.* The electron is a particle with large charge and small mass, so it is easily deflected, giving scopes their rapid response rates.

Typically, the horizontal deflection is driven by a linear time signal, which is called the *time base* or x axis. The vertical deflection is driven by the variable under study and is called the y axis. Often the beam itself may be turned off and on; this is called z-axis modulation. Some dual-beam scopes have two electron guns in one glass envelope so that two independent signals can be displayed simultaneously.

The scope's time base may be free-running; that is, the trace moves across the scope at a known rate of speed and snaps back to sweep again. This way repetitive phenomena can be viewed as a succession of superimposed plots. However, frequently an experiment is not repetitive, so a triggered single sweep is used. With the single sweep there is no sweep until a signal appears at the vertical input. At this time the sweep is triggered, and the horizontal time base makes one sweep across the tube, plotting a single event versus time. If the scope has "storage," this trace remains displayed until manually removed. If the scope does not have storage, the trace can be photographed, as described in Sec. 5.11. Often the experimenter needs neither storage nor a photograph to interpret the trace because human persistence of vision may permit adequate interpretation of the result.

Modern scopes contain internal circuits for calibrating both vertical traces and time bases and can be set up in minutes for reliable quantitative measurements. Refer to Fig. 5.69.

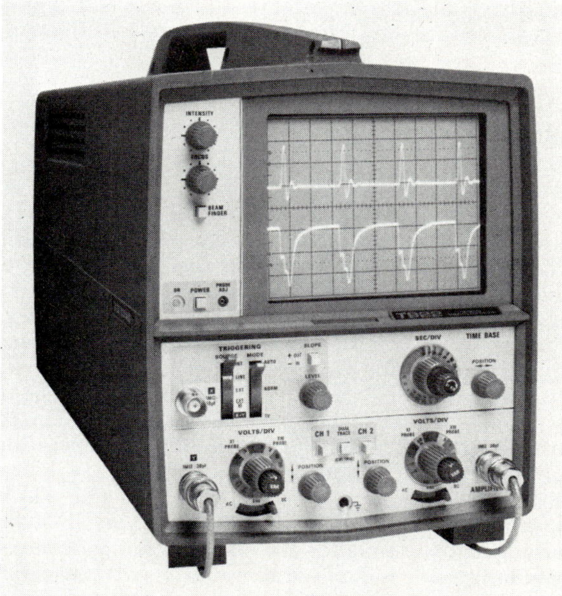

FIG. 5.69 Oscilloscope, two-channel. *(Tektronix Inc.)*

FIG. 5.70 Oscilloscope camera. *(Hewlett-Packard Co.)*

Photographic Recording of Oscilloscope Displays

It is often necessary to photograph scope displays for record keeping or publication. This is a well-developed and simple procedure. Most laboratory scopes use a 5-in (130-mm)-round scope tube and have a concentric and integral mounting sleeve suitable for a viewing hood or camera. Conventional photographic techniques can be used but involve the usual delays and inconvenience of darkroom processing. The most popular photorecording method involves Polaroid film, which presents visible experimental results in minutes. The scope or experiment can be readjusted for optimum results. Figure 5.70 shows a modern oscilloscope camera.

5.11 COUNTERS

Simple mechanical and electromechanical counters are cheap and reliable for slow-acting events up to perhaps 10 operations per second. However, much mechanical phenomena occur at faster repetition rates; the electronic counter must be used for these applications. Originally these counters were developed for nuclear instrumentation and provided only binary display. (Binary number systems are based upon powers of 2 rather than powers of 10, which is the decimal system.) Modern electronic counters are still designed with internal logic based upon the binary system, but the display is in decimal form (see Fig. 5.71).

When an electronic counter is used, typically the event to be counted is converted into electric-signal form and connected to the input. The counter is manually reset and then displays zero. The total number of operations or "events" can then be read at any later instant.

Another method involves an events per unit time (EPUT) meter. Such a meter is simply a counter with a built-in time base. The EPUT meter typically resets itself to zero, counts events for 1 s, displays that count for 1 s, and resets to zero. This rate indicator can be used to indicate such measurements as revolutions per minute and pounds per second. Generally, both the time base and display time are adjustable. Most EPUT meters can be used as simple counters,

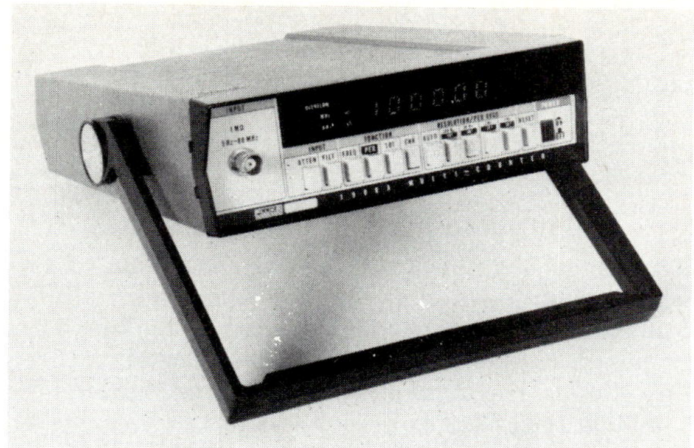

FIG. 5.71 Electronic counter, model 1900A. *(John Fluke Mfg. Co., Inc.)*

so they are more versatile than regular electronic counters. An EPUT meter is preferred for general-purpose applications.

5.12 POSITION SENSING

The position of an object with respect to another, or with respect to some reference coordinate, is a basic measurement factor in mechanical technology. Position measurement can be divided into the two general categories of linear and rotational. Because one form of position can be readily converted into the other for measuring by familiar means, we do not distinguish here between linear and rotational position.

Several position-sensing methods involving conversion to electric signals were discussed in Sec. 5.3 but are reviewed briefly here.

The potentiometric method uses a fine resistance wire wound upon an insulated mandrel. A wiper arm is the position-sensitive element, the wire-wound bobbin the fixed or reference element. If a known electric signal is impressed upon the winding, the signal appearing on the wiper arm measures its position. This principle is also described in Sec. 5.3 under "Pressure Transducers."

The synchro, sometimes called the selsyn, is a pair of small synchronous motors whose rotors are connected electrically and their stators connected to a common ac power source. With such a system, if the rotor of motor *A* is rotated 15°, the rotor of motor *B* rotates the same amount. Such a system represents mechanical-electrical-mechanical conversion, but a single synchro can be used to provide electric signals from mechanical position. These synchros are widely used in sophisticated electromechanical instrumentation systems.

The linear variable differential transformer (LVDT) is described in Sec. 5.3 as a pressure transducer but can also be a position transducer. A capacitance transducer is described in Sec. 5.3. Its principle applies to position sensing but only for limited motion because as the separation of the capacitor plates becomes large, the sensitivity of the device decreases.

Pneumatic gaging is a nonelectrical position-sensing method. The principle involved is the same as that used in the pneumatic pressure transmitter described in Sec. 5.3.

The optical interferometer is a position-sensing method capable of measuring to a few millionths of an inch. Monochromatic light is reflected between two optically flat and parallel reflecting surfaces. As these surfaces move with respect to one another, an observer first sees reinforcement of light (brightness) and then cancellation of light (darkness). This alternation of brightness and darkness occurs at each increment of one-half the wavelength of light. For visible light, this is on the order of 0.0000010 in. There are position-sensing devices that count the alternating pulses of light and indicate position to the fine precision described.

A position-sensing method now widely used for machine-tool control involves the digital encoder, which looks like a checkerboard, with alternating strips of clear and opaque material. The encoder reads the moving element optically much the way a punch card is read, and the resulting digital information is a measure of the position of the machine-tool element. Digital encoders also have been applied to analog recorders to obtain digital information for transmission or data processing.

5.13 LEAK DETECTION

In any hermetically sealed system there is always the possibility of leaks. If the leak is large, it becomes evident through loss of pressure or vacuum over a reasonable period of time. But small leaks might not be evident for years by conventional means. The helium mass spectrometer is an extraordinarily sensitive leak-detection method, on the order of 10^{-11} atmospheric cm^3/s (10^{-17} m^3/s). These systems are based upon two principles. (1) The helium molecule is very small and flows through microscopic holes at a greater rate than almost any other gas. Hydrogen also has the ability to flow through fine holes, but it is unsuitable for use in leak detectors because of its combustibility and its presence in water vapor (detecting hydrogen as distinct from the ever-present water in the atmosphere is difficult at best). Helium is a chemically inert, easily available material and present in the atmosphere in only microscopic quantities. It is the ideal medium for leak detection. (2) The mass spectrometer can separate and detect chemical elements. A mass spectrometer ionizes a gas and accelerates the ions under electrostatic and magnetic forces. Material of a particular atomic weight has its own characteristic path, so by properly locating a target, the ions of a particular element can be collected to the exclusion of all others.

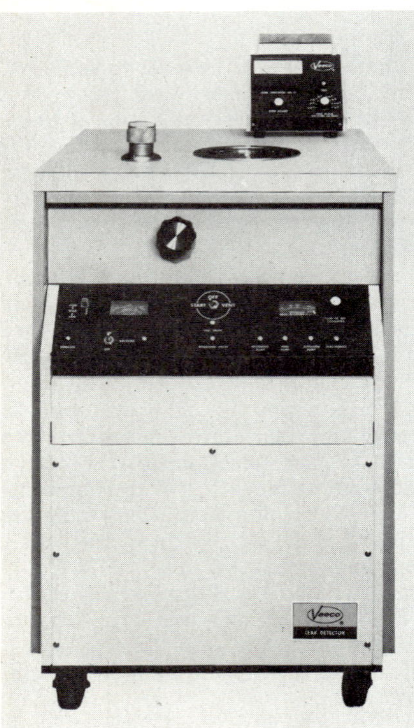

FIG. 5.72 Leak detector, model MS-17 AB. *(Veeco Instruments, Inc.)*

A typical leak detector is shown in Fig. 5.72. The cabinet houses a high-vacuum pumping system and a mass spectrometer. The vessel to be tested is mounted at the test port and pumped down to a high vacuum. The degree of vacuum obtainable and the time of pump down required provide a general idea of the pressure/vacuum integrity of the testpiece. The operator then directs a small jet of helium at the testpiece and explores the entire outside surface. If the helium stream is directed at or near a leak, helium atoms are swept into the vacuum system and find their way to the mass spectrometer. Not only is it thus possible to detect the presence of a leak, but the flow rate through the leak is inferred by the current in the mass spectrometer.

If the component to be tested cannot be made part of the vacuum system, it can be pressurized with gas containing helium and a "snifter" probe connected to the leak-detector vacuum system to explore for a leak. When the probe passes near a leak, the helium issuing from the leak is pulled into the leak detector with the surrounding air and is readily sensed by the mass spectrometer.

To test small sealed objects, seal in a sample of helium and then place the object to be tested under a bell jar on the leak-tester vacuum system. With this technique it is possible to determine the presence of a leak and its size but not the location of the leak.

Figure 5.73 is a schematic drawing of the more important elements of a mass spectrometer leak detector. The testpiece is fastened to the test port, usually with a quick-disconnect seal. The cryogenic trap is filled with a cold material such as liquid nitrogen, which condenses out volatile materials like hydrocarbons and water. The diffusion pump circulates a special oil that sweeps gas molecules toward the forepump. The forepump is the last element in the system and exhausts gases to the atmosphere. The use of these components represents standard high-vacuum technology. The mass spectrometer is designed to have high sensitivity to helium atoms, which, when they enter the test port, diffuse

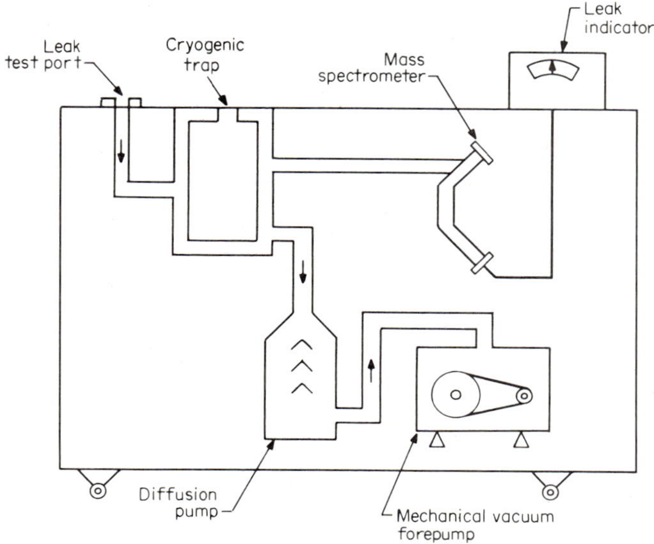

FIG. 5.73 Leak-detector schematic.

throughout the system and are seen as an electric current detected by the leak indicator. The leak detector in Fig. 5.72 is a general-purpose device capable of detecting leaks as small as 6×10^{-11} atmospheric cm^3/s (6×10^{-17} m^3/s).

5.14 ROTATIONAL-SPEED MEASUREMENT

Rotational speed is measured with an instrument called a *tachometer*. Tachometers are hand-held or permanent-mounting. Most laboratories use a hand-held instrument because of its versatility; the noncontact type is best. Figure 5.74 shows a modern noncontact, digital readout tachometer.

FIG. 5.74 Hand-held, noncontact, digital readout tachometer. *(Meylan Stopwatch Corp.)*

The noncontact instruments receive a reflective signal with each revolution of the part whose rotational speed is being measured. A light beam, usually originating from the instrument itself, is aimed at the rotating object (one part of the object is reflective). A small piece of self-adhesive reflective tape on the rotating object may be required to provide an adequately reflective area. Digital readout is almost instantaneous, and the instruments have wide speed-range-measuring capability (100 to 30,000 rpm or 10 to 3000 rad/s) and good accuracy because of their several range selections. Rotational speed is most often measured in revolutions per minute or sometimes in revolutions per second (rps); the SI unit is radians per second (rad/s).

Contact, hand-held instruments require access to the center of the rotating piece. The spindle of the instrument is placed and held against the center of rotation so that the spindle rotates with the rotating piece without slipping. Typically a rubber tip is placed on the end of the spindle to facilitate the technique. Dial or digital readouts are available with these instruments.

chapter 6

Instrument Calibration and Standards

Robert C. Knauer

6.1 GENERAL CONSIDERATIONS

The calibration process in some way must be carried out during the manufacture of any instrument. Calibration indicates the scale of the instrument; to establish this, the scale is compared with a "standard" scale. However, during the comparison there are many opportunities for error, and these errors can accumulate, rendering the calibrated scale less accurate than the standard. Clearly the standard must be of greater accuracy than that required of the instrument; and the calibration method must be carefully devised to minimize error and ensure reproducibility. An amazing degree of precision has been achieved by the use of primary standards; these standards must be carefully maintained. Secondary standards, calibrated against a primary standard, are usually adopted for engineering work and provide adequate precision. As mentioned, all instruments need calibration at the time of manufacture to establish a degree of accuracy, and periodic recalibration is required to verify that the accuracy is being maintained or to provide a correction to the original reading or scale.

6.2 NATIONAL BUREAU OF STANDARDS TRACEABILITY

Standards are useful only if everyone follows them. Plainly, a unit of length in one laboratory should mean the same thing in every other laboratory, no matter where. To this end, all industrial nations have established standards laboratories, whose responsibilities are to maintain standards of measurement and to disseminate this information to the industrial and scientific communities. A vital part of their function is to compare their standards with those of the International Bureau and with each other. In this country the organization charged with this responsibility is the National Bureau of Standards (NBS). Its main facility

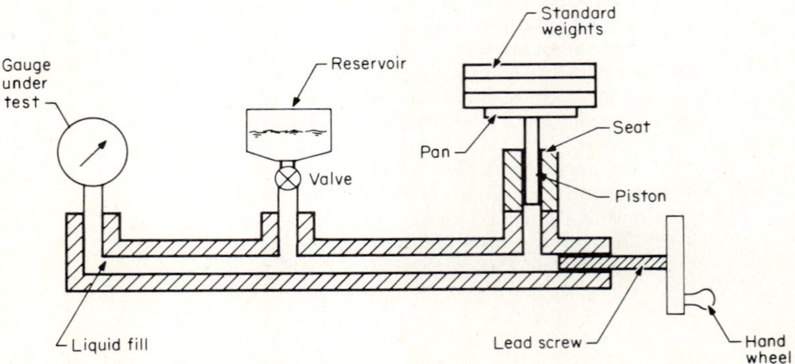

FIG. 6.1 Simple piston deadweight tester.

is located at Gaithersburg, Maryland, with several branch facilities at other locations.

When writing specifications for measurement and calibration equipment, the measurements have to be "traceable to NBS." This does not mean that every piece of equipment has to be shipped to the NBS for calibration, just that there is a documented history of calibration which started at the NBS and reached the ultimate user through a recognizable and fully documented chain of secondary standards, each of which is shown to have adequate accuracy for the transfer of measurements. The new accuracy is specified at each step in the process. This procedure is not hard to implement because there are many private organizations that specialize in just such work.

6.3 PRESSURE CALIBRATION

Pressure is expressed in force per unit area (pounds per square inch or newtons per square meter,* for example). This common quantity is of fundamental importance in much engineering and scientific work. Standards of pressure are usually based upon the definition of pressure; that is, a known force is exerted upon a known area, and the pressure thus generated is applied to the transducer or gage under calibration, which is the familiar deadweight tester (described in detail in the following section). Another form of pressure standard is the liquid-filled manometer (described in detail in Sec. 5.3), which when used under carefully controlled conditions achieves such good accuracy that it can be regarded as a primary standard. This type standard is also described in this chapter.

Hydraulic Deadweight Tester

The elements of the hydraulic deadweight tester are shown in Fig. 6.1; Fig. 6.2 is a photograph of one widely used tester. In Fig. 6.2, the pressure gage on the left is the instrument to be calibrated. This gage is attached to the tester, and the end valve is opened to admit the test fluid (usually oil) to the gage. The

*The SI unit for pressure is the pascal (Pa), which is equal to 1 newton per square meter.

FIG. 6.2 Deadweight tester. (*Amthor Testing Instrument Company, Inc.*)

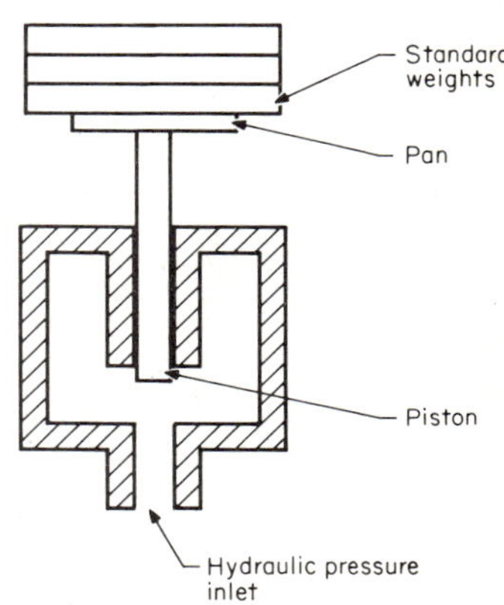

FIG. 6.3 Reentrant cylinder deadweight tester.

reservoir valve is then opened, and the hand wheel is backed off to admit more oil to the system. The reservoir valve is closed, and the hand wheel is turned inward until the pan/weight combination lifts off its seat by the hydraulic pressure on the piston. Under these conditions a uniform hydraulic pressure is established throughout the system and transmitted to both the gage and the pan. The piston diameter, carefully controlled by the manufacturer, is often of some convenient size, such as $\frac{1}{10}$ in^2 or $\frac{1}{10,000}$ m^2. In this example the weights exert a pressure of 10 or 10,000 times their true weight. In any event, the designer of the tester has calculated the relationship of weight values, piston areas, and so on; the user finds stamped on the weight the particular pressure increment to be applied when that particular weight is placed on the pan. The pan itself contributes to the total applied pressure, and this value is commonly marked upon the pan/piston combination. This value represents the smallest pressure that can be applied with the particular tester.

When the desired weights are placed on the pan and the pan lifted off its seat, the weights are given a rotational spin to reduce frictional effects. For very precise work, readings are taken with first a clockwise and then a counterclockwise rotation of the pan, to average out any effect caused by possible microscopic scratches on the piston that might tend to lift or depress the pan. A properly operated tester permits a small fluid leakage past the piston, and this loss in volume allows the piston to sink very slowly.

Oil is an ideal fluid for most deadweight tester work, but if the gage is to be used on an oxygen or other flammable system, the tester may be set up to use water or some other nonflammable liquid.

The simple piston/cylinder configuration in Fig. 6.1 is accurate and linear only up to a moderate pressure level. As the pressure increases, the cylinder diameter begins to grow and the effective area begins to increase. One simple way to handle this situation is to use the reentrant cylinder shown in Fig. 6.3. The fluid pressure is applied to both the inner and outer walls of the cylinder, greatly reducing cylinder distortion under pressure. This form of compensation is easily provided, and most deadweight testers do employ such a design. However, although the cylinder wall tends to be stable, there remains some small dimensional change. A further refinement in design is the controlled-clearance cylinder, shown in Fig. 6.4. Pressure from a separate controllable source is admitted

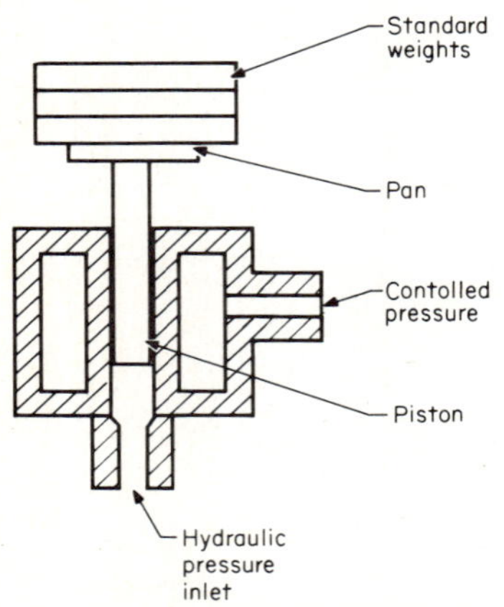

FIG. 6.4 Controlled-clearance deadweight tester.

FIG. 6.5 Primary pressure standard. (*CEC Division/Bell & Howell.*)

to the outside of the cylinder and can be adjusted to compensate fully for the pressure within the cylinder. The clearance between the piston and cylinder is judged by the leakage rate, and the controlled pressure is set to maintain a small but finite leakage. Such a deadweight tester is relatively complex to build and use, but it is one way to achieve high accuracy at pressures above 2000 psi (15 MPa).

Pneumatic Deadweight Tester

In the range between 0.3 and 15 psi (2 to 100 kPa), the oil-filled deadweight tester loses its applicability. (The pan and piston are too heavy to permit such low pressures.) In this range, and up to 500 psi (3500 kPa), the pneumatic tester is often the preferred standard. The principle is the same as for the oil-filled deadweight tester, that is, a gravity-actuated weight working against a piston of known area. An example of this design is the CEC-type 6-201 primary pressure standard shown in Fig. 6.5. The working fluid is usually dry nitrogen, although air can be used. The bell jar permits referencing to a vacuum for absolute pressures or to atmosphere for gage pressures. The piston in the cylinder automatically rotates, and the air or gas film between the cylinder wall and the piston provides a near frictionless condition.

Precision Mercury Manometer

Mercury is well-suited to manometry for many reasons. First, it is a liquid at ordinary temperatures and can be readily handled. Second, it is relatively dense (known to a precision of 1 ppm; see Table 6.1 for the density of mercury versus

TABLE 6.1 The Density of Mercury as a Function of Temperature

Temperature, °C	Density, kg/m³	Temperature, °C	Density, kg/m³
0	13595.08	21	13543.41
1	13592.61	22	13540.96
2	13590.15	23	13538.51
3	13587.68	24	13536.06
4	13585.21	25	13533.61
5	13582.75	26	13531.16
6	13580.29	27	13528.71
7	13577.82	28	13526.26
8	13575.36	29	13523.81
9	13572.90	30	13521.36
10	13570.44	31	13518.91
11	13567.98	32	13516.47
12	13565.52	33	13514.02
13	13563.06	34	13511.58
14	13560.60	35	13509.13
15	13558.14	36	13506.69
16	13555.69	37	13504.25
17	13553.23	38	13501.81
18	13550.78	39	13499.36
19	13548.32	40	13496.92
20	13545.87		

SOURCE: Data based on A. H. Cook, *Philos. Trans. R. Soc. London*, **A254:** 125 (1961); James A. Beattie et al., *Proc. Am. Acad. Arts Sci.*, **74:** 371 (1941); and IPTS-68.

temperature), and therefore the apparatus does not become overly large. To illustrate the difference: In a manometer to measure 30 psi (200 kPa), a mercury column is only about 5 ft (1.5 m) high, but a water column is 69 ft (21 m) high. A third advantage is that mercury's characteristics are well-known, having been studied by many investigators over the years. Fourth, mercury is readily available in a pure form, and when it becomes contaminated it can be easily cleaned by distillation.

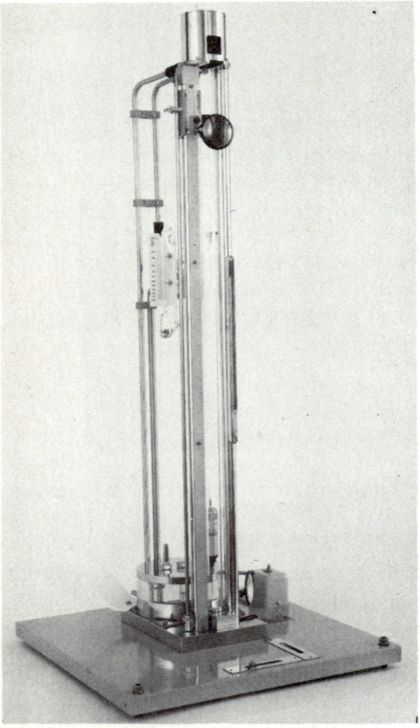

FIG. 6.6 Precision mercury manometer. (*Hass Instrument Co.*)

Figures 6.6 to 6.8 show a typical primary pressure standard, model MS-3 of the Hass Instrument Company. It is a true primary standard in that it does not require comparison with another barometer. To establish a pressure value with this standard, it is necessary to have only available standards of length and temperature and to know the local value of gravity.

In principle, the instrument is a U-tube manometer; when the pressure difference increases, one leg rises and the other leg drops. Here the ratio of the two legs is 100:1, so if the column rises 1 in (or 1 mm), the cistern level drops only 0.01 in (or 0.01 mm). This simplifies reading the cistern level because it has a relatively small effect on instrument accuracy. In fact, this device approximately cuts in half the uncertainty of observation of the measurement.

To illustrate the usage of the standard, a generated pressure is introduced above the cistern (Fig. 6.8); this pressure forces the mercury up the column to a height of, for example, 29 in (737 mm). The mercury level in the cistern is then lowered about 0.29 in (7.37 mm). In any case, the exact ratio between the column height and the lowering of the cistern level is not important. The observation of the actual difference in column height is the sum of the column above the zero reference and the lowering of the cistern level. The column height is observed optically and referred to a graduated vernier scale, which can be checked by stacking gage blocks on a reference plate. The temperature of the apparatus is noted, as indicated on the thermometer mounted on a supporting column. The apparatus is generally used in a temperature-controlled room. If ambient-temperature changes are kept small, the temperature of the mercury is only slightly different than that of the apparatus.

Periodically check the vacuum-reference pressure because it is potentially a source of error if allowed to rise above about 1 μm. A pressure of 1 μmHg (0.001 mmHg) is readily achieved with a conventional vacuum pump and a leaktight system. A vacuum of better than 1 μm is not attainable under ordinary circumstances because this is about the vapor pressure of mercury at room temperature.

FIG. 6.7 Precision mercury manometer. (*Hass Instrument Co.*)

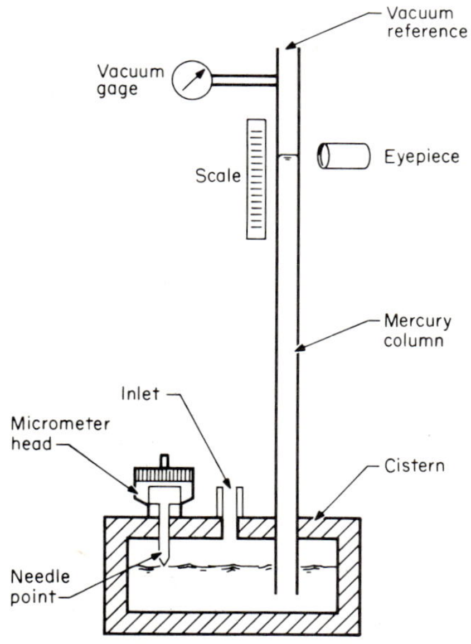

FIG. 6.8 Precision mercury manometer.

If greater accuracy is required, the pressure above the column can be applied as an additional correction to the observed column height.

Here is an example of how to calculate pressure from observing the mercury column height. First write down these observations:

1. Column height	29.6510 in
2. Cistern depression	0.2965 in
3. Net column height (add steps 1 and 2)	29.9475 in
4. Metric (convert by multiplying step 3 by 2.54×10^{-2})	0.760666 m
5. Temperature	22.0°C
6. Gravity (accel. of)	9.80300 m/s²
7. Vacuum reference	1.0 μm

Now substitute in the pressure formula:

$$P = dgh$$

where P = pressure, pascals (Pa)
d = density of mercury, kg/m³ (Table 6.1)
g = local acceleration of gravity, m/s²
h = height of mercury column, m

$$P = 13.540.96 \times 9.80300 \times 0.76066$$
$$= 100{,}972.6 \text{ Pa}$$

To convert to inHg, from Table 6.2:

$$1029.634 \times 0.028959 = 29.8172 \text{ inHg}$$

To convert to pounds per square inch, from Table 6.2:

$$1029.634 \times 0.014223 = 14.6445 \text{ lb/in}^2$$

TABLE 6.2 Conversion Factors for Pressure Units

Pressure-unit value	mb	mmHg, 0°C	inHg, 0°C	g/cm²	lb/in²	lb/ft²	cm water, 20°C	in water, 20°C
1 atm	1013.250	760.000	29.9213	1033.227	14.69595	2116.22	1035.08	407.513
1 mb	1	0.75006	0.029530	1.0197	0.014504	2.0885	1.0215	0.40218
1 mmHg	1.3332	1	0.03937	1.3595	0.019337	2.7845	1.3619	0.53620
1 inHg	33.864	25.400	1	34.531	0.49115	70.726	34.593	13.619
1 gm/cm²	0.98067	0.73556	0.028959	1	0.014223	2.0482	1.0018	0.39441
1 lb/in²	68.9476	51.715	2.0360	70.307	1	144	70.433	27.730
1 lb/ft²	0.47880	0.35913	0.014139	0.48824	0.0069444	1	0.48912	0.19257
1 cm water, 20° C	0.97891	0.73424	0.028907	0.99821	0.014198	2.0444	1	0.3937
1 in water, 20°C	2.4864	1.8650	0.073424	2.5354	0.036063	5.1930	2.5400	1
1 Pa	0.01000	0.0075008	0.0002953	0.010197	0.00014505	0.020885	0.010197	0.0040147

1 atm = 10332.3 kg/m² = 1.03323 kg/cm²
1 bar = 1000 mb
1 in = 2.54 cm
1 lb = 0.45359237 kg
Density of water at 20°C = 0.998207 gm/cm³
SOURCE: National Bureau of Standards Monograph No. 8, May 20, 1960, plus conversion for Pascals.

The pressure above the column of 1 μm is a typical working pressure in a vacuum system containing mercury. One micron is a negligible sum because the column height is readable to only the fourth decimal point.

In the example, the observations were noted in a mixture of metric and English units. All English units were converted to the metric equivalent, the calculation was made in the metric system, and the pressure was then converted back to English units.

If one wishes to work in all English units, an example follows:

1. Column height	29.1234 in
2. Cistern depression	0.2912 in
3. Net column height	29.4146 in
4. Temperature	68.0°F
5. Acceleration of gravity	32.1710 ft/s^2
6. Acceleration of gravity on an inch basis	386.052 in/s^2
7. Vacuum reference	2.0 μm

Again, the applicable formula is

$$P = dgh$$

where P = pressure, lb/in^2
d = density, lb/in^3
g = accel. of gravity, in/s^2
h = column height, in

Table 6.1 for the density of mercury shows values for 60 and 70°F. The density at 68°F is determined by *interpolation.* Assume that the density varies linearly with temperature; the proportional change for 8°F is calculated as follows:

Density of mercury at 60°F	0.48977 lb/in^3
Density of mercury at 70°F	−0.48928 lb/in^3
Difference for 10°F	0.00049 lb/in^3
8/10 of the difference	0.00039 lb/in^3
Density at 60°F	0.48977 lb/in^3
Difference at 8°F	−0.00039 lb/in^3
Density at 68°F	0.48938 lb/in^3

Proceeding with the calculation:

$$P = 0.48938 \times 32.1710 \times 29.4146$$
$$= 463.099 \text{ poundals}$$

But, because 1 poundal is defined as 1/32.1740 lb, dividing by this fraction gives

$$P = 14.3936 \text{ lb/in}^2$$

Either calculation method can be used routinely for determining pressure from the observed mercury column height, but both are laborious. In most practical cases, the gravity at the laboratory location does not change and the temperature of the apparatus is held to within ±1°. When these conditions are met, it is most convenient to devise a chart relating column height to pressure directly for specific use in the laboratory.

Gravity Determination

Generally, the value of the acceleration of gravity depends upon geographical location and elevation above sea level. The necessity for accuracy when determining gravity depends upon the precision desired in the pressure measurement.

As a first approximation, one may accept the standard value of 9.80665 m/s^2 or 32.1740 ft/s^2, which is the value of gravity at sea level at a 45° latitude. A refinement can be obtained by using Table 6.3, which provides accuracy of about 0.01 percent. Look for the latitude of the location in question: If the latitude differs from an exact multiple of 10°, interpolate from Table 6.3 to find the sea level gravity and then apply a suitable correction for the elevation above sea level.

Here is an example of a gravity calculation:

Location: Denver, Colorado
Latitude: 39°40′

(The latitude here is obtained from a road map, but this information is usually available from other sources as well.)

Elevation: 5280 ft (786.4 m)

(Elevation can be determined from topographic maps, from the city engineer, or from nearby bench marks.) Interpolation from Table 6.3 yields a sea level value of gravity of 32.157 ft/s^2. The elevation correction is thus $-0.003 \times 5.280 = -0.016$ ft/s^2. The resulting value is then

$$\begin{array}{r} 32.157 \\ -0.016 \\ \hline 32.141 \end{array} \text{ ft/s}^2 \text{ or } 9.7966 \text{ m/s}^2$$

If greater precision is required, consult the U.S. Coast & Geodetic Survey and the U.S. Geological Survey publications. These groups have measured the accel-

TABLE 6.3 Acceleration of Gravity

Latitude, deg.	g* cm/s^2	g* ft/s^2	$\left(\frac{g}{g^\circ}\right)$†
0	978.0	32.088	0.9974
10	978.2	32.093	0.9975
20	978.6	32.108	0.9980
30	979.3	32.130	0.9987
40	980.2	32.158	0.9995
50	981.1	32.187	1.0005
60	981.9	32.215	1.0013
70	982.6	32.238	1.0020
80	983.1	32.253	1.0025
90	983.2	32.258	1.0027

*Correction for altitude above sea level: -0.31 cm/s^2 for each 1000 m; -0.0033 ft/s^2 for each 1000 ft.

†Standard acceleration of gravity is g° = 980.619 cm/s^2 or 32.1725 ft/s^2, and g° is assumed to be the value of g at sea level and latitude 45°.

SOURCE: National Geodetic Survey, National Ocean Survey, NOAA, 1981.

eration of gravity at many points in the United States; if there is a published value at a nearby point, latitude and elevation corrections can be applied to transfer the value to the location in question. For the ultimate precision, the gravity value must be measured at the location. This procedure requires specialized equipment and skill and is not a requirement for most engineering laboratory work.

A more general computation of gravity can be obtained by formula. Taking as an example Islip Airport, for which we have an observed and published value from the U.S. Coast & Geodetic Survey,

Latitude: 40° 47.40′
Elevation: 24.4 m
Observed value of g = 9.80250 m/s^2

We can then compare this value with the calculated value as follows and find excellent agreement:

$$g = 9.80616\,(1 - 0.0026373 \cos 2\phi + 5.9 \times 10^{-6} \cos^2 2\phi)$$

$$\begin{aligned} \text{where } \phi &= 40.79° \\ 2\phi &= 81.58° \\ \cos 2\phi &= 0.146428 \\ \cos^2 2\phi &= 0.02144 \end{aligned}$$

$$\begin{aligned} g &= 9.80616\,(1 - 0.00038618 + 0.00000013) \\ &= 9.80616\,(0.999614) \\ &= 9.802374 \end{aligned}$$

Elevation correction = $-0.003 \times 24.4 \times 0.001 = -0.000073$
Calculated value of g = 9.802301 m/s^2

Servomanometer

One instrument widely accepted in the pressure-calibration field is a servo-driven mercury manometer (Fig. 6.9). This manometer combines the inherent accuracy of the mercury column with great operational convenience and speed.

The manometer positioning system, which is responsible for its pressure accuracy, consists of a precision lead screw driven by a servomotor (Fig. 6.10). The position of the lead screw, which moves a cistern of mercury, is measured by a pair of precision control transformers (synchros) geared to the lead screw. Pressure null sensing is done by a capacitive technique, as will be described.

The moving cistern can be positioned to ±0.0003 in (0.0075 mm); repeatability of positioning is ±0.0001 in (0.0025 mm). By preprogramming compensation of known errors, a position accuracy of better than ±0.0001 in (0.0025 mm) is achieved. The standard condition to which the manometer is corrected is that standard pressure relative to a column height of mercury at 0°C (273 K) with a standard value of gravity of 9.80665 m/s^2.

The manometer (Fig. 6.9) provides the user with a primary pressure standard whose accuracy can be verified anywhere, anytime, with gage blocks, the value of local gravity, and a reliable measurement of temperature. This automatic system, which can generate pressure within the accuracy of the basic manometer [0.0003 inHg (0.0075 mmHg) ±0.003 percent of reading], represents a significant advance in pressure-generation techniques.

The mercury column of the manometer system is contained in a flexible steel tube (stainless-steel seamless bellows). Each end of the tube terminates in a large-diameter cistern, one end of which is fixed. Mercury column height within the flexible line is established by elevating the moving cistern. The moving and fixed cisterns are connected to a rough sensing reservoir to form a U-tube configuration (Fig. 6.10).

Input pressures are measured as a function of mercury column height between the mercury level in the fixed cistern and the other end of the U tube, which is the mercury level in the moving cistern.

The moving cistern is supported by a carriage guided by rollers on a precision column. The carriage is elevated by a circulating ball nut in contact with a precision lead screw. The lead screw is rotated by a drive gear connected through appropriate gearing to an electric-drive motor and a Manual Drive knob. The electric-drive motor is actuated to advance the moving cistern by the Up/Down controls on the front control panel. As the lead screw is rotated to advance the moving cistern, the height of the mercury column (difference between the fixed and the moving cistern's mercury levels) is denoted on the front panel's readout counter. Fine adjustments for the mercury column height (readout-counter setting), representative of a desired pressure, are made by the Manual Drive knob. Pressures can be set and read within 0.0001 inHg (0.34 Pa) by using the 0.0001-in (0.0025-mm) indicator mounted below and geared to the front panel's readout counter.

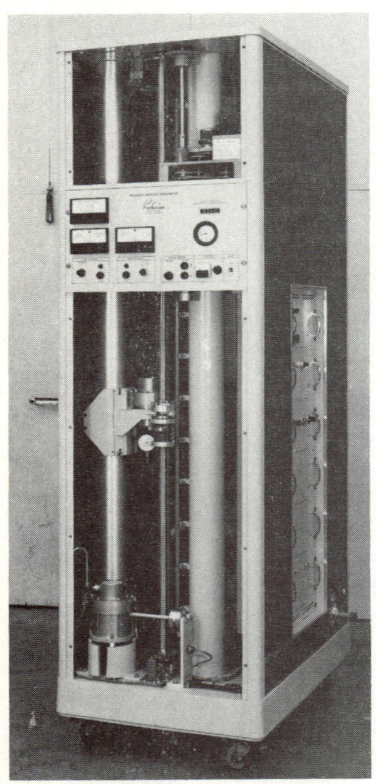

FIG. 6.9 Precision servomanometer. (*Schwien Engineering Inc.*)

Cistern elevation is maintained and gear backlash is eliminated by weight loading. A unique pantograph lever system supports the mercury column tube and reference vacuum line to the moving cistern. This pantograph lever system is extended by cables and pulleys when the moving cistern is elevated.

The weight of the moving cistern's carriage and the pantograph assembly is constant and continually supported by the screw/ball nut assembly. This constant downward load ensures exact positioning of the cistern and accurate registry with the front panel's readout counter through the loaded gear train.

A lapped plate mounted on the manometer base beneath the moving cistern is provided for use with gage blocks to verify, at anytime, the elevation system's accuracy. Five-second spirit levels, installed and adjusted on the manometer base, verify that the lead screw is vertical and the manometer level. An accuracy of 0.0001-in (0.0025-mm) elevation between the fixed and moving cisterns can be verified.

Circular steel plates are supported on glass insulators above the mercury surface in each cistern, to provide variable capacitors for sensing a capacitance null

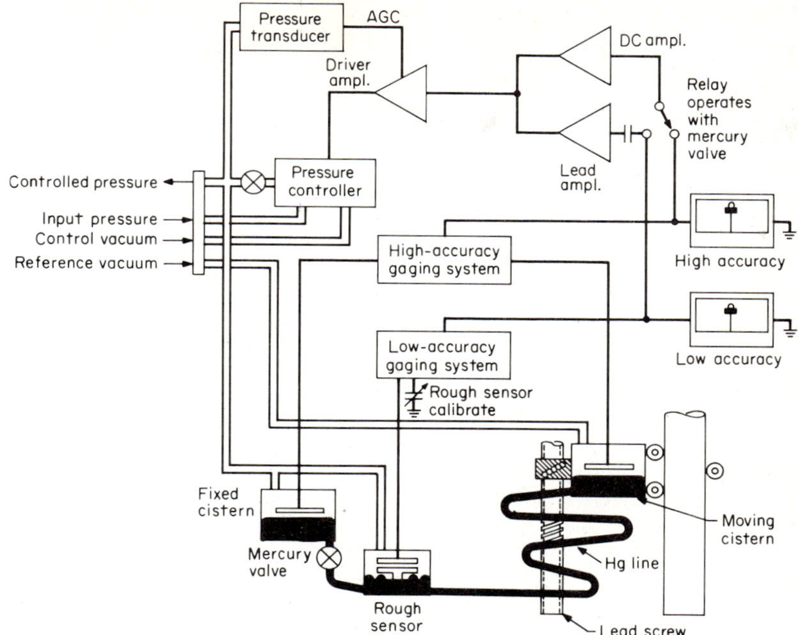

FIG. 6.10 Servomanometer schematic. (*Schwien Engineering Inc.*)

when input pressure exactly balances the mercury column height. The capacitors formed by this 0.045-in (1.14-mm) plate-to-mercury gap are identical at null, and their capacitance values can be verified anytime by a calibrated reference capacitor (self-test feature).

Circuitry identical for the high-accuracy gaging system is duplicated in the low-accuracy gaging system, which uses the rough sensor. The rough sensor contains a pressure diaphragm that changes a capacitance gap as a function of mercury head pressure relative to input pressure. This capacitance is compared with a fixed capacitor in the rough-sensor calibration unit, which simulates a second capacitance gage to complete a two-legged capacitance bridge similar to the bridge and oscillator in the high-accuracy system. A low-accuracy capacitance null is read on the front panel's null meter only when the mercury column height is balanced by the input pressure as seen by the rough-sensor diaphragm and capacitance gage (low-accuracy null meter on the front panel).

The automatic pressure controller generates the pressure necessary to balance the mercury column height sufficiently to effect a capacitance null. When a null condition does not exist, error signals from the low- and high-accuracy gaging systems are used as inputs to the pressure controller's amplifier system. These error signals are summed with appropriate signal conditioning to provide voltages at the output of the controller's drive amplifier, to operate valves connected to pressure and vacuum sources. This causes the proper nulling pressure at the pressure controller's pneumatic output, which is in fact the manometer system's pressure input when operating in the automatic pressure-controlling mode.

This pressure is available at the manometer's escutcheon-plate controlled-pressure port and is exactly that pressure dialed in by number (in inHg) on the front control panel's readout counter within the basic accuracy of the manometer.

The compensation system operates continually to algebraically add to the moving cistern's height the correction values necessary to make the front panel's readout and pressure output a standardized pressure. The compensation-system circuitry uses temperature information derived from temperature probes (one in the flex mercury line, about midlength, and one in each cistern), column height data derived from a height potentiometer geared to the lead screw, and gravity information to provide the appropriate signal at the input of the compensation system's servoamplifier.

The compensation system's servomotor then advances the lead screw through the differential gearing in the compensation computer, to provide the total column correction necessary to compensate for the decrease in the density of mercury (temperature higher than 0°C or 273 K,) the elongation or contraction of the lead screw, and the nonstandard value of local gravity. The servomotor also drives a follow-up potentiometer, which is geared to a compensation system's readout counter. Manual calculations for the total correction necessary for a given temperature and gravity value can be compared with the compensation system's readout counter to verify the total correction being made.

When absolute pressures are being measured or generated, the moving cistern must be evacuated. A thermocouple vacuum-gage tube, which provides a reading from 1000 to 1 μm, is installed in the moving cistern to monitor this reference pressure.

When generating a desired pressure, note the following sequences: The desired column height is established by slewing the moving cistern to the desired pressure number. The column height can be changed with a slew rate of up to 40 in/min (or 0.017 m/s) (slew-rate control). After the exact pressure desired is set at the front panel's readout counter, the pressure controller operates to null automatically the low-accuracy gaging system. The logic operates automatically to open the mercury valve, and a high-accuracy null results within 30 s nominal. During this time, the compensation system establishes the proper column height correction. The pressure output thus established is accurate to within 0.0003 inHg (1 Pa) $\pm$0.003 percent of reading.

A Secondary Standard

Many pressure-calibration tasks in engineering laboratories do not justify the cost and complexity of the primary pressure standards just described. Certain precision Bourdon tubes and capsule pressure gages offer greater convenience, only moderately sacrificing accuracy as compared to a primary standard such as a precision mercury column. Figure 6.11 shows such a precision pressure gage; it has a full-scale range of 150 psig and uses a capsule as the pressure-sensitive element. Bourdon tubes are used for higher pressure ranges. Available ranges are from a low 0 to 4.5 psig (30 kPa) to as high as 0 to 500 psig (3500 kPa). Accuracy claimed is 0.066 percent of full scale; readability is 0.02 percent of full scale. These gages have a long scale of about 45 in (1 m) because the pointer covers the full scale in 2 revolutions. The instruments are manufactured with a calibration that can be checked at appropriate intervals against a primary standard like a precision mercury manometer or a pneumatic piston gage.

Typical usage of such secondary standards might involve daily calibration of pressure transducers on a production line. The secondary standards might then go to the standards laboratory monthly for checking against the primary standard.

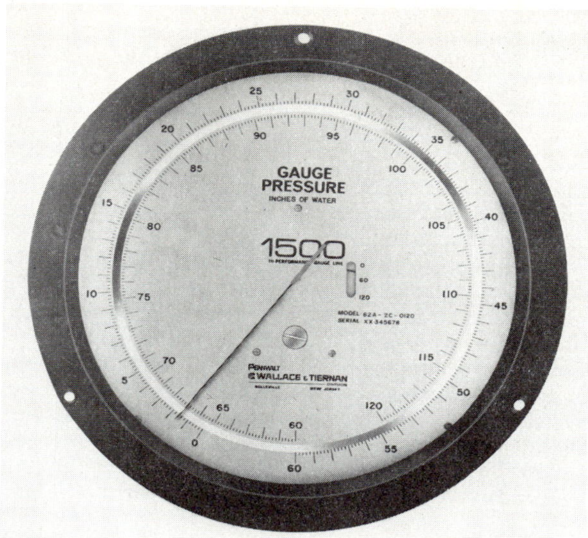

FIG. 6.11 Secondary pressure standard. (*Pennwalt/Wallace & Tiernan.*)

6.4 TEMPERATURE CALIBRATION

Concept of a Temperature Scale

The concept of heat is well-known to us. For example, we touch an object and say it is hot or cold, and we are also able to discern the object's degree of heat or coolness. However, this perception is not suitable for a quantitative definition of a temperature scale. To define such a scale, at least two reproducible fixed points must be defined, and some scheme of interpolation (gradations between the fixed points) and extrapolation (gradations beyond the fixed points) must be developed. Historically, many fixed points were based upon such a factor as the temperature of the human body. It is now obvious that bodily temperature is a variable and so not suitable for precise reference thermometry.

Most early experimenters recognized the suitability of the melting point of ice as a reliable and reproducible fixed-temperature point. Some thermometric pioneers also recognized that the boiling point of water at normal atmospheric pressure was another suitable fixed point. These two points have survived the test of time, and the modern centigrade scale (now known as Celsius) defines the freezing point of water as 0° and the boiling point as 100°. For many years interpolation on this scale was based upon the expansion of a fluid; that is, a liquid-in-glass thermometer was made, and the space between the freezing and boiling points was divided into 100 equal parts. The range was extended both above and below the fixed points by the same method, thus creating an extended thermometric scale.

It became evident that two problems had been overlooked in establishing such a scale: (1) There was an absolute zero, a temperature below which no experiment could be carried; (2) the liquid-expansion scale did not precisely fit the behavior of gases. These problems were solved by the creation of the so-called thermodynamic temperature scale, which can be used to interpolate and extrapolate from the two fixed points. The ideal gas law says that for 1 mol of gas,

$$PV = RT$$

where P = absolute pressure
V = volume
R = universal gas constant
T = temperature (absolute)

The SI unit for temperature is the kelvin (K); the Celsius and Kelvin scales are related by a constant such that K = °C + 273.15.

From this relationship it is possible to measure pressure and volume to thereby establish a temperature scale.

It is not convenient to conduct gas-pressure or volume tests to measure temperatures, but there is an alternative. A platinum resistance thermometer has a characteristic very close to the thermodynamic scale over a broad range of practical interest. (See the "Resistance Thermometers" and "Resistance-Thermometer Readout Systems" discussions in Sec. 5.2.) The modern temperature scale is based upon fixed points and interpolation with the platinum resistance thermometer.

International Practical Temperature Scale (IPTS)

An international committee in 1927 adopted a temperature scale based upon the principles just discussed. The scale has undergone several minor modifications (the most recent was in 1968) and is now known as IPTS-68. It defines six primary points:

1. Boiling point of liquid oxygen
2. Triple point of water
3. Boiling point of water
4. Boiling point of sulfur
5. Freezing point of silver
6. Freezing point of gold

In addition, several secondary fixed points are defined. Table 6.4 summarizes the more important fixed-temperature points. The terms freezing point and melting point are synonymous; they denote that temperature at which the solid and liquid are in equilibrium. The boiling point of a substance can also be called the condensation point, and it is that temperature at which the vapor and liquid phases are in equilibrium. The triple point is that temperature at which all three phases—solid, liquid, gas—are in equilibrium. Interpolation between points is by the platinum resistance thermometer up to 630.74°C (903.89 K), the freezing point of antimony. From 630.74 to 1064.43°C (903.89 to 1337.58 K), interpolation is by the platinum/10 percent rhodium versus platinum thermocouple (see the "Thermocouples" and "Thermocouple Readout Systems" discussions in Sec. 5.2 for details of thermocouple measurements). Extrapolation above 1064.43°C (1337.58 K) is by optical pyrometer (see the "Radiation Pyrometers" discussion in Sec. 5.2 for details of optical pyrometry).

TABLE 6.4 Fixed-Temperature Points

Defining fixed point	IPTS-68*	
	°C	K
tp† hydrogen	−259.34	13.81
bp hydrogen, 25/76 atm	−256.108	17.042
bp hydrogen	−252.87	20.28
bp neon	−246.048	27.102
tp oxygen	−218.789	54.361
bp oxygen	−182.962	90.188
fp water	0.0	273.15
tp water	+0.01	273.16
bp water	100	373.15
fp zinc	419.58	692.73
bp sulfur	444.674	717.824
fp silver	961.93	1235.08
fp gold	1064.43	1337.58

*IPTS-68 = international practical temperature scale of 1968.
†tp = triple point, bp = boiling point, fp = freezing point.
SOURCE: Data from ANSI/ASME PTC 19.3–1974, reproduced by permission of The American Society of Mechanical Engineers.

Primary Standards

Generally, the typical temperature-calibration laboratory does not actually determine temperatures from the published fixed points. Some of these points require sophisticated apparatus and skilled handling. Calibration of a primary standard like the platinum resistance thermometer or the platinum-rhodium thermometer is best carried out by the NBS or other qualified laboratories. These primary standards are then used to calibrate working secondary standards. The one exception is the triple point of water, for which a special device is manufactured, and which is a relatively straightforward measurement for the average laboratory.

Triple Point of Water

The freezing point of water (0.0°C or 273.15 K) is a fixed point readily duplicated in any laboratory. Add distilled water to crushed ice made from distilled water, and place the mixture in an insulated container. If no impurities enter the mixture and the water is air-saturated, this is a simple way to set up the freezing point of water. Errors are no more than ±0.1°C (±0.1 K) with this method.

However, the triple-point cell is capable of accuracies better than 0.001°C (0.001 K), and for precise work this method is preferred to the conventional ice bath. Figure 6.12 shows a triple-point cell for water. It consists of a double-walled glass tube containing a sealed-off volume containing ultrapure water and a small evacuated space. Once the volume is sealed off, a minute amount of the water fills the evacuated space as water vapor. The glass assembly fits into an insulated holder with cover, and concentrically down the center is a well that permits insertion of a temperature-sensing device, usually a platinum resistance thermometer. To use the cell, circulate a refrigerant through the well until the water is partly frozen in a cylindrical pattern (called an ice mantle.) Next, insert a heating rod into the well briefly to form a film of water between the ice mantle and

FIG. 6.12 Triple-point cell. (*Foxboro/Transsonics.*)

the well. At this stage the mantle of ice is in equilibrium with water and water vapor; the triple point is established, which is defined as +0.01°C (273.16 K). Accuracy claimed is better than 5×10^{-4} °C or K. The manipulation of the cell is not overly complicated, and the method offers the average laboratory a very precise fixed-temperature point.

Temperature Comparators

To compare a temperature standard with a working thermometer or sensor, there must be an environment which ensures that the devices to be compared will be held very nearly at the same temperature. This is most often accomplished with a bath of some sort: The bath usually consists of an insulated outer container, controlled heat input, and a stirring device. For the range from room temperature to its boiling point, water is ordinarily a suitable medium. From this point up to perhaps 500°F (960 K), certain oils are suitable. For safety, the oil should have a high flashpoint and its temperature should be held well below that point. Molten salts are often used in the next temperature range, say, up to about 1000°F (1500 K). From 1000 to about 2000°F (1500 to 2500 K), air furnaces are used.

Fluidized-Solids Bath

An alternative to the oil and salt baths is the fluidized-solids bath, a device useful over the range −100 to 2000°F (360 to 2500 K). If a reservoir containing finely divided inert particles is constructed so that a gas (e.g., air, steam, nitrogen) can be fed through the "bed" of particles, a "fluidization" state can be achieved, in which the individual particles become microscopically separated from each other by the moving gas. This fluidized bed of particles has unusual properties that differ markedly from those of either the gas or the particles. The fluidized bed behaves very much like a liquid, exhibiting characteristics generally attributable to a liquid. For example, the fluidized bed can be agitated and bubbled; it always seeks a common level; light-density materials float whereas those with densities greater than the equivalent fluidized bed density sink; and, most important, the heat-transfer characteristics between the fluidized bed and a solid interface can have an efficiency approaching that of an agitated liquid.

In addition, the fluidized-solids phase has a most unusual physical behavior in that its basic characteristics change only slightly over very large temperature ranges; it has no melting point and no boiling point. The lowest temperature available is the liquefaction point of the gas used for fluidization; the highest temperature level is the usable temperature of the inert solids. The metal oxide beds commonly used (e.g., aluminum oxide) are nonflammable, nonexplosive, nontoxic, and stable to over 2000°F (1400 K). The most commonly used fluidizing gas is compressed air obtained from a blower or compressor. For situations where a nonoxidizing environment is required, nitrogen is used; for a reducing environment, steam or cracked gas is used.

The unique characteristic of gas-fluidized solids is the relatively high rate of heat transfer within the phase, which yields highly isothermal conditions and excellent heat transfer to the solid surfaces submerged in the phase. This characteristic is the result of the turbulent motion and rapid circulation rate of the solid particles in conjunction with the extremely high solid-gas interfacial area. Therefore, even though the gas-solid interfaces normally yield low heat-transfer coefficients and the solids normally used have low thermal conductivities, the overall heat-transfer characteristics of a fluidized-solids phase approach those of a liquid.

The combination of excellent heat-transfer characteristics and high heat capacity make fluidized solids an excellent medium for providing an isothermal

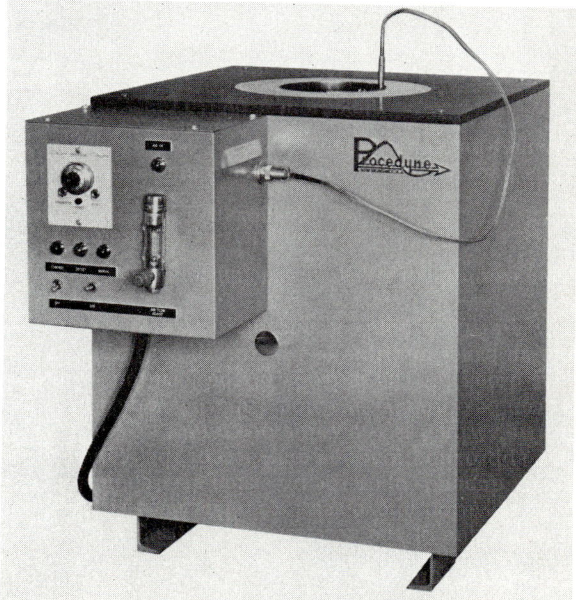

FIG. 6.13 Fluidized-solids temperature bath. (*Procedyne Corp.*)

environment for temperature calibration, heat treating, and testing materials over a very wide temperature range. Figure 6.13 is a commercially available bath. These baths provide many advantages for temperature calibrations: The fluidized aluminum oxide solids are odorless, fumeless, nontoxic, dry, inert, nonflammable, noncorrosive, and nonabrasive.

Secondary Standards

Much routine temperature calibration does not require the fine accuracy of the platinum resistance thermometer. Certain instruments, notably the liquid-in-glass thermometer and base-metal thermocouples, can be used for the calibration of many devices when properly compared to primary standards. See Sec. 5.2 for a discussion of liquid-filled glass thermometers and thermocouples.

6.5 FLOW CALIBRATION

Flow is a derived quantity, and hence there is no primary flow standard. Flow is expressed as weight per unit time or volume per unit per unit time and is therefore derived from the standards of mass, length, and time. The typical calibration laboratory sets up a "flow stand" or flow calibrator, and the three measurements are then referred to the primary standards of mass (weight), length (volume), and time.

Flow calibrations can be carried out on either a weight basis or a volumetric basis. For most applications the weight basis is preferred, even if the flowmeter under calibration is a volumetric unit.

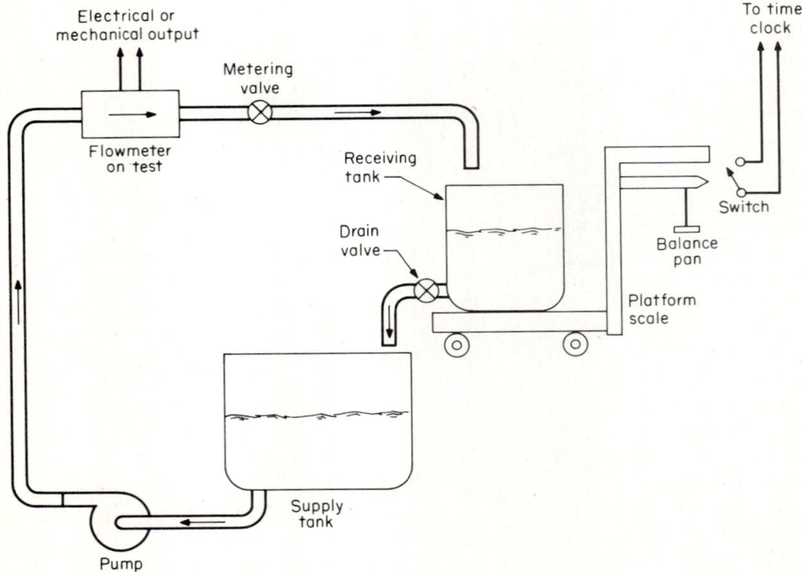

FIG. 6.14 Gravimetric-flow calibrator.

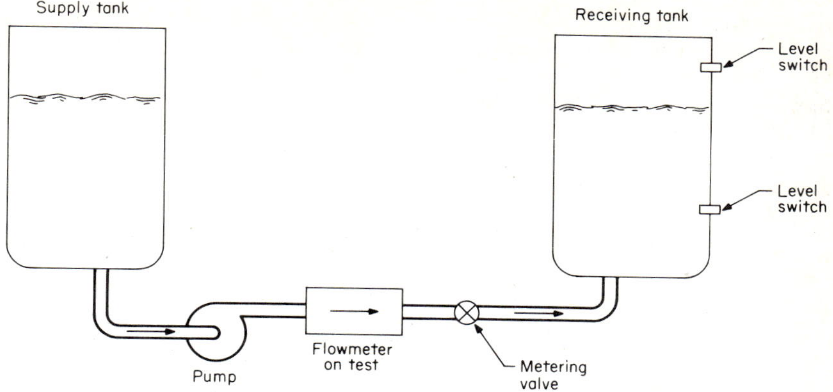

FIG. 6.15 Volumetric-flow calibrator.

Gravimetric-Flow Calibration

Figure 6.14 shows a typical flow calibrator. Operation of the stand is as follows. Initially leave the drain valve open, and set some flow rate with the metering valve. Then close the drain valve; the receiving tank begins to fill. The balance pan should have been previously adjusted to exceed the tare weight of tank and small amount of fluid. As the receiving tank fills, the balance pan rises and the time clock starts. Now add a known weight to the balance pan; it assumes the lower position. Suitable circuitry keeps the clock running until the pan again rises, at which time it stops the clock. Calculate the gravimetric flow rate by dividing the added weight by the elapsed time. If the flowmeter being tested is of the volumetric type, a measurement of specific gravity is also required. This is customarily done by a simple hydrometer test. Open the drain valve and remove the added weight from the pan. When the receiving tank is empty, another point can be measured by repeating the steps.

Volumetric-Flow Calibration

Volumetric-flow stands usually are not as convenient as gravimetric stands. Figure 6.15 shows the elements of such a system. In operation it is similar to that of the gravimetric stand, but the time clock is started and stopped by the level switches in the receiving standpipe. It is necessary to know the precise dimensions of the receiving standpipe and the spacing of the level switches. Not shown on the drawing is a return flow path. This can be accomplished with a separate pump or suitable valving at the pump shown.

The reader is cautioned about using pumps in liquid-flow measuring systems. Often, because of cavitation in the pump, the issuing liquid contains vapor bubbles that reduce the true density of the fluid. These bubbles collapse with time, but there must be adequate distance between the pump and instrument under calibration for this to occur. Alternatively, a settling tank or accumulator can be used to provide the necessary time for the vapor bubbles to disappear from the liquid.

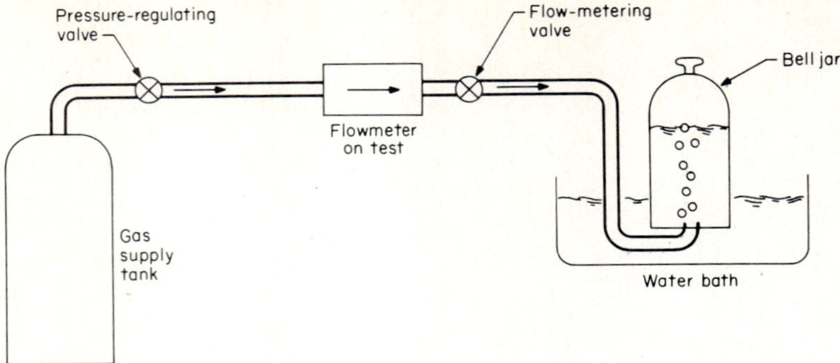

FIG. 6.16 Gas-flow calibrator.

Gas-Flow Calibration

In principle, either a volumetric or gravimetric method can be used to calibrate gas flowmeters. However, in practice the mass of material involved, and therefore its weight, is small compared to the tare weight of such items as tanks, piping, and valving. For this reason gas-flow calibration is usually carried out on a volume basis, as shown in Fig. 6.16. The volume of the bell jar must be known, and the typical technique is to time the lowering of the gas-water interface between two marks on the bell jar. In any calibration it is necessary to measure and control both temperature and pressure, parameters that are always specified for gas flowmeters.

Secondary Standards

The acceptance of flowmeters as secondary standards is somewhat controversial because their behavior depends upon many external conditions. Nevertheless, some authorities have accepted orifices, turbine types, and rotameters as secondary standards. Generally, the best practice is to use one of the flow stands described in this chapter, unless there is some compelling reason to go to a secondary standard.

6.6 FORCE CALIBRATION

Standards of force are based upon standards of mass, or weight. See Sec. 5.7 for a discussion of the interrelationship of these terms.

Standard Weights

The prototype standard of mass is the kilogram weight kept at the international bureau in Paris. The NBS has several copies of this weight, and one copy, No. 20, has been designated the United States national standard. This standard is compared regularly with the Paris standard and others at NBS. From these standards, several deadweight machines have been calibrated; these machines serve as force calibrators.

Until about 1965, the maximum capacity available from these deadweight machines was 111,000 lb (50,000 kg). For loads greater than this, force-multiplying techniques were used. These force multipliers are simple in principle, consisting of unequal lever arms and fine pivots of flexures. Unfortunately, this "bootstrapping" technique involves a loss in precision.

In response to the needs of the space program, which required precision-thrust calibration to unprecedented accuracy and force levels, the NBS embarked upon a development program of large deadweight force calibrators. This effort culminated in the construction of three testers of 1,000,000-, 300,000-, and 112,000-lb (450,000-, 136,000-, and 50,800-kg) capacities. Uncertainty is 0.002 percent. Obviously, it is not practical to submit 1,000,000-lb weights for calibration, so at these high load levels, secondary force standards are submitted. In the 50- to 10,000-lb (20 to 4500 kg), range, it is customary to use some other calibration facility that has traceability to NBS.

Secondary Standards

Two principal secondary standards used to transfer force measurements are the electric load cell (see Section 5.7), and the proving ring (Fig. 6.17). The ring is a product of the Morehouse Instrument Company and has a capacity of 250,000-lb (113,400-kg) compression. When used for either calibration or force testing, the ring is loaded at the projecting buttons and becomes distorted from a circular to an oval shape. This distortion is readily measurable as a change in diameter that is determined by adjusting the micrometer mounted across the throat of the instrument. Calibration of the device consists of establishing the relationship between force and deflection. When the device is used to measure an unknown force, the force-versus-deflection curve is consulted.

FIG. 6.17 Proving ring. (*Morehouse Instrument Co.*)

The design of such a proving ring mainly involves a suitable material; this material is generally a steel forging designed to operate within the elastic limit. A good proving ring shows a linear response and has negligible hysteresis.

There are units for tension and compression loading, in ranges up to 1,000,-000-lb capacity. In principle, they can be made for any capacity.

chapter 7

Safety Techniques and Devices

Maurice J. Webb

7.1 GENERAL SAFETY PRACTICES AND PRECAUTIONS

The technician's broad spectrum of duties coupled with the originality of much of the work, where no established failure modes or known hazards exist, call for an above-average demand for knowledge and awareness of potential hazards. It is *always* necessary to exercise great care. When doubt exists as to the safety of an operation, there can be no other course but to take the appropriate preventive measures. Once a potential hazard has been identified, yet does not occur after repeated operations, the temptation to relax safety procedures must be positively rejected. Most accidents occur not from new and unrecognized hazards but from established ones. Whether these accidents occur from complacency, ignorance, or carelessness, most can be prevented by habitual safe practices and by the use of safety devices.

Every technician should develop a safety-conscious attitude by repeatedly looking for hazardous situations and correcting them and by acquiring certain work habits. Good housekeeping is a prerequisite to good safety. Keep machines, benches, desks, and floors clear of items not immediately required for work. Always keep aisles, stairs, and walkways clear, and properly stack and secure materials so they cannot fall. Where possible, keep unneeded items in cabinets, drawers, or cupboards. Each individual's personal conduct in these work areas creates a safety-awareness atmosphere that will extend to other work activities. Practical jokes or buffoonery are *never* conducive to good safety practices in any technician's workplace or laboratory.

The supervisory technician has a special responsibility to educate and train new personnel in the safety requirements of the facility. In particular, the special hazards and proper procedures for that specific facility should be thoroughly explained for the type of work performed there. Not only do different facilities have different safety procedures, but the nature of the work may well be very different, too. Never assume that safety was taught and practiced elsewhere.

7.2 PRESSURE SAFETY

The existence of pressure in a system represents the most potentially serious general hazard in the laboratory. Even low pressures, especially if the fluid is gaseous, can result in serious accidents, and the possible consequences multiply greatly as the dimensions of the system (piping, vessels, etc.) increase. But pressurized systems are such a necessary part of modern technology that the only recourse is to observe a number of fundamental steps in design and handling.

Fluid-pressure hazards are often greatly increased by the presence of elevated temperature. Temperature structurally weakens the fluid-containment system by reducing material strength and intensifies material stresses from thermal expansion. In a closed system, increasing temperature may raise pressures enormously. Temperature protection alone is treated in Sec. 7.5, but temperature that intensifies pressure is primarily a pressure hazard.

The first requirement for pressure safety is to attach to the system some kind of pressure-measuring device. Usually this will be required for other reasons, but in any case it is needed for safety. The simplest device is a common Bourdon pressure gage. It is important to know the pressure in a system at all times to guard against initial overpressurization and to indicate what conditions continue to prevail in a closed container or system. Pressure gages or other pressure-measuring and indicating devices must be periodically calibrated to verify their accuracy.

Hydrostatic Testing

To verify that a pressure-containment component or system is capable of working safely at the intended pressure level, perform a hydrostatic test. This test serves several purposes. It proves out the design, verifies the integrity of the materials, confirms the adequacy of the workmanship, and locates the places where corrective action may be required (e.g., leaks).

Hydrostatic means that the pressurizing medium is a liquid and that there is essentially no flow into or out of the system. Hydrostatic (liquid) testing is done because it is much safer than pneumatic or gaseous testing. Liquids are essentially incompressible so that if, during the course of the test, a vessel ruptures the system loses pressure extremely quickly. If a gas is used as a pressurizing medium, then in the event of a vessel rupture the gas will continue to expand, pushing the ruptured vessel fragments with enormous force. These fragments may reach fantastic speeds, as does bomb shrapnel. In fact, any pneumatically pressurized system should be considered a potential bomb. This is not to say that there is no stored energy in a hydrostatic system. The walls of the containment vessel, which are usually metal, behave in an elastic manner so that upon rupture and pressure release these walls rapidly spring back. Fortunately, this spring-back movement is very small and fragments are unlikely. There are still certain precautions that must be taken with hydrostatic testing, but the severity of the hazard is much reduced.

To hydrostatically test a system, first fill it with liquid, usually water but not necessarily so, taking care to vent all air. Pockets of trapped air negate the advantages of the hydrostatic approach. Once the system is completely filled, connect a pump of small-volume capacity to the system. The small-volume capacity is important because the system pressure will rise rapidly if all air has been vented. The pump should be capable of fine control, and a hand-operated one is preferred where pressures permit. The pump operator must be able to see the pressure gage so that the test pressure can be achieved slowly.

The pressure selected for test purposes must provide an adequate margin over the maximum operating pressure. "Adequate margin" is certainly difficult to determine and few general rules can be given, so it is best to consult an engineer or approved specification for the specific item to obtain a value. The margin, however, considers factors such as the yield stresses for the construction materials (it is unlikely that one wants to cause permanent deformation), the possible eventual corrosion allowance and consequent thinning of vessel walls, and the operating temperature. Knowing the construction-material behavior at elevated operating temperature, the room-temperature test pressure may be increased by a factor such that operating stress margins are reproduced. A test pressure three times the maximum working pressure is typical after all factors are considered.

The time for which the test pressure is maintained depends upon the purposes of the test. If only structural integrity is of concern, then only a few minutes (10 to 15) will be required. If the test is also to locate leaks, then the test may take several hours to permit complete inspection of the system. A system under pressure should be well-labeled so that all can see that a potential hazard exists.

It is a wise precaution to take vessel or component dimensions before and after testing. For example, record pressure-vessel diameter or valve-body width. If permanent deformation results, report this to an engineer as it may be an indication of future trouble.

It was already mentioned that hydrostatic testing is not without some hazard. If a system is being tested for the first time, then it is wise to keep at a distance, preferably with some protective barrier between the system and operating personnel. If the system is complex, many of the individual items, especially the "one-of-a-kind" components, are best tested individually before assembling into the entire system. This reduces the risks while testing the complete system.

All pneumatic systems should first be hydrostatically tested before they are charged with gas, even if they are only intended for low-pressure service. If pneumatic testing is essential from considerations of liquid contamination, then perform it with adequate protection. Use a separate test area fitted with an indirect viewing system and outside pressurizing controls. Another safe way for pneumatic testing of small items is to submerge them in a large tank of water. If the component ruptures, the water will arrest the shrapnel.

If water is objectionable as the fluid for hydrostatic testing, then select hydraulic oil or an inert liquid. Never use volatile hydrocarbons because they may form an explosive mixture with unvented compressed air.

Once a hydrostatic test is complete, vent the pressure and then remove the notices and labels posted to advise that the test was in progress. Do not leave tests in progress without adequate restriction of the area.

Pressure Safety Valves

While it is generally a wise precaution to place a safety valve on any pressurized system, it is absolutely necessary to use one if the system pressure can exceed its maximum working pressure as a result of a component malfunction or human error. For example, a pressure vessel may have to be charged with a gas by using a regulator from a gas reservoir (cylinder) maintained at a pressure much greater than that required by the vessel. Obviously, a malfunction of the regulator or inadvertent use of the regulator can result in a pressure much above that intended. In vessels where chemical reactions take place or heat is added, the pressure may also rise unduly.

The safety valve must be sized to take care of the worst possible situation. That is, it must be capable of venting fluid at the maximum rate that fluid can accumulate or pressure increase in the system. In this way, once the valve opens, it can keep pace with the pressure generation to relieve it fast enough to prevent further increase.

If a safety valve is activated, first bring the system under control and investigate the cause. Repeated operations of a safety valve indicate improper operation, which must be corrected. Safety valves are neither designed nor intended to be pressure control valves.

Safety valves are typically adjustable. Once set, never reset them without the approval of the cognizant engineer. Periodic verification of the set pressure is in order, but the valve is often removed and "bench" tested for this purpose. Safety valves often have locking devices on the adjustable components. It is then readily apparent if the valve has been readjusted or tampered with because the lock is broken.

A venting safety valve can be a hazard in itself unless full consideration is given to the potential discharge. The discharge or blast from a high-pressure source can be dangerous, and adequate baffling or ducting should be installed to prevent problems. However, no restriction to the discharge flow is acceptable for a safety valve.

Frangible Diaphragms

A frangible diaphragm is a thin metal disk clamped at its edges. Pressure is applied to one side while the other side is exposed to the atmosphere. Usually the disk is curved, with the high pressure against the concave side of the disk (Fig. 7.1). These diaphragms are manufactured to fail at a stated pressure, and they do so with surprising precision. The lowest practical value that the disk can be designed to break at is 20 percent above the working pressure. The diaphragm material can fatigue with repeated pressure fluctuations and then break at a lower-than-intended pressure. However, with the 20 percent margin this occurrence is minimized.

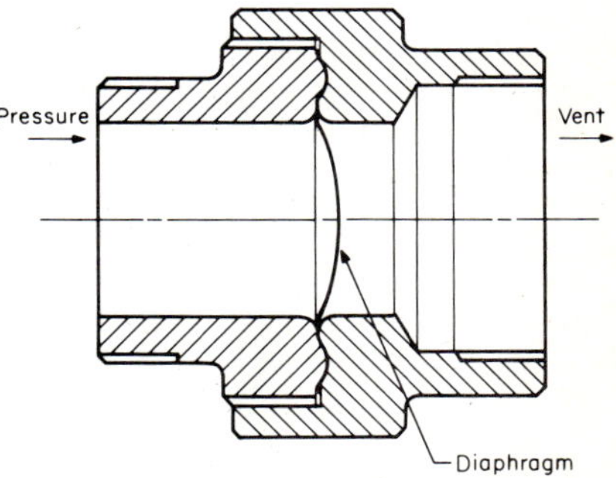

FIG. 7.1 Pressure safety diaphragm.

Frangible diaphragms are used alone or with safety valves. They are sized to provide adequate pressure relief. Once the diaphragm fractures, of course, it continues to discharge until the system pressure is zero, whereas a safety valve reseals once the pressure has decayed to the design value. When used with a safety valve, it is set at a pressure above that of the safety valve and provides added relief should the pressure continue to climb after the safety valve has opened.

The discharge from a frangible or burst diaphragm also needs safety considerations. First, the thin metallic disk should be retained. Second, very high thrust can be produced, especially from high-pressure systems, unless the discharge is equally dispersed in opposite directions. Figure 7.2 shows a simple attachment to a burst diaphragm that retains the disk and eliminates the thrust. Third, the discharge blast, like that of a safety valve, requires attention. The device of Fig. 7.2 is also effective in distributing the discharge and minimizing appreciably the dangers of a single blast of gas or liquid.

Pressure Gages

Pressure gages may be both an asset or a detriment if certain precautions are not taken. All pressure gages should be maintained in calibration by a well-disciplined retest program. An incorrectly reading gage may be worse than no gage at all for it can be misleading. Periodic calibrations are often necessary and insti-

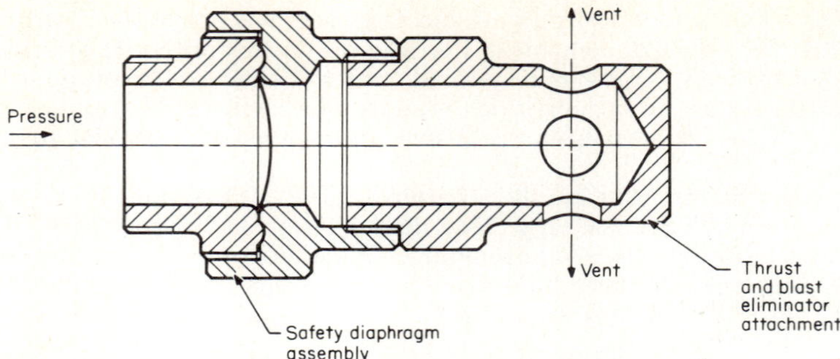

FIG. 7.2 Safety diaphragm with thrust and blast eliminator.

tuted to maintain accuracy for reasons of quality data, but there is also an important safety aspect, too. It is good practice to record the date of its last calibration right on the front of the gage.

On occasion, the internal mechanism of a gage may break so that the fluid pressurizes the inside of the gage. The glass front of the gage can be fractured by this internal pressure and can be dangerous to a person situated in front of the gage. This type of accident can be eliminated by using gages with "blow-out" backs. On these gages, the back blows out before the glass fractures, which is certainly the preferred design.

When measuring the pressure of a hazardous fluid, the fluid itself should not be brought in proximity to personnel who will be monitoring the system. The pressure data are paramount and readily obtained by one of two means. (1) Locate a pressure transducer (see Sec. 5.3) where the pressure measurement is required and locate an electrical readout device elsewhere, where personnel can safely read the information. (2) Use a standard Bourdon-type pressure gage but insert a diaphragm-seal device into the pressure line. With this device a flexible rubber or metal diaphragm separates the process fluid from the gage fluid. Figure 7.3 shows a diaphragm assembly with a gage, and Fig. 7.4 shows the device schematically. The instrument side of the diaphragm is completely filled with an inert fluid of minimal temperature expansion.

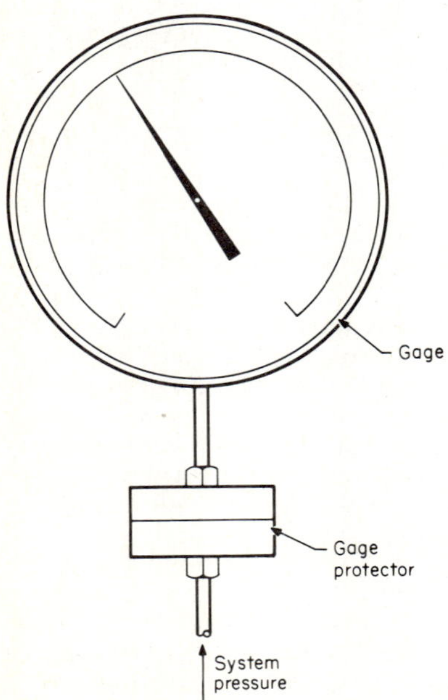

FIG. 7.3 Gage protector and gage.

Not only do diaphragm seals protect personnel but also they protect the instrument itself from exposure to corrosive and otherwise unsuitable fluids (e.g., slurries). In this application they can, of course, be used to protect pressure-measuring instruments other than gages (e.g., transducers).

Another device which has safety aspects and is used with pressure gages is the gage snubber. A snubber imposes restrictions to fluid flow in the gage connecting line either by a short length of capillary or by a porous plug in the line. Snubbers are used to dampen large-amplitude pressure fluctuations that are likely to cause gage failure and poor accuracy. Apart from this safety aspect there is also the added danger that the capillary or porous plug will become completely blocked, causing an entirely erroneous gage reading.

That a system requires snubbers in the gage lines should raise questions regarding structural components in the system and their ability to withstand the fatigue effects of cyclic pressures. Is the system adequately designed and tested for these phenomena?

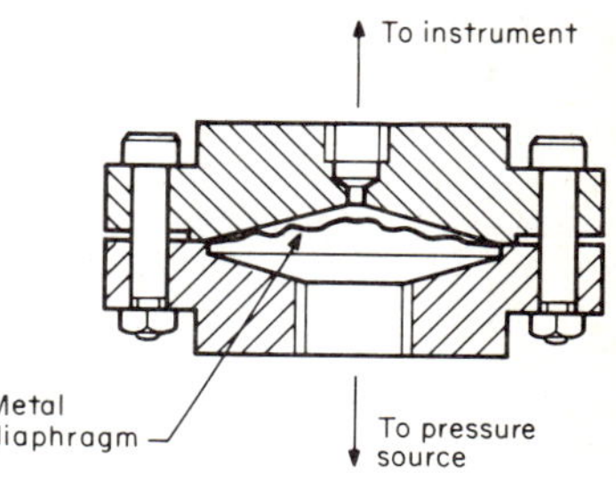

FIG. 7.4 Diaphragm gage protector.

Pressure Switches

Pressure switches are described in Sec. 4.11. They are very effective in reducing pressure hazards and in many situations are the preferred method because systems are often too complex to permit individual monitoring of every gage in a timely manner. Furthermore, pressure switches can trigger automatic corrective action faster than, and in the absence of, humans. They should be regarded as a primary safety device.

There are many ways that a pressure switch can add to safety. The simplest is to have it operate a warning light or alarm to attract the attention of operating personnel to a potentially hazardous situation. The switch may be set at a pressure that permits sufficient advance warnings so that corrections can be made before a safety valve must operate. Alternatively the pressure switch can be directly coupled to valves to either dump excess fluid or to prevent additional fluid from entering the system. Another use would be to turn off pumps, compressors, or electric heaters to remove the cause of pressurization.

As for other pressure safety devices, pressure switches and their associated electric circuitry and controls should be periodically checked to provide the necessary assurance that they will operate correctly when called upon to do so.

Overpressurizing Containers and Vessels

Often it is necessary to fill containers with liquids or gases either to transport them from one place to another or to use small quantities for a particular project. These filling procedures often cause serious accidents because simple precautionary measures were not taken. This is particularly true for metal vessels because it is not easy to judge the thickness of the vessel walls or to know, for example, how much liquid is contained in a vessel. (Note that the Interstate Commerce Commission has regulations governing the safe transportation of liquids and gases between states, and many states have adopted those same regulations for intrastate transportation.)

The first thing to determine is the safe working pressure for the vessel. It may already be stamped on the vessel. However, unless the condition of the vessel is known, the maximum working pressure may no longer be applicable. It is not unusual in laboratories to use things differently from the way they were originally designed to be used. For example, the vessel may have been previously used for corrosive service and the inside severely eroded. When the prior history for a pressure vessel is unknown, it must be subjected to a hydrostatic pressure test to verify its suitability. Vessels being used frequently should be retested frequently if there is any suspicion that their use may deteriorate the structural integrity through corrosion. Carefully record the date of each retest; preferably right on the vessel itself.

It should also be noted that the strength of a vessel can be adversely affected by welding, for example, with the attachment of supports, lifting lugs, and pipes. In general, make such attachments only with the knowledge and concurrence of an engineer. The welding may not only affect the immediate strength of the vessel at the weld heat-affected zone, but it may also change the crystalline structure of the material so as to render those zones more susceptible to future corrosion. This is especially true of stainless-steel vessels.

Having determined that the vessel is safe for its planned use there are still other safety precautions to be observed. When filling the vessel from a pressure source, be sure to have a gage and a relief valve in the fill line. The gage monitors the pressure, and the relief valve prevents overpressurization during filling.

For liquid filling, it is important to know how much liquid to put into the vessel. If the vessel is completely filled, then any temperature increase is likely to cause the vessel to rupture from overpressurization. Since liquids expand much more rapidly than metals, even small temperature changes can result in extremely high pressures because liquids are essentially incompressible. Always leave sufficient ullage in a liquid-filled vessel to allow for this expansion. One way to determine the amount of liquid in a vessel is simply to measure it before pouring it in. Another is to weigh the vessel as it is being filled, but this has some

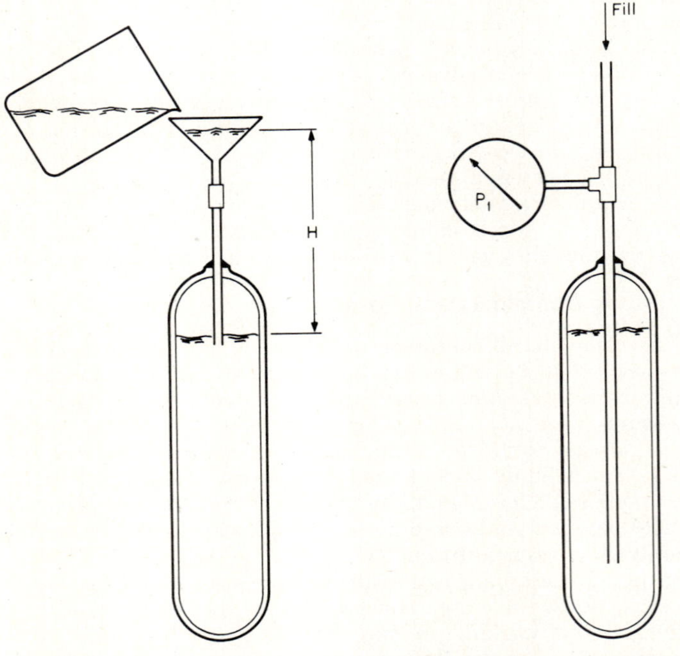

(a) Gravity fill. Ullage pressure = H.

(b) Pressure fill. Ullage pressure = P_1.

FIG. 7.5 Vessels with dip tubes to prevent overfilling.

practical difficulties, especially if the weight of liquid is small compared with the weight of the vessel. A third method is to use a dip tube.

A dip tube is a tube inserted into the vessel through which liquid passes during filling. The length of the tube determines the amount of liquid that can be put in the vessel (Fig. 7.5*a*). Once the liquid level in the vessel reaches the tube, the air in the ullage is trapped, and the addition of further liquid compresses the air until the ullage pressure equals the height of the column of liquid in the dip tube (typically less than 1 psi, or 7 kPa, pressure). At this point it is impossible to add more liquid. If the vessel must be charged under pressure because of the nature of the liquid, then use a longer dip tube and no further liquid will enter the vessel once the ullage pressure equals the filling pressure, neglecting the height of the liquid column in the dip tube, which is generally negligible (Fig. 7.5*b*).

In addition to the expansion of a liquid with temperature, consider also the vapor pressure of the liquid. Vapor pressure increases rapidly with temperature and can easily result in overpressurization. Remember, too, that vapor pressure is independent of the quantity of liquid in the vessel.

Plastics are particularly deceptive in their failure modes. Short-term, hydrostatic proving tests do not foretell future potential problems. Under stress, and especially cyclic stresses, the materials are prone to stress cracking, and such cracks typically propagate from surface discontinuities (e.g., scratches and notches). For these reasons, use plastic piping cautiously for corrosive or dangerous fluids or high pressures.

7.3 COMPRESSED GASES IN CYLINDERS

Many different gases are readily available in cylinders that have been charged to high pressure. While vessels containing high-pressure gases are inherently dangerous, their widespread use with such an excellent safety record is ample testimony that they may be safely handled when certain simple rules are followed. In fact, the few accidents that do occur can invariably be traced to improper handling. The following rules should be committed to memory for anyone handling compressed gases.

Operating Safety Rules

1. Never use a cylinder whose gas content is not clearly identified by the markings.
2. If a protective cap is on the cylinder, it should remain there until the connection is to be made.
3. Use a suitable trolley for moving cylinders. Do not drag, roll, slide, or drop cylinders.
4. Be sure the cylinder is securely fastened at its required location.
5. Do not locate the cylinder near combustible materials.
6. Do not locate the cylinder where ambient temperature might exceed 125°F (585 K).
7. Connect only to suitable pressure-regulated and pressure-safety-protected systems (see Sec. 7.2).

8. Be sure the connecting system is suitable for use with the particular gas, for example, oxygen-cleaned for oxygen gas.
9. Do not use force on the threaded cylinder connection. It should make up easily. Check for correct thread series, and if difficulty is still experienced, return the cylinder to the supplier, clearly indicating the reason.
10. Do not attempt to repair, modify, or change any part of the cylinder connection, valve, or safety devices to effect a connection or for any other reason.
11. Open the cylinder valve slowly, using only the prescribed equipment (handwheel or special handle). Do not force. Return to the supplier any cylinder whose valve does not function freely.
12. Close the cylinder valve immediately if any leaks become apparent.
13. Do not connect the cylinder to a system that may eventually have a pressure greater than the cylinder pressure (as cylinder pressure decays) because reverse flow may contaminate the cylinder and its remaining contents.
14. Keep the cylinder valve closed when gas is not required.
15. Always close the cylinder valve and reduce the system to atmospheric pressure before disconnecting the cylinder.

In addition to these operating rules, store cylinders in a dry, cool area, away from combustibles and protected from unauthorized personnel. It is best to segregate the full and empty cylinders in the storage area. Check also the state and local codes regarding the storage of gas cylinders because the codes often have specific regulations covering these items.

Every cylinder has stamped on it the name of the manufacturer, the date, capacity, retest dates and retest facility, etc. This information is placed there for specific safety-related reasons, and it is illegal to remove or change it. Never remove other markings such as labels and tags used to identify the contents because someone else may not exercise adequate identification care before using and may cause an accident. (See rule 1 above.)

To reduce the possibility of the wrong gas being connected to a system, different cylinder-threaded outlet connections are used for different gases. The threaded connections may be either male or female and left or right hand. Generally the combustible gases have left-hand threads, and similar gases (e.g., methane, ethane, ethylene, hydrogen) will have the same connection. Left-hand threads are identified by a groove in the center of the hexagonal nut.

7.4 CLEANING FOR OXYGEN SERVICE

Oxygen gas is readily available in steel cylinders at a pressure of about 1800 psi (125 MPa). It is commonly used in the workshop for welding and flame cutting and in the laboratory for a whole variety of purposes. It also has many medical applications. However, oxygen can be dangerous, especially when in contact with organic materials and at elevated pressures. The reaction between oxygen and many organics is rapid and accompanied by the release of energy that easily leads to explosions. With high pressure the reaction can be so rapid that detonations are not uncommon. Furthermore, there is no need for a flame or spark to initiate these reactions. Consequently it is important to clean all items that will come into contact with oxygen to remove all residues of organics, such as oil, grease, and other hydrocarbons that are readily combustible in the gas.

Never use oxygen gas as a substitute for compressed air or shop air supplies.

For oxygen use in the laboratory or pilot plant, the piping, valves, fittings, pressure gages, regulators, vessels, and similar components must first be cleaned to remove all traces of organic materials. Often these items may be purchased "cleaned for oxygen service." If this is done, then there is no need to reclean them. Often, however, components purchased for oxygen service have previously been used for other service anyway. If there is the slightest doubt about the cleanliness of items for oxygen service, clean them before use.

Proper cleaning requires the disassembly of valves, regulators, etc., to their component parts so that all traces of organics (lubricants, for example) are removed. Wash nonmetallic parts in hot detergent water. Dishwashing detergents are suitable. Following this, thoroughly rinse them in clean hot water before traces of detergent are permitted to dry on the surfaces. Following the water rinse, dry the parts in a stream of nitrogen gas. The nitrogen gas must be oilfree and may be purchased that way.

First clean metallic parts like the nonmetallic parts unless they are exceptionally oily or greasy, in which case first degrease them in an industrial vapor degreaser or other commercial liquid degreaser. Following the detergent cleaning, rinse each component in unused acetone or Freon-113 and then dry in an oilfree nitrogen gas stream. (*Note:* Shop air invariably contains oil vapor and cannot be used.)

Check the effectiveness of the cleaning with an ultraviolet lamp in a darkened room. With the lamp placed close (about 1 ft, or 30 cm, away) to the part, visually inspect it on all surfaces. The presence of hydrocarbon will be noted by fluorescence on the surface. Any fluorescence is cause for recleaning. Carefully protect acceptable parts against subsequent contamination. Handle parts to be reassembled with white, lintfree gloves. If lubrication is necessary, a silicone oil or grease is preferred, but never a hydrocarbon.

The cleaning of cylinders and other vessels for oxygen service is particularly difficult because the ultraviolet light check usually cannot be performed on internal surfaces. For these items, it is best to have them cleaned by specialized services, who verify the cleanliness through analysis of the used cleaning fluids for traces of organic contaminants.

Bourdon-type pressure gages are also difficult to clean because the complete length of the Bourdon tube must be cleaned. This may be done with a flexible (plastic) capillary tube attached to a syringe. Fill the syringe with a liquid degreaser such as Freon-113 and insert the capillary at the gage connection and push to the end of the Bourdon tube. The syringe then flushes the Bourdon tube for its entire length. Then put the gage into a vacuum chamber to remove all the degreaser. (*Note:* Do not calibrate oxygen-cleaned pressure gages with an oil-filled deadweight tester; instead calibrate them against a secondary standard, using oilfree nitrogen for pressurization).

7.5 TEMPERATURE SAFETY

Temperature hazards manifest themselves in various ways. The most common are elevated or lowered temperatures that can cause severe burns to individuals who make actual contact with the hot or cold item. The second most common temperature hazard results from reduced structural integrity at other than normal temperature. Others include significantly increased chemical reactivity, accelerated deterioration, and changes in physical properties or states. The nature of some of these effects, either alone or in combination, is often such that it leads to deceptive assurances of safety, when, in fact, there may be a developing catastrophe. It is neither easy to detect nor to measure many of these effects. Rigorous design, retest, and inspection procedures and early detection and corrective action programs have collectively proven to be effective. At the turn of the century, boiler explosions were not uncommon and deaths were numerous, but today such accidents are rare. New technology does not have the benefit of analysis of a history of previous failures, and safety must benefit from the proven techniques and procedures that at times must, of necessity, err on the side of overcautiousness.

The physical hazard from burns, from either high- or low-temperature surfaces, may be prevented with adequate insulation. Insulating materials are available for all temperatures, and adequate thickness or circulating coolants can bring temperatures to safe levels. If this cannot be done, then erect barriers and post adequate warning signs. A disadvantage of signs is that residual temperature may easily be underestimated and insufficient time elapsed before adjustments to equipment are made. To indicate the actual temperature, it is generally good practice to locate direct-reading surface thermometers adjacent to the warning signs.

The mechanical properties of all materials are affected by temperature change. At elevated temperature, materials lose strength. They may lose so much strength that they fail structurally; the hazard is compounded if the heated zone is also pressurized. Other types of failure mechanisms may also be initiated. For example, some materials creep at elevated temperature while under stress. The creep, or increasing permanent deformation, results in dimensional changes that can increase the forces on the part or cause other parts to assume a greater, and perhaps unplanned, share of the structural load. Another material-failure mechanism results from thermal expansion. The thermal growth induces higher forces on certain portions of the structure. A single structural member may be so hot at one point and relatively cool at another adjacent point that it enters a plastic flow regime locally, which ultimately will reduce its strength capabilities, especially after several thermal cycles.

Many of these incipient mechanical failures can be detected by recording at frequent intervals the dimensional stability of parts subjected to thermal loading. Continuing deformation is an early warning sign for future problems. The complexity of the structure and accessibility often preclude such measurements, however. At low temperatures, especially those in cryogenic work, materials may also lose their strength. A few improve in strength. More frequently, materials lose their ductility, and brittle failures occur. Some metal alloys retain their ductility, and these make obvious choices for cryogenic applications.

Material degradation resulting from high- or low-temperature operation is not easily tested for, as was the case for pressure. The best approach for safety is to erect suitable protective barriers between operating personnel and equipment. These barriers should be adequate for the worst anticipated problems. In addition, prevent unscheduled temperature excursion beyond planned operating conditions by temperature measurements at selected points in the system. If the excursions are likely to be too rapid or complex for manual system correction, then use automated devices, such as temperature switches or thermostats, to provide the necessary system control. Some detailed knowledge of the system is required to select the appropriate control devices.

7.6 SAFETY WEARING APPAREL

Technicians should not think that they are immune to accidents because they are out of the mainstream factory environment. Many accidents are preventable simply by wearing appropriate protective apparel, which is almost always available and can invariably be obtained if the need is recognized. The protective items are applicable to all personnel who may be exposed to hazards, if only momentarily.

Eyeglasses (Safety Glasses)

Probably the most common accidents involve foreign objects in the eye. Most can be prevented by wearing eyeglasses provided with splinterproof lenses. They are important for any machinists and for everyone around the machining area because small chips can fly in all directions. If there are a lot of high-velocity chips, such as with grinding, then more complete protection may be provided with goggles or face shields.

Personnel conducting hydrostatic tests should always wear safety glasses. High-pressure leaks sometimes result in high-velocity streams that can be most injurious to the eye.

Safety Shoes and Boots

Falling objects can inflict severe damage to the feet, especially to the toes. Most of these injuries can be eliminated and many reduced in severity by wearing safety shoes or boots that have steel-reinforced toe caps. As a guide, personnel normally moving items weighing 5 lb (2.3 kg) or over that can fall from the height of the average desk or bench should wear safety footwear.

For special hazards, especially electrostatic discharges, there are other types of safety footwear. Where explosive dust might be in the air, sparks from electrostatic discharges are particularly dangerous, and special floors and clothing are required to prevent them.

Hard Hats

Fortunately, hard hats have become commonplace and prevent enormous numbers of head injuries. Personnel walking beneath overhead operations, navigating low structures, or around loose overhead materials must always wear hard hats.

Hearing Protection

Loud noises and high frequencies can damage the ears. Many devices are available that attenuate and filter the noise and yet still permit adequate hearing for speech and warning signals. The devices include earmuffs and reusable or disposable earplugs.

Gloves

Industrial gloves are available to resist abrasion, chemicals, and extremes of temperature and to prevent skin oils and moisture from contaminating cleaned parts. Many nicks and cuts can be eliminated by using proper gloves. Choose ones that fit well and are suited to the particular work. Gloves with fingers or thumbs that are too long may be a detriment rather than an asset in working with equipment where the loose ends may get caught.

Face Masks

It is a wise precaution to eliminate inhaling airborne particles, chemical vapors, and offensive odors, regardless of the currently known or suspected toxicity, especially if working in that environment for long periods. Testing equipment is available to detect traces of all solids and vapors, and if any are suspect a determination can be made. Lightweight, slip-on-type respirators and masks are available for specific occasions, such as working near toxic particles, nontoxic particles, paint vapors, and odors. Remember that masks are not a substitute for good ventilation.

Where particularly dangerous gases are being used (e.g., carbon monoxide or chlorine), masks with a self-contained pure air supply (compressed air) should be readily available in case of a leak.

General Attire

Loose clothing, particularly ties and sleeves, are likely to get caught on rotating equipment. They should be well-secured or avoided when operating lathes, millers, drilling machines, etc. Likewise, long hair can be caught in the same manner, and a hat should be worn that will prevent this. Wedding bands and other rings remain a continuing cause for accidents when moving objects get caught in them. Always wear gloves or remove the rings while working.

7.7 MERCURY

Since mercury is used for many purposes in the laboratory, spillage occasionally occurs. Once spilled, mercury breaks up into small beads that roll in every direction, disappearing into cracks in floors and benches and behind cabinets and equipment. Very soon it is completely out of sight. The large surface area of the dissipated spill leads to easy vaporization. Because the vapor pressure of mercury is low, it remains for a long while, maintaining vapor in the atmosphere. Mercury vapor may be harmful to one's health and also to various manufacturing processes, especially those concerned with the metals aluminum and copper and their alloys. The liquid form itself is not poisonous.

Once mercury has been spilled, thoroughly clean the area. Small mercury beads are difficult to get up. The best way is to use a vacuum cleaner, but the vacuum cleaner should be one with a disposable bag, not a reusable cloth bag. Dispose of the bag when the cleaning is finished instead of waiting until the bag is full. Following cleaning, the area must be well-ventilated for several hours.

It is very difficult to know how well the cleanup was performed. When there is doubt because the spill was large (several hundred milliliters) and very dispersed, test the atmosphere. Specialized services are available in most localities to test the air for mercury contamination. These same services can advise if the level is harmful.

7.8 WELDING, BRAZING, AND SOLDERING VESSELS

Vessels not intended for pressure service and generally constructed from sheet metal sometimes require repair or alteration, and the fabrication techniques used typically include welding, brazing, or soldering. However, there are certain dangers involved in this type of work, and special safety precautions are necessary.

The most dangerous situation occurs when work is performed upon used vessels that once contained liquid hydrocarbons. Even dry tanks have exploded, causing fatalities, when heat applied by the fabrication technique vaporized residues that mixed with air in the vessel to form an explosive mixture. Kerosene tanks that have been empty and vented to the atmosphere for years have exploded under such work circumstances. Do not work on such tanks if it can be avoided. When it must be done, use the following procedure.

First steam clean or clean the vessel with a hot detergent and water solution and thoroughly rinse with hot water. Using a volatile liquid degreaser, flush the tank and again clean with detergent and rinse with hot water. Purge with a stream of oilfree air (most shop air supplies contain oil) for at least 12 h. Whenever possible, the purging should leave no dead spaces. More than one purge port and alternating the vent exits help. Figure 7.6 illustrates how and where dead spaces can occur and be avoided.

Before performing any welding, brazing, or soldering, purge the tank with an inert gas to remove all air. Nitrogen is best and most readily available. All dead-air spaces in the vessel must be purged. If this cannot be done by a suitable selection of purge-gas inlet and vent locations,

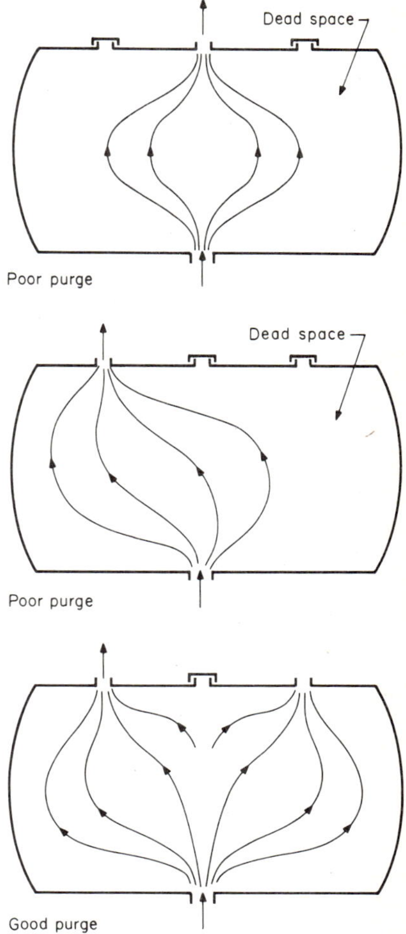

FIG. 7.6 Selecting inlets and outlets on vessel to maximize effectiveness of purge.

then it must be done by using a wand. With the nitrogen flowing steadily through the best purge locations of inlet and vent, insert a wand or tube, bent if necessary, having nitrogen flowing through it through the vent and manipulate it to purge the dead spaces of air. Figure 7.7 shows how this may be accomplished.

Without interrupting the nitrogen purge-gas flow, lest air reenter the tank and the preceding air-purge procedure be negated, perform the fabrication work. Maintain only a small pressure, 2 or 3 in H_2O (500 to 700 Pa), in the vessel during purge. Continue the nitrogen purge until the work is finished and the vessel has cooled.

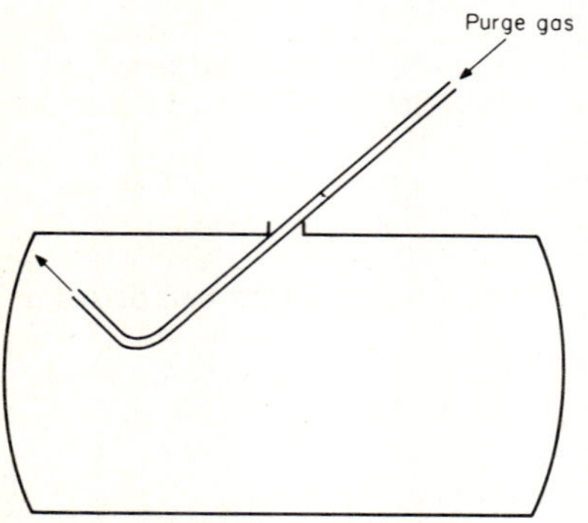

FIG. 7.7 Using a wand to purge vessel with limited-access ports.

Treat vessels whose histories are unknown or uncertain as though they had contained hydrocarbons.

Another dangerous situation that can occur is to have the vessel completely sealed (i.e., no vent) when it is heated. As the air or gas inside becomes heated, which is much slower than for the metal, the pressure will rise if completely contained and may reach a level sufficient to burst the vessel.

7.9 OPERATING PROCEDURES

Safety is often improved by using formalized written operating procedures. The primary intent of these written procedures is usually to assure repeatability of methods and uniformity of standards regardless of the personnel conducting the work. These documents are an ideal place to describe the methods that will provide the safest operation and can include specific cautionary safety notes as the work dictates. It is usual and preferred to have the technician participate in the preparation of operating procedures to contribute that special knowledge gained from the construction, testing, and checkout of the equipment. The completed document should be reviewed and approved by higher authority such as the laboratory supervisor or engineer. The collective talents of these individuals should benefit the final document.

If the system to be operated is especially complex or hazardous, it is best to include in the operating procedures a checklist of events arranged in the order of sequence, starting with all the preparatory requirements through start-up, run, shutdown, and final security. The checklist serves as a reminder of all the operations that must be performed. It is a significant help in achieving both correct operations and meeting safety requirements.

As a further step toward safety in operations, events can be arranged so that the operation cannot proceed unless others have first been correctly performed. For example, if a system becomes potentially hazardous as a result of pressurization, then the doors or other accesses, which for safety reasons are intended to be closed to exclude unauthorized personnel, may be fitted with a switch that controls the pressurization means (i.e., solenoid valve). If the operator inadvertently forgets to close the doors, the system cannot be pressurized until the operator goes back and completes the safety steps.

7.10 OBSERVATION WINDOWS

Reference has been made to the use of protective barriers between operating personnel and hazardous equipment. Typically these barriers are masonry walls or, if the hazard warrants, reinforced concrete. In either case, it is often necessary to observe equipment while the test is in progress to study failure modes in destructive tests and to detect visually any untoward event. While this may be done by closed-circuit television, it is more usual to make these observations directly through glass windows. Obviously these windows must provide the observer with protection at least equal to the dividing wall or barrier for consistent safety.

Observation windows are kept small in size to minimize the possibility of damage from flying debris. Also the smaller window has much greater strength. Typically these windows are constructed from laminated plate glass or Plexiglas. The construction is simple, as shown in Fig. 7.8*a*. A window that comprises five sheets of ordinary window glass is enormously strong.

For severe hazards, direct viewing is undesirable. Instead the window is protected from direct impact of flying objects with an angled steel plate, one side of

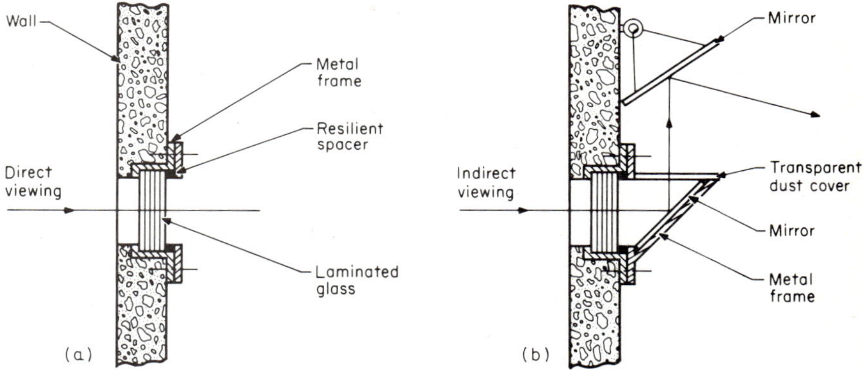

FIG. 7.8 Observation safety windows.

which is a mirror (Fig. 7.8*b*). Another mirror is suspended above the first and adjusted to provide the requisite view of equipment. Usually, the fixed mirror is fitted with a Plexiglas dust cover to eliminate mirror cleaning. If the adjustable mirror is curved or is a combination of several flat mirrors, then wide-angle viewing covering a large area can be achieved with only a small wall window.

7.11 LEAK DETECTION

Leak detection, as it pertains to safety, has two aspects. The first relates to the structural integrity or performance of a component, joint, seal, etc., where leakage is an indication of inadequacy in design or manufacture or of actual deterioration. The second is the detection of small amounts of toxic vapors or gases that may be present in the atmosphere for a variety of reasons, such as accidental discharges, system leaks, mechanical failures, or accumulation from many small leaks that are not individually detectable. Obviously there is some connection between the two aspects inasmuch as system leaks will be reduced through first checking component and system capability. But there is always a limit to the degree to which a single leak can be detected.

Some fluids leak more easily than others. For example, a system which is leak-tight with water to a certain pressure may leak with air at the same pressure. Similarly, some gases leak more easily than others. Recognition of these facts has led to various methods for leak-detection testing and specifying leakage rates.

Individual leaks are most commonly detected by a hydrostatic test. Leak rates are typically specified by observing for drops over a specified period of time. Leak rates may be as large as a few drops or milliliters per hour to zero drops per hour or per day. Zero drops over a period of time is most typical, but if the time becomes too long, not only is the observation difficult to make but the leak may be so slow that the water evaporates as it leaks, leaving no visible indication. Nevertheless, such a standard is quite strict and adequate for many applications.

A more stringent leak test is to pressurize with a gas, usually air or helium, and use a soap solution to detect the escaping gas when it forms soap bubbles. The only practical standard is to have no soap bubbles. Unfortunately soap solutions only remain effective for short periods of time so that this method is limited in its sensitivity as a leak-detection method.

Fortunately more sensitive and absolute methods for leak detection are available. The commonest one is helium leak detection using a mass spectrometer. Section 5.13 covers this instrument. It measures a wide range of specific leak values easily and quickly. The maximum sensitivity is 1×10^{-8} mL helium/s,

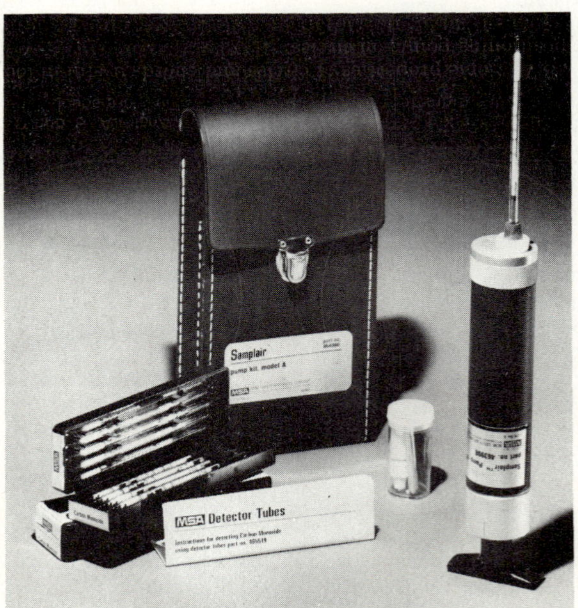

FIG. 7.9 Leak-detection kit. *(Mine Safety Appliances Co.)*

which is used especially for critical applications such as radioactive and toxic fluids.

Having obtained the necessary leak integrity for the joints and components of a system, the problem of atmospheric contamination must be considered. Some particularly useful and simple leak detectors have been developed for this purpose. A typical detector with accessories and carrying case is shown in Fig. 7.9. It is capable of sampling air to detect a broad range of toxic gases, vapors, and mists. This detector comprises a variable-volume, piston-type, hand-held pump to which is affixed a glass detector tube (Fig. 7.10). Each detector tube is filled with a chemical solution whose color changes when it reacts with specific airborne contaminants as listed on the packing box.

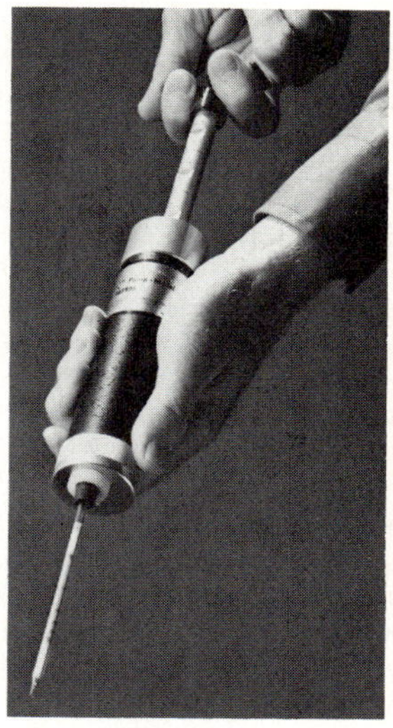

FIG. 7.10 Leak-detector hand-held pump and detector tube. *(Mine Safety Appliances Co.)*

To use, break off the glass tips at each end of the detector tube and push one end into the pump seal. By withdrawing the pump handle to the desired position, draw a specific volume (25, 50, 75, or 100 mL) of air and suspected contaminants into the pump through the detector tube. Depending upon the concentration of the specific contaminant and the volume of sample drawn through the tube, the chemical in the tube changes color for a certain length only. Using the calibrations on the tube, measure the concentration of contaminant in the sample. Use a new tube for each sample. Tubes have been developed and calibrated for a large number of contaminants including the dangerous and odorless gas carbon monoxide.

A difficulty can arise if several contaminants are present simultaneously. However, this is seldom the case. One ordinarily knows what particular chemical is in use that could leak and contaminate the atmosphere, and its presence should be checked often as part of a normal operating and safety procedure.

7.12 ELECTRICAL SAFETY

The following rules for electrical equipment should be adhered to for safety and, in some cases, may even prevent damage to equipment.

1. Verify that the power source (voltage, current, phases) is suitable for the equipment to be operated.
2. Be sure the current-carrying capacity (wire size) is adequate for the intended use when using an extension cord.
3. Make sure the ground connection (pigtail) is properly grounded when using a three-pin grounded adapter in a two-pin outlet.

4. Never place an extension cord where someone might trip over it.
5. Never permit extension cords to be near heaters.
6. Do not reset or replace a circuit breaker when it trips or a fuse when it blows until the cause for the overload has been investigated and corrected.
7. Always use a ground fault interrupter (GFI) adapter at the outlet when using portable electric tools while standing in damp or wet places or working on water lines.
8. Always disconnect equipment before exposing any wiring for examination or adjustment.
9. Be sure to specify explosionproof wiring, switches, controllers, etc., for all areas that may have explosive gases or vapors released to the atmosphere.
10. Call an electrician to connect equipment not provided with a plug or where no suitable receptacle for the plug exists.
11. Read carefully the manufacturer's literature regarding use of electric equipment.

7.13 HOISTS

Before using a hoist, be sure that its capacity is high enough to lift the intended load. Inspect chains, ropes, wires, and slings regularly to determine condition. Discard lines that are worn, frayed, corroded, or in any way suspect. Never assume that lines in poor condition will suffice for the lighter loads.

Whenever possible, stand well clear of the object being lifted. Raise and lower slowly and smoothly. When finished with a hoist, do not leave hooks hanging where someone might walk into them.

chapter 8

Useful Formulas

Maurice J. Webb

8.1 AREAS AND VOLUMES

The formulas for determining areas of many standard or basic shapes are shown in Fig. 8.1 in terms of measured linear dimensions. As discussed below, areas are usually the starting points for finding volumes and weights of objects. Often the required area is not one of the basic shapes but can usually be divided into basic shapes. Two examples of irregular shapes that comprise several basic shapes are shown in Fig. 8.2. Complex shapes such as those are often encountered in sheet-metal work.

Formulas for determining volumes for basic shapes are shown in Fig. 8.3. Of particular interest is the oblate spheroid because it is an ellipse rotated about its minor axis and is a shape commonly used for the ends of vessels (such as boilers and tanks) whose capacity or volume is often needed. Figure 8.4 shows such a vessel and the formula for its volume. For many applications the volume is determined by first finding the area and then multiplying by the thickness or length of the item. Typical examples are shown in Fig. 8.5.

Just as areas of actual shapes often have to be divided into a number of basic shapes for analysis, so, too, do volumes of actual objects. The example of Fig. 8.4 is typical of the technique, and the volume of that vessel comprises two semioblate spheroids (one at each end) and a cylinder.

8.2. WEIGHTS

The weight of an item is simply its volume multiplied by the density of the material. An important thing to remember is that the units must be consistent. Typically densities are expressed in pounds per cubic inch (lb/in^3), grams per milliliter (g/mL), or kilograms per cubic meter (kg/m^3). The volume value must be in appropriate units (cubic inches, milliliters, or cubic meters). The tables of Chap. 2 include values of density for many materials; others are in Tables 11.7 to 11.9. The formula for determining the density of gases is given in Sec. 8.5.

8.3 GEOMETRY

Triangles and circles are the most common actual or reference shapes required in checking the dimensional accuracy of components or in setting up machines to fabricate parts. The extensive use of numerical readout tables on machines and of small hand-held calculators appreciably simplifies the setup procedure and mathematical computations. The need still exists to have ready reference to a few basic geometric rules to determine a specific dimension not included on a drawing because it was not necessary to describe the part but is necessary to check or make the part.

Pythagoras' theorem relates the lengths of the sides of a triangle where one of the angles is a right angle (90°). Figure 8.6*a* gives the formula and shows the application. In practice not all triangles are right-angle triangles, and a consequence of Pythagoras' theorem is the simple extension shown in Fig. 8.6*b* that applies only for acute-angle triangles. (The sum of all angles in a triangle is 180° so that there must always be two acute angles, and the formula can be appropriately applied as required.) Thus these formulas enable the length of the third side to be determined from the lengths of the other two.

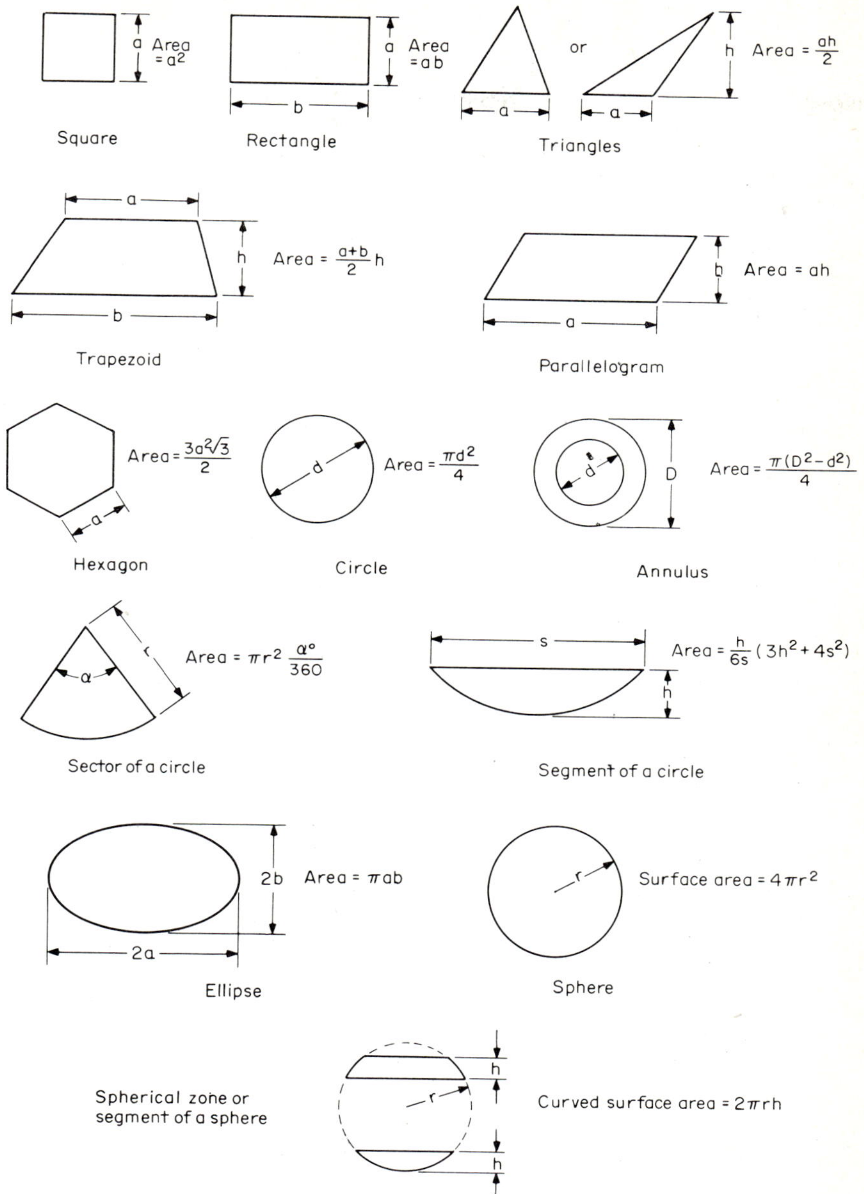

FIG. 8.1 Formulas for the areas of basic shapes.

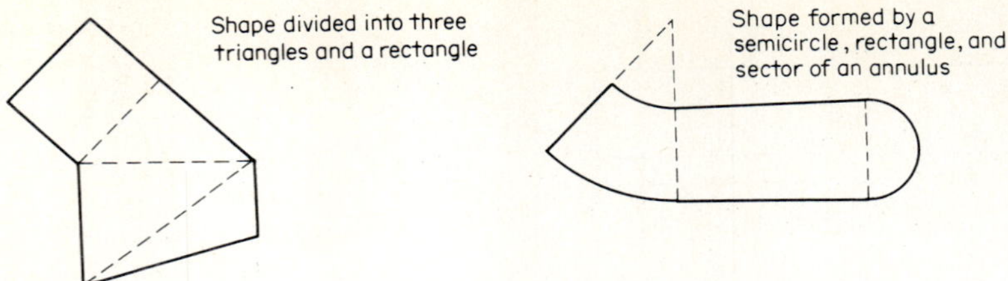

FIG. 8.2 Dividing complex shapes into basic shapes to determine area.

For circles, there are some simple relationships relating to chords that may help, for example, in determining and positioning bolt-circle diameters. In Fig. 8.7*a* the center of a circle lies on the perpendicular bisector of a chord so that given a portion of a circle the diameter and its location may be determined. The relationship between the lengths of intersecting chords, shown in Fig. 8.7*b*, may be used to determine the diameter of the circle or the rectilinear positions of locations around a circle (e.g., bolt-circle centers). The angles subtended by any chord are the same for all points on the circle on each side of the chord (Fig. 8.7*c*). For the special case when the chord is also a diameter, the angles are right angles.

8.4 TRIGONOMETRY

When measurements for two sides of a triangle are available then the methods above may be used. Often, though, only one side and an angle are given, and it is necessary to determine the length of another side. Also it may be preferred to set up an item from a known angle but only linear dimensions are known, not the required angle. Three simple trigonometric functions can be used, depending upon the available data, as shown in Fig. 8.8. Values for these functions are given in Tables 10.2 to 10.4.

8.5 THE DENSITY AND WEIGHT OF GASES

The density of solids and of most liquids varies little with temperature and pressure, but the density of gas varies greatly with these same factors. The density of gas, ρ, is found from the formula

$$\rho = \frac{P}{RT}$$

where P = pressure
R = constant for the particular gas
T = temperature (absolute)

For each specific gas the value of R is different but constant for that gas. Values of R for common gases are listed in Table 11.10. The units in that table are foot pounds per pound per degree Rankine [ft·lb/(lb·°R)]. When the pressure is expressed in pounds per square foot (lb/ft^2) and the temperature in degrees Rankine (°R = °F + 460), then the density will be in pounds per cubic foot (lb/ft^3). Conversions to other units are simply made, using the data from tables in Chap. 10.

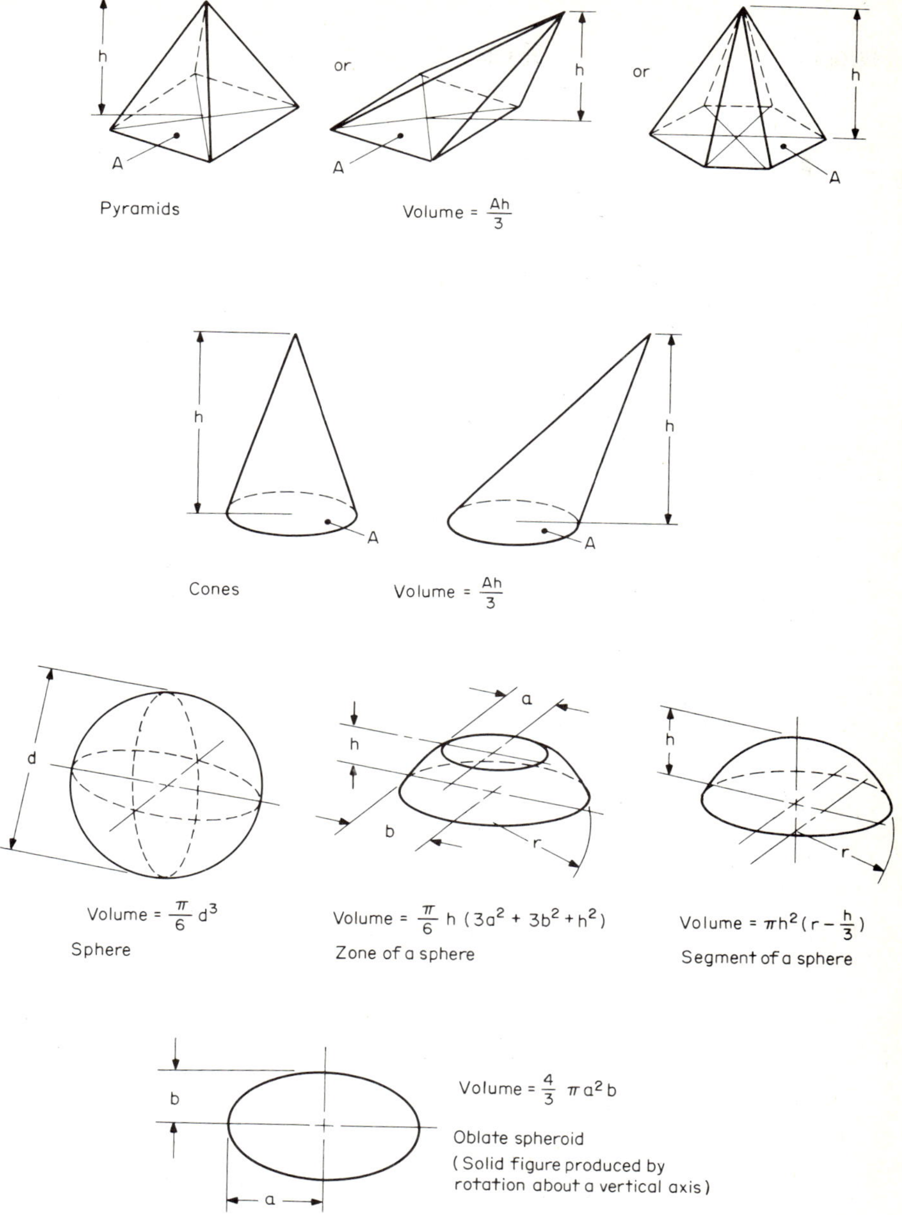

FIG. 8.3 Formulas for the volumes of basic solid shapes.

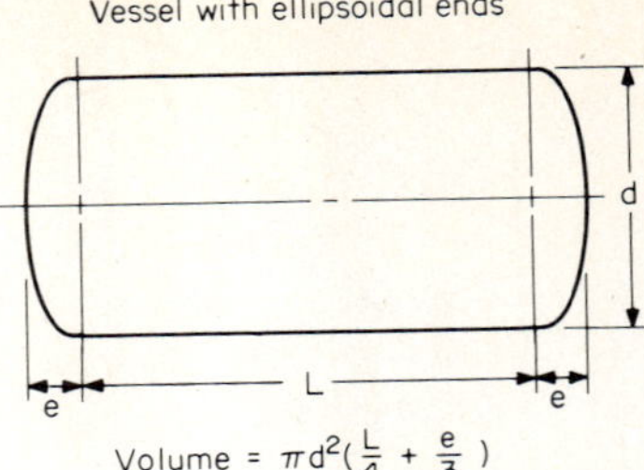

Fig. 8.4 Volume of a vessel obtained by combining volumes of basic solids.

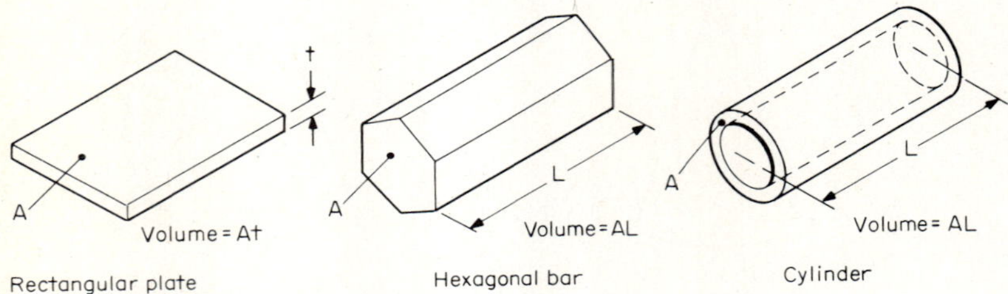

FIG. 8.5 Volumes of plates and bars.

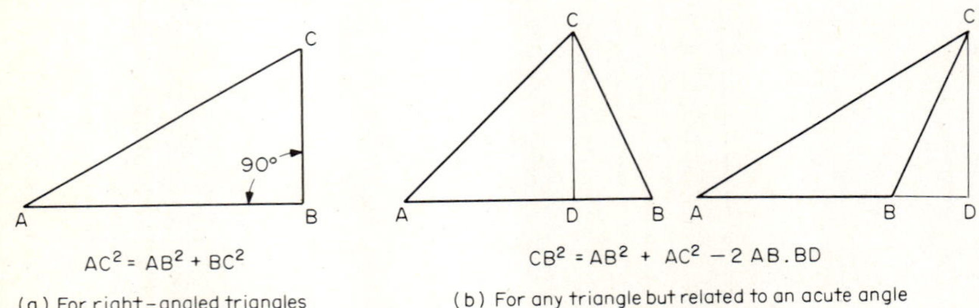

FIG. 8.6 The relationships between lengths of the sides of triangles.

For example, if the density of air is required at a pressure of 1000 psi and a temperature of 75°F, then first convert the pressure to lb/ft² and the temperature to °R.

$$\text{Pressure of 1000 psi} = 1000 \times 144 = 144{,}000 \text{ lb/ft}^2$$
$$\text{Temperature of 75°F} = 460 + 75 = 535\text{°R}$$

The value of R for air from Table 11.10 = 53.3; thus

$$\text{Density } \rho = \frac{144{,}000}{535 \times 53.3}$$
$$= 5.05 \text{ lb/ft}^3, \text{ or } 0.0029 \text{ lb/in}^3$$

The weight or mass of gas is $m = PV/RT$, where V = volume, ft³, to be consistent with the units above.

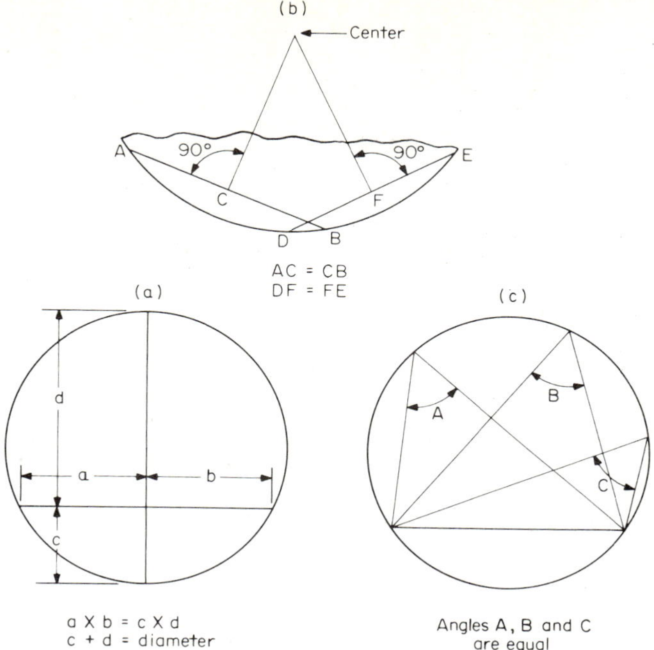

FIG. 8.7 Some properties of circles and chords useful in locating and positioning points on circles.

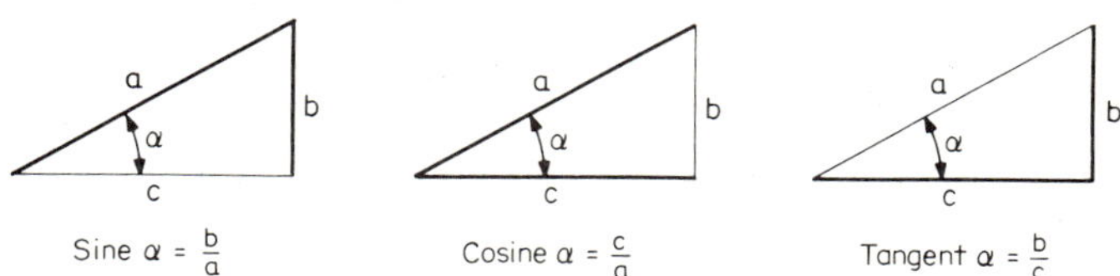

FIG. 8.8 The formulas and meaning for three basic trigonometric functions.

Values of R not given in Table 11.10 may be calculated from the molecular weight of the gas by the relationship

$$R = \frac{1545.4}{\text{mol. wt.}} \text{ ft}\cdot\text{lb}/(\text{lb}\cdot{}^\circ\text{R})$$

This relationship is especially useful for dealing with gas mixtures when the average molecular weight for the mixture is used.

8.6 RELATIONSHIP BETWEEN PRESSURE, TEMPERATURE, AND VOLUME OF GAS

For any fixed mass of gas the relationship that applies is

$$\frac{PV}{T} = \text{const.}$$

or

$$\frac{P_1 V_1}{T_1} = \frac{P_2 V_2}{T_2}$$

where subscript 1 refers to initial conditions and subscript 2 to the final conditions. For example, a vessel containing gas at 200 psi at room temperature (70°F) is heated to 500°F, and it is required to know the pressure then. In this case the volume $V_1 = V_2$ since it has not changed, and by rearranging the formula above,

$$\begin{aligned} P_2 &= P_1 \frac{T_2}{T_1} \\ &= 200 \times \frac{500 + 460}{70 + 460} \\ &= 362 \text{ psi} \end{aligned}$$

8.7 FLOW RATE

The relationship between flow rate, velocity, and density is an important one. The volume flow rate Q (e.g., cubic feet per minute or gallons per minute) depends upon the velocity of the fluid and the area of the duct or channel through which it is flowing and is determined by

$$Q = Av$$

where A = area
v = velocity

The most frequently used consistent units are to express the area in square feet (or square meters) and the velocity in feet per second (or meters per second). For example, if the water velocity in a pipeline is 20 ft/s (6.1 m/s) and the pipe has an internal diameter of 1.9 in (48.3 mm), then quantity of water flowing is

$$\frac{\pi(1.9)^2}{4 \times 144} \text{ ft}^2 \times 20 \text{ ft/s} = 0.39 \text{ ft}^3\text{/s, or } 23.63 \text{ ft}^3\text{/min } (0.0112 \text{ m}^3\text{/s})$$

This formula may also be used for gases if the velocities are low (up to about 100 ft/s, or 30 m/s, for air); beyond that it becomes increasingly inaccurate.

The mass (or weight) flow is provided by the formula

$$\dot{m} = \rho A v$$

where $\dot{m}$ = mass flow rate
ρ = density of the fluid

8.8 PRESSURE DROP IN PIPING AND COMPONENTS

The pressure drop in piping and fluid system components is required to determine, for example, the capabilities of a pump to meet pressure and flow requirements. The pressure drop through piping and components increases with the square of the velocity (or nearly so). Therefore if the flow rate is doubled the pressure drop increases by four times; if the flow is tripled the pressure drop increases by nine times. Considerations such as these may lead one to consider the use of larger pipe sizes and larger components rather than to use a more powerful pump and drive motor. The final analysis may be made on an economical basis after examining piping runs of several sizes to optimize the system.

Straight runs of pipe, bends, and changes in pipe size all contribute to pressure loss that must be overcome by a pump. For straight runs of pipe including long-radius curves, the pressure loss is often predicted from the Williams and Hazen formula. While this formula is an approximation of the more accurate calculation, its usefulness stems from its relative simplicity. It may, however, underestimate or overestimate the actual pressure drop and should only be used as a guide. Where expensive equipment is to be purchased based upon pressure requirements, the technician should refer the problem to an engineer for an accurate assessment. The Williams and Hazen formula is

$$\Delta P = \frac{4.52Q^{1.85}}{C^{1.85}d^{4.87}}$$

where ΔP = pressure drop, lb/in² per foot length of pipe
Q = flow rate through the pipe, gal/min
d = internal diameter of pipe, in
C = coefficient for pipe surface

Values of C for new steel pipe are typically 120; a value of 100 is applicable for pipe that has been in service for some years and subject to roughening from corrosion and scale. For tubing, where the surface is typically smoother, the value of C is from 140 to 150.

An advantage of the Williams and Hazen formula is that it illustrates dramatically the strong dependency of pressure drop on a pipe's internal diameter. Other things being equal, by doubling the pipe size the pressure drop declines by a factor of almost 30. Therefore small changes in diameter have large effects on pressure drop.

The total pressure drop for the system is the sum of pressure drops in all the individual components and piping. The pressure drop for valves is given in Sec. 4.5 and is based upon the flow rate and valve flow coefficient. For pipe fittings (elbows, tees, etc.) it is usual to assign to them a length of straight pipe that would provide the same pressure loss. The lengths for all the fittings are added together (for the same-diameter pipe) along with the lengths of straight pipe runs. This total length is used in the Williams and Hazen formula above. Table 8.1 gives the equivalent lengths for a number of fitting configurations expressed

TABLE 8.1 Equivalent Pipe Lengths for Fittings

Fitting	L/D
90° elbow	30
45° elbow	15
90° street elbow	50
45° street elbow	25
Tee:	
Flow straight through	20
Flow turning a 90° angle	60

as a length-to-diameter (L/D) ratio. Simply multiply the ratio by the nominal diameter of the fitting (for example, 4-in ell) to obtain the equivalent pipe length, remembering to express this length in appropriate units (feet) for use with the equation above.

8.9 PUMP AND MOTOR SIZING

Use the procedure outlined in Sec. 8.8 to estimate the total system pressure drop for piping and components. However, to size the pump, there is one more pressure drop to be considered. This drop is the one at the end of the flow line and may be the greatest of all. For example, the final discharge may be from an open pipe or special atomizing nozzle. For a special nozzle, the manufacturer will provide flow data and the pressure drop will be available for the required discharge (Q gal/min). For a smoothly contoured orifice or discharge from a straight pipe, the formula is

$$\Delta P = \frac{Q^2 \rho}{2A^2 g}$$

where ρ = fluid density, lb/ft^3
Q = flow, ft^3/s
A = area, ft^2
g = gravitational constant = 32.2 ft/s^2

Add this pressure-drop requirement to all the other drops to obtain the total drop (ΔP_{total}). Then select the pump that provides the requisite flow (Q) and delivers the requisite pressure (ΔP_{total} + atmospheric pressure of 14.7 psi). Calculate the size or horsepower of the motor to drive the pump from

$$\text{Horsepower} = \frac{\Delta P_{\text{total}} Q}{1715\eta}$$

where η = pump efficiency
Q = flow, gal/min
ΔP_{total} = total drop, lb/in^2

8.10 PRESSURE MEASUREMENT

When measuring the pressure of a flowing fluid there are two pressures of importance. One is the static pressure P, the pressure that a gage would sense if it were able to move along in the fluid at the same speed as the fluid is traveling. The other is the total, stagnation, or impact pressure P_{total}, the pressure measured when the flow is stagnated or stopped. That is, the dynamic pressure or ram effect of the moving fluid in bringing it to rest is added to its static pressure. Either pressure may be measured as shown in Fig. 8.9, and some care is required not to measure some value between the two, as may easily occur. The probe used to measure total pressure is called a *pitot tube*. The relationship between the two pressures is given by the formula

$$P_{\text{total}} = P + \frac{\rho v^2}{2g}$$

where ρ = fluid density
v = fluid velocity
g = gravitational constant

It is the total pressure that one is ordinarily interested in because the power required to move or pump a fluid depends upon the total energy required of the system. Unfortunately, it is easier to measure static pressure. However, all the pressure-drop formulas above are losses of total pressure.

The above formula is used for all liquids and for gases where the velocity is relatively low (less than 100 ft/s, or 30 m/s). If large pressure losses and high speeds are present with gas flows, then the gas density ρ also changes and adds considerable complication to the analyses and is beyond the scope of this text.

The pressure-measuring arrangement shown in Fig. 8.9 is often used to determine the velocity of flow. The above formula is simply rearranged to

$$v = \sqrt{\frac{2g(P_{\text{total}} - P)}{\rho}}$$

Air velocity or the speed of a vehicle traveling in air may be determined from this formula by measuring the total or impact pressure with a pitot tube. This has been done for sea-level (atmospheric pressure 14.7 psia, or 101.3 kPa) conditions, and the pressures for various speeds are shown in Table 11.11.

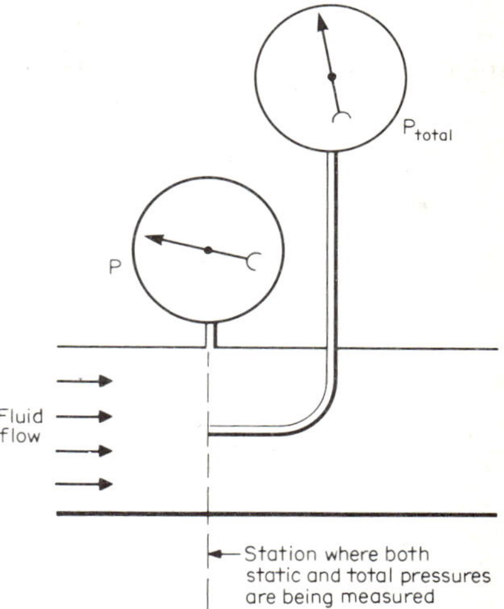

FIG. 8.9 Measuring pressures for use in the fluid-velocity formula.

8.11 THERMAL EXPANSION

Most materials expand when heated and contract when cooled. This fact is used to advantage in press fitting components together, and often it is a disadvantage because the thermal growth, if restrained, causes increase in component stresses, misalignments, and incorrect measurements. Fortunately this thermal movement is easily calculated and can be taken into consideration. Thermal expansion or contraction is calculated from

$$\epsilon = L\alpha\,\Delta T$$

where ϵ = expansion or contraction
L = length of part
α = coefficient of linear expansion
ΔT = change in temperature

Values of the linear expansion coefficient α are available for most materials (Table 11.13). The units are expressed as the change in length per unit length per degree temperature change. In this way the expansion is expressed in the same units as the length. For example, to find the increase in length of a steel tube that is 120 in long when heated from room temperature (70°F) to 500°F, then from Table 11.13, α is 7.8×10^{-6} and

$$\begin{aligned}\epsilon &= 120 \times 7.8 \times 10^{-6}(500 - 70)\\ &= 0.40 \text{ in } (10.16 \text{ mm})\end{aligned}$$

Similarly, the changes in area or volume can be determined from the initial area or volume of an item. The values of coefficients to be used, however, are, respectively, two and three times that for the linear expansion coefficient of the same material. For example, for a steel vessel of 12-ft^3 capacity heated from 40 to 350°F, the increase in volume is

$$\begin{aligned}\epsilon &= 3 \times 7.8 \times 10^{-6} \times 12(350 - 40)\\ &= 0.087 \text{ ft}^3\text{, or } 150 \text{ in}^3 \ (2.46 \times 10^{-3} \text{ m}^3)\end{aligned}$$

8.12 STRESS AND STRAIN

Materials subjected to stress or load expand or contract. When the stress is removed, they return to their original size if the stress has not exceeded a certain value called the *elastic limit*. It is usual to keep stresses low enough so as not to exceed the elastic limit, otherwise permanent deformation occurs. The relationship between the applied stress and the resulting expansion or contraction, called *strain*, is easily calculated. First, however, the values of stress and strain require exploration.

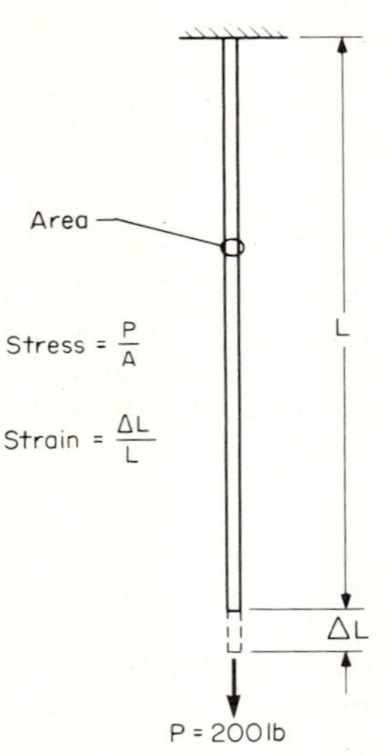

FIG. 8.10 Stress and strain in a rod.

Stress is the load per unit area that a part or component is subjected to. It is found from

$$\text{Stress } f = \frac{P}{A}$$

where P = force
A = area normal to the force

For example, a rod of ⅒-in diameter supports a load of 200 lb (Fig. 8.10). Now the stress everywhere along the rod is 200 per area, or

$$\frac{200}{(\pi/4)(0.1)^2} \text{ psi} = 25{,}500 \text{ psi (176 MPa)}$$

Strain is the increase (or decrease) in length as compared with the original length. Thus strain S is

$$S = \frac{\Delta l}{l}$$

where Δl = change in dimension
l = original dimension

Stress and strain are related through the modulus of elasticity E by

$$E = \frac{f}{S}$$

Values of the modulus of elasticity for many materials are given in several tables in Chap. 2.

If in the example above the rod was made of steel, then E is 30×10^6 psi. Therefore the strain in the rod is

$$S = \frac{f}{E} = \frac{25{,}500}{30 \times 10^6} = 0.85 \times 10^{-3}$$

If the original length of the rod was 120 in, then the increase in length Δl is

$$0.85 \times 10^{-3} \times 120 = 0.10 \text{ in (2.54 mm)}$$

While this extension is small, high-strength materials which permit stresses in the range 100,000 to 150,000 psi (700 to 1000 MPa) also permit larger strains and extensions to the extent that they cannot often be ignored.

It is normal to design with a safety factor; that is, the full strength (stress) capability of the material is seldom used. The reasons for this are that material properties may vary (although manufacturers strive to maintain consistency) and the true loads to be applied may not be accurately known, thus subjecting the material to various unintentional overloads. To protect against failures, safety factors are selected based upon the anticipated variations. The safety factor (S.F.) is equal to

$$\text{S.F.} = \frac{\text{material stress capability}}{\text{actual or design stress}}$$

Safety factors are based upon two material stresses: the material yield stress (the maximum stress within the elastic range) and the ultimate (failure) stress. Typical values of safety factors based upon yield are 1.2 to 2.0, and factors based on ultimate stress are 3 to 5.

8.13 FRICTION

When one body slides over another, the force at the contacting points or surfaces that opposes the motion is called *friction.* Friction depends upon the force between the surfaces and the nature of the materials in contact. The relationship of the force between the surfaces (W in Fig. 8.11) and the force F to slide one over the other is

$$F = \mu W$$

where μ is the coefficient of friction. The value of μ cannot exceed unity and for well-lubricated surfaces is typically 0.1.

Actually there are two values of μ. One is called the *coefficient for static friction* and is the value just to start to move one surface over the other. The other is the value when the surfaces are moving, called the *coefficient of sliding friction.* The former is always greater than the latter; that is, it takes a little more force to start movement than it does to maintain it. Values of these coefficients are given in Table 11.12.

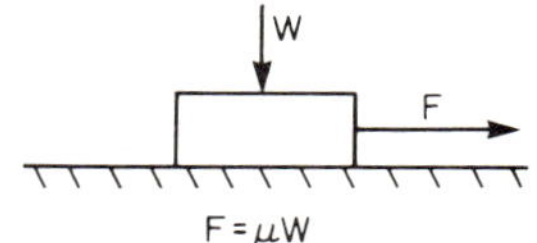

FIG. 8.11 Frictional force between two surfaces.

For the most part frictional force is independent of the area of contact. Of course, if the force between the surfaces is high or if one is softer than the other, then the surfaces may be deformed and appreciably greater effort is required to slide one over the other.

8.14 THERMAL CONDUCTIVITY AND INSULATION

Heat losses from reactor vessels, environmental chambers, pipelines, and the like often need to be quantified or reduced. If the loss is allowed, then it may add too much heat to a room or building or require too great an energy input to maintain the desired temperatures inside vessels and pipes. If fluid temperatures drop too much while passing along pipelines, then insulation must be added to minimize it. These kinds of problems may be readily approached with reasonable accuracy by the use of simple formulas, some temperature measurements, and the data in Table 11.14 for the thermal conductivities of various materials.

The heat conducted through a wall may be calculated from

$$q = kA\frac{\Delta t}{x}$$

where q = heat flow through wall, Btu h
k = thermal conductivity for wall material, Btu/(h · ft·°F)
Δt = temperature drop through wall, °F
x = thickness of wall, ft
A = surface area of wall, ft^2

Figure 8.12*a* shows how the temperature through the wall varies for Δt and x.

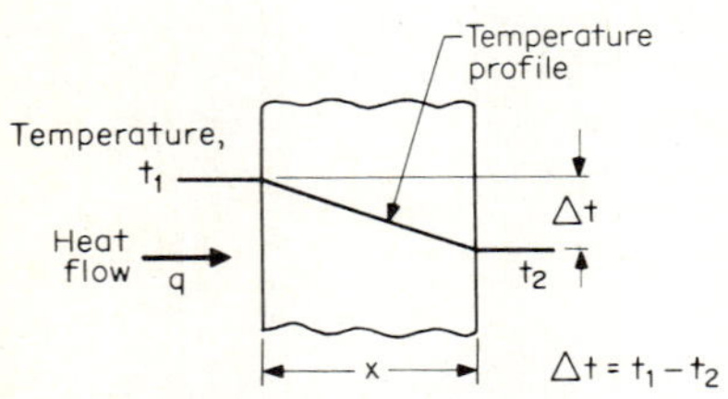

(a) Heat conduction through a single wall

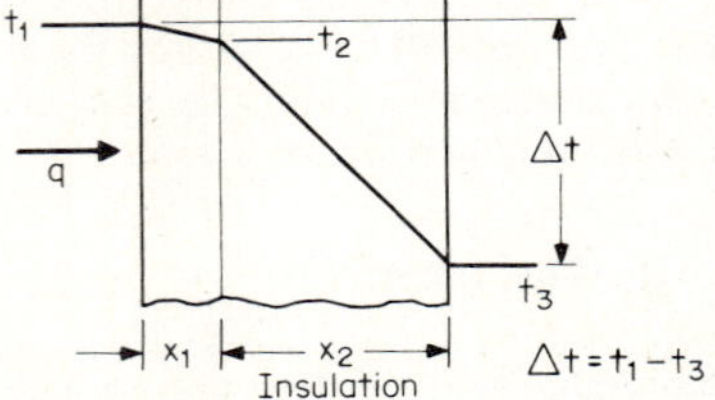

(b) Heat conduction through a composite wall

Fig. 8.12 Thermal conductivity and insulation for a uniform wall.

Often the situation of interest involves a composite wall of two materials. For example, a steel vessel with insulation material on the outside is typical. In this situation the heat transmitted is

$$q = \frac{A\,\Delta t}{x_1/k_1 + x_2/k_2}$$

where subscript 1 applies to one material and subscript 2 to the other. Figure 8.12*b* shows how the temperature would probably vary through the two walls. One is thin and of high thermal conductivity (e.g., steel), while the other is thick and of low thermal conductivity (e.g., insulation material). The effect of reducing the surface temperature and heat loss in this way becomes very apparent.

As an example, the heat loss per unit area (q/A) from a steel reactor vessel whose reactants are at 300°F and whose outer surface is measured to be 250°F is

$$\frac{q}{A} = \frac{(300 - 250)26}{\frac{1}{2} \times \frac{1}{12}}$$
$$= 31{,}200 \text{ Btu/(ft}^2\cdot\text{h), or } 98 \text{ kW/m}^2$$

The thermal conductivity of steel is 26 Btu/(h·ft·°F). If the same vessel is now insulated with 4-in-thick insulation with a thermal conductivity of 0.03 Btu/(hr · ft · °F) and the surface temperature is now measured to be 90°F, then the heat loss is

$$\frac{q}{A} = \frac{300 - 90}{1/(2 \times 12 \times 26) + 4/(12 \times 0.03)}$$
$$= 19 \text{ Btu/(ft}^2\cdot\text{h), or } 60 \text{ W/m}^2$$

This example illustrates the capability and importance of insulation for surfaces that are flat or nearly so. It also shows that the steel wall is insignificant in reducing heat loss compared with the insulation and can, in fact, be ignored for estimating purposes. When dealing with pipelines the steel or metal pipe can

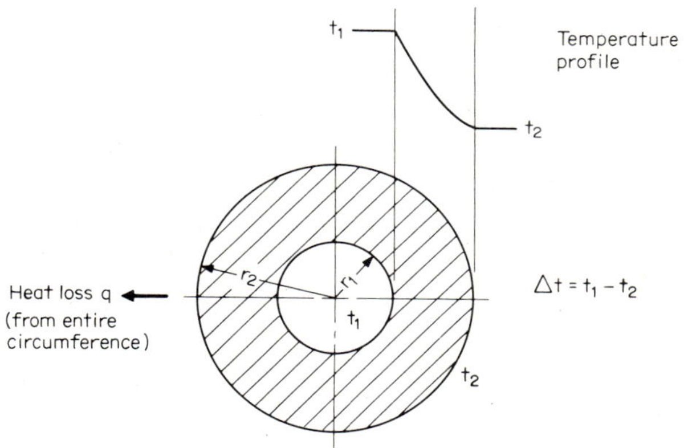

FIG. 8.13 Heat conduction through circular insulation.

also be ignored. However, the equation for determining the heat loss must be modified because the area through which the heat flows increases as the diameter of insulation increases. Figure 8.13 illustrates this. The formula to be used is

$$q = \frac{2\pi kL\,\Delta t}{\ln(r_2/r_1)}$$

where k, Δt = same as above
L = length of pipe, ft
r_1, r_2 = inner and outer radii of insulation, respectively
ln = natural logarithm (see Table 10.1)

8.15 TEMPERATURE CONVERSIONS

Absolute temperatures must be used for certain scientific disciplines (e.g., thermodynamics) because the zeros on these scales represent a base or absolute condition from which certain phenomena may be regarded to start. Two common absolute scales are in use: the Rankine scale (°R), which corresponds to the same basic intervals as the Fahrenheit scale, and the Kelvin scale (K), which corresponds to the Celsius (°C) scale. To convert to the absolute scale the formulas are

$$°R = °F + 459.67$$
$$K = °C + 273.15$$

These numbers are typically approximated as 460 and 273, respectively, with sufficient accuracy for most work.

To convert from Celsius to Fahrenheit, and vice versa, then

$$°C = \frac{°F - 32}{1.8}$$
$$°F = 1.8°C + 32$$

Conversion data for °F to °C are given in Table 10.17.

8.16 INTERPOLATIONS

The tables of Chap. 10 will provide the requisite degree of accuracy for most practical situations the technician must deal with. However, when using these or other tables where improved precision is required beyond that printed in the table, then one may interpolate. For example, if the area of circle whose diameter is 3.68 in is required, it will be adequate for most purposes to use the value for a 3.7-in diameter (10.752 from Table 10.7). For greater accuracy linear interpolation may be used.

The process of linear interpolation is to take two known (table) values that bracket the number of interest. In this case they are 3.6 and 3.7. The difference between the corresponding functional values for those two numbers selected (in this case the difference between the areas given for 3.6 and 3.7, which is 0.573) is proportioned according to the interval required [that is, (0.08/0.10)(0.573) = 0.458]. This is then added to the functional value corresponding to the first selected interval (that is, 0.458 + 10.179 = 10.637).

In this example the true area corresponding to a diameter of 3.68 in is 10.636 in^2 so that there is a small error. This is because linear interpolation is not strictly correct for the functional relationship considered (area varies with d^2, not d). However, for all practical purposes it is adequate.

chapter 9

Miscellaneous Techniques

Maurice J. Webb

9.1 IDENTIFYING METALS AND ALLOYS

Metals and alloys, especially rods, bars, pipe, and tubing stored in racks, can easily be confused. While it is good practice to stamp the end of each piece of metal for permanent identification, this procedure is seldom done rigorously, with the result that the various pieces racked together become mixed. Many metals and alloys once mixed cannot be visually identified because their colors are identical. For example, the various stainless-steel alloys and nickel and monel are identical in appearance, as are many aluminum alloys and steel alloys. Yet they have distinctly different properties, and serious consequences can result from error.

One way to determine the type of material is to cut a sample, being careful to identify the remainder, and send it to a specialized laboratory for analysis. Quicker methods are available, however, especially when the choices may be narrowed by knowing prior purchases, shape, etc. One of these methods is the emery-wheel spark test.

Although experienced testers can identify well over 100 metals and alloys from the spark characteristics alone, the technician can best use the technique by referring to a known material and by close observation. The wheel must always be clean and free from previous ground deposits. A high wheel speed (about 800 surface ft/min, or 40 m/s) is preferred to a slow one because it provides more sparks and a longer stream to observe. Spark observations are best made in subdued light with ample room to see the entire spark plume. There are several characteristics to watch for:

1. Spark color
2. Spark length
3. Spark bursts
4. Spark trajectories
5. Spark plume
6. Wheel following

First observe the color of the spark itself as well as that of the trajectory trail. Ordinarily, the colors for hardened or annealed alloy steels are the same, but the hardened material often provides longer trajectories and a greater abundance of sparks. The color, then, is important, and spark intensity and length are functions of hardness and other factors, such as wheel pressure, wheel speed, and wheel grit size. When the cause for spark-length variation is uncertain (hardness or different alloy particularly), then make independent hardness measurements of the various specimens.

The spark bursts often have distinct features. Figure 9.1 shows various spark characteristics that may be present alone or in combination. There are often preliminary bursts along the trajectory, with a main burst or spark at or near the end of the trajectory. Observe the size, frequency, and type of spark closely. The end of the trajectory is especially noteworthy for it may terminate in a simple spark, a branching spark, a spearlike point, a tongue, or forked tongues, as shown in Fig. 9.2.

The plume or stream of sparks that is formed may vary from a narrow cone to a wide array reminiscent of a Fourth of July pinwheel. While the wide arrays are usually accompanied by sparks being carried around the circumference of the

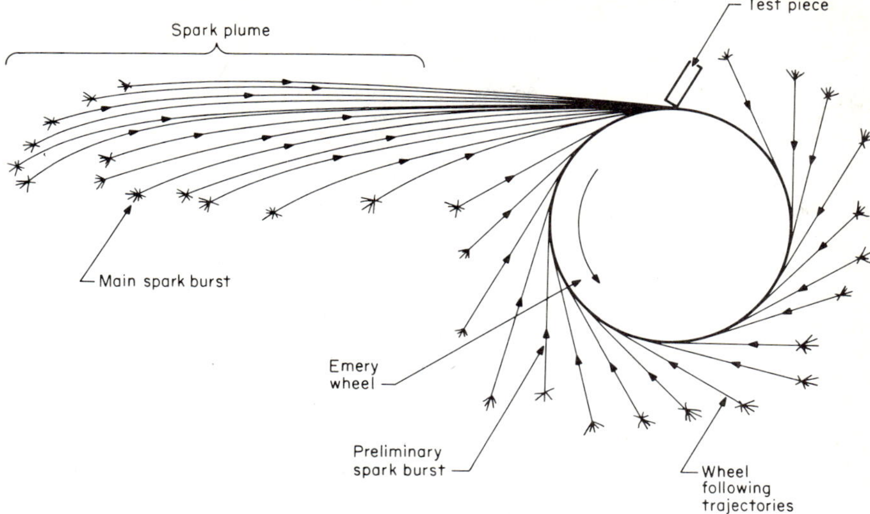

FIG. 9.1 Spark characteristics used for identifying metals.

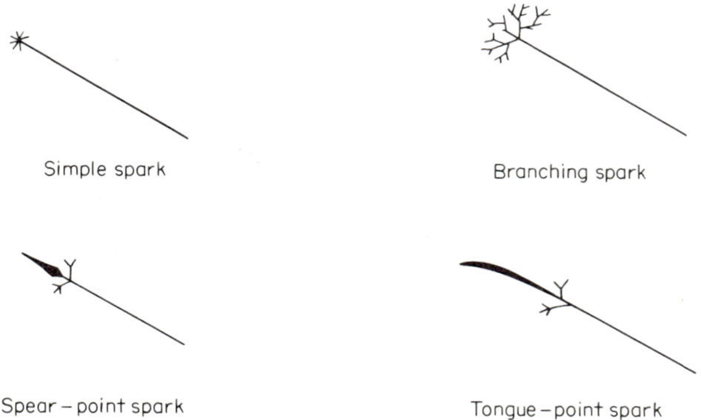

FIG. 9.2 Spark configurations at the end of the trajectories.

grinding wheel (i.e., wheel following), this phenomenon is not always present and should be judged independently of the array.

It is strongly recommended that technicians keep with their tool kits samples of materials (used toolbits, bar ends, etc.) that they use frequently. These samples should be permanently labeled (e.g., etched) as to their type and used as a reference for spark testing. It takes some practice to recognize the various spark-test characteristics, and some preliminary trials followed by occasional reuse is a good habit.

A magnetic test is another mechanical test that can be used to segregate materials. A simple bar magnet can determine if materials are magnetic or not. Table 9.1 lists some common magnetic and nonmagnetic alloys.

TABLE 9.1 Magnetic and Nonmagnetic Alloys

Magnetic	Nonmagnetic
Carbon steel Cast iron Low-alloy steel Stainless steel types 410, 420, 430, 431, 440, 443	Aluminum alloys Zinc alloys Brass Magnesium alloys Monel Inconel Stainless steel types 302, 303, 304, 316, 321, 347

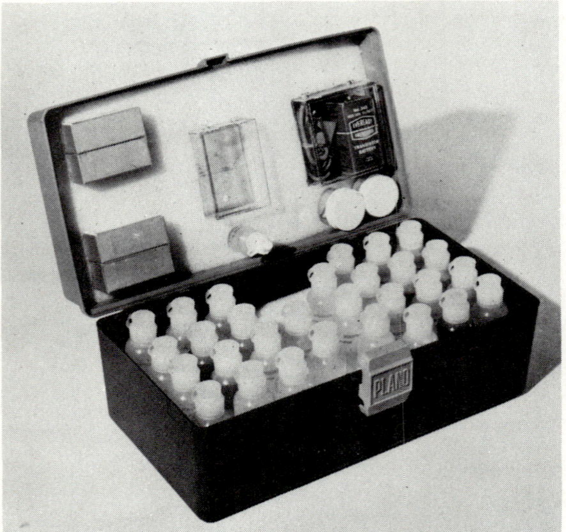

FIG. 9.3 Kit for identifying metals and alloys. *(Systems Scientific Laboratories.)*

Another technique for identifying metals, alloys, and electroplated or galvanized coatings is the chemical spot test. Place selected chemicals on the metal to be identified and observe the reaction and coloration. By systematically applying the chemicals, a wide variety of metals and alloys can be identified. For laboratory and workshop convenience the chemicals and complete instructions are available in kits. Figure 9.3 shows such a kit. The kit is equipped with identified metal samples so that the user can verify the exact reaction.

The chemical approach readily identifies stainless steels, 17-4PH steel, monel, Inconel, cobalt alloys, and Stellite. It separates alloy steels with nickel from those without nickel and separates low-sulfur from high-sulfur steels. Aluminum alloys 1100, 2011, 2024, 3003, 5086, 6061, and 7075 may be segregated, too. Coatings of copper, gold, silver, nickel, chromium, cadmium, zinc, and tin are easily distinguished.

9.2 FREEZING MIXTURES

Ice is usually available to any laboratory and may be used to keep equipment at the freezing point of water (0°C, 32°F, or 273 K). However, occasions arise when lower temperatures are required for relatively short periods (a few hours). For longer periods, use special refrigerators or cold boxes. Short-duration low-temperature environments can be obtained by using a mixture of ice and a salt. The best salt for this purpose is calcium chloride crystals. To maintain the low temperature, a well-insulated box should surround the freezing-mixture container. Figure 9.4 shows a typical arrangement.

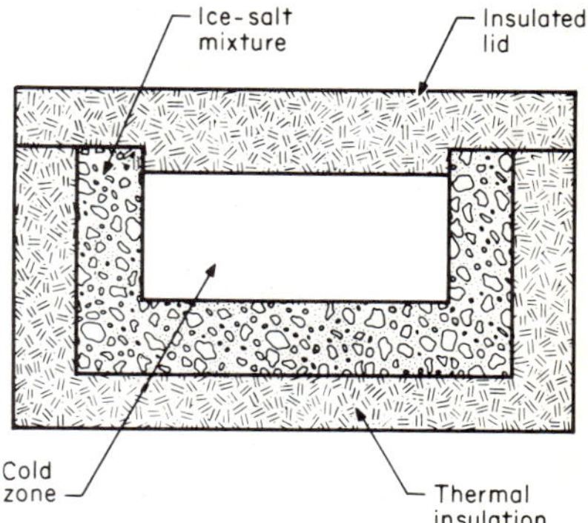

Fig. 9.4 Cold box using freezing mixtures.

The control over the temperature is obtained from the ratio of ice to salt. The ice must first be crushed and then mixed with the salt and packed around the chamber or vessel to be cooled. The temperatures that may be obtained through various mixtures are shown in Fig. 9.5; temperatures as low as −67°F (218 K) may be obtained in this manner when the ratio of salt to ice is 1.43. Adding more salt does not lower the temperature further but raises it.

While other salts may be used (e.g., common table salt), they do not produce such low-temperature freezing mixtures.

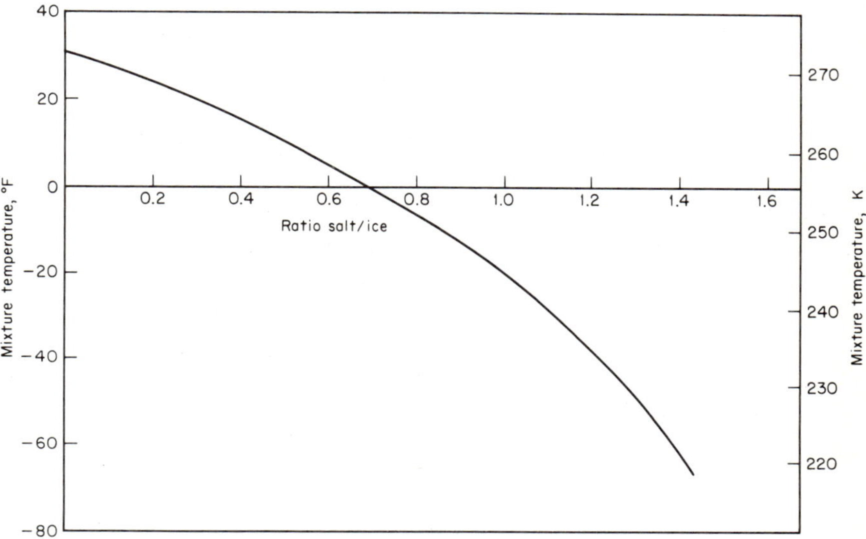

FIG. 9.5 Ratio of salt to ice for various freezing mixture temperatures.

9.3 CLEANING MERCURY

The use of mercury in manometers for measuring pressure, differential pressure, and vacuum is common. The mercury, however, becomes contaminated with oil and dust and leaves deposits on the manometer glass tubing, and measurements become imprecise. The mercury and the system need cleaning.

Mercury is easily cleaned in a three-step process. The apparatus is shown in Fig. 9.6. Break the mercury into small droplets by using a stainless-steel or glass capillary. The droplets pass through various washing solutions for cleaning. As the mercury accumulates in the lower vessel, remove it by decanting through the bottom stopcock as necessary. The column of washing solution should be about 3 ft (0.9 m) in length.

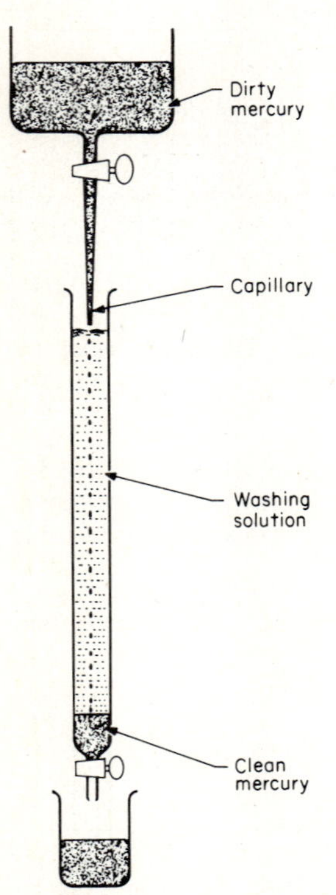

Fig. 9.6 Apparatus for cleaning mecury.

For mercury heavily contaminated with oil and grease, the first washing should be through a degreasing fluid such as 1,1,1-trichloroethane or alcohol. For mildly oil-contaminated mercury, use a 10% solution of potassium hydroxide. Following the oil and grease removal, next wash in a 15% nitric acid solution. Perform a final rinse with distilled water. Repeat any of the washing cycles as required. It is, in any case, good practice to rinse twice in distilled water, using clean rinsing water each time. Remove traces of the water by pulling a vacuum of at least 1 mmHg.

9.4 CLEANING GLASS

Glass, especially glass tubing used for manometers, rotameters, etc., requires cleaning from time to time. Oil and grease are common contaminants, having entered the system by oil-lubricated pumps and compressors. Remove oil and grease by flushing with alcohol or 1, 1, 1-trichloroethane. In addition pull a plug of cotton through with a wire. If the tube is especially dirty it may be necessary to replace the plug several times.

If this procedure proves inadequate, then use stronger cleaning agents. However, it is first generally necessary to remove the tube from the metallic parts of the instrument. Flushing may then be done with 15% nitric acid or chromic acid (sodium dichromate dissolved in sulfuric acid). Thoroughly rinse with distilled water and dry in warm air before reassembling the glass to its metallic components.

9.5 CRACK DETECTION

Fine cracks, invisible to the naked eye, may be present in bars, sheets, forgings, castings, and weldments. The cracks represent sites for possible material failure when subject to normal stress. Where such conditions can lead to severe consequences (e.g., safety-related), then inspection for them is necessary. The process for investigating the presence of such cracks is called *penetrant inspection* and is applicable only to cracks that propagate to the surface of the material. Cracks that are entirely within the body of the material may be detected by other means, such as ultrasonic inspection, but ultrasonic inspection does not necessarily detect surface cracks. Surface cracks, however, are usually more significant because typically the maximum material stresses occur at the surface boundaries.

A technique developed for easy or laboratory detection of surface cracks is the penetrant method; it can be applied to either machined or unmachined surfaces. A simple four-step process using prepackaged aerosol containers is all that is required. The Ardrox system (Brent Chemical Corp.) is typical. Three aerosol propellants are used.

Spray the surface to be inspected with an aerosol cleaner and wipe with a clean, lintfree cloth. Next, spray a dye penetrant on the surface from a second aerosol can. This penetrant is able to seep into the very tiny crevices formed by the cracks. After a few minutes, wipe the excess penetrant from the surface and apply the contents of the third aerosol can, the so-called "developer" that covers the surface with a white powder. Penetrant (dye) that seeped into a crack is absorbed by the developer, and the crack becomes visible as a red line on the white surface.

Not every crack detected by this means renders the part useless. If the crack is very short and in an area of low stress, it may be determined acceptable by an engineer. The crack may have very little depth, particularly if it is on an unmachined surface, and after removing (e.g., with a hand grinder) a small amount of material the crack may disappear upon reexamination.

Internal cracks that do not emerge at a surface may be detected by ultrasonic means. Specialized equipment and trained personnel are required to interpret the ultrasonic measurements, and the technique must be modified and developed for each new part geometry. Specialized laboratories can provide this service when the component use justifies it.

9.6 LUBRICANTS

Most mechanical equipment requires lubricant of one kind or another. Because of the many lubricants available, most manufacturers specify what should be used with their equipment. These specifications should be strictly adhered to, for the correct lubricant not only extends life but provides the correct performance. Lubricants vary greatly in their coefficients of friction so that for mechanical reasons alone the correct one is important. For example, a simple screw jack can lift twice the load for the same effort with a lubricant of one-half the coefficient of friction but also must withstand twice the stress, a stress for which it may not have been designed.

There are four classes of lubricant: oils, greases, waxes, and dry films in order of their ability to withstand pressure between the moving parts. Lubricants tend to be squeezed out from between the surfaces they are lubricating. Certain surface treatments provide greater retention of the lubricant. Retentive surfaces are provided by phosphating or liquid honing. The coefficients of friction generally increase as the lubricant changes from oil through dry film, but not always so. Some dry-film lubricants under low pressure (e.g., graphite) provide very low friction.

The viscosity of oils and greases determines to a large extent the ability of the lubricant to perform. The higher the viscosity (heavier oils) the thicker the lubrication film and the greater the friction. However, viscosity decreases markedly with increasing temperature, and for this reason lubricants are often cooled to retain their properties. Other oils are specially blended and have additives to minimize the effect of temperature over certain temperature ranges. Oils and greases have a maximum useful operating temperature to about 500°F (530 K). The temperature will be lower if they cannot conveniently be replaced as they drip from the lubricated surfaces.

Dry-film lubricants and solid lubricants enormously extend the lubrication possibilities. Typical dry lubricants are Teflon, graphite, molybdenum disulfide, and certain metal coatings. The principal disadvantage to the dry lubricants is that they cannot easily be renewed while the surfaces are in use. Their application then may be limited to components that do not move continuously (e.g., valve stems). This is not true of Teflon and other plastics that provide excellent life where the bearing pressures are low. For the ultimate in bearing pressure, electrodeposited coatings of silver or cadmium are used but the number of operating cycles available before the coating fails is usually few. The dry lubricants are the only approach for temperatures above 500°F (530 K) and may go as high as 1500°F (1100 K) and for temperatures below −60°F (220 K). Another area of application for the dry lubricants is high vacuum, where the vapor pressure of oils and greases would prove objectionable.

For oxygen service, lubricants are limited because of the potential hazard from fire and explosion. The halocarbons and silicones, available in oils and greases, are well-suited as are most of the dry lubricants. However, powdered graphite and molybdenum sulfide should be avoided because they can react with oxygen especially under conditions of elevated temperature and pressure.

9.7 RECORDING DATA AND REPORT WRITING

Recording Data

It is good practice and mandatory in many laboratories for each technician to maintain a daily record of activities and results obtained. Even if the work is routine and there is a specific form for recording test results, a separate log should be kept, stating the number of tests performed, the number of forms completed (which will typically carry some form of serial number), and any unusual events either of the results themselves or of the equipment used. The log should positively show any changes in procedure, the introduction of new or modification of existing equipment, and the start and finish of any particular series of tests. It is preferable to maintain this log in notebooks rather than in loose-leaf binders because pages are easily lost from the latter.

When fabricating apparatus or prototypes, reference specific drawing numbers in the log. If construction is from sketches, then paste the sketch itself or a copy of it into the log. Note on the sketch actual dimensions, materials used, finishes, etc. Where an assembly is a combination of purchased items and fabricated parts, show on a sketch or schematic diagram the particular manufacturer and corresponding identifying model or serial number of each purchased item. Drawings, sketches, lists of equipment, photocopies from manufacturers' brochures, etc., should figure prominently in the log. Photographs, especially those of the instant variety, are most valuable in recording test arrangements and apparatus. Each photograph must be dated and titled.

In deciding what information to include and what to leave out, it is better to err on the side of too much rather than too little. What seems obvious and logical on the day of reporting will not seem that way when reviewed months or years later. It is a valuable exercise to go back to earlier log entries—2 or 3 years old—and attempt to reconstruct exactly what was done and use that as a basis for improving current entries. If you have no difficulty in reconstructing your own earlier entries, then try the same technique with a log written by someone else.

Report Writing

Report writing is sometimes the duty of the technician, often a requirement for a laboratory supervisor, and always a useful skill for anyone. Reports should not be unnecessarily detailed, and it is reasonable to expect the reader, typically one's supervisor, to have some knowledge and understanding of the subject matter. It is neither necessary nor expected to educate the uninitiated in the subject matter, but it must be clear why the report is being written. It must be meaningful and logical to a reader years later and not leave a host of unanswered questions. Yet, the current reader, who is probably very familiar with the subject, will not want to read through pages of detailed test data but will only want to know what the results were. The apparent antithesis between being complete, precise, and historically meaningful on the one hand and concise and results-oriented on the other can be achieved by adopting the following outline format:

1. Title
   (*a.* Background)
2. Purpose or Objective
3. Conclusions or Summary
4. Procedures and Techniques
5. Detailed Results
6. Analysis
7. References and Bibliography

This format satisfies the many needs of the report. The familiar reader need only read the first three items in the format, which should be brief, and therefore can easily be kept abreast of progress, developments, or findings. The more inquisitive reader will want to go through the entire report to be fully satisfied.

The Title should be carefully chosen to give the reader as much information as possible regarding the subject matter of the report. If the report is one of a series, it must be clear that this is so (e.g., The Development of the Self-Erasing Ink Pen, Report No. 4).

The Background is seldom required. It is used only when it would not be clear to the reader why the work was done. It might be included, for example, in the first of a series of reports and synopsize prior history to explain the starting point for the current activity, or to explain why the activity would even be considered or fit the present organization. It would explain what may be viewed as an unusual departure from the organization's normal activity or unnecessary piece of work.

The Purpose or Objective is a clear and concise statement as to what is sought. A single sentence will usually suffice and should seldom be more than a single paragraph. If it is one of a series of reports, it must show the objective for that reporting period which, in all probability, will be different from but complementary to the overall program.

The Conclusions or Summary must clearly relate to the purpose of the program. No doubt should be left as to whether the objectives were achieved or not. It should answer the question, Where do we stand now? If reconsideration of the program direction, or additional work, or new approaches, etc., are required in the best judgment of the writer, then such recommendations should be made here. The detailed reasoning for these conclusions will then be apparent in the material that follows.

If the easiest and most concise way to express the Conclusion is with a bar chart, graph, or figure, then it may be included here, although such material is most likely to be found in the body of the report.

Under Procedures and Techniques should come a description of the equipment used, supported with drawings (or reference to them if they are filed elsewhere) and figures for clarification and brevity, not for padding. Included here should be information on the calibrations for the instruments (e.g., frequency), the measuring range of the instruments used, and any special reasons for making their selection. Someone should be able to repeat exactly what was done at some future date from the descriptions in this section of the report.

The Detailed Results provide the actual data recorded plus any observations made, especially if they were unexpected, inconsistent, or in any way unusual for this type of work for which you, the technician, are probably the best judge. These data are best presented in tabular form.

The Analysis should show what was done with the test results, if anything, to obtain the answers sought. It should note if calibration or other corrections were made to the actual data, record any formulas used, and show all techniques used in analyzing the observations. The calculations themselves should not be shown, but there must be sufficient information so that anyone can take the same results and arrive at the same answers by using the techniques described. The final results should be presented in the simplest way possible, and graphical or other pictorial representation is recommended.

In the References and Bibliography, include those sources from which formulas were taken or that refer to techniques, experimental or analytical, used in the work. Each reference should be numbered and that number used in the body of the report, in parentheses, immediately following its use. The reference should include the author's name, publication title, date of publication, name of publisher, and the page number on which the referenced material may be found.

9.8 HEAT TREATMENT

Metals and alloys are subjected to various heating and cooling cycles to change their mechanical properties and resistance to corrosion. The most common heat-treat processes are annealing, stress relieving, hardening, and tempering.

Annealing

Annealing is a heat-treating process that provides the metal with increased ductility and softness and with properties that are more like those of wrought material. Typically, annealing reduces hardness and strength, facilitates machining, and changes the crystalline structure (grain size). Most materials harden and lose ductility when cold-worked, and while this is sometimes beneficial, it may also preclude further cold working because of cracking. An annealing step restores the ductility and permits further cold work. In some processes, where cold working is extensive, in deep drawing, for example, several annealing cycles may be required.

To anneal, heat the material to an elevated temperature, hold it at that temperature for a period of time, and then cool it. The initial heating cannot be too rapid because distortion occurs from nonuniform expansions. The time for which the material is held at the selected temperature is important. It must be kept there long enough to permit necessary recrystallization and for certain elements and compounds to dissolve. However, keeping the material at the annealing temperature also promotes crystal growth (increased grain size), which is usually undesirable from considerations of additional cold working and reduced mechanical properties.

Thin sections attain the furnace temperature more quickly than heavier ones, yet all the material must be raised to the required temperature. Therefore the thinner sections are held at temperature for a longer period of time, and this factor should be considered for items having widely different thickness sections. Cooling time is also important. Rapid cooling tends to cause distortion, but for some alloys it is important to maintain (freeze) certain compounds in solid solution.

Table 9.2 provides a guide for annealing some metals and alloys. The times given are for durations at the specified temperature. For large workpieces with heavy sections it may be necessary to employ longer times to reach uniform temperatures.

Stress Relieving

Stress relieving, as the name implies, is a heat-treating process that reduces internal stresses which have been imparted either mechanically (e.g., cold working) or thermally (e.g., quenching). The pressure of internal stresses often accelerates corrosion. In fact, tests to determine the adequacy or need for a stress-relieving process are based upon the rapid deterioration of some materials in selected corrosive media (see Sec. 9.9). Stress relieving is similar to annealing, but in some cases no loss of material strength occurs. Table 9.3 provides a guide for stress-relieving heat treatments for a number of alloys.

Hardening and Tempering

Many materials, especially steels, may be heat-treated to improve their hardness and strength. In its simplest form, heat the item to be hardened to a specific temperature and then cool or quench it. Although the process appears simple, it becomes increasingly complex because factors such as maximum temperature, cooling rate, quench medium (oil, water, air), quench temperature, quench agitation, and geometry of the part all contribute to the final result. The temperature required and the maximum hardness attainable are shown in Table 9.4 for some common alloys. Manufacturers of tool steels usually provide specific data for their heat treatment.

TABLE 9.2 Annealing Heat Treatment for Some Metals and Alloys

Material	Anneal temperature, °F (K)	Time	Cooling
Aluminum and aluminum alloys	800 ± 25 (700 ± 15)	2 h	Slow (50°F/h) to 500°F (30 K/h to 530 K)
Carbon steel 1018	1600 ± 25 (1145 ± 15)	30 min	Slow (50°F/h) to 1300°F (30 K/h to 980 K)
Carbon steel 1060	1500 ± 50 (1090 ± 30)	30 min	Slow (50°F/h) to 1200°F (30 K/h to 920 K)
Stainless steel 304, 316	1950 ± 50 (1340 ± 30)	1 h	Water quench
Stainless steel 321	1800 ± 50 (1245 ± 30)	1 h	Air cool
Stainless steel 304L and 316L	1950 ± 50 (1340 ± 30)	1 h	Air cool
Copper	1000 ± 100 (810 ± 55)	5 min	As desired
Brass	1000 ± 100 (810 ± 55)	15 min	As desired
Phospher bronze	1000 ± 100 (810 ± 55)	15 min	As desired
Aluminum bronze	1200 ± 50 (920 ± 30)	1 h	As desired
Beryllium copper	1600 ± 100 (1145 ± 55)	1 h	As desired
Magnesium alloys	700 ± 50 (645 ± 30)	1 h	As desired
Nickel 200	1400 ± 100 (1030 ± 55)	30 min	As desired
Duranickel 301*	1700 ± 100 (1200 ± 55)	5 min	Water quench
Monel 400*	1700 ± 50 (1200 ± 30)	5 min	As desired
Inconel 625	2100 ± 100 (1420 ± 55)	1 h	As desired

*Registered trademarks of the International Nickel Co.

TABLE 9.3 Stress-Relieving Heat Treatment for Some Materials

Material	Stress-relieve temperature, °F (K)	Time	Cooling
Aluminum and aluminum alloys	650 ± 15 (615 ± 10)	10 min	Slow cool in furnace
Stainless steel 304 and 316	2000 ± 50 (1370 ± 30)	1 h	Water quench
Stainless steel 321	2000 ± 50 (1370 ± 30)	1 h	Slow cool
Stainless steel 304L and 316L	2000 ± 50 (1370 ± 30)	1 h	Slow cool
Brass	500 ± 25 (530 ± 15)	1 h	As desired
Phospher bronze	375 ± 25 (460 ± 15)	1 h	As desired
Magnesium alloy	450 ± 25 (505 ± 15)	30 min	As desired
Nickel 200	1000 ± 50 (810 ± 30)	3 h	Air cool
Monel 400*	1100 ± 100 (870 ± 55)	1 h	Slow cool
Inconel 625*	1600 ± 50 (1140 ± 30)	1 h	As desired

*Registered trademarks of the International Nickel Co.

TABLE 9.4 Hardening and Tempering Temperatures for Some Common Steel Alloys

Steel	Hardening temperature, °F (K)	Maximum hardness, Rc	Tempering temperature, °F (K)
1030	1575 ± 25 (1130 ± 15)	50	400–1000 (480–810)
1050	1500 ± 25 (1090 ± 15)	60	400–1000 (480–810)
1060	1500 ± 25 (1090 ± 15)	62	400–1000 (480–810)
420 stainless	1900 ± 50 (1310 ± 30)	52	400–800 (480–700)
440C stainless	1900 ± 50 (1340 ± 30)	60	400–800 (480–700)

Tempering is a heat-treating process performed after hardening steels to lower the hardness and reduce brittleness (improve ductility). The hardening process itself leaves little variation in the resulting hardness, and this often results in unacceptable mechanical properties. The tempering or drawing process permits a range of lower hardness values to be obtained. In some cases, tempering can improve ductility without materially affecting the hardness. Table 9.4 gives some typical times and temperatures.

Tempering is performed by heating the workpiece to a lower temperature than for hardening and quenching. The effect of increased time is typically similar to using a higher temperature, and to a large degree one may be traded against the other to obtain specific mechanical properties.

In general, it is better to seek the services of a heat-treating facility than to do one's own work. These facilities normally have a variety of equipment to provide controlled heating rates to suit the geometry and size of the workpiece and, if necessary, controlled atmospheres to eliminate oxidation and scaling. Additionally, various quench baths are usually available to provide the optimum cooling.

9.9 STRESS-CORROSION DETECTION

Internal stresses left in cold-worked materials (e.g., from rolling, drawing, machining) or thermally induced (e.g., casting shrinkage) result in more rapid corrosion and cracking than for annealed material. The need for an annealing operation can often be judged through an accelerated corrosion test of a representative piece. If this test piece shows cracking to occur, then all similar pieces should be subject to an anneal cycle to remove the internal or residual stresses. Applied stresses, resulting from mechanical loads in the normal use of a part, also cause more rapid corrosion than for unstressed material. However, the aggregate effect of stresses can certainly be reduced by removing the internal stresses, and perhaps the effect can be negated if the applied stresses are minimal.

Copper and its alloys are especially prone to stress-corrosion cracking, and a quick test has been devised to detect such cracks. Immerse the test piece in a mercurous nitrate solution for half an hour, then wash and dry the piece and visually examine it with a magnifying glass for surface cracks. The presence of any crack indicates the need for annealing.

Thoroughly clean the test piece before immersing it. This is typically done by rinsing in a grease solvent followed by an acid etch to remove oxides. The acid etch is a solution of nitric acid (40 volume percent).

To make the mercurous nitrate solution, dissolve 10 g mercurous nitrate in 1 liter distilled water to which 10 mL nitric acid (1.42 sp gr) has been added.

Stainless steels are also subject to stress-corrosion cracking, and because they are often selected for their corrosion resistance it is usually desirable to provide the ultimate corrosion capability and perform an anneal cycle if necessary. The test used is a polythionic acid test and is applicable to iron alloys containing nickel and chromium.

To verify that the acid solution is suitable for the test, use a controlled specimen of type 302 stainless steel. Heat a small strip of 302 stainless to 1200°F (920 K) for 4 h and allow to cool in the air. Bend the strip to a U shape with the radius of the U portion being about four times the thickness. Thoroughly degrease the specimen before immersing it in the polythionic acid solution. A specimen prepared in this way should crack with less than 1-h exposure.

Prepare similar specimens from the actual piece for which a determination is to be made and also immerse them along with a controlled specimen of type 302 stainless steel prepared as above. Exposure can continue for several hours. Inspect under a low-power (×20) microscope to determine the susceptibility of the material to cracking, although the cracks are often severe enough that no visual aid is required to detect them. The reason for including the control specimen (302 stainless) is to maintain a check on the effectiveness of the solution.

Another factor that leads to fast corrosion of stainless steels is the presence of carbides between crystal boundaries. The presence of these carbides is an indication of improper heat treat (anneal) or absence of an anneal cycle. A test may be used for types 304, 304L, 316, 316L, 317, 321, and 347 stainless steels.

The test method is to subject specimens to a boiling solution of acidified copper sulfate and copper metal for 24 h. Bend the specimens 180° over a bar of diameter equal to the thickness. The presence of cracks (using ×20 magnification) shows that intergranular corrosion has taken place.

To make the solution, dissolve 100 g copper sulfate crystals in 500 mL water to which add 100 mL sulfuric acid (1.84 sp gr) and enough distilled water to make 1 liter. Add enough copper (granulated) to the vessel to cover completely the specimens to be tested. Place a condenser on the vessel outlet so that, as boiling occurs, the vapors are condensed and returned to the vessel and no makeup solution is needed.

9.10 COMBUSTION

The combustion of fuels, especially gaseous fuels with air, may be required for heating or other laboratory processes. In sizing equipment and choosing instruments for measuring the flow of fuel and air, it is necessary to know the ratios of each. The maximum temperature attainable with a specific combination of fuel and air occurs approximately at the theoretical amount of air (oxygen) required to react all the hydrogen in the fuel to water ($H_2 + \frac{1}{2}O_2 \rightarrow H_2O$) and all the carbon to carbon dioxide ($C + O_2 \rightarrow CO_2$). Actually, because of dissociation at

TABLE 9.5 The Properties for Combustion of Some Fuels in Air

Fuel	Theoretical air/fuel ratio by volume	Combustion temperature, °F (K)	Flammability limits, air/fuel ratio by volume
Hydrogen	2.38	4390 (2694)	0.35 to 24
Methane	9.52	3480 (2189)	5.7 to 19
Acetylene	11.90	4770 (2906)	0.23 to 39
Ethane	16.66	3540 (2222)	7 to 32
Ethylene	14.28	4250 (2617)	2.5 to 35
Propane	23.8	3570 (2239)	8.9 to 47
Propylene	21.4	4090 (2528)	8.0 to 49
Butane	30.9	3580 (2244)	10.9 to 53
Isobutane	30.9	3580 (2244)	10.8 to 55

high temperatures for various combustion products, the maximum temperature occurs with slightly excess fuel (about 10 percent) than the theoretical ratio. Another factor that must be considered is that not all ratios of fuel and air provide a combustible mixture and that certain limits, called *flammability limits,* exist. Only for mixtures within the flammability limits can combustion be sustained. Table 9.5 provides data for some common fuels when combusted in air.

chapter 10

Tables and Conversions

Maurice J. Webb

TABLE 10.1 Natural Logarithms

											Mean differences								
	0	1	2	3	4	5	6	7	8	9	1	2	3	4	5	6	7	8	9
1.0	0.00000	00995	01980	02956	03922	04879	05827	06766	07696	08618	95	191	286	381	477	572	667	762	858
1.1	.09531	10436	11333	12222	13103	13976	14842	15700	16551	17395	87	174	261	348	435	522	609	696	783
1.2	.18232	19062	19885	20701	21511	22314	23111	23902	24686	25464	80	160	240	320	400	480	560	640	720
1.3	.26236	27003	27763	28518	29267	30010	30748	31481	32208	32930	74	148	222	296	371	445	519	593	667
1.4	.33647	34359	35066	35767	36464	37156	37844	38526	30204	39878	69	138	207	276	345	414	483	552	621
1.5	.40547	41211	41871	42527	43178	43825	44469	45108	45742	46373	65	129	194	258	323	387	452	516	581
1.6	.47000	47623	48243	48858	49470	50078	50682	51282	51879	52473	61	121	182	242	303	364	424	485	546
1.7	.53063	53649	54232	54812	55389	55962	56531	57098	57661	58222	57	114	172	229	286	343	400	458	515
1.8	.58779	59333	59884	60432	60977	61519	62058	62594	63127	63658	54	108	162	216	271	325	379	433	487
1.9	.64185	64710	65232	65752	66269	66783	67294	67803	68310	68813	51	103	154	205	257	308	359	410	462
2.0	.69315	69813	70310	70804	71295	71784	72271	72755	73237	73716	49	98	146	195	244	293	342	390	439
2.1	.74194	74669	75142	75612	76081	76547	77011	77473	77932	78390	47	93	140	180	233	278	326	372	419
2.2	.78846	79300	79751	80200	80648	81093	81536	81978	82418	82855	45	89	133	178	223	267	312	356	401
2.3	.83291	83725	84157	84587	85015	85442	85866	86289	86710	87129	43	85	128	171	214	255	298	341	383
2.4	.87547	87963	88377	88789	89200	89609	90016	90422	90826	91228	41	82	123	164	205	244	285	326	367
2.5	.91629	92028	92426	92822	93216	93609	94001	94391	94779	95166	39	78	117	156	197	236	275	314	353
2.6	.95551	95935	96317	96698	97078	97456	97833	98208	98582	98954	38	75	113	150	189	227	264	302	339
2.7	.99325	99695	1.00063	1.00430	1.00796	1.01160	1.01523	1.01885	1.02246	1.02604	36	73	109	146	182	218	255	291	328
2.8	1.02962	03318	03674	04028	04380	04732	05082	05431	05779	06126	35	70	106	141	176	210	245	281	316
2.9	1.06471	06816	07158	07500	07841	08181	08519	08856	09192	09527	34	68	102	136	170	203	237	271	305
3.0	1.09861	10194	10526	10856	11186	11514	11841	12168	12493	12817	33	66	99	132	165	196	229	262	295
3.1	1.13140	13462	13783	14103	14422	14740	15057	15373	15688	16002	32	63	95	127	160	191	223	255	286
3.2	1.16315	16627	16938	17248	17557	17865	18173	18479	18784	19089	31	62	92	123	154	185	216	246	277
3.3	1.19392	19695	19996	20297	20597	20896	21194	21491	21788	22083	30	60	89	119	150	180	210	239	269
3.4	1.22378	22671	22964	23256	23547	23837	24127	24415	24703	24990	29	58	87	116	145	174	203	232	261

3.5	1.25276	25562	25846	26130	26413	26695	26976	27257	27536	27815	28	56	85	113	141	169	197	226	254
3.6	1.28093	28371	28647	28923	29198	29473	29746	30019	30291	30563	27	55	82	109	138	165	192	219	247
3.7	1.30833	31103	31372	31641	31909	32176	32442	32708	32972	33237	27	53	80	106	134	161	187	214	240
3.8	1.33500	33763	34025	34286	34547	34807	35067	35325	35584	35841	26	52	78	104	130	156	182	208	234
3.9	1.36098	36355	36609	36864	37118	37372	37624	37877	38128	38379	25	50	76	101	127	152	177	203	228
4.0	1.38629	38879	39128	39377	39624	39872	40118	40364	40610	40854	25	49	74	99	124	148	173	198	222
4.1	1.41100	41342	41585	41828	42070	42311	42552	42792	43031	43270	24	48	72	96	121	145	169	193	217
4.2	1.43508	43746	43984	44220	44456	44692	44927	45161	45395	45629	23	47	70	94	118	141	165	188	212
4.3	1.45862	46094	46326	46557	46787	47018	47247	47476	47705	47933	23	46	69	92	115	138	161	184	207
4.4	1.48160	48387	48614	48840	49065	49290	49516	49739	49962	50185	23	45	68	90	113	135	158	180	203
4.5	1.50408	50630	50851	51072	51294	51513	51732	51951	52170	52388	22	44	66	88	110	132	154	176	198
4.6	1.52606	52823	53039	53256	53471	53687	53902	54116	54330	54543	22	43	65	86	108	129	151	172	194
4.7	1.54756	54969	55181	55393	55604	55814	56025	56235	56444	56653	21	42	63	84	106	127	148	169	190
4.8	1.56862	57070	57278	57485	57691	57898	58104	58309	58515	58719	21	41	62	82	103	124	144	165	185
4.9	1.58924	59127	59331	59534	59737	59939	60141	60342	60543	60744	20	40	61	81	101	121	141	162	182
5.0	1.60944	61144	61343	61542	61741	61939	62137	62334	62531	62728	20	40	59	79	99	119	139	158	178
5.1	1.62924	63120	63315	63511	63705	63900	64094	64287	64481	64673	19	39	58	78	97	116	136	155	175
5.2	1.64866	65058	65250	65441	65632	65823	66013	6620	66393	66582	19	38	57	76	96	115	134	153	172
5.3	1.66771	66959	67147	67335	67523	67710	67896	68083	68269	68455	19	37	56	75	94	112	131	150	168
5.4	1.68640	68825	69010	69194	69378	69562	69745	69928	70111	70293	18	37	55	74	92	110	129	147	166
5.5	1.70475	70657	70838	71019	71199	71380	71560	71740	71919	72098	18	36	54	72	90	108	126	144	162
5.6	1.72277	72456	72633	72811	72988	73166	73342	73519	73695	73871	18	35	53	71	89	106	124	142	159
5.7	1.74047	74222	74397	74572	74746	74920	75094	75267	75440	75613	17	35	52	70	87	104	122	139	157
5.8	1.75786	75958	76130	76302	76473	76644	76815	76985	77156	77326	17	34	51	68	86	103	120	137	154
5.9	1.77495	77665	77834	78002	78171	78339	78507	78675	78842	79009	17	34	50	67	84	101	118	134	151
6.0	1.79176	79342	79509	79675	79840	80006	80171	80336	80500	80665	17	33	50	66	83	99	116	132	149
6.1	1.80829	80993	81156	81319	81482	81645	81808	81970	82132	82294	16	33	49	65	82	98	114	130	147
6.2	1.82455	82616	82777	82938	83098	83258	83418	83578	83737	83896	16	32	48	64	80	96	112	128	144
6.3	1.84055	84214	84372	84530	84689	84845	85003	85160	85317	85473	16	32	47	63	79	95	111	126	142
6.4	1.85630	85786	85942	86097	86253	86408	86563	86718	86872	87026	16	31	47	62	78	93	109	124	140

TABLE 10.1 *(Continued)*

	0	1	2	3	4	5	6	7	8	9	Mean differences								
											1	2	3	4	5	6	7	8	9
6.5	1.87180	87334	87487	87641	87794	87947	88100	88251	88403	88555	15	31	46	61	77	92	107	122	138
6.6	1.88707	88858	89010	89160	89311	89462	89612	89762	89912	90061	15	30	45	60	76	91	106	121	136
6.7	1.90211	90360	90509	90658	90806	90954	91103	91250	91398	91545	15	30	44	59	74	89	104	118	133
6.8	1.91692	91839	91980	92132	92279	92425	92571	92716	92862	93007	15	29	44	58	73	88	102	117	131
6.9	1.93152	93297	93442	93586	93730	93874	94018	94162	94305	94448	14	29	43	58	72	86	101	115	130
7.0	1.94591	94734	94876	95019	95161	95303	95445	95586	95727	95869	14	28	43	57	71	85	99	114	128
7.1	1.96009	96150	96291	96431	96571	96711	96851	96991	97130	97269	14	28	42	56	70	84	98	112	126
7.2	1.97408	97547	97685	97824	97962	98100	98238	98376	98513	98650	14	28	41	55	69	83	97	110	124
7.3	1.98787	98924	99061	99198	99334	99470	99606	99742	99877	2.00013	14	27	41	54	68	82	95	109	122
7.4	2.00148	00283	00418	00553	00687	00821	00956	01089	01223	01357	13	27	40	54	67	80	94	107	121
7.5	2.01490	01624	01757	01890	02022	02155	02287	02419	02551	02683	13	27	40	53	67	80	93	106	120
7.6	2.02815	02946	03078	03209	03340	03471	03601	03732	03862	03992	13	26	39	52	66	79	92	105	118
7.7	2.04122	04252	04381	04511	04640	04769	04898	05027	05156	05284	13	26	39	52	65	77	90	103	116
7.8	2.05412	05540	05668	05796	05924	06051	06179	06306	06433	06560	13	25	38	51	64	76	89	102	114
7.9	2.06686	06813	06939	07065	07191	07317	07443	07568	07694	07819	13	25	38	50	63	76	88	101	113
8.0	2.07944	08069	08194	08318	08443	08567	08691	08815	08939	09063	12	25	37	50	62	74	87	99	112
8.1	2.09186	09310	09433	09556	09679	09802	09924	10047	10169	10291	12	25	37	49	62	74	86	98	111
8.2	2.10413	10535	10657	10779	10900	11021	11142	11263	11384	11505	12	24	36	48	61	73	85	97	109
8.3	2.11626	11746	11867	11986	12106	12226	12346	12465	12585	12704	12	24	36	48	60	72	84	96	108
8.4	2.12823	12942	13061	13180	13298	13417	13535	13653	13771	13889	12	24	35	47	59	71	83	94	106
8.5	2.14007	14124	14242	14359	14476	14593	14710	14827	14943	15060	12	23	35	47	59	70	82	94	105
8.6	2.15176	15292	15409	15524	15640	15756	15871	15987	16102	16218	12	23	35	46	58	70	81	93	104
8.7	2.16332	16447	16562	16677	16791	16905	17020	17134	17248	17361	11	23	34	46	57	68	80	91	103
8.8	2.17475	17589	17702	17816	17929	18042	18155	18267	18380	18493	11	23	34	45	57	68	79	90	102
8.9	2.18605	18717	18830	18942	19054	19165	19277	19389	19500	19611	11	22	34	45	56	67	78	90	101

9.0	2.19722	19834	19944	20055	20166	20276	20387	20497	20607	20717	11	22	33	44	56	67	78	89	100
9.1	2.20828	20937	21047	21157	21267	21375	21485	21594	21703	21812	11	22	33	44	55	66	77	88	99
9.2	2.21920	22029	22138	22246	22354	22462	22570	22678	22786	22894	11	22	32	43	54	65	76	86	97
9.3	2.23001	23109	23216	23324	23431	23537	23645	23751	23858	23965	11	21	32	43	54	64	75	86	96
9.4	2.24071	24178	24284	24390	24496	24601	24707	24813	24918	25024	11	21	32	42	53	64	74	85	95
9.5	2.25129	25234	25339	25444	25549	25654	25759	25863	25968	26072	11	21	32	42	53	63	74	84	95
9.6	2.26176	26280	26384	26488	26592	26696	26799	26903	27006	27109	10	21	31	42	52	62	73	83	94
9.7	2.27213	27316	27419	27521	27624	27727	27829	27931	28034	28130	10	21	31	41	52	62	72	82	93
9.8	2.28238	28340	28442	28544	28646	28747	28849	28950	29051	29152	10	20	31	41	51	61	71	82	92
9.9	2.29253	29354	29455	29556	29657	29757	29858	29958	30058	30158	10	20	30	40	51	61	71	81	91
10.0	2.3026																		

For values 10^{+n} times those listed, add the following:

n	1	2	3	4	5	6	7	8	9
$\log_e 10^n$	2.30259	4.60517	6.90776	9.21034	11.51293	13.81551	16.11810	18.42068	20.72327

For values 10^{-n} times those listed, subtract the following:

n	1	2	3	4	5	6	7	8	9
$\log_e 10^{-n}$	$\bar{3}.69741$	$\bar{5}.39483$	$\bar{7}.09224$	$\overline{10}.78966$	$\overline{12}.48707$	$\overline{14}.18449$	$\overline{17}.88190$	$\overline{19}.57932$	$\overline{21}.27673$

For the roots of number factors of 10^{2n} greater or smaller than these, simply move the decimal point by 10^n for positive or negative integers of n.

TABLE 10.2 Sines of Angles

Deg.	0′ 0.0°	6′ 0.1°	12′ 0.2°	18′ 0.3°	24′ 0.4°	30′ 0.5°	36′ 0.6°	42′ 0.7°	48′ 0.8°	54′ 0.9°	Mean differences 1	2	3	4	5
0	.00000	00175	00349	00524	00698	00873	01047	01222	01396	01571	29	58	87	117	145
1	.01745	01920	02094	02269	02443	02618	02792	02967	03141	03316	29	58	87	117	145
2	.03490	03664	03839	04013	04188	04362	04536	04711	04885	05059	29	58	87	116	145
3	.05234	05408	05582	05756	05931	06105	06279	06453	06627	06802	29	58	87	116	145
4	.06976	07150	07324	07498	07672	07846	08020	08194	08368	08542	29	58	87	116	145
5	.08716	08889	09063	09237	09411	09585	99758	09932	10106	10279	29	58	87	116	145
6	.10453	10626	10800	10973	11147	11320	11494	11667	11840	12014	29	58	87	116	145
7	.12187	12360	12533	12706	12880	13053	13226	13399	13572	13744	29	58	87	115	144
8	.13917	14090	14263	14436	14608	14781	14954	15126	15299	15471	29	58	86	115	144
9	.15643	15816	15988	16160	16333	16505	16677	16849	17021	17193	29	57	86	115	144
10	.17365	17537	17708	17880	18052	18224	18395	18567	18738	18910	29	57	86	114	143
11	.19081	19252	19423	19595	19766	19937	20108	20279	20450	20620	29	57	86	114	143
12	.20791	20962	21132	21303	21474	21644	21814	21985	22155	22325	28	57	85	114	142
13	.22495	22665	22835	23005	23175	23345	23514	23684	23853	24023	28	57	85	113	141
14	.24192	24362	24531	24700	24869	25038	25207	25376	25545	25713	28	56	85	113	141
15	.25882	26050	26219	26387	26556	26724	26892	27060	27228	27396	28	56	84	112	140
16	.27564	27731	27899	28067	28234	28402	28569	28736	28903	29070	28	56	84	112	139
17	.29237	29404	29571	29737	29904	30071	30237	30403	30570	30736	28	56	83	111	139
18	.30902	31068	31233	31399	31565	31730	31896	32061	32227	32392	28	55	83	110	138
19	.32557	32722	32887	33051	33216	33381	33545	33710	33874	34038	27	55	82	110	137
20	.34202	34366	34530	34694	34857	35021	35184	35347	35511	35674	27	55	82	109	136
21	.35837	36000	36162	36325	36488	36650	36812	36975	37137	37299	27	54	81	108	135
22	.37461	37622	37784	37946	38107	38268	38430	38591	38752	38912	27	54	81	107	134
23	.39073	39234	39394	39555	39715	39875	40035	40195	40355	40514	27	53	80	107	133
24	.40674	40833	40992	41151	41310	41469	41628	41787	41945	42104	27	53	79	106	132
25	.42262	42420	42578	42736	42894	43051	43209	43366	43523	43680	26	53	79	105	131
26	.43837	43994	44151	44307	44464	44620	44776	44932	45088	45243	26	52	78	104	130

27	.45399	45554	45710	45865	46020	46175	46330	46484	46639	46793	26	52	77	103	129
28	.46947	47101	47255	47409	47562	47716	47869	48022	48175	48328	26	51	77	102	128
29	.48481	48634	48786	48938	49090	49242	49394	49546	49697	49849	25	51	76	101	126
30	.50000	50151	50302	50453	50603	50754	50904	51054	51204	51354	25	50	75	100	125
31	.51504	51653	51803	51952	52101	52250	52399	52547	52696	52844	25	50	74	99	124
32	.52992	53140	53288	53435	53583	53730	53877	54024	54171	54317	25	49	74	98	123
33	.54464	54610	54756	54902	55048	55194	55339	55484	55630	55775	24	49	73	97	121
34	.55919	56064	56208	56353	56497	56641	56784	56928	57071	57215	24	48	72	96	120
35	.57358	57501	57643	57786	57928	58070	58212	58354	58496	58637	24	48	71	95	118
36	.58778	58920	59061	59201	59342	59482	59622	59763	59902	60042	23	47	70	94	117
37	.60181	60321	60460	60599	60738	60876	61015	61153	61291	61429	23	46	69	92	115
38	.61566	61704	61841	61978	62115	62251	62388	62524	62660	62796	23	46	68	91	114
39	.62932	63068	63203	63338	63473	63608	63742	63877	64011	64145	23	45	67	90	112
40	.64279	64412	64546	64679	64812	64945	65077	65210	65342	65474	22	44	66	88	111
41	.65606	65738	65869	66000	66131	66262	66393	66523	66653	66783	22	44	65	87	109
42	.66913	67043	67172	67301	67430	67559	67688	67816	67944	68072	22	43	64	86	107
43	.68200	68327	68455	68582	68709	68835	68962	69088	69214	69340	21	42	63	84	106
44	.69466	69591	69717	69842	69966	70091	70215	70339	70463	70587	21	41	62	83	104
45	.70711	70834	70957	71080	71203	71325	71447	71569	71691	71813	20	41	61	82	102
46	.71934	72055	72176	72297	72417	72537	72657	72777	72897	73016	20	40	60	80	100
47	.73135	73254	73373	73491	73610	73728	73846	73963	74080	74198	20	39	59	79	98
48	.74314	74431	74548	74664	74780	74896	75011	75126	75241	75356	20	39	58	77	96
49	.75471	75585	75700	75813	75927	76041	76154	76267	76380	76492	19	38	57	76	94
50	.76604	76717	76828	76940	77051	77162	77273	77384	77494	77605	19	37	56	74	93
51	.77715	77824	77934	78043	78152	78261	78369	78478	78586	78694	18	36	54	72	91
52	.78801	78908	79015	79122	79229	79335	79441	79547	79653	79758	18	35	53	71	89
53	.79864	79968	80073	80178	80282	80386	80489	80593	80696	80799	17	35	52	69	87
54	.80902	81004	81106	81208	81310	81412	81513	81614	81714	81815	17	34	51	68	84
55	.81915	82015	82115	82214	82314	82413	82511	82610	82708	82806	16	33	49	66	82
56	.82904	83001	83098	83195	83292	83389	83485	83581	83676	83772	16	32	48	64	80
57	.83867	83962	84057	84151	84245	84339	84433	84526	84619	84712	16	31	47	62	78
58	.84805	84897	84989	85081	85173	85264	85355	85446	85536	85627	15	31	46	61	76
59	.85717	85806	85896	85985	86074	86163	86251	86340	86427	86515	15	30	44	59	74

TABLE 10.2 *(Continued)*

Deg.	0′ 0.0°	6′ 0.1°	12′ 0.2°	18′ 0.3°	24′ 0.4°	30′ 0.5°	36′ 0.6°	42′ 0.7°	48′ 0.8°	54′ 0.9°	Mean differences 1	2	3	4	5
60	.86603	86690	86777	86863	86949	87036	87121	87207	87292	87377	14	29	43	57	72
61	.87462	87546	87631	87715	87798	87882	87965	88048	88130	88213	14	28	42	56	69
62	.88295	88377	88458	88539	88620	88701	88782	88862	88942	89021	14	27	40	54	67
63	.89101	89180	89259	89337	89415	89493	89571	89649	89726	89803	13	26	39	52	65
64	.89879	89956	90032	90108	90183	90259	90334	90408	90483	90557	13	25	38	50	63
65	.90631	90704	90778	90851	90924	90996	91068	91140	91212	91283	12	24	36	48	60
66	.91355	91425	91496	91566	91636	91706	91775	91845	91914	91982	12	23	35	46	58
67	.92050	92119	92186	92254	92321	92388	92455	92521	92587	92653	11	22	33	45	56
68	.92718	92784	92849	92913	92978	93042	93106	93169	93232	93295	11	21	32	43	54
69	.93358	93420	93483	93544	93606	93667	93728	93789	93849	93909	10	20	31	41	51
70	.93969	94029	94088	94147	94206	94264	94322	94380	94438	94495	10	19	29	39	49
71	.94552	94609	94665	94721	94777	94832	94888	94943	94997	95052	9	18	28	37	46
72	.95106	95159	95213	95266	95319	95372	95424	95476	95528	95579	9	17	26	35	44
73	.95630	95681	95732	95782	95832	95882	95931	95981	96029	96078	8	17	25	33	41
74	.96126	96174	96222	96269	96316	96363	96410	96456	96502	96547	8	16	23	31	39
75	.96593	96638	96682	96727	96771	96815	96858	96902	96945	96987	7	16	22	29	36
76	.97030	97072	97113	97155	97196	97237	97278	97318	97358	97398	7	14	20	27	34
77	.97437	97476	97515	97553	97592	97630	97667	97705	97742	97778	6	13	19	25	32
78	.97815	97851	97887	97922	97958	97992	98027	98061	98096	98129	6	12	17	23	29
79	.98163	98196	98229	98261	98294	98325	98357	98388	98420	98450	5	11	16	21	27
80	.98481	98511	98541	98570	98600	98629	98657	98686	98714	98741	5	10	14	19	24
81	.98769	98796	98823	98849	98876	98902	98927	98953	98978	99002	4	9	13	17	22
82	.99027	99051	99075	99098	99122	99144	99167	99189	99211	99233	4	8	11	15	19
83	.99255	99276	99297	99317	99337	99357	99377	99396	99415	99434	3	7	10	13	16
84	.99452	99470	99488	99506	99523	99540	99556	99572	99588	99604	3	6	8	11	14
85	.99619	99635	99649	99664	99678	99692	99705	99719	99731	99744	2	5	7	9	11
86	.99756	99768	99780	99792	99803	99813	99824	99834	99844	99854	2	4	5	7	9
87	.99863	99872	99881	99889	99897	99905	99912	99919	99926	99933	1	3	4	5	6
88	.99939	99945	99951	99956	99961	99966	99970	99974	99978	99982	1	2	2	3	4
89	.99985	99988	99990	99993	99995	99996	99998	99999	99999	1.0000	0	1	1	1	1
90	1.000														

TABLE 10.3 Cosines of Angles

Deg.	0′ 0.0°	6′ 0.1°	12′ 0.2°	18′ 0.3°	24′ 0.4°	30′ 0.5°	36′ 0.6°	42′ 0.7°	48′ 0.8°	54′ 0.9°	Mean differences* 1	2	3	4	5
0	1.000	99999	99999	99999	99998	99996	99995	99993	99990	99988	0	1	1	1	1
1	.99985	99982	99978	99974	99970	99966	99961	99956	99951	99945	1	2	2	3	4
2	.99939	99933	99926	99919	99912	99905	99897	99889	99881	99872	1	3	4	5	6
3	.99863	99854	99844	99834	99824	99813	99803	99792	99780	99768	2	4	5	7	9
4	.99756	99744	99731	99719	99705	99692	99678	99664	99649	99635	2	5	7	9	11
5	.99619	99604	99588	99572	99556	99540	99523	99506	99488	99470	3	6	8	11	14
6	.99452	99434	99415	99396	99377	99357	99337	99317	99297	99276	3	7	10	13	16
7	.99255	99233	99211	99189	99167	99144	99122	99098	99075	99051	4	8	11	15	19
8	.99027	99002	98978	98953	98927	98902	98876	98849	98823	98796	4	9	13	17	22
9	.98769	98741	98714	98686	98657	98629	98600	98570	98541	98511	5	10	14	19	24
10	.98481	98450	98420	98388	98357	98325	98204	98261	98229	98196	5	11	16	21	27
11	.98163	98129	98096	98061	98027	97992	97958	97922	97887	97851	6	12	17	23	29
12	.97815	97778	97742	97705	97667	97630	97592	97553	97515	97476	6	13	19	25	32
13	.97437	97398	97358	97318	97278	97237	97196	97155	97113	97072	7	14	20	27	34
14	.97030	96987	96945	96902	96858	96815	96771	96727	96682	96638	7	16	22	29	36
15	.96593	96547	96502	96456	96410	96363	96316	96269	96222	96174	8	16	23	31	39
16	.96126	96078	96029	95981	95931	95882	95832	95782	95732	95681	8	17	25	33	41
17	.95630	95579	95528	95476	95424	95372	95319	95266	95213	95159	9	17	26	35	44
18	.95106	95052	94997	94943	94888	94832	94777	94721	94665	94609	9	18	28	37	46
19	.94552	94495	94438	94380	94322	94264	94206	94147	94088	94029	10	19	29	39	49
20	.93969	93909	93849	93789	93728	93667	93606	93544	93483	93420	10	20	31	41	51
21	.93358	93295	93232	93169	93106	93042	92978	92913	92849	92784	10	21	32	43	54
22	.92718	92653	92587	92521	92455	92388	92321	92254	92186	92119	11	22	33	45	56
23	.92050	91982	91914	91845	91775	91706	91636	91566	91496	91425	11	23	35	46	58
24	.91355	91283	91212	91140	91068	90996	90924	90851	90778	90704	12	24	36	48	60
25	.90631	90557	90483	90408	90334	90259	90183	90108	90032	89956	13	25	38	50	63
26	.89879	89803	89726	89649	89571	89493	89415	89337	89259	89180	13	26	39	52	65
27	.89101	89021	88942	88862	88782	88701	88620	88539	88458	88377	13	27	40	54	67
28	.88295	88213	88130	88048	87965	87882	87798	87715	87631	87546	14	28	42	56	69
29	.87462	87377	87292	87207	87121	87036	86949	86863	86777	86690	14	29	43	57	72
30	.86603	86515	86427	86340	86251	86163	86074	85985	85896	85806	15	30	44	59	74
31	.85717	85627	85536	85446	85355	85264	85173	85081	84989	84897	16	30	46	61	76
32	.84805	84712	84619	84526	84433	84339	84245	84151	84057	83962	16	31	47	62	78
33	.83867	83772	83676	83581	83485	83389	83292	83195	83098	83001	16	32	48	64	80
34	.82904	82806	82708	82610	82511	82413	82314	82214	82115	82015	16	33	49	66	82
35	.81915	81815	81714	81614	81513	81412	81310	81208	81106	81004	17	34	51	68	84
36	.80902	80799	80696	80593	80489	80386	80282	80178	80073	79968	17	35	52	69	87
37	.79864	79758	79653	79547	79441	79335	79229	79122	79015	78908	18	35	53	71	89
38	.78801	78693	78586	78478	78369	78261	78152	78043	77934	77824	18	36	54	72	91
39	.77715	77605	77494	77384	77273	77162	77051	76940	76828	76717	18	37	56	74	93
40	.76604	76492	76380	76267	76154	76041	75927	75813	75700	75585	19	38	57	76	94
41	.75471	75356	75241	75126	75011	74896	74780	74664	74548	74431	19	39	58	77	96
42	.74314	74198	74080	73963	73846	73728	73610	73491	73373	73254	20	39	59	79	98
43	.73135	73016	72897	72777	72657	72537	72417	72297	72176	72055	20	40	60	80	100
44	.71934	71813	71691	71569	71447	71325	71203	71080	70957	70834	20	41	61	82	102
45	.70711	70587	70463	70339	70215	70091	69966	69842	69717	69591	21	41	62	83	104
46	.69466	69340	69214	69088	68962	68835	68709	68582	68455	68327	21	42	63	84	106
47	.68200	68072	67944	67816	67688	67559	67430	67301	67172	67043	21	43	64	86	107
48	.66913	66783	66653	66523	66393	66262	66131	66000	65869	65738	22	44	65	87	109
49	.65606	65474	65342	65210	65077	64945	64812	64679	64546	64412	22	44	66	88	111
50	.64279	64145	64011	63877	63742	63608	63473	63338	63203	63068	22	45	67	90	112
51	.62932	62796	62660	62524	62388	62251	62115	61978	61841	61704	23	45	68	91	114
52	.61566	61429	61291	61153	61015	60876	60738	60599	60460	60321	23	46	69	92	115
53	.60181	60042	59902	59763	59622	59482	59342	59201	59061	58920	23	47	70	94	117
54	.58779	58637	58496	58354	58212	58070	57928	57786	57643	57501	24	47	71	95	118
55	.57358	57215	57071	56928	56784	56641	56497	56353	56208	56064	24	48	72	96	120
56	.55919	55775	55630	55484	55339	55194	55048	54902	54756	54610	24	49	73	97	121
57	.54464	54317	54171	54024	53877	53730	53583	53435	53288	53140	24	49	74	98	123
58	.52992	52844	52696	52547	52399	52250	52101	51952	51803	51653	25	50	74	99	124
59	.51504	51354	51204	51054	50904	50754	50603	50453	50302	50151	25	50	75	100	125

TABLE 10.3 *(Continued)*

Deg.	0′ 0.0°	6′ 0.1°	12′ 0.2°	18′ 0.3°	24′ 0.4°	30′ 0.5°	36′ 0.6°	42′ 0.7°	48′ 0.8°	54′ 0.9°	Mean differences* 1	2	3	4	5
60	.50000	49849	49697	49546	49394	49242	49090	48938	48786	48634	25	51	76	101	126
61	.48481	48328	48175	48022	47869	47716	47562	47409	47255	47101	26	51	77	102	128
62	.46947	46793	46639	46484	46330	46175	46020	45865	45710	45554	26	52	77	103	129
63	.45399	45243	45088	44932	44776	44620	44464	44307	44151	43994	26	52	78	104	130
64	.43837	43680	43523	43366	43209	43051	42894	42736	42578	42420	26	53	79	105	131
65	.42262	42104	41945	41787	41628	41469	41310	41151	40992	40833	26	53	79	106	132
66	.40674	40514	40355	40195	40035	39875	39715	39555	39394	39234	27	53	80	107	133
67	.39073	38912	38752	38591	38430	38268	38107	37946	37784	37622	27	54	81	107	134
68	.37461	37299	37137	36975	36812	36650	36488	36325	36162	36000	27	54	81	108	135
69	.35837	35674	35511	35347	35184	35021	34857	34694	34530	34366	27	54	82	109	136
70	.34202	34038	33874	33710	33545	33381	33216	33051	32887	32722	27	55	82	110	137
71	.32557	32392	32227	32061	31896	31730	31565	31399	31233	31068	28	55	83	110	138
72	.30902	30736	30570	30403	30237	30071	29904	29737	29571	29404	28	56	83	111	139
73	.29237	29070	28903	28736	28569	28402	28234	28067	27899	27731	28	56	84	112	139
74	.27564	27396	27228	27060	26892	26724	26556	26387	26219	26050	28	56	84	112	140
75	.25882	25713	25545	25376	25207	25038	24869	24700	24531	24362	28	56	85	113	141
76	.24192	24023	23853	23684	23514	23345	23175	23005	22835	22665	28	57	85	113	141
77	.22495	22325	22155	21985	21814	21644	21474	21303	21132	20962	28	57	85	114	142
78	.20791	20620	20450	20279	20108	19937	19766	19595	19423	19252	29	57	86	114	143
79	.19081	18910	18738	18567	18395	18224	18052	17880	17708	17537	29	57	86	114	143
80	.17365	17193	17021	16849	16677	16505	16333	16160	15988	15816	29	57	86	115	144
81	.15643	15471	15299	15126	14954	14781	14608	14436	14263	14090	29	58	86	115	144
82	.13917	13744	13572	13399	13226	13053	12880	12706	12533	12360	29	58	87	115	144
83	.12187	12014	11840	11667	11494	11320	11147	10973	10800	10626	29	58	87	116	145
84	.10453	10279	10106	09932	09758	09585	09411	09237	09063	08889	29	58	87	116	145
85	.08716	08542	08368	08914	08020	07846	07672	07498	07432	07150	29	58	87	116	145
86	.06976	06802	06627	06453	06279	06105	05931	05756	05582	05408	29	58	87	116	145
87	.05234	05059	04885	04711	04536	04362	04188	04013	03839	03664	29	58	87	116	145
88	.03490	03316	03141	02967	02792	02618	02443	02269	02094	01920	29	58	87	117	145
89	.01745	01571	01396	01222	01047	00873	00698	00524	00349	00175	29	58	87	117	145
90	.0000														

*Numbers in difference columns to be subtracted.

TABLE 10.4 Tangents of Angles

Deg.	0′ 0.0°	6′ 0.1°	12′ 0.2°	18′ 0.3°	24′ 0.4°	30′ 0.5°	36′ 0.6°	42′ 0.7°	48′ 0.8°	54′ 0.9°	Mean differences 1	2	3	4	5
0	.00000	00175	00349	00524	00698	00873	01047	01222	01396	01571	29	58	87	116	146
1	.01746	01920	02095	02269	02444	02619	02793	02968	03143	03317	29	58	87	116	146
2	.03492	03667	03842	04016	04191	04366	04541	04716	04891	05066	29	58	87	117	146
3	.05241	05416	05591	05766	05941	06116	06291	06467	06642	06817	29	58	88	117	146
4	.06993	07168	07344	07519	07695	07870	08046	08221	08397	08573	29	59	88	117	146
5	.08749	08925	09101	09277	09453	09629	09805	09981	10158	10334	29	59	88	117	147
6	.10510	10687	10863	11040	11217	11394	11570	11747	11924	12101	29	59	88	118	147
7	.12278	12456	12633	12810	12988	13165	13343	13521	13698	13876	30	59	89	118	148
8	.14054	14232	14410	14588	14767	14945	15124	15302	15481	15660	30	59	89	119	149
9	.15838	16017	16196	16376	16555	16734	16914	17093	17273	17453	30	60	90	120	150
10	.17633	17813	17993	18173	18353	18534	18714	18895	19076	19257	30	60	90	120	150
11	.19438	19619	19801	19982	20164	20345	20527	20709	20891	21073	30	60	91	121	152
12	.21256	21438	21621	21804	21986	22169	22353	22536	22719	22903	30	61	92	122	153
13	.23087	23271	23455	23639	23823	24008	24193	24377	24562	24747	31	61	93	124	155
14	.24933	25118	25304	25490	25676	25862	26048	26235	26421	26608	31	62	93	124	155
15	.26795	26982	27169	27357	27545	27732	27920	28109	28297	28486	31	63	94	125	157
16	.28675	28864	29053	29242	29432	29621	29811	30001	30192	30382	32	63	95	127	158
17	.30573	30764	30955	31147	31338	31530	31722	31914	32106	32299	32	64	96	128	160
18	.32492	32685	32878	33072	33266	33460	33654	33848	34043	34238	32	65	97	129	162
19	.34433	34628	34824	35019	35216	35412	35608	35805	36002	36199	33	66	98	131	164
20	.36397	36595	36793	36991	37190	37388	37588	37787	37986	38186	33	66	99	133	166
21	.38386	38587	38787	38988	39190	39391	39593	39795	39997	40200	34	67	101	134	168
22	.40403	40606	40809	41013	41217	41421	41626	41831	42036	42242	34	68	102	136	170
23	.42447	42654	42860	43067	43274	43481	43689	43897	44105	44314	34	69	104	138	173
24	.44523	44732	44942	45152	45362	45573	45784	45995	46206	46418	35	70	105	141	176

TABLE 10.4 *(Continued)*

Deg.	0′ 0.0°	6′ 0.1°	12′ 0.2°	18′ 0.3°	24′ 0.4°	30′ 0.5°	36′ 0.6°	42′ 0.7°	48′ 0.8°	54′ 0.9°	Mean differences 1	2	3	4	5
25	.46631	46843	47056	47270	47483	47698	47912	48127	48342	48557	36	71	107	143	179
26	.48773	48989	49206	49423	49640	49858	50076	50295	50514	50733	36	73	109	145	182
27	.50953	51173	51393	51614	51835	52057	52279	52501	52724	52947	37	74	111	148	185
28	.53171	53395	53620	53844	54070	54296	54522	54748	54975	55203	38	75	113	151	188
29	.55431	55659	55888	56117	56347	56577	56808	57039	57271	57503	38	77	115	154	192
30	.57735	57968	58201	58435	58670	58905	59140	59376	59612	59849	39	78	118	157	196
31	.60086	60324	60562	60801	61040	61280	61520	61761	62003	62245	40	79	120	160	200
32	.62487	62730	62973	63217	63462	63707	63953	64199	64446	64693	41	82	123	164	205
33	.64941	65189	65438	65688	65938	66189	66440	66692	66944	67197	42	84	126	167	209
34	.67451	67705	67960	68215	68471	68728	68985	69243	69502	69761	43	86	129	171	214
35	.70021	70281	70542	70804	71066	71329	71593	78157	72122	72388	44	88	132	176	219
36	.72654	72921	73189	73457	73726	73996	74267	74538	74810	75082	45	90	135	180	225
37	.75355	75629	75904	76180	76456	76733	77010	77289	77568	77848	46	92	139	185	231
38	.78129	78410	78692	78975	79259	79544	79829	80115	80402	80690	47	95	142	190	237
39	.80978	81268	81558	81849	82141	82434	82727	83022	83317	83613	49	98	147	195	244
40	.83910	84208	84507	84806	85107	85408	85710	86014	86318	86623	50	100	151	201	252
41	.86929	87236	87543	87852	88162	88473	88784	89097	89410	89725	52	103	155	207	259
42	.90040	90357	90674	90993	91313	91633	91955	92277	92601	92926	53	107	160	214	268
43	.93252	93578	93906	94235	94565	94896	95229	95562	95897	96232	55	111	165	221	276
44	.96569	96907	97246	97586	97927	98270	98613	98958	99304	99652	57	114	171	229	286
45	1.00000	00350	00701	01053	01406	01761	02117	02474	02832	03192	58	118	177	237	296
46	1.03553	03915	04279	04644	05010	05378	05747	06117	06489	06862	61	123	184	245	307
47	1.07237	07613	07990	08369	08749	09131	09514	09899	10285	10672	63	127	191	255	319
48	1.11061	11452	11844	12238	12633	13029	13428	13828	14229	14632	66	132	199	265	331
49	1.15037	15443	15851	16261	16672	17085	17500	17916	18344	18754	69	138	207	276	344
50	1.19175	19599	20024	20451	20879	21310	21742	22176	22612	23050	72	143	216	288	359
51	1.23490	23931	24375	24820	25268	25717	26169	26622	27077	27535	75	150	225	300	375
52	1.27994	28456	28919	29385	29853	30323	30795	31269	31745	32224	78	157	235	314	392
53	1.32704	33187	33673	34160	34650	35142	35637	36134	36633	37134	82	164	247	329	411
54	1.37638	38145	38653	39165	36679	40195	40714	41235	41759	42286	86	172	259	345	431

55	1.42815	43347	43881	44418	44958	45501	46046	46595	47146	47700	91	181	272	362	453
56	1.48256	48816	49378	49944	50512	51084	51658	52235	52816	53400	95	191	286	382	477
57	1.53987	54576	55170	55767	56366	56969	57575	58184	58797	59414	100	201	302	403	504
58	1.60033	60657	61283	61914	62548	63185	83826	64471	65120	65772	106	213	319	426	533
59	1.66428	67088	67752	68419	69091	69766	70446	71129	71817	72509	113	226	339	452	564
60	1.73205	73905	74610	75319	76032	76749	77471	78198	78929	79665	120	240	360	481	600
61	1.80405	81150	81900	82654	83413	84177	84946	85720	86500	87283	128	255	383	511	639
62	1.88073	88867	89667	90472	91282	92098	92920	93746	94579	95417	136	273	409	546	683
63	1.96261	97111	97967	98828	99695	2.00569	2.01449	2.02335	2.03227	2.04125	146	292	438	584	731
64	2.05030	05942	06860	07785	08716	09654	10600	11552	12511	13477	157	314	471	629	786
65	2.14451	15432	16420	17416	18419	19430	20449	21475	22510	23553	169	338	508	677	846
66	2.24604	25663	26730	27806	28891	29984	31086	32197	33317	34447	183	366	549	732	915
67	2.35585	36733	37891	39058	40235	41421	42618	43825	45043	46270	199	397	596	795	994
68	2.47509	48758	50018	51289	52571	53865	55170	56487	57815	59156					
69	2.60509	61874	63252	64642	66046	67462	66892	70335	71792	73263					
70	2.74748	76247	77761	79289	80833	82391	83965	85556	87161	88783					
71	2.90421	92076	93748	95437	97144	98868	3.00611	3.02372	3.04152	3.05950					
72	3.07768	09606	11464	13341	15240	17159	19100	21063	23048	25055					
73	3.27085	29139	31216	33317	35443	37594	39771	41973	44202	46458					
74	3.48741	51053	53393	55761	58160	60588	63048	65538	68061	70616					
75	3.73205	75828	78485	81177	83906	86671	89474	92316	95196	98117					
76	4.01078	04081	07127	10216	13350	16530	19756	23030	26352	29724					
77	4.33148	36623	40152	43735	47374	51071	54826	58641	62518	66458					
78	4.70463	74534	78673	82882	87162	91516	95945	5.00451	5.05037	5.09704					
79	5.14455	19293	24218	29235	34345	39552	44857	50264	55777	61397					
80	5.67128	72974	78938	85024	91236	97576	6.04051	6.10664	6.17419	6.24321					
81	6.31375	38587	45961	53503	61220	69116	77199	85475	93952	7.02637					
82	7.11537	20661	30018	39616	49465	59575	69957	80622	91582	8.02848					
83	8.14435	26356	38625	51259	64275	77689	91520	9.05789	9.20516	9.35724					
84	9.51436	9.6768	9.8448	10.019	10.199	10.385	10.579	10.780	10.988	11.205					
85	11.430	11.664	11.909	12.163	12.429	12.706	12.996	13.300	13.617	13.951					
86	14.301	14.669	15.056	15.464	15.895	16.350	16.832	17.343	17.866	18.464					
87	19.081	19.740	20.446	21.205	22.022	22.904	23.859	24.898	26.031	27.271					
88	28.636	30.145	31.821	33.694	35.801	38.188	40.917	44.066	47.740	52.081					
89	57.290	63.657	71.615	81.847	95.489	114.60	143.24	190.98	286.48	572.96					
90	∞														

TABLE 10.5 Square Roots for Numbers from 1 to 10

											Mean differences								
	0	1	2	3	4	5	6	7	8	9	1	2	3	4	5	6	7	8	9
1.0	1.0000	1.0050	1.0100	1.0149	1.0198	1.0247	1.0296	1.0344	1.0392	1.0440	5	10	15	20	24	29	34	39	44
1.1	1.0488	1.0536	1.0583	1.0630	1.0677	1.0724	1.0770	1.0817	1.0863	1.0909	5	9	14	19	23	28	33	37	42
1.2	1.0954	1.1000	1.1045	1.1091	1.1136	1.1180	1.1225	1.1269	1.1314	1.1358	4	9	13	18	22	27	31	36	40
1.3	1.1402	1.1446	1.1489	1.1533	1.1576	1.1619	1.1662	1.1705	1.1747	1.1790	4	9	13	17	22	26	30	34	39
1.4	1.1832	1.1874	1.1916	1.1958	1.2000	1.2042	1.2083	1.2124	1.2166	1.2207	4	8	13	17	21	25	29	33	37
1.5	1.2247	1.2288	1.2329	1.2369	1.2410	1.2450	1.2490	1.2530	1.2570	1.2610	4	8	12	16	20	24	28	32	36
1.6	1.2649	1.2689	1.2728	1.2767	1.2806	1.2845	1.2884	1.2923	1.2961	1.3000	4	8	12	16	19	23	27	31	35
1.7	1.3038	1.3077	1.3115	1.3153	1.3191	1.3229	1.3266	1.3304	1.3342	1.3379	4	8	11	15	19	23	27	30	34
1.8	1.3416	1.3454	1.3491	1.3528	1.3565	1.3601	1.3638	1.3675	1.3711	1.3748	4	7	11	15	18	22	26	29	33
1.9	1.3784	1.3820	1.3856	1.3892	1.3928	1.3964	1.4000	1.4036	1.4071	1.4107	4	7	11	14	18	22	25	29	32
2.0	1.4142	1.4177	1.4213	1.4248	1.4283	1.4318	1.4353	1.4387	1.4422	1.4457	4	7	11	14	18	21	24	28	31
2.1	1.4491	1.4526	1.4560	1.4595	1.4629	1.4663	1.4697	1.4731	1.4765	1.4799	3	7	10	14	17	20	24	27	31
2.2	1.4832	1.4866	1.4900	1.4933	1.4966	1.5000	1.5033	1.5067	1.5100	1.5133	3	7	10	13	17	20	24	27	30
2.3	1.5166	1.5199	1.5232	1.5264	1.5297	1.5330	1.5362	1.5395	1.5427	1.5460	3	7	10	13	16	20	23	26	29
2.4	1.5492	1.5524	1.5556	1.5588	1.5620	1.5652	1.5684	1.5716	1.5748	1.5780	3	6	10	13	16	19	22	26	29
2.5	1.5811	1.5843	1.5875	1.5906	1.5937	1.5969	1.6000	1.6031	1.6062	1.6093	3	6	9	16	16	19	22	25	28
2.6	1.6125	1.6155	1.6186	1.6217	1.6248	1.6279	1.6310	1.6340	1.6371	1.6401	3	6	9	12	15	18	22	25	28
2.7	1.6432	1.6462	1.6492	1.6523	1.6553	1.6583	1.6613	1.6643	1.6673	1.6703	3	6	9	12	15	18	21	24	27
2.8	1.6733	1.6763	1.6793	1.6823	1.6852	1.6882	1.6912	1.6941	1.6971	1.7000	3	6	9	12	15	18	21	24	27
2.9	1.7029	1.7059	1.7088	1.7117	1.7146	1.7176	1.7205	1.7234	1.7263	1.7292	3	6	9	12	15	18	20	23	26
3.0	1.7321	1.7349	1.7378	1.7407	1.7436	1.7464	1.7493	1.7521	1.7550	1.7578	3	6	9	11	14	17	20	23	26
3.1	1.7607	1.7635	1.7664	1.7692	1.7720	1.7748	1.7776	1.7804	1.7833	1.7861	3	6	9	11	14	17	20	23	25
3.2	1.7889	1.7916	1.7944	1.7972	1.8000	1.8028	1.8055	1.8083	1.8111	1.8138	3	6	8	11	14	17	19	22	25
3.3	1.8166	1.8193	1.8221	1.8248	1.8276	1.8303	1.8330	1.8358	1.8385	1.8412	3	5	8	11	14	16	19	22	25
3.4	1.8439	1.8466	1.8493	1.8520	1.8547	1.8574	1.8601	1.8628	1.8655	1.8682	3	5	8	11	13	16	19	22	24
3.5	1.8708	1.8735	1.8762	1.8788	1.8815	1.8841	1.8868	1.8894	1.8921	1.8947	3	5	8	11	13	16	19	21	24
3.6	1.8974	1.9000	1.9026	1.9053	1.9079	1.9105	1.9131	1.9157	1.9183	1.9209	3	5	8	10	13	16	18	21	24
3.7	1.9235	1.9261	1.9287	1.9313	1.9339	1.9365	1.9391	1.9416	1.9442	1.9468	3	5	8	10	13	16	18	21	23
3.8	1.9494	1.9519	1.9545	1.9570	1.9596	1.9621	1.9647	1.9672	1.9698	1.9723	3	5	8	10	13	15	18	21	23
3.9	1.9748	1.9774	1.9799	1.9824	1.9849	1.9875	1.9900	1.9925	1.9950	1.9975	3	5	8	10	13	15	18	20	23

4.0	2.0000	2.0025	2.0050	2.0075	2.0100	2.0125	2.0149	2.0174	2.0199	2.0224	2	5	7	10	12	15	17	20	22
4.1	2.0248	2.0273	2.0298	2.0322	2.0347	2.0372	2.0396	2.0421	2.0445	2.0469	2	5	7	10	12	15	17	20	22
4.2	2.0494	2.0518	2.0543	2.0567	1.0591	2.0616	2.0640	2.0664	2.0688	2.0712	2	5	7	10	12	15	17	19	22
4.3	2.0736	2.0761	2.0785	2.0809	2.0833	2.0857	2.0881	2.0905	2.0928	2.0952	2	5	7	10	12	14	17	19	22
4.4	2.0976	2.1000	2.1024	2.1048	2.1071	2.1095	2.1119	2.1142	2.1166	2.1190	2	5	7	9	12	14	17	19	21
4.5	2.1213	2.1237	2.1260	2.1284	2.1307	2.1331	2.1354	2.1378	2.1401	2.1424	2	5	7	9	12	14	16	19	21
4.6	2.1448	2.1471	2.1494	2.1517	2.1541	2.1564	2.1587	2.1610	2.1633	2.1656	2	5	7	9	12	14	16	19	21
4.7	2.1679	2.1703	2.1726	2.1749	2.1772	2.1794	2.1817	2.1840	2.1863	2.1886	2	5	7	9	12	14	16	18	21
4.8	2.1909	2.1932	2.1954	2.1977	2.2000	2.2023	2.2045	2.2068	2.2091	2.2113	2	5	7	9	11	14	16	18	20
4.9	2.2136	2.2159	2.2181	2.2204	2.2226	2.2249	2.2271	2.2293	2.2316	2.2338	2	5	7	9	11	14	16	18	20
5.0	2.2361	2.2383	2.2405	2.2428	2.2450	2.2472	2.2494	2.2517	2.2539	2.2561	2	4	7	9	11	13	16	18	20
5.1	2.2583	2.2605	2.2627	2.2650	2.2672	2.2694	2.2716	2.2738	2.2760	2.2782	2	4	7	9	11	13	15	18	20
5.2	2.2804	2.2825	2.2847	2.2809	2.2891	2.2913	2.2935	2.2956	2.2978	2.3000	2	4	7	9	11	13	15	17	20
5.3	2.3022	2.3043	2.3065	2.3087	2.3108	2.3130	2.3152	2.3173	2.3195	2.3216	2	4	6	9	11	13	15	17	19
5.4	2.3238	2.3259	2.3281	2.3302	2.3324	2.3345	2.3367	2.3388	2.3409	2.3431	2	4	6	9	11	13	15	17	19
5.5	2.3452	2.3473	2.3495	2.3516	2.3537	2.3558	2.3580	2.3601	2.3622	2.3643	2	4	6	8	11	13	15	17	19
5.6	2.3664	2.3685	2.3707	2.3728	2.3749	2.3770	2.3791	2.3812	2.3833	2.3854	2	4	6	8	11	13	15	17	19
5.7	2.3875	2.3896	2.3917	2.3937	2.2958	2.3979	2.4000	2.4021	2.4042	2.4062	2	4	6	8	10	12	15	17	19
5.8	2.4083	2.4104	2.4125	2.4145	2.4166	2.4187	2.4207	2.4228	2.4249	2.4269	2	4	6	8	10	12	14	16	19
5.9	2.4290	2.4310	2.4331	2.4352	2.4372	2.4393	2.4413	2.4434	2.4454	2.4474	2	4	6	8	10	12	14	16	18
6.0	2.4495	2.4515	2.4536	2.4556	2.4576	2.4597	2.4617	2.4637	2.4658	2.4678	2	4	6	8	10	12	14	16	18
6.1	2.4698	2.4718	2.4739	2.4759	2.4779	2.4799	2.4819	2.4839	2.4860	2.4880	2	4	6	8	10	12	14	16	18
6.2	2.4900	2.4920	2.4940	2.4960	2.4980	2.5000	2.5020	2.5040	2.5060	2.5080	2	4	6	8	10	12	14	16	18
6.3	2.5100	2.5120	2.5140	2.5159	2.5179	2.5199	2.5219	2.5239	2.5259	2.5278	2	4	6	8	10	12	14	16	18
6.4	2.5298	2.5318	2.5338	2.5357	2.5377	2.5397	2.5417	2.5436	2.5456	2.5475	2	4	6	8	10	12	14	16	18
6.5	2.5495	2.5515	2.5534	2.5554	2.5573	2.5593	2.5612	2.5632	2.5652	2.5671	2	4	6	8	10	12	14	16	18
6.6	2.5690	2.5710	2.5729	2.5749	2.5768	2.5788	2.5807	2.5826	2.5846	2.5865	2	4	6	8	10	12	14	16	17
6.7	2.5884	2.5904	2.5923	2.5942	2.5962	2.5981	2.6000	2.6019	2.6038	2.6058	2	4	6	8	10	12	14	15	17
6.8	2.6077	2.6096	2.6115	2.6134	2.6153	2.6173	2.6192	2.6211	2.6230	2.6249	2	4	6	8	10	11	13	15	17
6.9	2.6268	2.6287	2.6306	2.6325	2.6344	2.6363	2.6382	2.6401	2.6420	2.6439	2	4	6	8	10	11	13	15	17
7.0	2.6458	2.6476	2.6495	2.6514	2.6533	2.6552	2.6571	2.6589	2.6608	2.6627	2	4	6	8	9	11	13	15	17
7.1	2.6646	2.6665	2.6683	2.6702	2.6721	2.6739	2.6758	2.6777	2.6796	2.6814	2	4	6	7	9	11	13	15	17
7.2	2.6833	2.6851	2.6870	2.6889	2.6907	2.6926	2.6944	2.6963	2.6981	2.7000	2	4	6	7	9	11	13	15	17
7.3	2.7019	2.7037	2.7055	2.7074	2.7092	2.7111	2.7129	2.7148	2.7166	2.7185	2	4	6	7	9	11	13	15	17
7.4	2.7203	2.7221	2.7240	2.7258	2.7276	2.7295	2.7313	2.7331	2.7350	2.7368	2	4	5	7	9	11	13	15	16

TABLE 10.5 *(Continued)*

											Mean differences								
	0	1	2	3	4	5	6	7	8	9	1	2	3	4	5	6	7	8	9
7.5	2.7386	2.7404	2.7423	2.7441	2.7459	2.7477	2.7495	2.7514	2.7532	2.7550	2	4	5	7	9	11	13	15	16
7.6	2.7568	2.7586	2.7604	2.7622	2.7641	2.7659	2.7677	2.7695	2.7713	2.7731	2	4	5	7	9	11	13	14	16
7.7	2.7749	2.7767	2.7785	2.7803	2.7821	2.7839	2.7857	2.7875	2.7893	2.7911	2	4	5	7	9	11	13	14	16
7.8	2.7928	2.7946	2.7964	2.7982	2.8000	2.8018	2.8036	2.8054	2.8071	2.8089	2	4	5	7	9	11	13	14	16
7.9	2.8107	2.8125	2.8142	2.8160	2.8178	2.8196	2.8213	2.8231	2.8249	2.8267	2	4	5	7	9	11	12	14	16
8.0	2.8284	2.8302	2.8320	2.8337	2.8355	2.8373	2.8390	2.8408	2.8425	2.8443	2	4	5	7	9	11	12	14	16
8.1	2.8460	2.8478	2.8496	2.8513	2.8531	2.8548	2.8566	2.8583	2.8601	2.8618	2	4	5	7	9	11	12	14	16
8.2	2.8636	2.8653	2.8671	2.8688	2.8705	2.8723	2.8740	2.8758	2.8775	2.8792	2	3	5	7	9	10	12	14	16
8.3	2.8810	2.8827	2.8844	2.8862	2.8879	2.8896	2.8914	2.8931	2.8948	2.8965	2	3	5	7	9	10	12	14	16
8.4	2.8983	2.9000	2.9017	2.9034	2.9052	2.9069	2.9086	2.9103	2.9120	2.9138	2	3	5	7	9	10	12	14	15
8.5	2.9155	2.9172	2.9189	2.9206	2.9223	2.9240	2.9257	2.9275	2.9292	2.9309	2	3	5	7	9	10	12	14	15
8.6	2.9326	2.9343	2.9360	2.9377	2.9394	2.9411	2.9428	2.9445	2.9462	2.9479	2	3	5	7	9	10	12	14	15
8.7	2.9496	2.9513	2.9530	2.9547	2.9563	2.9580	2.9597	2.9614	2.9631	2.9648	2	3	5	7	9	10	12	14	15
8.8	2.9665	2.9682	2.9698	2.9715	2.9732	2.9749	2.9766	2.9783	2.9799	2.9816	2	3	5	7	8	10	12	13	15
8.9	2.9833	2.9850	2.9866	2.9883	2.9900	2.9917	2.9933	2.9950	2.9967	2.9983	2	3	5	7	8	10	12	13	15
9.0	3.0000	3.0017	3.0033	3.0050	3.0067	3.0083	3.0100	3.0116	3.0133	3.0150	2	3	5	7	8	10	12	13	15
9.1	3.0166	3.0183	2.0199	2.0216	2.0232	3.0249	3.0265	3.0282	3.0299	3.0315	2	3	5	7	8	10	12	13	15
9.2	3.0332	3.0348	3.0364	3.0381	3.0397	3.0414	3.0430	3.0447	3.0463	3.0480	2	3	5	7	8	10	11	13	15
9.3	3.0496	3.0512	3.0529	3.0545	3.0561	3.0578	3.0594	3.0610	3.0627	3.0643	2	3	5	7	8	10	11	13	15
9.4	3.0659	3.0676	3.0692	3.0708	3.0725	3.0741	3.0757	3.0773	3.0790	3.0806	2	3	5	7	8	10	11	13	15
9.5	3.0822	3.0838	3.0854	3.0871	3.0887	3.0903	3.9019	3.0935	3.0952	3.0968	2	3	5	6	8	10	11	13	15
9.6	3.0984	3.1000	3.1016	3.1032	3.1048	3.1064	3.1081	3.1097	3.1113	3.1129	2	3	5	6	8	10	11	13	14
9.7	3.1145	3.1161	3.1177	3.1193	3.1209	3.1225	3.1241	3.1257	3.1273	3.1289	2	3	5	6	8	10	11	13	14
9.8	3.1305	3.1321	3.1337	3.1353	3.1360	3.1385	3.1401	3.1417	3.1432	3.1448	2	3	5	6	8	10	11	13	14
9.9	3.1464	3.1480	3.1496	3.1512	3.1528	3.1544	3.1559	3.1575	3.1591	3.1607	2	3	5	6	8	9	11	13	14

TABLE 10.6 Square Roots for Numbers from 10 to 100

											Mean differences								
	0	1	2	3	4	5	6	7	8	9	1	2	3	4	5	6	7	8	9
10	3.1623	3.1780	3.1937	3.2094	3.2249	3.2404	3.2558	3.2711	3.2863	3.3015	15	31	46	62	77	92	108	123	139
11	3.3166	3.3317	3.3466	3.3615	3.3764	3.3912	3.4059	3.4205	3.4351	3.4496	15	30	44	59	74	89	104	118	133
12	3.4641	3.4785	3.4928	3.5071	3.5214	3.5355	3.5496	3.5637	3.5777	3.5917	14	28	43	57	71	85	99	114	128
13	3.6056	3.6194	3.6332	3.6469	3.6606	3.6742	3.6878	3.7014	3.7148	3.7283	14	27	41	54	68	82	95	109	122
14	3.7417	3.7550	3.7683	3.7815	3.7947	3.8079	3.8210	3.8341	3.8471	3.8601	13	26	39	52	66	79	92	105	118
15	3.8730	3.8859	3.8987	3.9115	3.9243	3.9370	3.9497	3.9623	3.9749	3.9875	13	25	38	51	64	76	89	102	114
16	4.0000	4.0125	4.0249	4.0373	4.0497	4.0620	4.0743	4.0866	4.0988	4.1110	12	25	37	49	62	74	86	98	111
17	4.1231	4.1352	4.1473	4.1593	4.1713	4.1833	4.1952	4.2071	4.2190	4.2308	12	24	36	48	60	72	84	96	108
18	4.2426	4.2544	4.2661	4.2778	4.2895	4.3012	4.3128	4.3243	4.3359	4.3474	12	23	35	46	58	70	81	93	104
19	4.3589	4.3704	4.3818	4.3932	4.4045	4.4159	4.4272	4.4385	4.4497	4.4609	11	23	34	45	57	68	79	90	102
20	4.4721	4.4833	4.4944	4.5056	5.5166	4.5277	4.5387	4.5497	4.5607	4.5717	11	22	33	44	56	67	78	89	100
21	4.5826	4.5935	4.6043	4.6152	4.6260	4.6368	4.6476	4.6583	4.6690	4.6797	11	22	32	43	54	65	76	86	97
22	4.6904	4.7011	4.7117	4.7223	4.7329	4.7434	4.7539	4.7645	4.7749	4.7854	11	21	32	42	53	63	74	84	95
23	4.7958	4.8062	4.8166	4.8270	4.8374	4.8477	4.8580	4.8683	4.8785	4.8888	10	21	31	41	52	62	72	82	93
24	4.8990	4.9092	4.9193	4.9295	4.9396	4.9497	4.9598	4.9699	4.9800	4.9900	10	20	30	40	51	61	71	81	91
25	5.0000	5.0100	5.0200	5.0299	5.0398	5.0498	5.0596	5.0695	5.0794	5.0892	10	20	30	40	50	59	69	79	89
26	5.0990	5.1088	5.1186	5.1284	5.1381	5.1478	5.1575	5.1672	5.1769	5.1865	10	19	29	39	49	58	68	78	87
27	5.1962	5.2058	5.2154	5.2249	5.2345	5.2440	5.2536	5.2631	5.2726	5.2820	10	19	29	38	48	57	67	76	86
28	5.2915	5.3009	5.3104	5.3198	5.3292	5.3385	5.3479	5.3572	5.3666	5.3759	9	19	28	38	47	56	66	75	85
29	5.3852	5.3944	5.4037	5.4129	5.4222	5.4314	5.4406	5.4498	5.4589	5.4681	9	18	28	37	46	55	64	74	83
30	5.4772	5.4863	5.4995	5.5045	5.5136	5.5227	5.5317	5.5408	5.5498	5.5588	9	18	27	36	46	55	64	73	82
31	5.5678	5.5767	5.5857	5.5946	5.6030	5.6125	5.6214	5.6303	5.6391	5.6480	9	18	27	36	45	53	62	71	80
32	5.6569	5.6657	5.6745	5.6833	5.6921	5.7009	5.7096	5.7184	5.7271	5.7359	9	18	26	35	44	53	62	70	79
33	5.7446	5.7533	5.7619	5.7706	5.7793	5.7879	5.7966	5.8052	5.8138	5.8224	9	17	26	34	43	52	60	69	77
34	5.8310	5.8395	5.8481	5.8566	5.8652	5.8737	5.8822	5.8907	5.8992	5.9076	9	17	26	34	43	51	60	68	77
35	5.9161	5.9245	5.9330	5.9414	5.9498	5.9582	5.9666	5.9749	5.9833	5.9917	8	17	25	34	42	50	59	67	76
36	6.0000	6.0083	6.0166	6.0249	6.0332	6.0415	6.0498	6.0581	6.0663	6.0745	8	17	25	33	42	50	58	66	75
37	6.0828	6.0910	6.0992	6.1074	6.1156	6.1237	6.1319	6.1400	6.1482	6.1563	8	16	25	33	41	49	57	66	74
38	6.1644	6.1725	6.1806	6.1887	6.1968	6.2048	6.2129	6.2209	6.2290	6.2370	8	16	24	32	41	49	57	65	73
39	6.2450	6.2530	6.2610	6.2690	6.2769	6.2849	6.2929	6.3008	6.3087	6.3166	8	16	24	32	40	48	56	64	72

TABLE 10.6 (*Continued*)

	0	1	2	3	4	5	6	7	8	9	Mean differences								
											1	2	3	4	5	6	7	8	9
40	6.3246	6.3325	6.3403	6.3482	6.3561	6.3640	6.3718	6.2797	6.3875	6.3953	8	16	24	32	40	47	55	63	71
41	6.4031	6.4109	6.4187	6.4265	6.4343	6.4420	6.4498	6.4576	6.4653	6.4730	8	16	23	31	39	47	55	62	70
42	6.4807	6.4885	6.4962	6.5038	6.5115	6.5192	6.5269	6.5345	6.5422	6.5498	8	15	23	31	39	46	54	62	69
43	6.5574	6.5651	6.5727	6.5803	6.5879	6.5955	6.6030	6.6106	6.6182	6.6257	8	15	23	30	38	46	53	61	68
44	6.6332	6.6408	6.6483	6.6558	6.6633	6.6708	6.6783	6.6858	6.6933	6.7007	8	15	23	30	38	45	53	60	68
45	6.7082	6.7157	6.7231	6.7305	6.7380	6.7454	6.7528	6.7602	6.7676	6.7750	7	15	22	30	37	44	52	59	67
46	6.7823	6.7897	6.7971	6.8044	6.8118	6.8191	6.8264	6.8337	6.8411	6.8484	7	15	22	29	37	44	51	58	66
47	6.8557	6.8629	6.8702	6.8775	6.8848	6.8920	6.8993	6.9065	6.9138	6.9210	7	15	22	29	37	44	51	58	66
48	6.9282	6.9354	6.9426	6.9498	6.9570	6.9642	6.9714	6.9785	6.9857	6.9929	7	14	22	29	36	43	50	58	65
49	7.0000	7.0071	7.0143	7.0214	7.0285	7.0356	7.0427	7.0498	7.0569	7.0640	7	14	21	28	36	43	50	57	64
50	7.0711	7.0781	7.0852	7.0922	7.0993	7.1063	7.1134	7.1204	7.1274	7.1344	7	14	21	28	35	42	49	56	63
51	7.1414	7.1484	7.1554	7.1624	7.1694	7.1764	7.1833	7.1903	7.1972	7.2042	7	14	21	28	35	42	49	56	63
52	7.2111	7.2180	7.2250	7.2319	7.2388	7.2457	7.2526	7.2595	7.2664	7.2732	7	14	21	28	35	41	48	55	62
53	7.2801	7.2870	7.2938	7.3007	7.3075	7.3144	7.3212	7.3280	7.3348	7.3417	7	14	20	27	34	41	48	54	61
54	7.3485	7.3553	7.3621	7.3689	7.3756	7.3824	7.3892	7.3959	7.4027	7.4095	7	14	20	27	34	41	48	54	61
55	7.4162	7.4229	7.4297	7.4364	7.4431	7.4498	7.4565	7.4632	7.4699	7.4766	7	13	20	27	34	40	47	54	60
56	7.4833	7.4900	7.4967	7.5033	7.5100	7.5166	7.5233	7.5299	7.5366	7.5432	7	13	20	27	34	40	47	54	60
57	7.5498	7.5565	7.5631	7.5697	7.5763	7.5829	7.5895	7.5961	7.6026	7.6092	7	13	20	26	33	40	46	53	59
58	7.6158	7.6223	7.6289	7.6354	7.6420	7.6485	7.6551	7.6616	7.6681	7.6746	7	13	20	26	33	39	46	52	59
59	7.6811	7.6877	7.6942	7.7006	7.7071	7.7136	7.7201	7.7266	7.7330	7.7395	7	13	20	26	33	39	46	52	59
60	7.7460	7.7524	7.7589	7.7653	7.7717	7.7782	7.7846	7.7910	7.7974	7.8038	6	13	19	26	32	38	15	51	58
61	7.8102	7.8166	7.8230	7.8294	7.8358	7.8422	7.8486	7.8549	7.8613	7.8677	6	13	19	26	32	38	45	51	58
62	7.8740	7.8804	7.8867	7.8930	7.8994	7.9057	7.9120	7.9183	7.9246	7.9310	6	13	19	25	32	38	44	50	57
63	7.9373	7.9436	7.9498	7.9561	7.9624	7.9687	7.9750	7.9812	7.9875	7.9937	6	13	19	25	32	38	44	50	57
64	8.0000	8.0062	8.0125	8.0187	8.0250	8.0312	8.0374	8.0436	8.0498	8.0561	6	12	19	25	31	37	43	50	56
65	8.0623	8.0685	8.0747	8.0808	8.0870	8.0932	8.0994	8.1056	8.1117	8.1179	6	12	19	25	31	37	43	50	56
66	8.1240	8.1302	8.1363	8.1425	8.1486	8.1548	8.1609	8.1670	8.1731	8.1792	6	12	18	24	31	37	43	49	55
67	8.1854	8.1915	8.1976	8.2037	8.2098	8.2158	8.2219	8.2280	8.2341	8.2401	6	12	18	24	31	37	43	49	55
68	8.2462	8.2523	8.2583	8.2644	8.2704	8.2765	8.2825	8.2885	8.2946	8.3006	6	12	18	24	30	36	42	48	54
69	8.3066	8.3126	8.3187	8.3247	8.3307	8.3367	8.3427	8.3487	8.3546	8.3606	6	12	18	24	30	36	42	48	54

70	8.3666	8.3726	8.3785	8.3845	8.3905	8.3964	8.4024	8.4083	8.4143	8.4202	6	12	18	24	30	36	42	48	54
71	8.4261	8.4321	8.4380	8.4439	8.4499	8.4558	8.4617	8.4676	8.4735	8.4794	6	12	18	24	30	35	41	47	53
72	8.4853	8.4912	8.4971	8.5029	8.5088	8.5147	8.5206	8.5264	8.5323	8.5381	6	12	18	24	30	35	41	47	53
73	8.5440	8.5499	8.5557	8.5615	8.5674	8.5732	8.5790	8.5849	8.5907	8.5965	6	12	17	23	29	35	41	46	52
74	8.6023	8.6081	8.6139	8.6197	8.6255	8.6313	8.6371	8.6429	8.6487	8.6545	6	12	17	23	29	35	41	46	52
75	8.6603	8.6660	8.6718	8.6776	8.6833	8.6891	8.6948	8.7006	8.7063	8.7121	6	12	17	23	29	35	41	46	52
76	8.7178	8.7235	8.7293	8.7350	8.7407	8.7464	8.7521	8.7579	8.7636	8.7693	6	11	17	23	29	34	40	46	51
77	8.7750	8.7807	8.7864	8.7920	8.7977	8.8034	8.8091	8.8148	8.8204	8.8261	6	11	17	23	29	34	40	46	51
78	8.8318	8.8374	8.8431	8.8487	8.8544	8.8600	8.8657	8.8713	8.8769	8.8826	6	11	17	22	28	34	39	45	50
79	8.8882	8.8938	8.8994	8.9051	8.9107	8.9163	8.9219	8.9275	8.9331	8.9387	6	11	17	22	28	34	39	45	50
80	8.9443	9.9499	8.9554	8.9610	8.9666	8.9722	8.9778	8.9833	8.9889	8.9944	6	11	17	22	28	34	39	45	50
81	9.0000	9.0056	9.0111	9.0167	9.0222	9.0277	9.0333	9.0388	9.0443	9.0499	6	11	17	22	28	33	39	44	50
82	9.0554	9.0609	9.0664	9.0719	9.0774	9.0830	9.0885	9.0940	9.0995	9.1049	6	11	17	22	28	33	39	44	50
83	9.1104	9.1159	9.1214	9.1269	9.1324	9.1378	9.1433	9.1488	9.1542	9.1597	6	11	17	22	28	33	39	44	50
84	9.1652	9.1706	9.1761	9.1815	9.1869	9.1924	9.1978	9.2033	9.2087	9.2141	5	11	16	22	27	32	38	43	49
85	9.2195	9.2250	9.2304	9.2358	9.2412	9.2466	9.2520	9.2574	9.2628	9.2682	5	11	16	22	27	32	38	43	49
86	9.2736	9.2790	9.2844	9.2898	9.2952	9.3005	9.3059	9.3113	9.3167	9.3220	5	11	16	22	27	32	38	43	49
87	9.3274	9.3327	9.3381	9.3434	9.3488	9.3541	9.3595	9.3648	9.3702	9.3755	5	11	16	21	27	32	37	42	48
88	9.3808	9.3862	9.3915	9.3968	9.4021	9.4074	9.4128	9.4181	9.4234	9.4287	5	11	16	21	27	32	37	42	48
89	9.4340	9.4393	9.4446	9.4499	9.4552	9.4604	9.4657	9.4710	9.4763	9.4816	5	11	16	21	27	32	37	42	48
90	9.4868	9.4921	9.4974	9.5026	9.5079	9.5131	9.5184	9.5237	9.5289	9.5341	5	11	16	21	27	32	37	42	48
91	9.5394	9.5446	9.5499	9.5551	9.5603	9.5656	9.5708	9.5760	9.5812	9.5864	5	11	16	21	26	31	36	42	47
92	9.5917	9.5969	9.6021	9.6073	9.6125	9.6177	9.6229	9.6281	9.6333	9.6385	5	11	16	21	26	31	36	42	47
93	9.6437	9.6488	9.6540	9.6592	9.6644	9.6695	9.6747	9.6799	9.6850	9.6902	5	10	16	21	26	31	36	42	47
94	9.6954	9.7005	9.7057	9.7108	9.7160	9.7211	9.7263	9.7314	9.7365	9.7417	5	10	15	20	26	31	36	41	46
95	9.7468	9.7519	9.7570	9.7622	9.7673	9.7724	9.7775	9.7826	9.7877	9.7929	5	10	15	20	26	31	36	41	46
96	9.7980	9.8031	9.8082	9.8133	9.8184	9.8234	9.8285	9.8336	9.8387	9.8438	5	10	15	20	26	31	36	41	46
97	9.8489	9.8539	9.8590	9.8641	9.8691	9.8742	9.8793	9.8843	9.8894	9.8944	5	10	15	20	26	31	36	41	46
98	9.8995	9.9045	9.9096	9.9146	9.9197	9.9247	9.9298	9.9348	9.9398	9.9448	5	10	15	20	25	30	35	40	45
99	9.9499	9.9549	9.9599	9.9649	9.9700	9.9750	9.9800	9.9850	9.9900	9.9950	5	10	15	20	25	30	35	40	45

TABLE 10.7 Areas of Circles: Diameters Increasing by Tenths

Diameter	0.0	0.1	0.2	0.3	0.4	0.5	0.6	0.7	0.8	0.9
0	0.0	0.0078	0.0314	0.0706	0.1256	0.1963	0.2827	0.3848	0.5027	0.6362
1	0.78540	0.95033	1.1310	1.3273	1.5394	1.7671	2.0106	2.2698	2.5447	2.8353
2	3.1416	3.4636	3.8013	4.1548	4.5239	4.9087	5.3093	5.7255	6.1575	6.6052
3	7.0686	7.5477	8.0425	8.5530	9.0792	9.6211	10.179	10.752	11.341	11.946
4	12.566	13.202	13.854	14.522	15.205	15.904	16.619	17.349	18.096	18.857
5	19.635	20.428	21.237	22.062	22.902	23.758	24.630	25.518	26.421	27.340
6	28.274	29.225	30.191	31.172	32.170	33.183	34.212	35.257	36.317	37.393
7	38.485	39.592	40.715	41.854	43.008	44.179	45.365	46.566	47.784	49.017
8	50.266	51.530	52.810	54.106	55.418	56.745	58.088	59.447	60.821	62.211
9	63.617	65.039	66.476	67.929	69.398	70.882	72.382	73.898	75.430	76.977
10	78.540	80.118	81.713	83.323	84.949	86.590	88.247	89.920	91.609	93.313
11	95.033	96.769	98.520	100.29	102.07	103.87	105.68	107.51	109.36	111.22
12	113.10	114.99	116.90	118.82	120.76	122.72	124.69	126.68	128.68	130.70
13	132.73	134.78	136.85	138.93	141.03	143.14	145.27	147.41	149.57	151.75
14	153.94	156.14	158.37	160.61	162.86	165.13	167.41	169.72	172.03	174.37
15	176.71	179.08	181.46	183.85	186.26	188.69	191.13	193.59	196.07	198.56
16	201.06	203.58	206.12	208.67	211.24	213.82	216.42	219.94	221.67	224.32
17	226.98	229.66	232.35	235.06	237.79	249.53	243.28	246.06	248.85	251.65
18	254.47	257.30	260.15	263.02	265.90	268.80	271.72	274.65	277.59	280.55
19	283.53	286.52	289.53	292.55	295.59	298.65	301.72	304.80	307.91	311.02
20	314.16	317.31	320.47	323.65	326.85	330.06	333.29	336.53	339.79	343.07
21	346.36	349.67	352.99	356.33	359.68	363.05	366.43	369.84	373.25	376.68
22	380.13	383.60	387.08	390.57	394.08	397.61	401.15	404.71	408.28	411.87
23	415.48	419.10	422.73	426.38	430.05	433.74	437.43	441.15	444.88	448.63
24	452.39	456.17	459.96	463.77	467.59	471.43	475.29	479.16	483.05	486.95
25	490.87	494.81	498.76	502.73	506.71	510.70	514.72	518.75	522.79	526.85
26	530.93	535.02	539.13	543.25	547.39	551.55	555.72	559.90	564.10	568.32
27	572.56	576.80	581.07	585.35	589.65	593.96	598.28	602.63	606.99	611.36
28	615.75	620.16	624.58	629.02	633.47	637.94	642.42	646.92	651.44	655.97
29	660.52	665.08	669.66	674.26	678.87	684.49	683.13	692.79	697.46	702.15

30	706.86	711.58	716.31	721.07	725.83	730.62	735.41	740.23	745.06	749.91
31	754.77	759.64	764.54	769.45	774.37	779.31	784.27	789.24	794.23	799.23
32	804.25	809.28	814.33	819.49	824.48	829.58	834.69	839.82	844.96	850.12
33	855.30	860.49	865.70	870.92	876.16	881.41	886.68	891.97	897.27	902.59
34	907.92	913.27	918.63	924.01	929.41	934.82	940.25	945.69	951.15	956.62
35	962.11	967.62	973.14	978.68	984.23	989.80	995.38	1000.98	1006.6	1012.2
36	1017.9	1023.5	1029.2	1034.9	1040.6	1046.3	1052.1	1057.8	1063.6	1069.4
37	1075.2	1081.0	1086.9	1092.7	1098.6	1104.5	1110.4	1176.3	1122.2	1128.1
38	1134.1	1140.1	1146.1	1152.1	1158.1	1164.2	1170.2	1176.3	1182.4	1188.5
39	1194.6	1200.7	1206.9	1213.0	1219.2	1225.4	1231.6	1237.6	1244.1	1250.4
40	1256.6	1262.9	1269.2	1275.6	1281.9	1288.2	1294.6	1301.0	1307.4	1313.8
41	1320.3	1326.7	1333.2	1339.6	1346.1	1352.6	1359.2	1365.7	1372.3	1378.8
42	1385.4	1392.0	1398.7	1405.3	1412.0	1418.6	1425.3	1432.0	1438.7	1445.4
43	1452.2	1459.0	1465.7	1472.5	1479.3	1486.2	1493.0	1499.9	1506.7	1513.6
44	1520.5	1527.4	1534.4	1541.3	1548.3	1555.3	1562.3	1569.3	1576.3	1583.4
45	1590.4	1597.5	1604.6	1611.7	1618.8	1626.0	1633.1	1640.3	1647.5	1654.7
46	1661.9	1669.1	1676.4	1683.6	1690.9	1698.2	1705.5	1712.9	1720.2	1727.6
47	1743.9	1742.3	1749.7	1757.2	1764.6	1772.0	1779.5	1787.0	1794.5	1802.0
48	1809.6	1817.1	1824.7	1832.2	1839.8	1847.4	1855.1	1862.7	1870.4	1878.0
49	1885.7	1893.4	1901.2	1908.9	1916.6	1924.4	1932.2	1940.0	1947.8	1955.6
50	1963.5	1971.4	1979.2	1987.1	1995.0	2003.0	2010.9	2018.8	2026.8	2034.8
51	2042.8	2050.8	2058.9	2066.9	2075.0	2083.1	2091.2	2099.3	2107.4	2115.6
52	2123.7	2131.9	2140.1	2148.3	2156.5	2164.7	2173.0	2181.3	2189.6	2197.9
53	2206.2	2214.5	2222.9	2231.2	2239.6	2248.0	2256.4	2264.8	2273.3	2281.7
54	2290.2	2298.7	2307.2	2315.7	2324.3	2332.8	2341.4	2350.0	2358.6	2367.2
55	2375.8	2384.5	2393.1	2401.8	2410.5	2419.2	2428.0	2436.7	2445.4	2454.2
56	2463.0	2471.8	2480.6	2489.5	2498.3	2507.2	2516.1	2525.0	2533.9	2542.8
57	2551.8	2560.7	2569.7	2578.7	2587.7	2596.7	2605.8	2614.8	2623.9	2633.0
58	2642.1	2651.2	2660.3	2669.5	2678.6	2687.8	2697.0	2706.2	2715.5	2724.7
59	2734.0	2743.2	2752.5	2761.8	2771.2	2780.5	2789.9	2799.2	2808.6	2818.0

TABLE 10.7 *(Continued)*

Diameter	0.0	0.1	0.2	0.3	0.4	0.5	0.6	0.7	0.8	0.9
60	2827.4	2836.9	2846.3	2855.8	2865.3	2874.7	2884.3	2893.8	2903.3	2912.9
61	2922.5	2932.1	2941.7	2951.3	2960.9	2070.6	2980.2	2989.9	2999.6	3009.3
62	3019.1	3028.8	3038.6	3048.4	3058.1	3068.0	3077.8	3087.6	3097.5	3107.4
63	3117.2	3127.1	3137.1	3147.0	3157.0	3166.9	3176.9	3186.9	3196.9	3206.9
64	3217.0	3227.1	3237.1	3247.2	3257.3	3267.4	3277.6	3287.7	3297.9	3308.1
65	3318.3	3328.5	3338.8	3349.0	3359.3	3369.6	3379.8	3390.2	3400.5	3410.8
66	3421.2	3431.6	3442.0	3452.4	3462.8	3473.2	3483.7	3494.1	3504.6	3515.1
67	3525.7	3536.2	3546.7	3557.3	3567.9	3578.5	3589.1	3599.7	3610.3	3621.0
68	3631.7	3642.4	3653.1	3663.8	3674.5	3685.3	3696.1	3706.8	3717.6	3728.4
69	3739.3	3750.1	3761.0	3771.9	3782.8	3793.7	3804.6	3815.5	3826.5	3837.5
70	3848.5	3859.4	3870.5	3881.5	3892.6	3903.6	3914.7	3925.8	3936.9	3948.0
71	3959.2	3970.3	3981.5	3992.7	4003.9	4015.1	4026.4	4037.6	4048.9	4060.2
72	4071.5	4082.8	4094.2	4105.5	4116.9	4128.2	4139.6	4151.1	4162.5	4173.9
73	4185.4	4196.9	4208.3	4219.9	4231.4	4242.9	4254.5	4266.0	4277.6	4289.2
74	4300.8	4312.5	4324.1	4335.8	4347.5	4359.2	4370.9	4382.6	4394.3	4406.1
75	4417.9	4429.6	4441.5	4453.3	4465.1	4477.0	4488.8	4500.7	4512.6	4524.5
76	4536.5	4548.4	4560.4	4572.3	4584.3	4596.3	4608.4	4620.4	4632.5	4644.5
77	4656.6	4668.7	4680.8	4693.0	4705.1	4717.3	4729.5	4741.7	4753.9	4766.1
78	4778.4	4790.6	4802.9	4815.2	4827.5	4839.8	4852.2	4864.5	4876.9	4889.3
79	4901.7	4914.1	4926.5	4939.0	4951.4	4963.9	4976.4	4988.9	5001.4	5014.0
80	5026.6	5039.1	5051.7	5064.3	5076.9	5089.6	5102.2	5114.9	5127.6	5140.3
81	5153.0	5165.7	5178.5	5191.2	5204.0	5216.8	5229.6	5242.4	5255.3	5268.1
82	5281.0	5293.9	5306.8	5319.7	5332.7	5345.6	5358.6	5371.6	5384.6	5397.6
83	5410.6	542.6	5436.7	5449.8	5462.9	5476.0	5489.1	5502.3	5515.4	5528.6
84	5541.8	5555.0	5568.2	5581.4	5594.7	5607.9	5621.2	5634.5	5647.8	5661.2
85	5674.5	5687.9	5701.2	5714.6	5728.0	5741.5	5754.9	5768.3	5781.8	5795.3
86	5808.8	5822.3	5835.8	5849.4	5863.0	5876.6	5890.1	5903.7	5917.4	5931.0
87	5944.7	5958.3	5972.0	5985.7	5999.5	6013.2	6027.0	6040.7	6054.5	6968.3
88	6982.1	6096.0	6109.8	6123.7	6137.5	6151.4	6165.3	6179.3	6193.2	6207.2
89	6221.1	6235.1	6249.1	6263.1	6277.2	6291.2	6305.3	6319.4	6333.5	6347.6

90	6361.7	6375.9	6390.0	6404.2	6418.4	6432.6	6446.8	6461.1	6475.3	6489.6
91	6503.9	6518.2	6532.5	6546.8	6561.2	6575.6	6589.9	6604.3	6618.7	6633.2
92	6647.6	6662.1	6676.5	6691.0	6705.5	6720.1	6734.6	6749.1	6793.7	6778.3
93	6792.9	6807.5	6822.2	6836.8	6851.5	6866.1	6880.8	6895.6	6910.3	6925.0
94	6939.8	6954.6	6969.3	6984.1	6999.0	7013.8	7028.7	7043.5	7058.4	7073.3
95	7088.2	7103.1	7118.1	7133.1	7148.0	7163.0	7178.0	7193.1	7208.1	7223.2
96	7218.2	7253.3	7268.4	7283.5	7298.7	7313.8	7329.0	7344.2	7359.4	7374.6
97	7389.8	7405.1	7420.3	7435.6	7450.9	7466.2	7481.5	7496.9	7512.2	7527.6
98	7543.0	7558.4	7573.8	7589.2	7604.7	7620.1	7635.6	7651.1	7666.6	7682.1
99	7697.7	7713.2	7728.8	7744.4	7760.0	7775.6	7791.3	7806.9	7822.6	7838.3
100	7854.0	7869.7	7885.4	7901.2	7916.9	7932.7	7948.5	7964.3	7980.1	7996.0

TABLE 10.8 Areas of Circles: Diameters Increasing by Eighths

Diameter	0	⅛	¼	⅜	½	⅝	¾	⅞
0	0.0	0.01227	0.04909	0.11045	0.19635	0.30680	0.44179	0.60132
1	0.78540	0.99402	1.2272	1.4849	1.7671	2.0739	2.4053	2.7612
2	3.1416	3.5466	3.9761	4.4301	4.9087	5.4119	5.9396	6.4918
3	7.0686	7.6699	8.2958	8.9462	9.6211	10.321	11.045	11.793
4	12.566	13.364	14.186	15.033	15.904	16.800	17.721	18.665
5	19.635	20.629	21.648	22.691	23.758	24.850	25.967	27.109
6	28.274	29.465	30.680	31.919	33.183	34.472	35.785	37.122
7	38.485	39.871	41.282	42.718	44.179	45.664	47.173	48.707
8	50.265	51.849	53.456	55.088	56.745	58.426	60.132	61.862
9	63.617	65.397	67.201	69.029	70.882	72.760	74.662	76.589
10	78.540	80.516	82.516	84.541	86.590	88.664	90.763	92.886
11	95.033	97.205	99.402	101.62	103.87	106.14	108.43	110.75
12	113.10	115.47	117.86	120.28	122.72	125.19	127.68	130.19
13	132.73	135.30	137.89	140.50	143.14	145.80	148.49	151.20
14	153.94	156.70	159.48	162.30	165.13	167.99	170.87	173.78
15	176.71	179.67	182.65	185.66	188.69	191.75	194.83	197.93
16	201.06	204.22	207.39	210.60	213.82	217.08	220.35	223.65
17	226.98	230.33	233.71	237.10	240.53	243.98	247.45	250.95
18	254.47	258.02	261.59	265.18	268.80	272.45	276.12	279.81
19	283.53	287.27	291.04	294.83	298.65	302.49	302.35	310.24
20	314.16	318.10	322.06	326.05	330.06	334.10	338.16	342.25
21	346.36	350.50	354.66	358.84	363.05	367.28	371.54	375.83
22	380.13	384.46	388.82	393.20	397.61	402.04	406.49	410.97
23	415.48	420.00	424.56	429.13	433.74	438.36	443.01	447.69
24	452.39	457.11	461.86	466.64	471.44	476.26	481.11	485.98
25	490.87	495.79	500.74	505.71	510.71	515.72	520.77	525.84
26	530.93	536.05	541.19	546.35	551.55	556.76	562.00	567.27
27	572.56	577.87	583.21	588.57	593.96	599.37	604.81	610.27
28	615.75	621.26	626.80	632.36	637.94	643.55	649.18	654.84
29	660.52	666.23	671.96	677.71	683.49	689.30	695.13	700.98

30	706.86	712.76	718.69	724.64	730.62	736.62	742.64	748.69
31	754.77	760.87	766.99	773.14	779.31	785.51	791.73	797.98
32	804.25	810.54	816.86	823.21	829.58	835.97	842.39	848.83
33	855.30	861.79	868.31	874.85	881.41	888.00	894.62	901.26
34	907.92	914.61	921.32	928.06	934.82	941.61	948.42	955.25
35	962.11	969.00	975.91	982.84	989.80	966.78	1003.8	1010.8
36	1017.9	1025.0	1032.1	1039.2	1046.3	1053.5	1060.7	1068.0
37	1075.2	1082.5	1089.8	1097.1	1104.5	1111.8	1119.2	1126.7
38	1134.1	1141.6	1149.1	1156.6	1164.2	1171.7	1179.3	1186.9
39	1194.6	1202.3	1210.0	1217.7	1225.4	1233.2	1241.0	1248.8
40	1256.6	1264.5	1272.4	1280.3	1288.2	1296.2	1304.2	1312.2
41	1320.3	1328.3	1336.4	1344.5	1352.7	1360.8	1369.0	1377.2
42	1385.4	1393.7	1402.0	1410.3	1418.6	1427.0	1435.4	1443.8
43	1452.2	1460.7	1469.1	1477.6	1486.2	1494.7	1503.3	1511.9
44	1520.5	1529.2	1537.9	1546.6	1555.3	1564.0	1572.8	1581.6
45	1590.4	1599.3	1608.2	1617.0	1626.0	1634.9	1643.9	1652.9
46	1661.9	1670.9	1680.0	1689.1	1698.2	1707.4	1716.5	1725.7
47	1734.9	1744.2	1753.5	1762.7	1772.1	1781.4	1790.8	1800.1
48	1809.6	1819.0	1828.5	1837.9	1847.5	1857.0	1866.5	1876.1
49	1885.7	1895.4	1905.0	1914.7	1924.4	1934.2	1943.9	1953.7
50	1963.5	1973.3	1983.2	1993.1	2003.0	2012.9	2022.8	2032.8
51	2042.8	2052.8	2062.9	2073.0	2083.1	2093.2	2103.3	2113.5
52	2123.7	2133.9	2144.2	2154.5	2164.8	2175.1	2185.4	2195.8
53	2206.2	2216.6	2227.0	2237.5	2248.0	2258.5	2269.1	2279.6
54	2290.2	2300.8	2311.5	2322.1	2332.8	2343.5	2354.3	2365.0
55	2375.8	2386.6	2397.5	2408.3	2419.2	2430.1	2441.1	2452.0
56	2463.0	2474.0	2485.0	2496.1	2507.2	2518.3	2529.4	2540.6
57	2551.8	2563.0	2574.2	2585.4	2596.7	2608.0	2619.4	2630.7
58	2642.1	2653.5	2664.9	2676.4	2687.8	2699.3	2710.9	2722.4
59	2734.0	2745.6	2757.2	2768.8	2780.5	2792.2	2803.9	2815.7

TABLE 10.8 *(Continued)*

Diameter	0	⅛	¼	⅜	½	⅝	¾	⅞
60	2827.4	2839.2	2851.0	2862.9	2874.8	2886.6	2898.6	2910.5
61	2922.5	2934.5	8946.5	2958.5	2970.6	2982.7	2994.8	3006.9
62	3019.1	3031.3	3043.5	3955.7	3068.0	3080.3	3092.6	3104.9
63	3117.2	3129.6	3142.0	3154.5	3166.9	3179.4	3191.9	3204.4
64	3217.0	3229.6	3242.2	3254.8	3267.5	3280.1	3292.8	3305.6
65	3318.3	3331.1	3343.9	3356.7	3369.6	3382.4	3395.3	3408.2
66	3421.2	3434.3	3447.2	3460.2	3473.2	3486.3	3499.4	3512.5
67	3525.7	3538.8	3552.0	3565.2	3578.5	3591.7	3605.0	3618.3
68	3631.7	3645.0	3658.4	3671.8	3685.3	3698.7	3712.2	3725.7
69	3793.3	3752.8	3766.4	3780.0	3793.7	3807.3	3821.0	3834.7
70	3848.5	3862.2	3876.0	3889.8	3903.6	3917.5	3931.4	3945.3
71	3959.2	3973.1	3987.1	4001.1	4015.2	4029.2	4043.3	4057.4
72	4071.5	4085.7	4099.8	4114.0	4128.2	4142.5	4156.8	4171.1
73	4185.4	4199.7	4214.1	4228.5	4242.9	4257.4	4271.8	4286.3
74	4300.8	4315.4	4329.9	4344.5	4359.2	4373.8	4388.5	4403.2
75	4417.9	4432.6	4447.4	4462.2	4477.0	4491.8	4506.7	4521.5
76	4536.5	4551.4	4566.4	4581.3	4596.3	4611.4	4626.4	4641.5
77	4656.6	4671.8	4686.9	4702.1	4717.3	4732.5	4747.8	4763.1
78	4778.4	4793.7	4809.0	4824.4	4839.8	4855.2	4870.7	4886.2
79	4901.7	4917.2	4932.7	4948.3	4963.9	4979.5	4995.2	5010.9

80	5026.5	5042.3	5058.0	5973.8	5089.6	5105.4	5121.2	5137.1
81	5153.0	5168.9	5184.9	5200.8	5216.8	5232.8	5248.9	5264.9
82	5281.0	5297.1	5313.3	5329.4	5345.6	5361.8	5378.1	5394.3
83	5410.6	5426.9	5443.3	5459.6	5476.0	5492.4	5508.8	5525.3
84	5541.8	5558.3	5574.8	5591.4	5607.9	5624.5	5641.2	5657.8
85	5674.5	5691.2	5707.9	5724.7	5741.5	5758.3	5775.1	5791.9
86	5808.8	5825.7	5842.6	5859.6	5876.5	5893.5	5910.6	5927.6
87	5944.7	5961.8	5978.9	5996.0	6013.2	6030.4	6047.6	6064.9
88	6082.1	6099.4	6116.7	6134.1	6151.4	6168.8	6186.2	6203.7
89	6221.1	6238.6	6256.1	6273.7	6291.2	6308.8	6326.4	6344.1
90	6361.7	6379.4	6397.1	6414.9	6432.6	6450.4	6468.2	6486.0
91	6503.9	6521.8	6539.7	6557.6	6575.5	6593.5	6611.5	6629.6
92	6647.6	6665.7	6683.8	6701.9	6720.1	6738.2	6756.4	6774.7
93	6792.9	6811.2	6829.5	6847.8	6866.1	6881.5	6902.9	6921.3
94	6939.8	6958.2	6976.7	6995.3	7013.8	7032.4	7051.0	7069.6
95	7988.2	7106.9	7125.6	7144.3	7163.0	7181.8	7200.6	7219.4
96	7238.2	7257.1	7276.0	7294.9	7313.8	7332.8	7351.8	7370.8
97	7389.8	7408.9	7428.0	7447.1	7466.2	7485.3	7504.5	7523.7
98	7543.0	7562.2	7581.5	7600.8	7620.1	7639.5	7658.9	7678.3
99	9697.7	7717.1	7736.6	7756.1	7775.6	7795.2	7814.8	7834.4
100	7854.0	7873.6	7893.6	7913.0	7932.7	7952.5	7972.2	7992.0

TABLE 10-9 Conversion of Inches into Millimeters in Increments of 1/16 Inch

In	0	1/16	1/8	3/16	1/4	5/16	3/8	7/16	1/2	9/16	5/8	11/16	3/4	13/16	7/8	15/16
0	0.0	1.6	3.2	4.8	6.4	7.9	9.5	11.1	12.7	14.3	15.9	17.5	19.1	20.6	22.2	23.8
1	25.4	27.0	28.6	30.2	31.8	33.3	34.9	36.5	38.1	39.7	41.3	42.9	44.5	46.0	47.6	49.2
2	50.8	52.4	54.0	55.6	57.2	58.7	60.3	61.9	63.5	65.1	66.7	68.3	69.9	71.4	73.0	74.6
3	76.2	77.8	79.4	81.0	82.6	84.1	85.7	87.3	88.9	90.5	92.1	93.7	95.3	96.8	98.4	100.0
4	101.6	103.2	104.8	106.4	108.0	109.5	111.1	112.7	114.3	115.9	117.5	119.1	120.7	122.2	123.8	125.4
5	127.0	128.6	130.2	131.8	133.4	134.9	136.5	138.1	139.7	141.3	142.9	144.5	146.1	147.6	149.2	150.8
6	152.4	154.0	155.6	157.2	158.8	160.3	161.9	163.5	165.1	166.7	168.3	169.9	171.5	173.0	174.6	176.2
7	177.8	179.4	181.0	182.6	184.2	185.7	187.3	188.9	190.5	192.1	193.7	195.3	196.9	198.4	200.0	201.6
8	203.2	204.8	206.4	208.0	209.6	211.1	212.7	214.3	215.9	217.5	219.1	220.7	222.3	223.8	225.4	227.0
9	228.6	230.2	231.8	233.4	235.0	236.5	238.1	239.7	241.3	242.9	244.5	246.1	247.7	249.2	250.8	252.4
10	254.0	255.6	257.2	258.8	260.4	261.9	263.5	265.1	266.7	268.3	269.9	271.5	273.1	274.6	276.2	277.8
11	279.4	281.0	282.6	284.2	285.8	287.3	288.9	290.5	292.1	293.7	295.3	296.9	298.5	300.0	301.6	303.2
12	304.8	306.4	308.0	309.6	311.2	312.7	314.3	315.9	317.5	319.1	320.7	322.3	323.9	325.4	327.0	328.6
13	330.2	331.8	333.4	335.0	336.6	338.1	339.7	341.3	342.9	344.5	346.1	347.7	349.3	350.8	352.4	354.0
14	355.6	357.2	358.8	360.4	362.0	363.5	365.1	366.7	368.3	369.9	371.5	373.1	374.7	376.2	377.8	379.4
15	381.0	382.6	384.2	385.8	387.4	388.9	390.5	392.1	393.7	395.3	396.9	398.5	400.1	401.6	403.2	404.8
16	406.4	408.0	409.6	411.2	412.8	414.3	415.9	417.5	419.1	420.7	422.3	423.9	425.5	427.0	428.6	430.2
17	431.8	433.4	435.0	436.6	438.2	439.7	441.3	442.9	444.5	446.1	447.7	449.3	450.9	452.4	454.0	455.6
18	457.2	458.8	460.4	462.0	463.6	465.1	466.7	468.3	469.9	471.5	473.1	474.7	476.3	477.8	479.4	481.0
19	482.6	484.2	485.8	487.4	489.0	490.5	492.1	493.7	495.3	496.9	498.5	500.1	501.7	503.2	504.8	506.4
20	508.0	509.6	511.2	512.8	514.4	515.9	517.5	519.1	520.7	522.3	523.9	525.5	527.1	528.6	530.2	531.8
21	533.4	535.0	536.6	538.2	539.8	541.3	542.9	544.5	546.1	547.7	549.3	550.9	552.5	554.0	555.6	557.2
22	558.8	560.4	562.0	563.6	565.2	566.7	568.3	569.9	571.5	573.1	574.7	576.3	577.9	579.4	581.0	582.6
23	584.2	585.8	587.4	589.0	590.6	592.1	593.7	595.3	596.9	598.5	600.1	601.7	603.3	604.8	606.4	608.0
24	609.6	611.2	612.8	614.4	616.0	617.5	619.1	620.7	622.3	623.9	625.5	627.1	628.7	630.2	631.8	633.4

25	635.0	636.6	638.2	639.8	641.4	642.9	644.5	646.1	647.7	649.3	650.9	652.5	654.1	655.6	657.2	658.8
26	660.4	662.0	663.6	665.2	666.8	668.3	669.9	671.5	673.1	674.7	676.3	677.9	679.5	681.0	682.6	684.2
27	685.8	687.4	689.0	690.6	692.2	693.7	695.3	696.9	698.5	700.1	701.7	703.3	704.9	706.4	708.0	709.6
28	711.2	712.8	714.4	716.0	717.6	719.1	720.7	722.3	723.9	725.5	727.1	728.7	730.3	731.8	733.4	735.0
29	736.6	738.2	739.8	714.4	743.0	744.5	746.1	747.7	749.3	750.9	752.5	754.1	755.7	757.2	748.8	760.4
30	762.0	763.6	765.2	766.8	768.4	769.9	771.5	773.1	774.7	776.3	777.9	779.5	781.1	782.6	784.2	785.8
31	787.4	789.0	790.6	792.2	793.8	795.3	796.9	798.5	800.1	801.7	803.3	804.9	806.5	808.0	809.6	811.2
32	812.8	814.4	816.0	817.6	819.2	820.7	822.3	823.9	825.5	827.1	828.7	830.3	831.9	833.4	835.0	836.6
33	838.2	939.8	841.4	843.0	844.6	846.1	847.7	849.3	850.9	852.5	854.1	855.7	857.3	858.8	860.4	862.0
34	863.6	865.2	866.8	868.4	870.0	871.5	873.1	874.7	876.3	877.9	879.5	881.1	882.7	884.2	885.8	887.4
35	889.0	890.6	892.2	893.8	895.4	896.9	898.5	900.1	901.7	903.3	904.9	906.5	908.1	909.6	911.2	912.8
36	914.4	916.0	917.6	919.2	920.8	922.3	923.9	925.5	927.1	928.7	930.3	931.9	933.5	935.0	936.6	938.2
37	939.8	941.4	943.0	944.6	946.2	947.7	949.3	950.9	952.5	954.1	955.7	957.3	958.9	960.4	962.0	963.6
38	965.2	966.8	968.4	970.0	971.6	973.1	974.7	976.3	977.9	979.5	981.1	982.7	984.3	985.8	987.4	989.0
39	990.6	992.2	993.8	995.4	997.0	998.5	1000.1	1001.7	1003.3	1004.9	1006.5	1008.1	1009.7	1011.2	1012.8	1014.4
40	1016.0	1017.6	1019.2	1020.8	1022.4	1023.9	1025.5	1027.1	1028.7	1030.3	1031.9	1033.5	1035.1	1036.6	1038.2	1039.8
41	1041.4	1043.0	1044.6	1046.2	1047.8	1049.3	1050.9	1052.5	1054.1	1055.7	1057.3	1058.9	1060.5	1062.0	1063.6	1065.2
42	1066.8	1068.4	1070.0	1071.6	1073.2	1074.7	1076.3	1077.9	1079.5	1081.1	1082.7	1084.3	1085.9	1087.4	1089.0	1090.6
43	1092.2	1093.8	1095.4	1097.0	1098.6	1100.1	1101.7	1103.3	1104.9	1106.5	1108.1	1109.7	1111.3	1112.8	1114.4	1116.0
44	1117.6	1119.2	1120.8	1122.4	1124.0	1125.5	1127.1	1128.7	1130.3	1131.9	1133.5	1135.1	1136.7	1138.2	1139.8	1141.4
45	1143.0	1144.6	1146.2	1147.8	1149.4	1150.9	1152.5	1154.1	1155.7	1157.3	1158.9	1160.5	1162.1	1163.6	1165.2	1166.8
46	1168.4	1170.0	1171.6	1173.2	1174.8	1176.3	1177.9	1179.5	1181.1	1182.7	1184.3	1185.9	1187.5	1189.0	1190.6	1192.2
47	1193.8	1195.4	1197.0	1198.6	1200.2	1201.7	1203.3	1204.9	1206.5	1208.1	1209.7	1211.3	1212.9	1214.4	1216.0	1217.6
48	1219.2	1220.8	1222.4	1224.0	1225.6	1227.1	1228.7	1230.3	1231.9	1233.5	1235.1	1236.7	1238.3	1239.8	1241.4	1243.0
49	1244.6	1246.2	1247.8	1249.4	1251.0	1252.5	1254.1	1255.7	1257.3	1258.9	1260.5	1262.1	1263.7	1265.2	1266.8	1268.4
50	1270.0	1271.6	1273.2	1274.8	1276.4	1277.9	1279.5	1281.1	1282.7	1284.3	1285.9	1287.5	1289.1	1290.6	1292.2	1293.8

TABLE 10.10 Conversion of Inches into Millimeters in Increments of 1⁄100 Inch

In	0.00	0.01	0.02	0.03	0.04	0.05	0.06	0.07	0.08	0.09
0.00	0.00	0.25	0.51	0.76	1.02	1.27	1.52	1.78	2.03	2.29
0.10	2.54	2.79	3.05	3.30	3.56	3.81	4.06	4.32	4.57	4.83
0.20	5.08	5.33	5.59	5.84	6.10	6.35	6.60	6.86	7.11	7.37
0.30	7.62	7.87	8.13	8.38	8.64	8.89	9.14	9.40	9.65	9.91
0.40	10.16	10.41	10.67	10.92	11.18	11.43	11.68	11.94	12.19	12.45
0.50	12.70	12.95	13.21	13.46	13.72	13.97	14.22	14.48	14.73	14.99
0.60	15.24	15.49	15.75	16.00	16.26	16.51	16.76	17.02	17.27	17.53
0.70	17.78	18.03	18.29	18.54	18.80	19.05	19.30	19.56	19.81	20.07
0.80	20.32	20.57	20.83	21.08	21.34	21.59	21.84	22.10	22.35	22.61
0.90	22.86	23.11	23.37	23.62	23.88	24.13	24.38	24.64	24.89	25.15

Note: By the use of Table 10.9 and simple addition, the metric equivalents up to 50.99 in are readily available. By manipulation of the decimal point, any magnitude number is readily converted.

TABLE 10.11 Hardness Scale Conversions

Brinell hardness no.	Shore scleroscope hardness no.	Rockwell		Brinell hardness no.	Shore scleroscope hardness no.	Rockwell		Brinell hardness no.	Shore scleroscope hardness no.	Rockwell	
		B scale	C scale			B scale	C scale			B scale	C scale
782	107	.....	72	277	39	.....	29	137	.....	75	
744	100	.....	69	269	38	.....	28	134	.....	74	
713	96	.....	67	262	37	.....	27	131	.....	72	
683	92	.....	65	255	36	.....	26	128	.....	71	
652	88	.....	63	248	36	.....	25	126	.....	70	
627	85	.....	61	241	35	100	24	124	.....	69	
600	81	.....	59	235	34	99	23	121	.....	67	
578	78	.....	58	228	33	98	22	118	.....	66	
555	75	.....	56	223	33	97	21	116	.....	65	
532	72	.....	54	217	32	96	20	114	.....	64	
512	70	.....	52	212	31	95	.....	112	.....	62	
495	68	.....	51	207	30	94	.....	109	.....	61	
477	66	.....	49	202	30	93	.....	107	.....	59	
460	64	.....	48	196	29	92	.....	105	.....	58	
444	61	.....	47	192	29	91	.....	103	.....	57	
430	59	.....	45	187	28	90	.....	101	.....	56	
418	57	.....	44	183	28	89	.....	99	.....	54	
402	55	.....	43	179	27	88	.....	97	.....	53	
387	53	.....	41	174	27	87	.....	96	.....	52	
375	52	.....	40	170	26	86	.....	95	.....	51	
364	50	.....	39	166	26	85	.....	93	.....	50	
351	49	.....	38	163	25	84	.....	92	.....	49	
340	47	.....	37	159	25	83	.....	90	.....	48	
332	46	.....	36	156	24	82	.....	88	.....	47	
321	45	.....	35	153	24	81	.....	87	.....	46	
311	44	.....	34	149	23	80	.....	86	.....	45	
302	42	.....	33	146	23	78	.....	85	.....	44	
293	41	.....	31	143	22	77	.....	83	.....	43	
286	40	.....	30	140	.....	76	.....	82	.....	42	

TABLE 10.12 Selected SI* Units and Symbols

Quantity	Unit	Symbol
Length	meter	m
Mass	kilogram	kg
Time	second	s
Electric current	ampere	A
Temperature	kelvin	K
Amount of substance	mole	mol
Frequency	hertz	Hz
Force	newton	N
Pressure and stress	pascal	Pa
Energy	joule	J
Power	watt	W
Electric potential	volt	V
Capacitance	farad	F
Electric resistance	ohm	Ω
Conductance	siemens	S
Inductance	henry	H

*The International System (SI) of units is that form of the metric system which has been adopted worldwide.

TABLE 10.13 SI Exponential Notation*

Exponential	Prefix	Symbol
10^{-18}	atto	a
10^{-15}	femto	f
10^{-12}	pico	p
10^{-9}	nano	n
10^{-6}	micro	μ
10^{-3}	milli	m
10^{3}	kilo	k
10^{6}	mega	M
10^{9}	giga	G
10^{12}	tera	T
10^{15}	peta	P
10^{18}	exa	E

*Prefixes used to describe larger or smaller values than those chosen as standard in Table 10.12.

TABLE 10.14 Conversion Factors to SI Units

To convert from	to	Multiply by
	Length	
foot (ft)	meter (m)	3.048×10^{-1}
inch (in)	meter (m)	2.54×10^{-2}
mile (mi)	meter (m)	1.609×10^{3}
	Area	
square foot (ft^2)	square meter (m^2)	9.290×10^{-2}
square inch (in^2)	square meter (m^2)	6.451×10^{-4}
square mile (mi^2)	square meter (m^2)	2.590×10^{6}
	Volume	
cubic foot (ft^3)	cubic meter (m^3)	2.831×10^{-2}
gallon (gal)	cubic meter (m^3)	3.785×10^{-3}
	Flow	
cubic foot per minute (ft^3/min)	cubic meter per second (m^3/s)	4.719×10^{-4}
cubic foot per second (ft^3/s)	cubic meter per second (m^3/s)	2.831×10^{-2}
gallon per minute (gal/min or gpm)	cubic meter per second (m^3/s)	6.309×10^{-5}
	Pressure	
bar	pascal (Pa)	1.000×10^{5}
foot of water	pascal (Pa)	2.989×10^{3}
inch of mercury (inHg)	pascal (Pa)	3.377×10^{3}
pounds per square inch (lb/in^2 or psi)	pascal (Pa)	6.895×10^{3}
torr (mmHg)	pascal (Pa)	1.333×10^{2}
	Velocity (speed)	
feet per second (ft/s)	meters per second (m/s)	3.048×10^{-1}
miles per hour (mi/h or mph)	meters per second (m/s)	4.470×10^{-1}
	Density	
pounds per cubic inch (lb/in^3)	kilograms per cubic meter (kg/m^3)	2.768×10^{4}
grams per cubic centimeter (g/cm^3)	kilograms per cubic meter (kg/m^3)	1.000×10^{3}
	Mass	
pound (lb)	kilogram (kg)	4.536×10^{-1}
gram (g)	kilogram (kg)	1.000×10^{-3}
	Torque	
pound inch ($lb \cdot in$)	newton-meter ($N \cdot m$)	1.130×10^{-1}
pound foot ($lb \cdot ft$)	newton-meter ($N \cdot m$)	1.356

TABLE 10.14 *(Continued)*

To convert from	to	Multiply by
	Thermal conductivity	
Btu-feet per hour per square foot per degree Fahrenheit [Btu·ft/(h·ft²·°F)]	watts per meter per kelvin [W/(m·K)]	1.731
	Force	
pound-force (lb)	newton (N)	4.448
kilogram-force (kg)	newton (N)	9.806
	Power	
Btu per hour (Btu/h)	watt (W)	2.931×10^{-1}
foot pound per minute (ft·lb/min)	watt (W)	2.260×10^{-2}
horsepower (hp)	watt (W)	7.457×10^{2}
	Energy	
Btu	joule (J)	1.055×10^{3}
calorie (cal)	joule (J)	4.187
foot pound (ft·lb)	joule (J)	1.356
	Energy per unit area and time	
Btu per square foot per hour [Btu/(ft²·h)]	watts per square meter (W/m²)	3.152

TABLE 10.15 Standard Gage Size in Inches for Wire, Sheet, Strip, and Twist Drills

Gage	Mfrs. steel sheet	USS steel sheet (old)	Birmingham or Stub	W & M or Roebling steel wire	AWG or B & S nonferrous wire or sheet	Numbered twist drills
0000000		0.500		0.4900		
000000		0.469		0.4615	0.580	
00000		0.438		0.4305	0.516	
0000		0.406	.454	0.3938	0.460	
000		0.375	.425	0.3625	0.410	
00		0.344	.380	0.3310	0.365	
0		0.313	.340	0.3065	0.325	
1		0.281	.300	0.2830	0.289	0.2280
2		0.266	.284	0.2625	0.258	0.2210
3	.2391	0.250	.259	0.2437	0.229	0.2130
4	.2242	0.234	.238	0.2253	0.204	0.2090
5	.2092	0.219	.220	0.2070	0.182	0.2055
6	.1943	0.203	.203	0.1920	0.162	0.2040
7	.1793	0.188	.180	0.1770	0.144	0.2010
8	.1644	0.172	.165	0.1620	0.128	0.1990
9	.1495	0.156	.148	0.1483	0.114	0.1960
10	.1345	0.141	.134	0.1350	0.102	0.1935
11	.1196	0.125	.120	0.1205	0.0907	0.1910
12	.1046	0.109	.109	0.1055	0.0808	0.1890
13	.0897	0.0937	.095	0.0915	0.0720	0.1850
14	.0747	0.0781	.083	0.0800	0.0641	0.1820
15	.0673	0.0703	.072	0.0720	0.0571	0.1800
16	.0598	0.0625	.065	0.0625	0.0508	0.1770
17	.0538	0.0562	.058	0.0540	0.0453	0.1730
18	.0478	0.0500	.049	0.0475	0.0403	0.1695
19	.0418	0.0437	.042	0.0410	0.0359	0.1660
20	.0359	0.0375	.035	0.0348	0.0320	0.1610
21	.0329	0.0344	.032	0.0318	0.0285	0.1590
22	.0299	0.0312	.028	0.0286	0.0253	0.1570
23	.0269	0.0281	.025	0.0258	0.0226	0.1540
24	.0239	0.0250	.022	0.0230	0.0201	0.1520
25	.0209	0.0219	.020	0.0204	0.0179	0.1495
26	.0179	0.0187	.018	0.0181	0.0159	0.1470
27	.0164	0.0172	.016	0.0173	0.0142	0.1440
28	.0149	0.0156	.014	0.0162	0.0126	0.1405
29	.0135	0.0141	.013	0.0150	0.0113	0.1360
30	.0120	0.0125	.012	0.0140	0.0100	0.1285
31	.0105	0.0109	.010	0.0132	0.0089	0.1200
32	.0097	0.0102	.009	0.0128	0.0080	0.1160
33	.0090	0.0094	.008	0.0118	0.0071	0.1130
34	.0082	0.0086	.007	0.0104	0.0063	0.1110
35	.0075	0.0078	.005	0.0095	0.0056	0.1100
36	.0067	0.0070	.004	0.0090	0.0050	0.1065
37	.0064	0.0066		0.0085	0.0045	0.1040
38	.0060	0.0062		0.0080	0.0040	0.1015

TABLE 10.15 (*Continued*)

Gage	Mfrs. steel sheet	USS steel sheet (old)	Birmingham or Stub	W & M or Roebling steel wire	AWG or B & S nonferrous wire or sheet	Numbered twist drills
39				0.0075	0.0035	0.0995
40				0.0070	0.0031	0.0980
41				0.0066	0.0028	0.0960
42				0.0062	0.0025	0.0935
43				0.0060	0.0022	0.0890
44				0.0058	0.0020	0.0860
45				0.0055	0.0018	0.0820
46				0.0052	0.0016	0.0810
47				0.0050	0.0014	0.0785
48				0.0048	0.0012	0.0760
49				0.0046	0.0011	0.0730
50				0.0044	0.0010	0.0700

TABLE 10.16 Other Useful Conversions

To convert from	to	Multiply by
	Length	
meters (m)	inches (in)	39.37
	feet (ft)	3.281
kilometers (km)	miles (mi)	0.6214
	feet (ft)	3281
miles (mi)	feet (ft)	5280
	Area	
acres	square feet (ft^2)	43560
	square meters (m^2)	4047
	Volume	
barrels (oil) (bbl)	gallons (gal)	42
cubic centimeters (cm^3)	cubic inches (in^3)	0.061
cubic feet (ft^3)	gallons (gal)	7.481
gallons (U.S. liquid) (gal)	cubic inches (in^3)	231
	cubic feet (ft^3)	0.1337
ounces, fluid (oz)	cubic centimeters (cm^3)	29.57
	cubic inches (in^3)	1.805
	Weight (mass)	
grams (g)	ounces (oz)	0.03527
	pounds (lb)	0.002205
kilograms (kg)	pounds (lb)	2.205
tons	pounds (lb)	2000
	kilograms (kg)	907.2
	Velocity	
feet per second (ft/s)	miles per hour (mph)	0.6818
	meters per second (m/s)	0.3048
miles per hour (mi/h or mph)	meters per second (m/s)	0.4470
	kilometers per hour (km/h)	1.609
	Pressure	
centimeters of mercury (cmHg)	inches of water	5.352
	pounds per square inch (psi)	0.1934
feet of water	pounds per square inch (psi)	0.4331
inches of mercury (inHg)	pounds per square inch (psi)	0.4912
inches of water	pounds per square inch (psi)	0.0361
	Density	
grams per cubic centimeter (g/cm^3)	pounds per cubic inch (lb/in^3)	0.0361
pounds per cubic inch (lb/in^3)	grams per cubic centimeter (g/cm^3)	27.70
pounds per cubic foot (lb/ft^3)	pounds per cubic inch (lb/in^3)	5.787×10^{-4}
	Energy	
Btu	joules (J)	1055

TABLE 10.17 Temperature Conversion from Degrees Fahrenheit to Degrees Celsius

°F	°C	°F	°C	°F	°C	°F	°C	°F	°C	°F	°C
−50	−45.6	70	21.1	280	138	530	277	780	416	1150	621
−45	−42.8	75	23.9	290	143	540	282	790	421	1200	649
−40	−40	80	26.7	300	149	550	288	800	427	1250	677
−35	−37.2	85	294	310	154	560	293	810	432	1300	704
−30	−34.4	90	32.2	320	160	570	299	820	438	1350	732
−25	−31.7	95	35	330	166	580	304	830	443	1400	760
−20	−28.9	100	37.8	340	171	590	310	840	449	1450	788
−15	−26.1	110	43	350	177	600	316	850	454	1500	816
−10	−23.3	120	49	360	182	610	321	860	460		
−5	−20.6	130	54	370	188	620	327	870	466		
0	−17.8	140	60	380	193	630	332	880	471		
5	−15	150	66	390	199	640	338	890	477		
10	−12.2	160	71	400	204	650	343	900	482		
15	−9.4	170	77	410	210	660	349	910	488		
20	−6.7	180	82	420	216	670	354	920	493		
25	−3.9	190	88	430	221	680	360	930	499		
30	−1.1	200	93	440	227	690	366	940	504		
32	0	210	99	450	232	700	371	950	510		
35	1.7	212	100	460	238	710	377	960	516		
40	4.4	220	104	470	243	720	382	970	521		
45	7.2	230	110	480	249	730	388	980	527		
50	10	240	116	490	254	740	393	990	532		
55	12.8	250	121	500	260	750	399	1000	538		
60	15.6	260	127	510	266	760	404	1050	566		
65	18.3	270	132	520	271	770	410	1100	593		

chapter 

Selected Properties of Materials and Physical Constants

Maurice J. Webb

TABLE 11.1 Abbreviated Table of the Elements with Their Key Properties

Element	Symbol	Atomic weight	Density, lb/in^3 (10^4 kg/m^3)*	Freezing point, °F (K)	Boiling point, °F (K)
Aluminum	Al	27.0	0.097 (0.269)	1219 (933)	3734 (2330)
Antimony	Sb	121.7	0.239 (0.662)	1167 (904)	2516 (1653)
Argon	A	39.9	0.111(g) (1.778)	−309 (84)	−302 (88)
Arsenic	As	74.9	0.207 (0.573)	1139 sublimes (888)	
Barium	Ba	137.4	0.136 (0.377)	1562 (1123)	2084 (1413)
Beryllium	Be	9.0	0.066 (0.183)	1586 (1137)	
Bismuth	Bi	209.0	0.352 (0.975)	510 (539)	2840 (1833)
Boron	B	10.8	0.120 (0.332)	4172 (2573)	4622 (2823)
Bromine	Br	79.9	0.113(l) (0.313)	19 (266)	138 (332)
Cadmium	Cd	112.4	0.313 (0.867)	610 (594)	1412 (1040)
Calcium	Ca	40.1	0.056 (0.155)	1548 (1116)	2264 (1513)
Carbon (graphite)	C	12.0	0.081 (0.224)	6620 sublimes (3933)	
Cesium	Cs	132.8	0.068 (0.188)	83 (302)	1238 (943)
Chlorine	Cl	35.5	0.200(g) (3.204)	−153 (171)	−30 (239)
Chromium	Cr	52.0	0.238 (0.659)	3436 (1768)	4496 (2753)
Cobalt	Co	58.9	0.314 (0.870)	2723 (1768)	5252 (3173)
Columbium (see niobium)					
Copper	Cu	63.6	0.321 (0.889)	981 (801)	4237 (2609)
Fluorine	F	19.0	0.106(g) (1.698)	−369 (51)	−305 (86)
Gallium	Ga	69.7	0.213 (0.590)	86 (303)	3601 (2256)
Germanium	Ge	72.6	0.197 (0.546)	1758 (1232)	4900 (2978)
Gold	Au	197.0	0.697 (1.931)	1945 (1336)	4712 (2873)
Hafnium	Hf	178.6	0.480 (1.330)	3092 (1973)	6000 (3589)
Helium	He	4.0	0.011(g) (0.176)	−458 (1)	−454 (3)
Hydrogen	H	1.0	0.0056(g) (0.090)	−435 (14)	−423 (21)

TABLE 11.1 *(Continued)*

Element	Symbol	Atomic weight	Density, lb/in^3 (10^4 kg/m^3)*	Freezing point, °F (K)	Boiling point, °F (K)
Indium	In	114.8	0.263 (0.729)	314 (430)	3632 (2273)
Iodine	I	126.9	0.178 (0.493)	237 (387)	363 (457)
Iridium	Ir	192.2	0.809 (2.241)	4449 (2727)	8670 (5072)
Iron	Fe	55.8	0.284 (0.787)	2795 (1808)	5432 (3273)
Krypton	Kr	83.8	0.231(g) (3.700)	−251 (116)	−241 (121)
Lead	Pb	207.2	0.409 (1.133)	621 (601)	2948 (1893)
Lithium	Li	6.9	0.019 (0.053)	367 (459)	2437 (1609)
Magnesium	Mg	24.3	0.063 (0.175)	204 (369)	2025 (1381)
Manganese	Mn	54.9	0.268 (0.742)	2300 (1583)	3452 (2173)
Mercury	Hg	200.6	0.491(l) (1.360)	−38 (234)	674 (630)
Molybdenum	Mo	96.0	0.325 (0.900)	4748 (2893)	8672 (5073)
Neon	Ne	20.2	0.056(g) (0.897)	−416 (24)	−411 (27)
Nickel	Ni	58.7	0.314 (0.870)	2651 (1728)	5252 (3173)
Niobium (columbium)	Nb	92.9	0.303 (0.839)	4532 (2773)	6692 (3973)
Nitrogen	N	14.0	0.078(g) (1.249)	−346 (63)	−320 (78)
Oxygen	O	16.0	0.089(g) (1.426)	−361 (55)	−297 (91)
Phosphorus	P	31.0	0.083 (0.230)	101 (312)	536 (553)
Platinum	Pt	195.1	0.771 (2.136)	3224 (2027)	7772 (4573)
Potassium	K	39.1	0.031 (0.086)	144 (386)	1400 (1033)
Radon	Rn	222.0	0.607(g) (9.724)	−96 (202)	79 (212)
Rhodium	Rh	102.9	0.449 (1.244)	3571 (2239)	4500 (2756)
Selenium	Se	79.0	0.159 (0.440)	420 (489)	1270 (961)
Silicon	Si	28.1	0.087 (0.241)	2588 (1693)	4271 (2628)
Silver	Ag	107.9	0.379 (1.050)	1761 (1234)	3542 (2223)
Sodium	Na	23.0	0.034 (0.094)	208 (371)	1616 (1153)
Strontium	Sr	87.6	0.090 (0.249)	1400 (1033)	2100 (1422)

TABLE 11.1 *(Continued)*

Element	Symbol	Atomic weight	Density, lb/in³ (10^4 kg/m³)*	Freezing point, °F (K)	Boiling point, °F (K)
Sulphur	S	32.0	0.072 (0.199)	235 (386)	832 (718)
Tantalum	Ta	181.0	0.065 (0.180)	5424 (3269)	7412 (4373)
Thorium	Th	232.1	0.411 (1.138)	3353 (2118)	8132 (4773)
Tin	Sn	118.7	0.264 (0.731)	450 (506)	4100 (2533)
Titanium	Ti	47.9	0.162 (0.449)	3272 (2073)	5450 (3283)
Tungsten	W	183.9	0.679 (1.881)	6098 (3643)	10,650 (6172)
Uranium	U	238.1	0.675 (1.870)	2071 (1406)	
Vanadium	V	51.0	0.205 (0.568)	3110 (1983)	5432 (3273)
Xenon	Xe	131.3	0.365(g) (5.847)	−169 (162)	−109 (195)
Zinc	Zn	65.4	0.256 (0.709)	536 (553)	
Zirconium	Zr	91.2	0.232 (0.643)	3374 (2130)	5250 (3172)

*Density of gases (g) expressed in pounds per cubic foot (lb/ft³) at atmospheric pressure and 68°F (kg/m³ at 293 K).

TABLE 11.2 Vapor Pressures of Some Liquids

Pressure, psia (kPa)

Substance	Temperature, °F (K)			
	0 (255.6)	100 (311.1)	200 (366.7)	300 (422.2)
Acetone	...	...	44.1 (304)	162 (1116)
Benzene	0.12 (0.83)	3.3 (22.8)	21.6 (149)	77.9 (537)
Ethyl alcohol	...	2.3 (15.9)	41.2 (284)	135 (930)
Methyl alcohol	...	4.5 (31.0)	40.3 (278)	184 (1268)
Propane	4.0 (27.6)	187 (1289)	720 (4761)	
Water	...	0.95 (6.6)	12.5 (86.2)	67.0 (462)

TABLE 11.3 Vapor Pressure of Water

Vapor pressure			Vapor pressure		
Absolute, psia	Vacuum, inHg	Temperature, °F	Absolute, psia	Gage, psig	Temperature, °F
0.20	29.51	53.14	14.696	0.0	212.00
0.25	29.41	59.30	15.0	0.3	213.03
0.30	29.31	64.47	16.0	1.3	216.32
0.35	29.21	68.93	17.0	2.3	219.44
0.40	29.11	72.86	18.0	3.3	222.41
			19.0	4.3	225.24
0.45	29.00	76.38	20.0	5.3	227.96
0.50	28.90	79.58	21.0	6.3	230.57
0.60	28.70	85.21	22.0	7.3	233.07
0.70	28.49	90.08	23.0	8.3	235.49
0.80	28.29	94.38	24.0	9.3	237.82
0.90	28.09	98.24	25.0	10.3	240.07
1.0	27.88	101.74	26.0	11.3	242.25
1.2	27.48	107.92	27.0	12.3	244.36
1.4	27.07	113.26	28.0	13.3	246.41
1.6	26.66	117.99	29.0	14.3	248.40
1.8	26.26	122.23	30.0	15.3	250.33
2.0	25.85	126.08	31.0	16.3	252.22
2.2	25.44	129.62	32.0	17.3	254.05
2.4	25.03	132.89	33.0	18.3	255.84
2.6	24.63	135.94	34	19.3	257.58
2.8	24.22	138.79	35.0	20.3	259.28
3.0	23.81	141.48	36.0	21.3	260.95
3.5	22.79	147.57	37.0	22.3	262.57
4.0	21.78	152.97	38.0	23.3	264.16
4.5	20.76	157.83	39.0	24.3	265.72
5.0	19.74	162.24	40.0	25.3	267.25
5.5	18.72	166.30	41.0	26.3	268.74
6.0	17.70	170.06	42.0	27.3	270.21
6.5	16.69	173.56	43.0	28.3	271.64
7.0	15.67	176.85	44.0	29.3	273.05
7.5	14.65	179.94	45.0	30.3	274.44
8.0	13.63	182.86	46.0	31.3	275.80
8.5	12.61	185.64	47.0	32.3	277.13
9.0	11.60	188.28	48.0	33.3	278.45
9.5	10.58	190.80	49.0	34.3	279.74
10.0	9.56	193.21	50.0	35.3	281.01
11.0	7.52	197.75	51.0	36.3	282.26
12.0	5.49	201.96	52.0	37.3	283.49
13.0	3.45	205.88	53.0	38.3	284.70
14.0	1.42	209.56	54.0	39.3	285.90

TABLE 11.3 *(Continued)*

Vapor pressure			Vapor pressure		
Absolute, psia	Gage, psig	Temperature, °F	Absolute, psia	Gage, psig	Temperature, °F
55.0	40.3	287.07	100.0	85.3	327.81
56.0	41.3	288.23	101.0	86.3	328.53
57.0	42.3	289.37	102.0	87.3	329.25
58.0	43.3	290.50	103.0	88.3	329.96
59.0	44.3	291.61	104.0	89.3	330.66
60.0	45.3	292.71	105.0	90.3	331.36
61.0	46.3	293.79	106.0	91.3	332.05
62.0	47.3	294.85	107.0	92.3	332.74
63.0	48.3	295.90	108.0	93.3	333.42
64.0	49.3	296.94	109.0	94.3	334.10
65.0	50.3	297.97	110.0	95.3	334.77
66.0	51.3	298.99	111.0	96.3	335.44
67.0	52.3	299.99	112.0	97.3	336.11
68.0	53.3	300.98	113.0	98.3	336.77
69.0	54.3	301.96	114.0	99.3	337.42
70.0	55.3	302.92	115.0	100.3	338.07
71.0	56.3	303.88	116.0	101.3	338.72
72.0	57.3	304.83	117.0	102.3	339.36
73.0	58.3	305.76	118.0	103.3	339.99
74.0	59.3	306.68	119.0	104.3	340.62
75.0	60.3	307.60	120.0	105.3	341.25
76.0	61.3	308.50	121.0	106.3	341.88
77.0	62.3	309.40	122.0	107.3	342.50
78.0	63.3	310.29	123.0	108.3	343.11
79.0	64.3	311.16	124.0	109.3	343.72
80.0	65.3	312.03	125.0	110.3	344.33
81.0	66.3	312.89	126.0	111.3	344.94
82.0	67.3	313.74	127.0	112.3	345.54
83.0	68.3	314.59	128.0	113.3	346.13
84.0	69.3	315.42	129.0	114.3	346.73
85.0	70.3	316.25	130.0	115.3	347.32
86.0	71.3	317.07	131.0	116.3	347.90
87.0	72.3	317.88	132.0	117.3	348.48
88.0	73.3	318.68	133.0	118.3	349.06
89.0	74.3	319.48	134.0	119.3	349.64
90.0	75.3	320.27	135.0	120.3	350.21
91.0	76.3	321.06	136.0	121.3	350.78
92.0	77.3	321.83	137.0	122.3	351.35
93.0	78.3	322.60	138.0	123.3	351.91
94.0	79.3	323.36	139.0	124.3	352.47
95.0	80.3	324.12	140.0	125.3	353.02
96.0	81.3	324.87	141.0	126.3	353.57
97.0	82.3	325.61	142.0	127.3	254.12
98.0	83.3	326.35	143.0	128.3	354.67
99.0	84.3	327.08	144.0	129.3	355.21

TABLE 11.3 *(Continued)*

Vapor pressure			Vapor pressure		
Absolute, psia	Gage, psig	Temperature, °F	Absolute, psia	Gage, psig	Temperature, °F
145.0	130.3	355.76	275.0	260.3	409.43
146.0	131.3	356.29	280.0	265.3	411.05
147.0	132.3	356.83	285.0	270.3	412.65
148.0	133.3	357.36	290.0	275.3	414.23
149.0	134.3	357.89	295.0	280.3	415.79
150.0	135.3	358.42	300.0	285.3	417.33
152.0	137.3	359.46	320.0	305.3	423.29
154.0	139.3	360.49	340.0	325.3	428.97
156.0	141.3	361.52	360.0	345.3	434.40
158.0	143.3	362.53	380.0	365.3	439.60
160.0	145.3	363.53	400.0	385.3	444.59
162.0	147.3	364.53	420.0	405.3	449.39
164.0	149.3	365.51	440.0	425.3	454.02
166.0	151.3	366.48	460.0	445.3	458.50
168.0	153.3	367.45	480.0	465.3	462.82
170.0	155.3	368.41	500.0	485.3	467.01
172.0	157.3	369.35	520.0	505.3	471.07
174.0	159.3	370.29	540.0	525.3	475.01
176.0	161.3	371.22	560.0	545.3	478.85
178.0	163.3	372.14	580.0	565.3	482.58
180.0	165.3	373.06	600.0	585.3	486.21
182.0	167.3	373.96	620.0	605.3	489.75
184.0	169.3	374.86	640.0	625.3	493.21
186.0	171.3	375.75	660.0	645.3	496.58
188.0	173.3	376.64	680.0	665.3	499.88
190.0	175.3	377.51	700.0	685.3	503.10
192.0	177.3	378.38	720.0	705.3	506.25
194.0	179.3	379.24	740.0	725.3	509.34
196.0	181.3	380.10	760.0	745.3	512.36
198.0	183.3	380.95	780.0	765.3	515.33
200.0	185.3	381.79	800.0	785.3	518.23
205.0	190.3	383.86	820.0	805.3	521.08
210.0	195.3	385.90	840.0	825.3	523.88
215.0	200.3	387.89	860.0	845.3	526.63
220.0	205.3	389.86	880.0	865.3	529.33
225.0	210.3	391.79	900.0	885.3	531.98
230.0	215.3	393.68	920.0	905.3	534.59
235.0	220.3	395.54	940.0	925.3	537.16
240.0	225.3	397.37	960.0	945.3	539.68
245.0	230.3	399.18	980.0	965.3	542.17
250.0	235.3	400.95	1000.0	985.3	544.61
255.0	240.3	402.70	1050.0	1035.3	550.57
260.0	245.3	404.42	1100.0	1085.3	556.31
265.0	250.3	406.11	1150.0	1135.3	561.86
270.0	255.3	407.78	1200.0	1185.3	567.22

TABLE 11.3 *(Continued)*

Vapor pressure			Vapor pressure		
Absolute, psia	Vacuum, inHg	Temperature, °F	Absolute, psia	Vacuum, inHg	Temperature, °F
1250.0	1235.3	572.42	2500.0	2485.3	668.13
1300.0	1285.3	577.46	2600.0	2585.3	673.94
1350.0	1335.3	582.35	2700.0	2685.3	679.55
1400.0	1385.3	587.10	2800.0	2785.3	684.99
1450.0	1435.3	591.73	2900.0	2885.3	690.26
1500.0	1485.3	596.23	3000.0	2985.3	695.36
1600.0	1585.3	604.90	3100.0	3085.3	700.31
1700.0	1685.3	613.15	3200.0	3185.3	705.11
1800.0	1785.3	621.03	3206.2	3191.5	705.40
1900.0	1885.3	628.58			
2000.0	1985.3	635.82			
2100.0	2085.3	642.77			
2200.0	2185.3	649.46			
2300.0	2285.3	655.91			
2400.0	2385.3	662.12			

TABLE 11.4 Properties of Cryogenic Fluids

Gas	Boiling point, °F (K)	Critical temp., °F (K)	Critical pressure, psia (MPa)	Liquid density, lb/ft^3 (kg/m^3)	Gas (NTP)/liquid (bp) ratio by volume
Acetylene	−118 (190)	97 (309)	911 (6.28)	24.9 (399)	371
Ammonia	−28 (240)	270 (406)	1639 (11.3)	42.6 (682)	964
Argon	−302 (87.8)	−189 (151)	705 (4.86)	86.8 (1891)	798
Carbon dioxide	−109 (199)	88 (304)	1073 (7.39)	48.0 (768)	421
Carbon monoxide	−310 (83.3)	−218 (134)	515 (3.55)	53.7 (8.59)	735
Chlorine	−30 (239)	291 (417)	1118 (7.70)	200.5 (3212)	1093
Ethylene	−155 (169)	50 (283)	748 (5.15)	13.1 (210)	180
Fluorine	−305 (197)	−200 (144)	808 (5.57)	94.4 (1510)	888
Helium	−454 (3.3)	−450 (5.6)	33.2 (0.23)	7.8 (125)	700
Hydrogen	−423 (20.6)	−400 (33.3)	188 (1.30)	4.4 (70.4)	700
Krypton	−241 (122)	−83 (209)	798 (5.50)	149.8 (2397)	694
Methane	−161 (166)	−117 (191)	672 (4.63)	26.5 (424)	578
Neon	−411 (27.2)	−380 (44.4)	395 (2.72)	74.9 (1198)	1340
Nitric oxide	−291 (93.9)	−137 (179)	956 (6.59)	91.7 (1467)	1176
Nitrogen	−320 (77.8)	−233 (126)	492 (3.39)	50.2 (803)	688
Oxygen	−297 (90.6)	−181 (155)	730 (5.03)	71.3 (1141)	850
Xenon	−109 (195)	62 (290)	855 (5.89)	193.5 (3096)	572

TABLE 11.5 Properties of Some Common Chemicals

Common name	Chemical name	State at 70°F (294 K)	Density at 70°F (294 K), lb/in^3 (kg/m^3)	Melting pt., °F (K)	Boiling pt., °F (K)	Solubility in water at 70°F (294 K), by wt.%
Aspirin	Acetylsalicylic acid	Solid	. . .	260 (400), decomposes	. . .	0.3
Baking powder	Sodium bicarbonate	Solid	0.079 (219)	Decomposes, releasing CO_2	. . .	6.9
Beryl	Beryllium aluminum silicate	Solid	0.112 (310)	. . .	. . .	Insoluble
Bleaching powder	Calcium chloride hypochlorite	Solid	. . .	Decomposes	. . .	Dissolves, releasing chlorine gas
Boracic acid	Boric acid	Solid	0.052 (144)	Decomposes above 365 (436)	. . .	2.0
Borax	Sodium tetraborate	Solid	0.085 (235)	1335 (997)	Decomposes, 2800 (1810)	1.5
Calcite	Calcium carbonate	Solid	. . .	Decomposes, releasing CO_2	. . .	Insoluble
Carbolic acid	Phenol	Solid	0.04 (110)	106 (314)	360 (456)	6.7
Carbonic acid gas	Carbon dioxide	Gas	6.6×10^{-5} (1.83)	. . .	−109 (195)	Very slight
Chalk	Calcium chloride	Solid	0.091 (252)	1390 (1028)	. . .	60
Epsom salt	Magnesium sulfate	Solid	0.060 (166)	. . .	. . .	71
Formaldehyde	Formaldehyde	Gas	4.5×10^{-5} (1.25)	−132 (182)	−6 (252)	Soluble
Gypsum (plaster of paris)	Calcium sulfate	Solid	0.084 (233)	. . .	. . .	0.3
Heavy water	Deuterium oxide	Liquid	0.040 (110)	. . .	. . .	All proportions
Hydrogen peroxide	Hydrogen peroxide	Liquid	0.053 (147)	−128 (184)	300 (420) (explodes)	All proportions
Lime	Calcium oxide	Solid	0.121 (335)	4680 (2860)	5160 (3120)	0.1
Moth balls	Napthalene	Solid	0.041 (114)	176 (353)	424 (491)	Very slight
Muriatic acid	Hydrochloric acid	Liquid	0.053 (147)	0 (256)	. . .	All proportions
Potassium permanganate	Potassium permanganate	Solid	0.100 (277)	460 (510), decomposes	. . .	2.8
Quartz	Silicon oxide	Solid	0.096 (266)	2640 (1722)	4050 (2500)	Insoluble
Salt	Sodium chloride	Solid	0.072 (199)	266 (403)	. . .	200
Sugar (cane sugar)	Sucrose	Solid	0.058 (161)	360 (456), decomposes	. . .	180
TNT	Trinitrotoluene	Solid	0.060 (166)	177 (354)	Explodes, 400 (480)	Very slight
Vinegar	Acetic acid	Liquid	0.038 (105)	62 (290)	244 (391)	All proportions
Water glass	sodium silicate	Solid	. . .	. . .	. . .	Soluble

TABLE 11.6 Hardness of Some Materials*

Material	Hardness scales			
	Mohs value	Knoop, 500 g and over	Rockwell C, 150 kg	Rockwell B, 100 kg
Diamond	10	7000	...	
Tungsten carbide	9	1800	...	
Topaz	8	1340	...	
Quartz	7	820	65	
Tool steel	7	850	66	
Glass	~5	530	49	
Copper	2.5–3.0	160	...	79
Silver	2.0–2.5	60	...	
Cadmium	2	35	...	

*See Table 10.11 for relative hardness scales.

TABLE 11.7 Densities of Miscellaneous Solids

Material	Density, lb/in³ (kg/m²)	Material	Density, lb/in³ (kg/m²)
Agate	0.094 (2600)	Leather	0.034 (941)
Aluminum alloy	0.100 (2770)	Limestone	0.087 (2410)
Asbestos	0.083 (2300)	Marble	0.094 (2600)
Beeswax	0.035 (969)	Nylon	0.040 (1110)
Brass	0.310 (8580)	Paper	0.032 (886)
Brick	0.065 (1800)	Phosphor bronze	0.318 (8800)
Chalk	0.087 (2410)	Plexiglas	0.043 (1190)
Coal, anthracite	0.058 (1610)	Quartz	0.094 (2600)
Concrete	0.083 (2300)	Rubber	0.042 (1160)
Cork, compressed	0.083 (2300)	Sandstone	0.080 (2210)
Diamond	0.119 (3300)	Steel, carbon	0.282 (7810)
Earth, dry and packed	0.054 (1490)	Steel, stainless	0.284 (7860)
Glass, flint	0.137 (3790)	Solder (70/30)	0.296 (8190)
Granite	0.098 (2710)	Tar	0.037 (1020)
Human fat	0.033 (913)	Teflon	0.079 (2190)
Ice	0.033 (913)	Wood, balsa, dry	0.006 (166)
Iron, cast	0.260 (7200)	Wood, oak, dry	0.028 (775)
Lead	0.412 (11,400)	Wood, pine, dry	0.016 (443)

TABLE 11.8 Densities and Boiling Points of Common Liquids

Liquid	Density, lb/in^3 (kg/m^3)	Boiling point, °F (K)
Acetone	0.029 (803)	134 (330)
Aniline	0.037 (1020)	364 (458)
Benzene	0.032 (886)	176 (353)
Carbon disulfide	0.043 (1190)	115 (319)
Carbon tetrachloride	0.058 (1610)	170 (350)
Chloroform	0.054 (1490)	142 (334)
Ethyl alcohol	0.028 (775)	173 (352)
Ethyl ether	0.026 (720)	94 (308)
Ethylene oxide	0.071 (1970)	51 (284)
Glycerol (glycerin)	0.045 (1250)	554 (563)
Glycol	0.040 (1110)	387 (471)
Heavy water	0.040 (1110)	215 (375)
Methyl alcohol	0.029 (803)	148 (338)
Octane	0.026 (720)	258 (399)
Sulfuric acid (98%)	0.066 (1830)	626 (603)
Toluene	0.031 (858)	231 (384)
Trichloroethane	0.047 (1300)	165 (347)
Water	0.036 (996)	212 (373)

TABLE 11.9 Densities and Molecular Weights of Gases

At atmospheric pressure and 58°F (288 K)

Gas	Molecular weight	Density, lb/in^3 (kg/m^3)
Acetylene	26	0.068 (1.09)
Air	29	0.075 (1.20)
Ammonia	17	0.044 (.705)
Argon	40	0.104 (1.67)
Butane	58	0.150 (2.40)
Carbon dioxide	44	0.114 (1.83)
Carbon monoxide	28	0.073 (1.17)
Ethylene	28	0.073 (1.17)
Helium	4	0.010 (.160)
Hydrogen	2	0.0052 (.0833)
Hydrogen sulfide	33	0.086 (1.38)
Krypton	83.8	0.217 (3.48)
Methane	16	0.042 (.673)
Neon	20.2	0.052 (.833)
Nitrogen	28	0.073 (1.17)
Nitric oxide	30	0.078 (1.25)
Nitrous oxide	44	0.114 (1.83)
Oxygen	32	0.083 (1.34)
Ozone	48	0.125 (2.00)
Propane	44	0.114 (1.83)
Sulfur dioxide	64	0.166 (2.66)
Xenon	54	0.140 (2.24)

TABLE 11.10 Gas Constant *R* for Various Gases

Gas	Constant, ft·lb/(lb·°R) [J/(kg·K)]
Acetylene	59.40 [319.6]
Air	53.30 [286.8]
Ammonia	90.77 [488.4]
Argon	38.70 [208.2]
Carbon dioxide	35.13 [189.0]
Carbon monoxide	55.19 [296.9]
Ethane	51.51 [277.1]
Ethylene	55.11 [296.5]
Helium	386.3 [2078]
Hydrogen	766.8 [4126]
Methane	96.37 [518.5]
Nitric oxide	51.52 [277.1]
Nitrogen	55.16 [296.8]
Oxygen	48.31 [259.9]

TABLE 11.11 Ram Air Pressure at Various Speeds

Adiabatic compressible flow at sea level

Air speed		Pressure	
mph	ft/s (m/s)	in water	psi (kPa)
20	29.3 (8.93)	0.197	0.007 (0.048)
40	58.7 (17.9)	0.788	0.028 (0.193)
60	88.0 (26.8)	1.744	0.064 (0.441)
80	117.3 (35.75)	3.158	0.114 (0.786)
100	146.7 (44.71)	4.942	0.178 (1.230)
120	176.0 (53.64)	7.130	0.257 (1.77)
140	205.3 (62.58)	9.726	0.351 (2.42)
160	234.7 (71.54)	12.736	0.460 (3.17)
180	264.0 (80.47)	16.167	0.584 (4.03)
200	293.3 (89.40)	20.025	0.723 (4.98)
300	440.0 (131.1)	46.033	1.66 (11.4)
400	566.7 (172.7)	84.22	3.04 (21.0)
500	733.0 (223.4)	136.87	4.94 (34.1)
600	880.2 (268.3)	206.37	7.45 (51.4)
700	1026.9 (313.00)	296.47	10.70 (73.8)

TABLE 11.12 Coefficients of Friction

Materials	Static	Sliding
Cast iron and bronze, lubricated	0.4	0.2
Mild steel and bronze, dry	0.7	0.5
Mild steel and bronze, lubricated	0.2	0.1
Teflon on steel	0.06	0.1
Steel on ice	0.2	0.1
Rubber on pavement	0.4	0.5*
Leather on wood	0.5	0.3
Stone on concrete	0.8	0.7

*Apparent coefficients of friction greater than 1.0 have been observed on vehicles.

TABLE 11.13 Coefficients of Linear Expansion per °F (per K)

Material	Coefficient $\times 10^{-6}$ at 70°F	(294 K)
Aluminum	13.2	(23.8)
Beryllium	6.8	(12.2)
Brass	10.5	(18.9)
Brick	1.2	(2.2)
Copper	7.8	(14.0)
Glass, flint	4.5	(8.1)
Glass, pyrex	2.0	(3.6)
Lead	16.3	(29.3)
Magnesium	14.4	(25.9)
Mercury	22.8	(41.0)
Molybdenum	2.7	(4.9)
Nickel	7.8	(14.0)
Quartz	4.4	(7.9)
Rubber	44.4	(79.9)
Steel, carbon	7.8	(14.0)
Steel, stainless	10.6	(19.1)
Tantalum	3.7	(6.7)
Tungsten	2.4	(4.3)

TABLE 11.14 Thermal Conductivities (*k*) of Some Common Materials

Material	Conductivity, Btu/(h·ft·°F) [W/(m·K)]
Aluminum	118 [204]
Asbestos	0.1 [0.17]
Brass	58 [100]
Brick, building	0.4 [0.69]
Brick, fire	0.1 [0.17]
Copper	220 [381]
Cardboard, corrugated	0.04 [0.069]
Cork board	0.03 [0.052]
Glass (typical)	0.5 [0.87]
Gold	170 [294]
Ice	1.3 [2.25]
Kapok	0.02 [0.035]
Leather	0.1 [0.17]
Magnesia, cast	0.3 [0.52]
Magnesium	92 [159]
Mercury	5.0 [8.65]
Mica	0.3 [0.52]
Mineral wool	0.02 [0.035]
Nickel	35 [60.6]
Paper	0.08 [0.14]
Platinum	40.0 [69.2]
Rubber	0.1 [0.17]
Silver	240 [415]
Steel, carbon	26 [45.0]
Steel, stainless 18/8	9.4 [16.3]
Snow	0.3 [0.52]
Wallboard	0.04 [0.069]
Wood, typical	0.1 [0.17]
Wool	0.02 [0.035]

TABLE 11.15 Atmospheric Pressure and Temperature at Altitude

Altitude, ft (m)	Temperature, °F (K)	Pressure, psia (kPa)
0	59 (288)	14.70 (101.3)
5,000 (1,500)	41.2 (278)	12.12 (83.56)
10,000 (3,050)	23.3 (268.5)	10.10 (69.64)
15,000 (4,600)	5.5 (258.6)	8.29 (57.2)
20,000 (6,100)	−12.3 (248.7)	6.75 (46.5)
25,000 (7,600)	−30.2 (238.8)	5.45 (37.6)
30,000 (9,100)	−48.0 (228.9)	4.36 (30.1)
35,000 (10,600)	−65.8 (219.0)	3.46 (23.9)
40,000 (12,200)	−67.0 (218.3)	2.72 (18.8)
45,000 (13,700)	−67.0 (218.3)	2.14 (14.8)
50,000 (15,200)	−67.0 (218.3)	1.69 (11.7)
75,000 (22,900)	−67.0 (218.3)	0.51 (3.5)
100,000 (30,500)	−51.0 (227.2)	0.16 (1.1)
150,000 (45,700)	19.0 (266.1)	0.020 (0.14)
200,000 (61,000)	−3.0 (253.9)	0.0029 (0.020)
250,000 (76,200)	−108.0 (195.6)	0.0003 (0.002)

Index